Honda CBR900RR FireBlade (CBR929RR, CBR954RR)
Service and Repair Manual

by Matthew Coombs

(4060 - 304)

Models covered
CBR900RR FireBlade. 929cc. 2000 and 2001 (Europe)
CBR929RR. 929cc. 2000 and 2001 (US and Canada)
CBR900RR FireBlade. 954cc. 2002 and 2003 (Europe)
CBR954RR. 954cc. 2002 and 2003 (US and Canada)

© Haynes Publishing 2003

ABCDE
FGHIJ
KLMNO
PQRST

A book in the Haynes Service and Repair Manual Series

All rights reserved. No part of this book may be reproduced or transmitted in any form or by any means, electronic or mechanical, including photocopying, recording or by any information storage or retrieval system, without permission in writing from the copyright holder.

ISBN 1 84425 060 1

British Library Cataloguing in Publication Data
A catalogue record for this book is available from the British Library

Library of Congress Control Number 2003108845

Printed in the USA

Haynes Publishing
Sparkford, Yeovil, Somerset, BA22 7JJ, England

Haynes North America, Inc
861 Lawrence Drive, Newbury Park, California 91320, USA

Editions Haynes S.A.
4, Rue de l'Abreuvoir, 92415 COURBEVOIE CEDEX, France

Haynes Publishing Nordiska AB
Box 1504, 751 45 UPPSALA, Sweden

Contents

LIVING WITH YOUR HONDA CBR900RR FIREBLADE

Introduction

The Birth of a Dream	Page	0•4
Acknowledgements	Page	0•8
About this manual	Page	0•8
Model development	Page	0•9
Bike spec	Page	0•9
Performance data	Page	0•11
Identification numbers	Page	0•12
Buying spare parts	Page	0•12
Safety first!	Page	0•13

Daily (pre-ride) checks

Engine/transmission oil level check	Page	0•14
Suspension, steering and drive chain checks	Page	0•14
Legal and safety checks	Page	0•14
Coolant level check	Page	0•15
Tyre checks	Page	0•15
Brake fluid level checks	Page	0•16

MAINTENANCE

Routine maintenance and servicing

Specifications	Page	1•1
Recommended lubricants and fluids	Page	1•2
Component locations	Page	1•3
Maintenance schedule	Page	1•4
Maintenance procedures	Page	1•6

Contents

REPAIRS AND OVERHAUL

Engine, transmission and associated systems

Engine, clutch and transmission	Page	2•1
Cooling system	Page	3•1
Fuel and exhaust systems	Page	4•1
Ignition system	Page	5•1

Chassis components

Frame, suspension and final drive	Page	6•1
Brakes, wheels and tyres	Page	7•1
Fairing and bodywork	Page	8•1

Electrical system

Page 9•1

Wiring diagrams

Page 9•28

REFERENCE

Tools and Workshop Tips	Page	REF•2
Security	Page	REF•20
Lubricants and fluids	Page	REF•23
Conversion factors	Page	REF•26
MOT Test Checks	Page	REF•27
Storage	Page	REF•32
Fault Finding	Page	REF•35
Fault Finding Equipment	Page	REF•44
Technical Terms Explained	Page	REF•48

Index

Page REF•52

The Birth of a Dream

by Julian Ryder

There is no better example of the Japanese post-war industrial miracle than Honda. Like other companies which have become household names, it started with one man's vision. In this case the man was the 40-year old Soichiro Honda who had sold his piston-ring manufacturing business to Toyota in 1945 and was happily spending the proceeds on prolonged parties for his friends. However, the difficulties of getting around in the chaos of post-war Japan irked Honda, so when he came across a job lot of generator engines he realised that here was a way of getting people mobile again at low cost.

A 12 by 18-foot shack in Hamamatsu became his first bike factory, fitting the generator motors into pushbikes. Before long he'd used up all 500 generator motors and started manufacturing his own engine, known as the 'chimney', either because of the elongated cylinder head or the smoky exhaust or perhaps both. The chimney made all of half a horsepower from its 50 cc engine but it was a major success and became the Honda A-type.

Less than two years after he'd set up in Hamamatsu, Soichiro Honda founded the Honda Motor Company in September 1948. By then, the A-type had been developed into the 90 cc B-type engine, which Mr Honda decided deserved its own chassis not a bicycle frame. Honda was about to become Japan's first post-war manufacturer of complete motorcycles. In August 1949 the first prototype was ready. With an output of three horsepower, the 98 cc D-type was still a simple two-stroke but it had a two-speed transmission and most importantly a pressed steel frame with telescopic forks and hard tail rear end. The frame was almost triangular in profile with the top rail going in a straight line from the massively braced steering head to the rear axle. Legend has it that after the D-type's first tests the entire workforce went for a drink to celebrate and try and think of a name for the bike. One man broke one of those silences you get when people are thinking, exclaiming 'This is like a dream!' 'That's it!' shouted Honda, and so the Honda Dream was christened.

> 'This is like a dream!'
> 'That's it'
> shouted Honda

Mr Honda was a brilliant, intuitive engineer and designer but he did not bother himself with the marketing side of his business. With hindsight, it is possible to see that employing Takeo Fujisawa who would both sort out the home market and plan the eventual expansion into overseas markets was a masterstroke. He arrived in October 1949 and in 1950 was made Sales Director. Another vital new name was Kiyoshi Kawashima, who along with Honda himself, designed the company's first four-stroke after Kawashima had told them that the four-stroke opposition to Honda's two-strokes sounded nicer and therefore sold better. The result of that statement was the overhead-valve 148 cc E-type which first ran in July 1951 just two months after the first drawings were made. Kawashima was made a director of the Honda Company at 34 years old.

The E-type was a massive success, over 32,000 were made in 1953 alone, a feat of mass-production that was astounding by the

Honda C70 and C90 OHV-engined models

standards of the day given the relative complexity of the machine. But Honda's lifelong pursuit of technical innovation sometimes distracted him from commercial reality. Fujisawa pointed out that they were in danger of ignoring their core business, the motorised bicycles that still formed Japan's main means of transport. In May 1952 the F-type Cub appeared, another two-stroke despite the top men's reservations. You could buy a complete machine or just the motor to attach to your own bicycle. The result was certainly distinctive, a white fuel tank with a circular profile went just below and behind the saddle on the left of the bike, and the motor with its horizontal cylinder and bright red cover just below the rear axle on the same side of the bike. This was the machine that turned Honda into the biggest bike maker in Japan with 70% of the market for bolt-on bicycle motors, the F-type was also the first Honda to be exported. Next came the machine that would turn Honda into the biggest motorcycle manufacturer in the world.

The C100 Super Cub was a typically audacious piece of Honda engineering and marketing. For the first time, but not the last, Honda invented a completely new type of motorcycle, although the term 'scooterette' was coined to describe the new bike which had many of the characteristics of a scooter but the large wheels, and therefore stability, of a motorcycle. The first one was sold in August 1958, fifteen years later over nine-million of them were on the roads of the world. If ever a machine can be said to have brought mobility to the masses it is the Super Cub. If you add in the electric starter that was added for the C102 model of 1961, the design of the Super Cub has remained substantially unchanged ever since, testament to how right Honda got it first time. The Super Cub made Honda the world's biggest manufacturer after just two years of production.

The CB250N Super Dream became a favorite with UK learner riders of the late seventies and early eighties

Honda's export drive started in earnest in 1957 when Britain and Holland got their first bikes, America got just two bikes the next year. By 1962 Honda had half the American market with 65,000 sales. But Soichiro Honda had already travelled abroad to Europe and the USA, making a special

The GL1000 introduced in 1975, was the first in Honda's line of GoldWings

Introduction

Carl Fogarty in action at the Suzuka 8 Hour on the RC45

An early CB750 Four

point of going to the Isle of Man TT, then the most important race in the GP calendar. He realised that no matter how advanced his products were, only racing success would convince overseas markets for whom 'Made in Japan' still meant cheap and nasty. It took five years from Soichiro Honda's first visit to the Island before his bikes were ready for the TT. In 1959 the factory entered five riders in the 125 class. They did not have a massive impact on the event being benevolently regarded as a curiosity, but sixth, seventh and eighth were good enough for the team prize. The bikes were off the pace but they were well engineered and very reliable.

The TT was the only time the West saw the Hondas in '59, but they came back for more the following year with the first of a generation of bikes which shaped the future of motorcycling – the double-overhead-cam four-cylinder 250. It was fast and reliable – it revved to 14,000 rpm – but didn't handle anywhere near as well as the opposition. However, Honda had now signed up non-Japanese riders to lead their challenge. The first win didn't come until 1962 (Aussie Tom Phillis in the Spanish 125 GP) and was followed up with a world-shaking performance at the TT. Twenty-one year old Mike Hailwood won both 125 and 250 cc TTs and Hondas filled the top five positions in both races. Soichiro Honda's master plan was starting to come to fruition, Hailwood and Honda won the 1961 250 cc World Championship. Next year Honda won three titles. The other Japanese factories fought back and inspired Honda to produce some of the most fascinating racers ever seen: the awesome six-cylinder 250, the five-cylinder 125, and the 500 four with which the immortal Hailwood battled Agostini and the MV Agusta.

When Honda pulled out of racing in '67 they had won sixteen rider's titles, eighteen manufacturer's titles, and 137 GPs, including 18 TTs, and introduced the concept of the modern works team to motorcycle racing. Sales success followed racing victory as Soichiro Honda had predicted, but only because the products advanced as rapidly as the racing machinery. The Hondas that came to Britain in the early '60s were incredibly sophisticated. They had overhead cams where the British bikes had pushrods, they had electric starters when the Brits relied on the kickstart, they had 12V electrics when even the biggest British bike used a 6V system. There seemed no end to the technical wizardry. It wasn't that the technology itself was so amazing but just like that first E-type, it was the fact that Honda could mass-produce it more reliably than the lower-tech competition that was so astonishing.

When in 1968 the first four-cylinder CB750 road bike arrived the world of motorcycling changed for ever, they even had to invent a new word for it, 'Superbike'. Honda raced again with the CB750 at Daytona and won the

World Endurance title with a prototype DOHC version that became the CB900 roadster. There was the six-cylinder CBX, the CX500T – the world's first turbocharged production bike, they invented the full-dress tourer with the GoldWing, and came back to GPs with the revolutionary oval-pistoned NR500 four-stroke, a much-misunderstood bike that was more a rolling experimental laboratory than a racer. Just to show their versatility Honda also came up with the weird CX500 shaft-drive V-twin, a rugged workhorse that powered a new industry, the courier companies that oiled the wheels of commerce in London and other big cities.

It was true, though, that Mr Honda was not keen on two-strokes – early motocross engines had to be explained away to him as lawnmower motors! However, in 1982 Honda raced the NS500, an agile three-cylinder lightweight against the big four-cylinder opposition in 500 GPs. The bike won in its first year and in '83 took the world title for Freddie Spencer. In four-stroke racing the V4 layout took over from the straight four, dominating TT, F1 and Endurance championships with the RVF750, the nearest thing ever built to a Formula 1 car on wheels. And when Superbike arrived Honda were ready with the RC30. On the roads the VFR V4 became an instant classic while the CBR600 invented another new class of bike on its way to becoming a best-seller. The V4 road bikes had problems to start with but the VFR750 sold world-wide over its lifetime while the VFR400 became a massive commercial success and cult bike in Japan. The original RC30 won the first two World Superbike Championships is 1988 and '89, but Honda had to wait until 1997 to win it again with the RC45, the last of the V4 roadsters. In Grands Prix, the NSR500 V4 two-stroke superseded the NS triple and became the benchmark racing machine of the '90s. Mick Doohan secured his place in history by winning five World Championships in consecutive years on it.

In yet another example of Honda inventing a new class of motorcycle, they came up with the astounding CBR900RR FireBlade, a bike with the punch of a 1000 cc motor in a package the size and weight of a 750. It became a cult bike as well as a best seller, and with judicious redesigns continues to give much more recent designs a run for their money.

When it became apparent that the high-tech V4 motor of the RC45 was too expensive to produce, Honda looked to a V-twin engine to power its flagship for the first time. Typically, the VTR1000 FireStorm was a much more rideable machine than its opposition and once accepted by the market formed the basis of the next generation of Superbike racer, the VTR-SP-1.

One of Mr Honda's mottos was that technology would solve the customers' problems, and no company has embraced cutting-edge technology more firmly than Honda. In fact Honda often developed new technology, especially in the fields of materials science and metallurgy. The embodiment of that was the NR750, a bike that was misunderstood nearly as much as the original NR500 racer. This limited-edition technological tour-de-force embodied many of Soichiro Honda's ideals. It used the latest techniques and materials in every component, from the oval piston, 32-valve V4 motor to the titanium coating on the windscreen, it was – as Mr Honda would have wanted – the best it could possibly be. A fitting memorial to the man who has shaped the motorcycle industry and motorcycles as we know them today.

The CX500 – Honda's first V-Twin and a favorite choice of dispatch riders

The CBR900RR FireBlade models

It took a long time for the original FireBlade to be knocked off its perch as king of the sportsbikes. The first 'Blade appeared in 1992 and until the R1 Yamaha appeared was the benchmark for big-bore sportsbikes. The original dictum of Tadeo Baba, the project leader, was the power of a 1000 in a bike the

The VFR400R was a cult bike in Japan and a popular grey import in the UK

Introduction

The 2001 929 cc RR-1 model

The 2002 954 cc RR-2 model

size of a 600, and that is the principle that still underlies the latest generation machine.

World beater it might have been, but the first 'Blade achieved all its objectives with very well understood technology – there was nothing ground breaking in the design of chassis or motor, they were just put together with lightness and power as the only things that mattered. The new FireBlade for the Millennium, the Y-model launched in 2000, can in many ways be called the first modern 'Blade: the first with fuel injection, the first with upsidedown front forks, and the first with unconventional chassis technology. It looked very different too, mainly thanks to the single, full width headlamp.

It didn't just look different from what had gone before; it was totally different in every department. The 58 mm stroke used on every 'Blade since its launch was massively shortened to 54 mm enabling the motor to be made more compact despite the taller valve gear. As Yamaha's patent on the Exup exhaust valve ran out in 2000, Honda introduced their version on the new Blade and used a similar variable geometry device inside the airbox on the inlet side as well. Naturally they had to have new acronyms: H-TEV for Honda Titanium Exhaust Valve, and H-VIX for Honda Variable Intake/Exhausts System. The software controlling these devices and the PGM-FI fuel injection system gives as much attention to emissions as performance with Honda spokesmen getting very Zen and talking about matching the requirements of humanity and engineering.

On the chassis side the 'Blade finally bid farewell to its 16-inch front wheel and said hello to a state-of-the-art 17-incher. This put paid to any lingering suspicions of skittishness that attached to early models especially. The pivotless frame design seen on smaller CBRs and the FireStorm V-twin didn't quite make it to the Millennium 'Blade. The main frame members stop short of the swinging arm pivot and the arm itself does pivot in castings in the rear of the crankcases. However, a very solid looking frame member runs under the motor and up either side to the swingarm pivots, but doesn't link with the main frame.

The result was a 'Blade that could once again compete with the R1 and the big GSX-R in the ultimate performance stakes. It wasn't the most powerful bike out there but it was a superb package. The bike that had once been a tank-slapping terror had turned into a user-friendly all-rounder you could ride to work or ride to Mediterranean sunshine. However, comparison tests with the Yamaha and Suzuki supersportsters unanimously damned the new FireBlade with faint praise. Criticism centred on the rough throttle response low down the rev range and harsh suspension.

Honda addressed these complaints with the 2002 model. Engine size was increased from 929 to 954 cc with a 1 mm overbore and the fuel injection was smoothed out considerably. Suspension modifications included a swinging arm that looked as if it had just been unbolted from a factory racer, lighter wheels and a lower fuel tank. The response to the modifications was unanimous and astonishing. The 'Blade still wasn't the most powerful or fastest bike in its very competitive market sector but most testers said it was the best. The reason goes back to Baba-san's words on the subject of usability (for an in-depth analysis see Rob Simmonds book on the FireBlade in the Haynes Great Bikes series).

Honda provided more than enough power in a superb chassis with sharp handling, amazing brakes – and all in a package that weighs less than contemporary Supersport 600s. How do they do that? Like all good tricks, it is devastatingly simple yet incredibly difficult to pull off. It relies on the bike being treated as a complete system not a collection of individual components. More than that, it relies on the designers understanding how every part of their design affects or will be used by the rider. Baba-san was the project leader and guiding light for the first FireBlade and has been involved with all ten iterations the bike has gone through. It shows, it really does.

Acknowledgements

Our thanks are due to Bransons Motorcycles of Yeovil who supplied the machines featured in the illustrations throughout this manual. We would also like to thank NGK Spark Plugs (UK) Ltd for supplying the colour spark plug condition photographs, the Avon Rubber Company for supplying information on tyre fitting and Draper Tools Ltd for some of the workshop tools shown.

Thanks are also due to Julian Ryder who wrote the introduction 'The Birth of a Dream' and to Honda (UK) Ltd. who supplied model photographs.

About this Manual

The aim of this manual is to help you get the best value from your motorcycle. It can do so in several ways. It can help you decide what work must be done, even if you choose to have it done by a dealer; it provides information and procedures for routine maintenance and servicing; and it offers diagnostic and repair procedures to follow when trouble occurs.

We hope you use the manual to tackle the work yourself. For many simpler jobs, doing it yourself may be quicker than arranging an appointment to get the motorcycle into a dealer and making the trips to leave it and pick it up. More importantly, a lot of money can be saved by avoiding the expense the shop must pass on to you to cover its labour and overhead costs. An added benefit is the sense of satisfaction and accomplishment that you feel after doing the job yourself.

References to the left or right side of the motorcycle assume you are sitting on the seat, facing forward.

We take great pride in the accuracy of information given in this manual, but motorcycle manufacturers make alterations and design changes during the production run of a particular motorcycle of which they do not inform us. No liability can be accepted by the authors or publishers for loss, damage or injury caused by any errors in, or omissions from, the information given.

Model development and Bike spec

Model development

CBR900RR-Y FireBlade (Europe 2000 model)
CBR929RR-Y (US and Canada 2000 model)

The new millennium version of Honda's legendary FireBlade, starting with the RR-Y (2000) model, is much more than an update of the previous year's model – it is an all-new motorcycle.

It retains the in-line four cylinder liquid-cooled engine. Drive to the double overhead camshafts which actuate the four valves per cylinder is by chain from the right-hand end of the crankshaft. The clutch is a conventional wet multi-plate unit and the gearbox is 6-speed. Drive to the rear wheel is by chain and sprockets.

The most significant change from previous models is the use of Honda's PGM-FI fuel injection system to replace the carburettors, and an engine management system that controls both the injection system and the ignition system.

The fuel system incorporates Honda's H-VIX (Honda variable intake and exhaust) system, which controls the flow of air through the air box using a two-position valve, and controls the flow of gases through the exhaust system using a three-position valve.

The engine sits in a twin-beam aluminium frame which uses the engine as a stressed member, with box-section tubing for the upper rails. Front suspension is by fully adjustable upside-down oil-damped 43 mm forks with cartridge type dampers. Rear suspension is by a fully adjustable single shock absorber via a three-way rising rate linkage. The swingarm pivots in a bracket which is bolted to the back of the engine.

The front brake system has two twin-opposed piston calipers acting on floating discs. The rear brake system has a single piston sliding caliper acting on a conventional disc.

Available in black, blue/yellow, and blue/red.

CBR900RR-1 FireBlade (Europe 2001 model)
CBR929RR-1 (US and Canada 2001 model)

There are no significant changes from the Y (2000) model.

Available in black, blue, blue/yellow, yellow/silver, blue/red and red/blue.

An additional models, the CBR929RE-1, was available in the US for 2001. There are no technical differences between the RR and the RE model. The only differences are cosmetic – the frame, engine bracket and swingarm are finished in black, and the fuel tank, front mudguard and bodywork are coloured to match. There are no colour options.

CBR900RR-2 FireBlade (Europe 2002 model)
CBR954RR-2 (US and Canada 2002 model)

The most significant change to this model is the larger engine capacity, increased from 929cc to 954cc, achieved by increasing the bore by 1mm.

Other changes include a revised swingarm, revised front fork internals, a different starter motor and a new headlight and instrument cluster. The bodywork is completely new, and there is no separate lower fairing – the fairing side panels now wrap round under the bike and join together.

Available in red, white/blue and yellow

CBR900RR-3 FireBlade (Europe 2003 model)
CBR954RR-3 (US and Canada 2003 model)

There are no significant changes from the 2 (2002) model.

Available in black/red, blue/black and black/yellow.

Bike spec

Engine
Type ....Four-stroke in-line four
Capacity
  RR-Y and RR-1 (2000 and 2001) models ....929 cc
  RR-2 and RR-3 (2002 and 2003) models ....954 cc
Bore
  RR-Y and RR-1 (2000 and 2001) models ....74.0 mm
  RR-2 and RR-3 (2002 and 2003) models ....75.0 mm
Stroke ....54.0 mm
Compression ratio
  RR-Y and RR-1 (2000 and 2001) models ....11.3 to 1
  RR-2 and RR-3 (2002 and 2003) models ....11.5 to 1
Cooling system ....Liquid cooled
Clutch ....Wet multi-plate
Transmission ....Six-speed constant mesh
Final drive ....Chain and sprockets
Camshafts ....DOHC, chain-driven
Fuel system ....PGM-FI fuel injection with H-VIX (Honda variable intake and exhaust)
Exhaust system ....Four-into-one titanium with variable valve system
Ignition system ....Computer-controlled digital transistorised with electronic advance

Chassis
Frame type ....Twin spar aluminium box section
Rake and Trail ....23°45', 97 mm
Fuel tank
  Capacity (including reserve) ....18.0 litres
  Reserve volume ....3.5 litres
Front suspension
  Type ....43 mm upside-down oil-damped telescopic forks
  Travel ....110 mm
  Adjustment ....Spring pre-load, rebound and compression damping
Rear suspension
  Type ....Single shock absorber, rising rate linkage, box section aluminium swingarm
  Travel (at axle) ....135 mm
  Adjustment ...Spring pre-load, rebound and compression damping
Wheels ....17 inch 3-spoke alloys
Tyres
  Front ....120/70-ZR17 (58W) Radial
  Rear ....190/50-ZR17 (73W) Radial
Front brake ....Twin 330 mm floating discs with twin opposed-piston calipers
Rear brake .....Single 220 mm disc with single piston sliding caliper

0•10 Dimensions and weights

Dimensions and weights

RR-Y and RR-1 models
Overall length . . . . .2065 mm
Overall width . . . . .680 mm
Overall height . . . . .1125 mm
Wheelbase
   UK and European models . . . . .1400 mm
   US and Canada models . . . . .1395 mm
Seat height . . . . .820 mm
Ground clearance . . . . .130 mm
Weight (dry)
   Germany models . . . . .171 kg
   UK and all other European models . . . . .170 kg
   US 49 States and Canada models . . . . .172 kg
   California models . . . . .174 kg
Weight (OTR)
   Germany models . . . . .199 kg
   UK and all other European models . . . . .198 kg
   US 49 States and Canada models . . . . .197 kg
   California models . . . . .199 kg
Maximum weight capacity
   UK and European models . . . . .189 kg
   US 49 States and California models . . . . .160 kg
   Canada models . . . . .164 kg

RR-2 and RR-3 models
Overall length . . . . .2025 mm
Overall width . . . . .680 mm
Overall height . . . . .1135 mm
Wheelbase
   UK and European models . . . . .1400 mm
   US and Canada models . . . . .1395 mm
Seat height . . . . .820 mm
Ground clearance . . . . .130 mm
Weight (dry)
   Germany models . . . . .169 kg
   UK and all other European models . . . . .168 kg
   US 49 States and Canada models . . . . .168 kg
   California models . . . . .170 kg
Weight (OTR)
   Germany models . . . . .193 kg
   UK and all other European models . . . . .194 kg
   US 49 States and Canada models . . . . .195 kg
   California models . . . . .197 kg
Maximum weight capacity
   UK and European models . . . . .189 kg
   US 49 States and California models . . . . .160 kg
   Canada models . . . . .164 kg

Performance data 0•11

Performance data

RR-Y and RR-1

Maximum power
Claimed . . . . .130 bhp (97 kW) @ 10,100 rpm

Maximum torque
Claimed . . . . .72.3 lbf ft (98 Nm) @ 7500 rpm

Top speed
Estimated . . . . .172 mph (276 km/h)

Acceleration
Time taken to cover a 1/4 mile from a standing start . . . . . .10.31 secs
Terminal speed after 1/4 mile . . . . .139 mph (224 km/h)

Average fuel consumption
Miles per Imp gal, miles per litre,
litres per 100 km .35 mpg, 8.3 mpl, 8.0 l/100km

Fuel tank range
Based on average fuel consumption rate . . . . .140 miles (225 km)

RR-2 and RR-3

Maximum power
Claimed . . . . .149 bhp (111 kW) @ 10,500 rpm

Maximum torque
Claimed . . . . .77 lbf ft (104 Nm) @ 8900 rpm

Top speed
Estimated . . . . .175 mph (282 km/h)

Acceleration
Time taken to cover a 1/4 mile from a standing start . . . . . . .10.8 secs
Terminal speed after 1/4 mile . . . . .137.5 mph (221 km/h)
0 to 60 mph . . . . .3.1 secs

Average fuel consumption
Miles per Imp gal, miles per litre,
  litres per 100 km . . . . .35 mpg, 8.3 mpl, 8.0 l/100km

Fuel tank range
Based on average fuel consumption rate . . . . .140 miles (225 km)

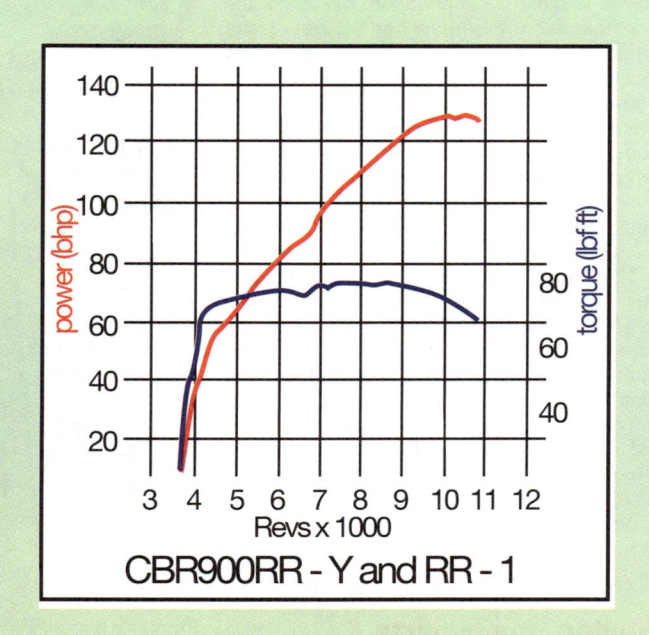

CBR900RR - Y and RR - 1

Performance data sourced from Motor Cycle News road test features.
See the MCN website for up-to-date biking news.

MCN www.motorcyclenews.com

0•12 Identification numbers

Frame and engine numbers

The frame serial number is stamped into the right-hand side of the steering head. The engine number is stamped into the top of the crankcase on the right-hand side of the engine. Both of these numbers should be recorded and kept in a safe place so they can be furnished to law enforcement officials in the event of a theft. There is also a colour code label on the storage compartment wall, visible after raising the passenger seat, and a VIN plate on the frame beam. The throttle bodies also have an ID number stamped into them on the intake side.

The frame serial number, engine serial number, and colour code should also be kept in a handy place (such as with your driver's licence) so they are always available when purchasing or ordering parts for your machine.

The procedures in this manual identify models by their code letter or number (e.g. RR-Y or RR-1) and production year (e.g. 2000 or 2001). The model code or production year is printed on the colour code label.

Model – Europe	Year	Initial engine no.	Initial frame no.
CBR900RR-Y	2000	SC44E-2000001	JH2SC44A/B/C*-YM000001
CBR900RR-1	2001	SC44E-2100001	JH2SC44A/B/C*-1M100001
CBR900RR-2	2002	SC50E-2000001	JH2SC50A/B/C*-2M000001
CBR900RR-3	2003	SC50E-2100001	JH2SC44A/B/C*-3M100001

*A – UK and all Europe except Germany – B, and France – C

Model – US 49-state	Year	Initial engine no.	Initial frame no.
CBR929RR-Y	2000	SC44E-2000001	JH2SC440-YM000001
CBR929RR-1	2001	SC44E-2100001	JH2SC440-1M100001
CBR929RE-1	2001	SC44E-2100001	JH2SC443-1M100001
CBR954RR-2	2002	SC50E-2000001	JH2SC500-2M200001
CBR954RR-3	2003	SC50E-2100001	JH2SC500-3M300001

Model – California	Year	Initial engine no.	Initial frame no.
CBR929RR-Y	2000	SC44E-2000001	JH2SC441-YM000001
CBR929RR-1	2001	SC44E-2100001	JH2SC441-1M100001
CBR929RE-1	2001	SC44E-2100001	JH2SC444-1M100001
CBR954RR-2	2002	SC50E-2000001	JH2SC501-2M200001
CBR954RR-3	2003	SC50E-2100001	JH2SC501-3M300001

Buying spare parts

Once you have found all the identification numbers, record them for reference when buying parts. Since the manufacturers change specifications, parts and vendors (companies that manufacture various components on the machine), providing the ID numbers is the only way to be reasonably sure that you are buying the correct parts.

Whenever possible, take the worn part to the dealer so direct comparison with the new component can be made. Along the trail from the manufacturer to the parts shelf, there are numerous places that the part can end up with the wrong number or be listed incorrectly.

The two places to purchase new parts for your motorcycle – the franchised or main dealer and the parts/accessories store – differ in the type of parts they carry. While dealers can obtain every single genuine part for your motorcycle, the accessory store is usually limited to normal high wear items such as chains and sprockets, brake pads, spark plugs and cables, and to tune-up parts and various engine gaskets, etc. Rarely will an accessory outlet have major suspension components, camshafts, transmission gears, or engine cases.

Used parts can be obtained from breakers yards for roughly half the price of new ones, but you can't always be sure of what you're getting. Once again, take your worn part to the breaker for direct comparison, or when ordering by mail order make sure that you can return it if you are not happy.

Whether buying new, used or rebuilt parts, the best course is to deal directly with someone who specialises in your particular make.

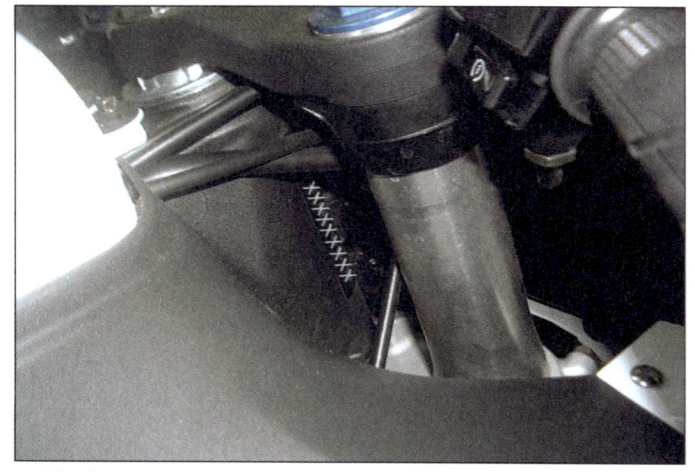

The frame number is stamped into the right-hand side of the steering head.

The engine number is stamped into the top of the crankcase on the right-hand side of the engine.

Safety first! 0•13

Professional mechanics are trained in safe working procedures. However enthusiastic you may be about getting on with the job at hand, take the time to ensure that your safety is not put at risk. A moment's lack of attention can result in an accident, as can failure to observe simple precautions.

There will always be new ways of having accidents, and the following is not a comprehensive list of all dangers; it is intended rather to make you aware of the risks and to encourage a safe approach to all work you carry out on your bike.

Asbestos

● Certain friction, insulating, sealing and other products - such as brake pads, clutch linings, gaskets, etc. - contain asbestos. Extreme care must be taken to avoid inhalation of dust from such products since it is hazardous to health. If in doubt, assume that they do contain asbestos.

Fire

● Remember at all times that petrol is highly flammable. Never smoke or have any kind of naked flame around, when working on the vehicle. But the risk does not end there - a spark caused by an electrical short-circuit, by two metal surfaces contacting each other, by careless use of tools, or even by static electricity built up in your body under certain conditions, can ignite petrol vapour, which in a confined space is highly explosive. Never use petrol as a cleaning solvent. Use an approved safety solvent.

● Always disconnect the battery earth terminal before working on any part of the fuel or electrical system, and never risk spilling fuel on to a hot engine or exhaust.
● It is recommended that a fire extinguisher of a type suitable for fuel and electrical fires is kept handy in the garage or workplace at all times. Never try to extinguish a fuel or electrical fire with water.

Fumes

● Certain fumes are highly toxic and can quickly cause unconsciousness and even death if inhaled to any extent. Petrol vapour comes into this category, as do the vapours from certain solvents such as trichloroethylene. Any draining or pouring of such volatile fluids should be done in a well ventilated area.
● When using cleaning fluids and solvents, read the instructions carefully. Never use materials from unmarked containers - they may give off poisonous vapours.
● Never run the engine of a motor vehicle in an enclosed space such as a garage. Exhaust fumes contain carbon monoxide which is extremely poisonous; if you need to run the engine, always do so in the open air or at least have the rear of the vehicle outside the workplace.

The battery

● Never cause a spark, or allow a naked light near the vehicle's battery. It will normally be giving off a certain amount of hydrogen gas, which is highly explosive.

● Always disconnect the battery ground (earth) terminal before working on the fuel or electrical systems (except where noted).
● If possible, loosen the filler plugs or cover when charging the battery from an external source. Do not charge at an excessive rate or the battery may burst.
● Take care when topping up, cleaning or carrying the battery. The acid electrolyte, evenwhen diluted, is very corrosive and should not be allowed to contact the eyes or skin. Always wear rubber gloves and goggles or a face shield. If you ever need to prepare electrolyte yourself, always add the acid slowly to the water; never add the water to the acid.

Electricity

● When using an electric power tool, inspection light etc., always ensure that the appliance is correctly connected to its plug and that, where necessary, it is properly grounded (earthed). Do not use such appliances in damp conditions and, again, beware of creating a spark or applying excessive heat in the vicinity of fuel or fuel vapour. Also ensure that the appliances meet national safety standards.
● A severe electric shock can result from touching certain parts of the electrical system, such as the spark plug wires (HT leads), when the engine is running or being cranked, particularly if components are damp or the insulation is defective. Where an electronic ignition system is used, the secondary (HT) voltage is much higher and could prove fatal.

Remember...

✗ **Don't** start the engine without first ascertaining that the transmission is in neutral.

✗ **Don't** suddenly remove the pressure cap from a hot cooling system - cover it with a cloth and release the pressure gradually first, or you may get scalded by escaping coolant.

✗ **Don't** attempt to drain oil until you are sure it has cooled sufficiently to avoid scalding you.

✗ **Don't** grasp any part of the engine or exhaust system without first ascertaining that it is cool enough not to burn you.

✗ **Don't** allow brake fluid or antifreeze to contact the machine's paintwork or plastic components.

✗ **Don't** siphon toxic liquids such as fuel, hydraulic fluid or antifreeze by mouth, or allow them to remain on your skin.

✗ **Don't** inhale dust - it may be injurious to health (see Asbestos heading).

✗ **Don't** allow any spilled oil or grease to remain on the floor - wipe it up right away, before someone slips on it.

✗ **Don't** use ill-fitting spanners or other tools which may slip and cause injury.

✗ **Don't** lift a heavy component which may be beyond your capability - get assistance.

✗ **Don't** rush to finish a job or take unverified short cuts.

✗ **Don't** allow children or animals in or around an unattended vehicle.

✗ **Don't** inflate a tyre above the recommended pressure. Apart from overstressing the carcass, in extreme cases the tyre may blow off forcibly.

✔ **Do** ensure that the machine is supported securely at all times. This is especially important when the machine is blocked up to aid wheel or fork removal.

✔ **Do** take care when attempting to loosen a stubborn nut or bolt. It is generally better to pull on a spanner, rather than push, so that if you slip, you fall away from the machine rather than onto it.

✔ **Do** wear eye protection when using power tools such as drill, sander, bench grinder etc.

✔ **Do** use a barrier cream on your hands prior to undertaking dirty jobs - it will protect your skin from infection as well as making the dirt easier to remove afterwards; but make sure your hands aren't left slippery. Note that long-term contact with used engine oil can be a health hazard.

✔ **Do** keep loose clothing (cuffs, ties etc. and long hair) well out of the way of moving mechanical parts.

✔ **Do** remove rings, wristwatch etc., before working on the vehicle - especially the electrical system.

✔ **Do** keep your work area tidy - it is only too easy to fall over articles left lying around.

✔ **Do** exercise caution when compressing springs for removal or installation. Ensure that the tension is applied and released in a controlled manner, using suitable tools which preclude the possibility of the spring escaping violently.

✔ **Do** ensure that any lifting tackle used has a safe working load rating adequate for the job.

✔ **Do** get someone to check periodically that all is well, when working alone on the vehicle.

✔ **Do** carry out work in a logical sequence and check that everything is correctly assembled and tightened afterwards.

✔ **Do** remember that your vehicle's safety affects that of yourself and others. If in doubt on any point, get professional advice.

● If in spite of following these precautions, you are unfortunate enough to injure yourself, seek medical attention as soon as possible.

Daily (pre-ride) checks

Note: *The daily (pre-ride) checks outlined in the owner's manual covers those items which should be inspected on a daily basis.*

Engine/transmission oil level check

Before you start:
✔ Take the motorcycle on a short run to allow it to reach normal operating temperature.
Caution: Do not run the engine in an enclosed space such as a garage or workshop.
✔ Stop the engine and support the motorcycle upright using an auxiliary stand. Allow it to stand undisturbed for a few minutes to allow the oil level to stabilise. Make sure the motorcycle is on level ground.

Bike care:
● If you have to add oil frequently, check whether you have any oil leaks from the engine joints, oil seals and gaskets. If not, the engine could be burning oil, in which case there will be white smoke coming out of the exhaust (see *Fault Finding*).

The correct oil
● Modern, high-revving engines place great demands on their oil. It is very important that the correct oil for your bike is used.
● Always top up with a good quality oil of the specified type and viscosity and do not overfill the engine.

Oil type	API grade SE, SF or SG
Oil viscosity*	SAE 10W40*

*If you are using the motorcycle constantly in extreme conditions of heat or cold, other more suitable viscosity ranges may be used – refer to the viscosity table to select the oil best suited to your conditions.

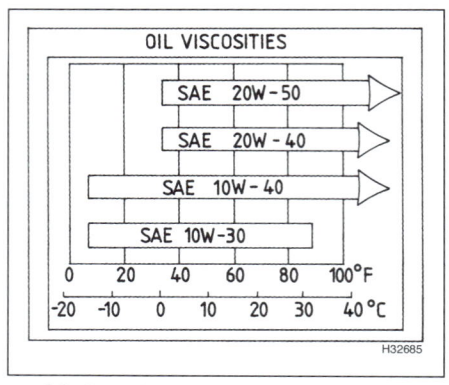

Oil viscosity table: select the oil best suited to your conditions

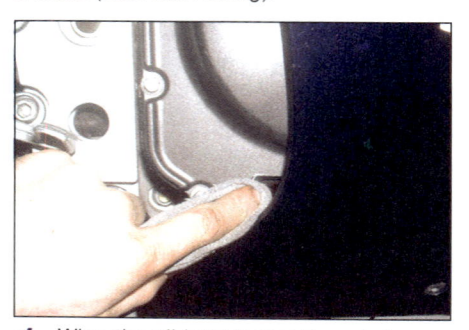

1 Wipe the oil level inspection window so that it is clean – it is located in the clutch cover on the right-hand side of the engine. With the motorcycle vertical, the oil level should lie between the maximum and minimum level lines on the window.

2 If the level is on or below the minimum line, unscrew the oil filler cap from the clutch cover.

3 Top up the engine with the recommended grade and type of oil to bring the level up to the maximum line on the inspection window. Do not overfill. On completion, make sure the filler cap is secure in the cover.

Suspension, steering and drive chain checks

Suspension and Steering:
● Check that the front and rear suspension operates smoothly without binding (see Chapter 1).
● Check that the suspension is adjusted as required (see Chapter 6).
● Check that the steering moves smoothly from lock-to-lock.

Drive chain:
● Check that the chain isn't too loose or too tight, and adjust it if necessary (see Chapter 1).
● If the chain looks dry, lubricate it (see Chapter 1).

Legal and safety checks

Lighting and signalling:
● Take a minute to check that the headlight, taillight, brake light, license plate light (where fitted), instrument lights and turn signals all work correctly.
● Check that the horn sounds when the button is pressed.
● A working speedometer, graduated in mph, is a statutory requirement in the UK.

Safety:
● Check that the throttle grip rotates smoothly when opened and snaps shut when released, in all steering positions. Also check for the correct amount of freeplay (see Chapter 1).
● Check that the engine shuts off when the kill switch is operated. Check the starter interlock circuit (see Chapter 1).
● Check that sidestand return springs hold the stand up securely when retracted.

Fuel:
● This may seem obvious, but check that you have enough fuel to complete your journey. If you notice signs of fuel leakage – rectify the cause immediately.
● Ensure you use the correct grade unleaded fuel – see Chapter 4 Specifications.

Daily (pre-ride) checks 0•15

Coolant level check

> **Warning: DO NOT remove the radiator pressure cap to add coolant. Topping up is done via the coolant reservoir tank filler. DO NOT leave open containers of coolant about, as it is poisonous.**

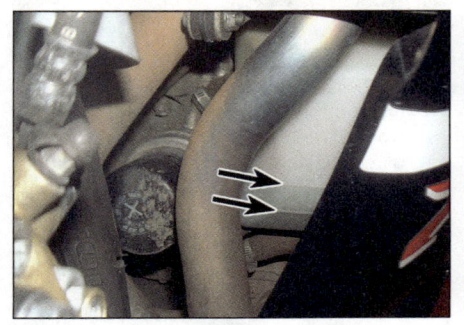

Before you start:
✔ Make sure you have a supply of coolant available (a mixture of 50% distilled water and 50% corrosion inhibited ethylene glycol anti-freeze is needed).
✔ Always check the coolant level when the engine is at normal working temperature. Take the motorcycle on a short run to allow it to reach normal temperature.
Caution: Do not run the engine in an enclosed space such as a garage or workshop.
✔ Stop the engine and support the motorcycle upright on an auxiliary stand. Make sure the motorcycle is on level ground.
✔ For best visibility and access remove the left-hand fairing side panel (see Chapter 8).

Bike care:
● Use only the specified coolant mixture. It is important that anti-freeze is used in the system all year round, and not just in the winter. Do not top the system up using only water, as the system will become too diluted.
● Do not overfill the reservoir tank. If the coolant is significantly above the UPPER level line at any time, the surplus should be siphoned or drained off to prevent the possibility of it being expelled out of the overflow hose.
● If the coolant level falls steadily, check the system for leaks (see Chapter 1). If no leaks are found and the level continues to fall, it is recommended that the machine is taken to a Honda dealer for a pressure test.

1 The coolant reservoir is located at the front of the engine on the left-hand side. The UPPER and LOWER level lines (arrowed) are marked on the front and side of the reservoir. If the marks are obscured by dirt, clean the reservoir. The coolant level should lie between the UPPER and LOWER level lines.

2 If the coolant level is on or below the LOWER line, remove the reservoir filler cap, then insert a suitable funnel into the reservoir and top the reservoir up with the recommended coolant mixture to the UPPER level line. If you do not have a funnel, it is best to remove the left-hand fairing side panel (see Chapter 8). Fit the cap securely. Install the fairing side panel if removed (see Chapter 8).

Tyre checks

The correct pressures:
● The tyres must be checked when **cold**, not immediately after riding. Note that incorrect tyre pressures will cause abnormal tread wear and unsafe handling. Low tyre pressures may cause the tyre to slip on the rim or come off.
● Use an accurate pressure gauge. Many forecourt gauges are wildly inaccurate. If you buy your own, spend as much as you can justify on a quality gauge.
● Proper air pressure will increase tyre life and provide maximum stability and ride comfort.

Front	Rear
36 psi (2.50 Bar)	42 psi (2.90 Bar)

Tyre care:
● Check the tyres carefully for cuts, tears, embedded nails or other sharp objects and excessive wear. Operation of the motorcycle with excessively worn tyres is extremely hazardous, as traction and handling are directly affected.
● Check the condition of the tyre valve and ensure the dust cap is in place.
● Pick out any stones or nails which may have become embedded in the tyre tread. If left, they will eventually penetrate through the casing and cause a puncture.
● If tyre damage is apparent, or unexplained loss of pressure is experienced, seek the advice of a tyre fitting specialist without delay.

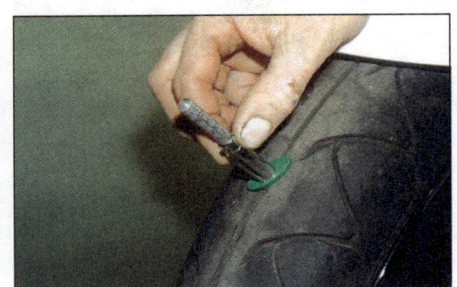

Tyre tread depth:
● At the time of writing UK law requires that tread depth must be at least 1 mm over 3/4 of the tread breadth all the way around the tyre, with no bald patches. Many riders, however, consider 2 mm tread depth minimum to be a safer limit. Honda recommend a minimum of 1.5 mm on the front and 2 mm on the rear, but note that German law requires a minimum of 1.6 mm.
● Many tyres now incorporate wear indicators in the tread. Identify the location marking on the tyre sidewall to locate the indicator bar and replace the tyre if the tread has worn down to the bar.

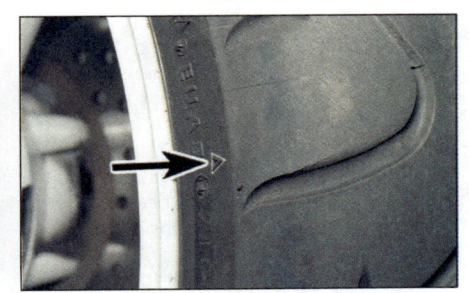

1 Remove the dust cap from the valve and check the tyre pressures when **cold**. Do not forget to fit the cap after checking the pressure.

2 Measure tread depth at the centre of the tyre using a depth gauge.

3 Tyre tread wear indicator location marking (arrowed) on the sidewall.

Daily (pre-ride) checks

Brake fluid level checks

> ⚠️ **Warning:** Brake hydraulic fluid can harm your eyes and damage painted surfaces, so use extreme caution when handling and pouring it and cover surrounding surfaces with rag. Do not use fluid that has been standing open for some time, as it is hygroscopic (absorbs moisture from the air) which can cause a dangerous loss of braking effectiveness.

Before you start:

✔ The front master cylinder reservoir is on the right-hand handlebar. The rear master cylinder reservoir is located under the seat on the right-hand side.
✔ Make sure you have the correct hydraulic fluid. DOT 4 is recommended.
✔ Wrap a rag around the reservoir being worked on to ensure that any spillage does not come into contact with painted surfaces.
✔ When checking the fluid in the front reservoir turn the handlebars as required so the reservoir is level. When checking the fluid in the rear reservoir support the motorcycle upright, if available using an auxiliary stand.

Note 1: When working on the front reservoir, access to the cover screws is partially restricted by the windshield, so if a short or angled screwdriver is not available, remove the windshield (see Chapter 8).

Note 2: When working on the rear reservoir, there is not enough clearance above it to remove the cap with the reservoir mounted, so unscrew the mounting bolt and displace the reservoir round the back of the sub-frame – make sure that you cover the area with plenty of rag and take care not to spill any fluid.

Bike care:

● The fluid in the front and rear brake master cylinder reservoirs will drop as the brake pads wear down. If the fluid level is low check the brake pads for wear (see Chapter 1).
● If either fluid reservoir requires repeated topping-up there could be a leak somewhere in the system, which must be investigated immediately.
● Check for signs of fluid leakage from the hydraulic hoses and/or brake system components – if found, rectify immediately (see Chapter 7).
● Check the operation of both brakes before taking the machine on the road; if there is evidence of air in the system (spongy feel to lever or pedal), it must be bled (see Chapter 7).

Front

1 The front brake fluid level is visible through the window in the reservoir body – it must be between the UPPER and LOWER level lines (arrowed).

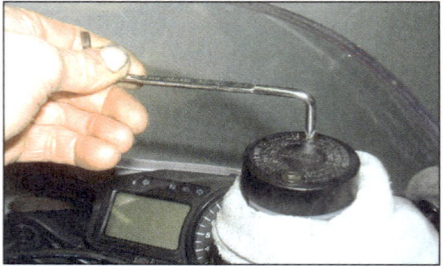

2 If the level is on or below the LOWER line, undo the two reservoir cap screws (see **Note 1**), and remove the cap, diaphragm plate and diaphragm.

3 Top up with new clean DOT 4 hydraulic fluid, until the level is up to the UPPER line. Do not overfill and take care to avoid spills (see **Warning** above).

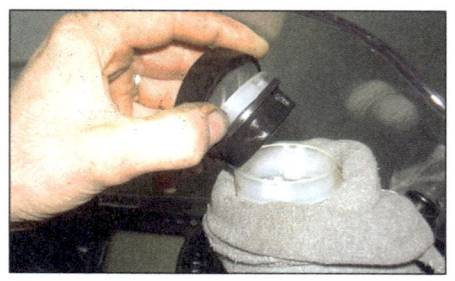

4 Ensure that the diaphragm is correctly seated before installing the plate and cap. Secure the cap with the screws.

Rear

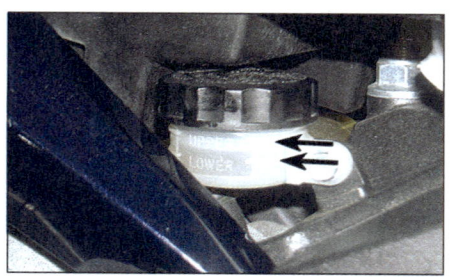

5 The rear brake fluid level is visible through the reservoir body – it must be between the UPPER and LOWER level lines (arrowed).

6 If the level is on or below the LOWER line, unscrew the reservoir mounting bolt and manoeuvre the reservoir round the back of the rear sub-frame.

7 Unscrew the reservoir cap, and remove the diaphragm plate and diaphragm.

8 Top up to the upper line with new clean DOT 4 hydraulic fluid. Do not overfill. Ensure that the diaphragm is correctly seated before installing the plate and cap. Install the reservoir and tighten its bolt.

Chapter 1
Routine maintenance and servicing

Contents

Air filter – renewal 21	Fuel filter – cleaning and renewal 36
Battery – charging ....see Chapter 9	Fuel system – check 9
Battery – check 10	Headlight aim – check and adjustment 14
Battery – removal, installation, inspection	H-VIX system – check and cable adjustment 25
and maintenance ....see Chapter 9	Idle speed – check and adjustment 2
Brake caliper and master cylinder seals – renewal 32	Nuts and bolts – tightness check 18
Brake fluid – change 23	Pulse secondary air injection (PAIR) system – check 20
Brake hoses – renewal 33	Sidestand and starter interlock circuit – check 15
Brake pads – wear check 3	Sidestand, lever pivots and cables – lubrication 7
Brake system – check 13	Spark plugs – check 5
Clutch – check and adjustment 4	Spark plugs – renewal 6
Cooling system – check 12	Starter interlock circuit – check 15
Cooling system – draining, flushing and refilling 26	Steering head bearings – freeplay check and adjustment ... 17
Cylinder compression – check 27	Steering head bearings – re-greasing 30
Drive chain and sprockets – check, adjustment, cleaning and	Suspension – check 16
lubrication 1	Swingarm and suspension linkage bearings – re-greasing ... 31
Engine oil pressure – check 28	Throttle cables – check 11
Engine – oil and oil filter change 8	Throttle body starter valve – synchronisation 37
Evaporative emission control (EVAP) system –	Valve clearances – check and adjustment 24
check (California models) 22	Wheels and tyres – general check 19
Front forks – oil change 35	Wheel bearings – check 29
Fuel system hoses – renewal 34	

Degrees of difficulty

Easy, suitable for novice with little experience	Fairly easy, suitable for beginner with some experience	Fairly difficult, suitable for competent DIY mechanic	Difficult, suitable for experienced DIY mechanic	Very difficult, suitable for expert DIY or professional

Specifications

Engine

Cylinder numbering	1 to 4 from left to right
Spark plug type	
RR-Y and RR-1 (2000 and 2001) models	
Standard	Denso IUH27D
For cold climate (below 5°C)	Denso IUH24D
RR-2 and RR-3 (2002 and 2003) models	
Standard	NGK IMR9C-9H or Denso VUH27D
For cold climate (below 5°C)	NGK IMR8C-9H or Denso VUH24D
Spark plug electrode gap	0.8 to 0.9 mm
Engine idle speed	1200 ± 100 rpm
Valve clearances (COLD engine)	
Intake valves	0.13 to 0.19 mm
Exhaust valves	0.24 to 0.30 mm
Cylinder compression	
RR-Y and RR-1 (2000 and 2001) models	178 psi (12.2 Bar) @ 350 rpm
RR-2 and RR-3 (2002 and 2003) models	174 psi (12.0 Bar) @ 350 rpm
Oil pressure (at main gallery plug, with engine warm)	71 psi (5.0 Bar) @ 5400 rpm, oil @ 80°C

Routine maintenance and servicing

Miscellaneous

Drive chain slack	40 to 50 mm
Throttle cable freeplay	2 to 6 mm
Clutch cable freeplay	10 to 20 mm
Tyre pressures (cold)	see Daily (pre-ride) checks
Steering head bearing pre-load (see text)	
RR-Y and RR-1 (2000 and 2001) models	10 to 15 N
RR-2 (2002) models	
Europe models	10 to 15 N
US and Canada models	11 to 16 N
RR-3 (2003) models	16 to 23 N

Recommended lubricants and fluids

Engine/transmission oil type	API grade SE, SF or SG motor oil
Engine/transmission oil viscosity	SAE 10W40 (but see Daily (pre-ride) checks)
Engine/transmission oil capacity	
Oil change	3.5 litres
Oil and filter change	3.7 litres
Following engine overhaul – dry engine, new filter	4.0 litres
Coolant type	50% distilled water, 50% corrosion inhibited ethylene glycol anti-freeze
Coolant capacity	
Radiator and engine	
RR-Y and RR-1 (2000 and 2001) models	3.1 litres
RR-2 and RR-3 (2002 and 2003) models	3.2 litres
Reservoir	0.4 litres
Brake fluid	DOT 4
Drive chain	SAE 80 or 90 gear oil or chain lubricant suitable for O-ring chains
Steering head bearings	Multi-purpose grease with EP2 rating
Swingarm pivot bearings	Multi-purpose grease with EP2 rating
Suspension linkage bearings	Multi-purpose grease with EP2 rating
Bearing seal lips	Multi-purpose grease
Gearchange lever/rear brake pedal/footrest pivots	Multi-purpose grease
Clutch lever pivot	Multi-purpose grease
Sidestand pivot	Multi-purpose or molybdenum disulphide grease
Throttle twistgrip	Molybdenum disulphide grease or dry film lubricant
Front brake lever pivot and piston tip	Silicone grease
Cables	Cable lubricant

Torque settings

Engine/transmission oil drain plug	29 Nm
Engine/transmission oil filter	26 Nm
Exhaust control valve pulley cover bolts	12 Nm
Handlebar end weight screw	10 Nm
Main oil gallery plug	30 Nm
Rear axle nut	113 Nm
Spark plugs	12 Nm
Steering head bearing adjuster nut	
RR-Y and RR-1 (2000 and 2001) models	
Initial setting	29 Nm
Final setting	29 Nm
RR-2 (2002) models with caged ball bearings	
Initial setting	30 Nm
Final setting	15 Nm + 90° (see Text)
RR-2 (2002) models with taper roller bearings	
Initial setting	40 Nm
Final setting	18 Nm
RR-3 (2003) models	
Initial setting	30 Nm
Final setting	15 Nm + 90° (see Text)
Steering stem nut	103 Nm
Timing inspection cap	18 Nm
Top yoke fork clamp bolts	
RR-Y and RR-1 (2000 and 2001) models	22 Nm
RR-2 and RR-3 (2002 and 2003) models	23 Nm

Routine maintenance and servicing

Component locations on the right side

1 Rear brake fluid reservoir
2 Engine oil filler cap
3 Clutch cable lower adjuster
4 Throttle cable upper adjuster
5 Front brake fluid reservoir
6 Radiator pressure cap
7 Engine oil filter
8 Engine oil drain plug
9 Engine oil level inspection window
10 Rear brake light switch
11 Rear brake pedal height adjuster

Component locations on the left side

1 Clutch cable upper adjuster
2 Steering head bearing adjuster
3 Air filter
4 Fuel filter
5 Battery
6 Drive chain adjuster
7 Drive chain slider
8 Idle speed adjuster
9 Coolant drain bolt
10 Coolant reservoir
11 Front fork seal

Maintenance schedule – European models

Note: *The daily (pre-ride) checks outlined in the owner's manual covers those items which should be inspected on a daily basis. Always perform the pre-ride inspection at every maintenance interval (in addition to the procedures listed). The intervals listed below are the intervals recommended by the manufacturer for each particular operation during the model years covered in this manual. Your owner's manual may have different intervals for your model.*

Daily (pre-ride)
See *'Daily (pre-ride) checks'* at the beginning of this manual.

After the initial 600 miles (1000 km)
Note: *This check is usually performed by a Honda dealer after the first 600 miles (1000 km) from new. Thereafter, maintenance is carried out according to the following intervals of the schedule.*

Every 600 miles (1000 km)
- Check, adjust and lubricate the drive chain (Section 1)

Every 4000 miles (6000 km) or 6 months (whichever comes sooner)
- Check and adjust the idle speed (Section 2)
- Check the brake pads (Section 3)
- Check the clutch (Section 4)

Every 8000 miles (12,000 km) or 12 months (whichever comes sooner)
Carry out all the items under the 4000 mile (6000 km) check, plus the following:
- Check the spark plugs – RR-Y and RR-1 (2000 and 2001) models (Section 5)
- Lubricate the clutch, gearchange and brake levers, brake pedal, sidestand pivot, and the throttle and choke cables (Section 7)
- Change the engine oil and fit a new filter (Section 8)
- Check the fuel system and hoses (Section 9)
- Check the battery terminals (Section 10)
- Check and adjust the throttle cables (Section 11)
- Check the cooling system (Section 12)
- Check the brake system and brake light switch operation (Section 13)
- Check and adjust the headlight aim (Section 14)
- Check the sidestand and starter interlock circuit (Section 15)
- Check the suspension (Section 16)
- Check and adjust the steering head bearings (Section 17)
- Check the tightness of all nuts, bolts and fasteners (Section 18)
- Check the condition of the wheels and tyres (Section 19)
- Check the pulse secondary air injection (PAIR) system (Section 20)

Every 12,000 miles (18,000 km) or 18 months (whichever comes first)
Carry out all the items under the 4000 mile (6000 km) check, plus the following:
- Fit a new air filter element (Section 21)

Every 12,000 miles (18,000 km) or two years (whichever comes first)
Carry out all the items under the 4000 mile (6000 km) check, plus the following:
- Change the brake fluid (Section 23)

Every 16,000 miles (24,000 km)
Carry out all the items under the 8000 mile (12,000 km) check, plus the following:
- Check and adjust the valve clearances (Section 24)
- Check the spark plugs – RR-2 and RR-3 (2002 and 2003) models (Section 5)
- Fit new spark plugs – RR-Y and RR-1 (2000 and 2001) models (Section 6)
- Check the H-VIX system (Section 25)

Every 24,000 miles (36,000 km) or two years (whichever comes sooner)
Carry out all the items under the 12,000 mile (18,000 km) and 8000 mile (12,000 km) checks, plus the following:
- Change the coolant (Section 26)

Every 32,000 miles (48,000 km)
Carry out all the items under the 16,000 mile (224,000 km) checks, plus the following:
- Fit new spark plugs – RR-2 and RR-3 (2002 and 2003) models (Section 6)

Non-scheduled maintenance
- Check the cylinder compression (Section 27)
- Check the engine oil pressure (Section 28)
- Check the wheel bearings (Section 29)
- Re-grease the steering head bearings (Section 30)
- Re-grease the swingarm and suspension linkage bearings (Section 31)
- Fit new brake master cylinder and caliper seals (Section 32)
- Fit new brake hoses (Section 33)
- Fit new fuel system hoses (Section 34)
- Change the front fork oil (Section 35)
- Clean/renew the fuel filters (Section 36)
- Synchronise the throttle body starter valve (Section 37)

Maintenance schedule – US and Canada models

Note: *The daily (pre-ride) checks outlined in the owner's manual covers those items which should be inspected on a daily basis. Always perform the pre-ride inspection at every maintenance interval (in addition to the procedures listed). The intervals listed below are the intervals recommended by the manufacturer for each particular operation during the model years covered in this manual. Your owner's manual may have different intervals for your model.*

Daily (pre-ride)
See '*Daily (pre-ride) checks*' at the beginning of this manual.

After the initial 600 miles (1000 km)
Note: *This check is usually performed by a Honda dealer after the first 600 miles (1000 km) from new. Thereafter, maintenance is carried out according to the following intervals of the schedule.*

Every 500 miles (800 km)
- Check, adjust and lubricate the drive chain (Section 1)

Every 4000 miles (6000 km) or 6 months (whichever comes sooner)
- Check and adjust the idle speed (Section 2)
- Check the brake pads (Section 3)
- Check the clutch (Section 4)
- Check the spark plugs – RR-Y and RR-1 (2000 and 2001) models (Section 5)

Every 8000 miles (12,000 km) or 12 months (whichever comes sooner)
Carry out all the items under the 4000 mile (6000 km) check, plus the following:
- Fit new spark plugs – RR-Y and RR-1 (2000 and 2001) models (Section 6)
- Lubricate the clutch, gearchange and brake levers, brake pedal, sidestand pivot, and the throttle and choke cables (Section 7)
- Change the engine oil and fit a new filter (Section 8)
- Check the fuel system and hoses (Section 9)
- Check the battery terminals (Section 10)
- Check and adjust the throttle cables (Section 11)
- Check the cooling system (Section 12)
- Check the brake system and brake light switch operation (Section 13)
- Check and adjust the headlight aim (Section 14)
- Check the sidestand and starter interlock circuit (Section 15)
- Check the suspension (Section 16)
- Check and adjust the steering head bearings (Section 17)
- Check the tightness of all nuts, bolts and fasteners (Section 18)
- Check the condition of the wheels and tyres (Section 19)
- Check the pulse secondary air injection (PAIR) system (Section 20)

Every 12,000 miles (18,000 km) or 18 months (whichever comes first)
Carry out all the items under the 4000 mile (6000 km) check, plus the following:
- Fit a new air filter element (Section 21)
- Check the evaporative emission control (EVAP) system hoses – California models (Section 22)

Every 12,000 miles (18,000 km) or two years (whichever comes first)
Carry out all the items under the 4000 mile (6000 km) check, plus the following:
- Change the brake fluid (Section 23)

Every 16,000 miles (24,000 km)
Carry out all the items under the 8000 mile (12,000 km) check, plus the following:
- Check and adjust the valve clearances (Section 24)
- Check the spark plugs – RR-2 and RR-3 (2002 and 2003) models (Section 5)
- Check the H-VIX system (Section 25)

Every 24,000 miles (36,000 km) or two years (whichever comes sooner)
Carry out all the items under the 12,000 mile (18,000 km) and 8000 mile (12,000 km) checks, plus the following:
- Change the coolant (Section 26)

Every 32,000 miles (48,000 km)
Carry out all the items under the 16,000 mile (24,000 km) checks, plus the following:
- Fit new spark plugs – RR-2 and RR-3 (2002 and 2003) models (Section 6)

Non-scheduled maintenance
- Check the cylinder compression (Section 27)
- Check the engine oil pressure (Section 28)
- Check the wheel bearings (Section 29)
- Re-grease the steering head bearings (Section 30)
- Re-grease the swingarm and suspension linkage bearings (Section 31)
- Fit new brake master cylinder and caliper seals (Section 32)
- Fit new brake hoses (Section 33)
- Fit new fuel system hoses (Section 34)
- Change the front fork oil (Section 35)
- Clean/renew the fuel filters (Section 36)
- Synchronise the throttle body starter valve (Section 37)

Introduction

1 This Chapter is designed to help the home mechanic maintain his/her motorcycle for safety, economy, long life and peak performance.

2 Deciding where to start or plug into the routine maintenance schedule depends on several factors. If your motorcycle has been maintained according to the warranty standards and has just come out of warranty, start routine maintenance as it coincides with the next mileage or calendar interval. If you have owned the machine for some time but have never performed any maintenance on it, start at the nearest interval and include some additional procedures to ensure that nothing important is overlooked. If you have just had a major engine overhaul, then start the maintenance routine from the beginning. If you have a used machine and have no knowledge of its history or maintenance record, combine all the checks into one large service initially and then settle into the specified maintenance schedule.

3 Before beginning any maintenance or repair, the machine should be cleaned thoroughly, especially around the oil filter, spark plugs, valve covers, body panels, carburettors, etc. Cleaning will help ensure that dirt does not contaminate the engine and will allow you to detect wear and damage that could otherwise easily go unnoticed.

4 Certain maintenance information is sometimes printed on labels attached to the motorcycle. If the information on the labels differs from that included here, use the information on the label.

1 Drive chain and sprockets – check, adjustment, cleaning and lubrication

Every 500 miles (800 km) – US and Canada models

Every 600 miles (1000 km) – European models

Check

1 A neglected drive chain won't last long and will quickly damage the sprockets. Routine chain adjustment and lubrication isn't difficult and will ensure maximum chain and sprocket life.

2 To check the chain, place the bike on its sidestand and shift the transmission into neutral.

3 Push up on the bottom run of the chain and measure the slack midway between the two sprockets, then compare your measurement to that listed in this Chapter's Specifications **(see illustration)**. As the chain stretches with wear, adjustment will periodically be necessary (see below). Since the chain will rarely wear evenly, roll the bike forward so that another section of chain can be checked (having an assistant to do this makes the task a lot easier); do this several times to check the entire length of chain, and mark the tightest spot.

Caution: Riding the bike with excess slack in the chain could lead to damage.

4 In some cases where lubrication has been neglected, corrosion and galling may cause the links to bind and kink, which effectively shortens the chain's length and makes it tight. Thoroughly clean and work free any such links, then highlight them with a marker pen or paint. After the bike has been ridden, repeat the measurement for slack in the highlighted area. If the chain has kinked again and is still tight, replace it with a new one. A rusty, kinked or worn chain will damage the sprockets and can damage transmission bearings. If in any doubt as to the condition of a chain, it is far better to install a new one than risk damage to other components and possibly yourself.

5 Check the entire length of the chain for damaged rollers, loose links and pins, and missing O-rings and replace it with a new one if necessary. **Note:** *Never install a new chain on old sprockets, and never use the old chain if you install new sprockets – replace the chain and sprockets as a set.*

6 Unscrew the bolts securing the front sprocket cover and remove the cover and the guide plate **(see illustrations)**. Check the teeth on the front sprocket and the rear sprocket for wear **(see illustration)**.

7 Inspect the drive chain slider on the front of the swingarm for excessive wear and damage. There are wear limit lines marked by arrows on the top and bottom of the slider at the front. Replace the slider with a new one if it has worn down to the lines or the tips of the arrows (see Chapter 6).

Adjustment

8 Move the bike so that the chain is positioned with the tightest point at the centre of its bottom run, then put it on the sidestand.

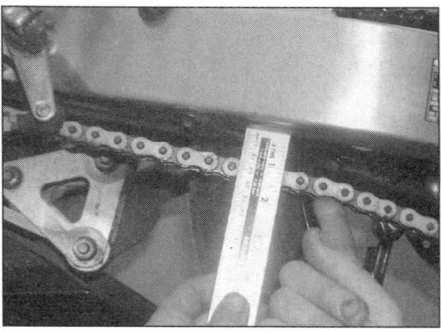

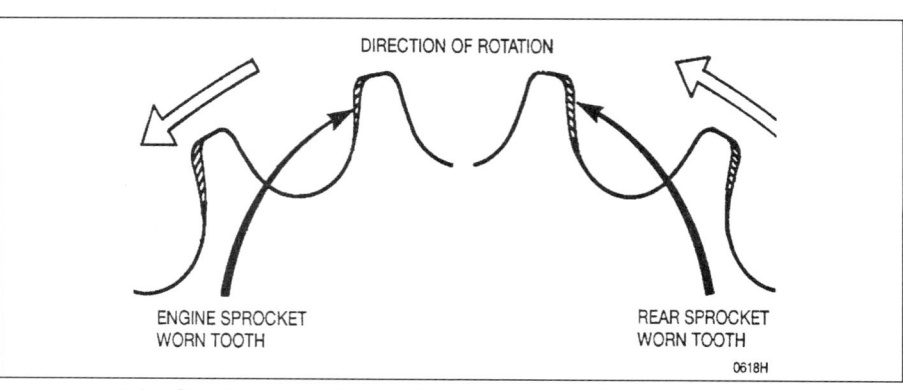

1.3 Push up on the chain and measure the slack

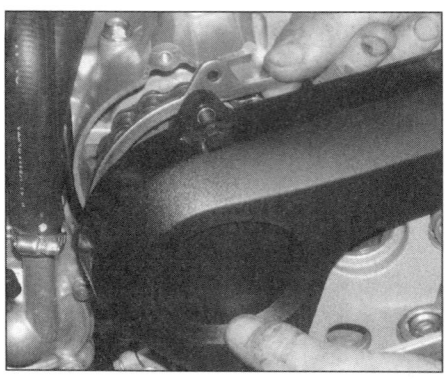

1.6a Unscrew the bolts (arrowed) . . .

1.6b . . . and remove the cover and guide plate

1.6c Check the sprockets in the areas indicated to see if they are worn excessively

Routine maintenance and servicing 1•7

1.9a Slacken the axle nut . . .

1.9b . . . and the adjuster locknut (arrowed) on each side

9 Slacken the rear axle nut **(see illustration)**. Slacken the locknut on each adjuster bolt **(see illustration)**.

10 Turn the adjuster bolt on each side of the swingarm evenly until the amount of freeplay specified at the beginning of the Chapter is obtained at the centre of the bottom run of the chain **(see illustration)**. Following adjustment, check that the index point on each chain adjustment marker is in the same position in relation to the marks on the swingarm **(see illustration)**. It is important the alignment is the same on each side of the swingarm; if not, the rear wheel will be out of alignment with the front. Always make sure that the front edge of each marker is butted against the end of the adjuster bolt.

11 If there is a discrepancy in the chain adjuster positions, adjust one of them so that its position is exactly the same as the other. Check the chain freeplay as described above and readjust if necessary.

12 Also check the alignment of the wear decal on the left-hand adjustment marker with the index line on the swingarm (do not use the rear edge of the swingarm as the index). When the index line meets the red REPLACE CHAIN zone, the drive chain has stretched excessively and must be replaced with a new one.

13 When adjustment is complete, tighten the adjuster bolt locknuts **(see illustration 1.9b)**. Now tighten the axle nut to the torque setting specified at the beginning of the Chapter. Recheck the adjustment as above, then place the machine on an auxiliary stand and spin the wheel to make sure it runs freely.

Cleaning and lubrication

14 If required, wash the chain in paraffin (kerosene) or a suitable non-flammable or high flash-point solvent that will not damage the O-rings, using a soft brush to work any dirt out if necessary. Wipe the cleaner off the chain and allow it to dry, using compressed air if available. If the chain is excessively dirty remove it from the machine and allow it to soak in the paraffin or solvent (see Chapter 6). **Caution: Don't use petrol (gasoline), an unsuitable solvent or other cleaning fluids which might damage the internal sealing properties of the chain. Don't use high-pressure water to clean the chain. The entire process shouldn't take longer than ten minutes, otherwise the O-rings could be damaged.**

15 For routine lubrication, the best time to lubricate the chain is after the motorcycle has been ridden. When the chain is warm, the lubricant will penetrate the joints between the sideplates better than when cold. **Note:** *Honda specifies SAE 80 to SAE 90 gear oil or an aerosol chain lube that it is suitable for O-ring or X-ring (sealed) chains; do not use any other chain lubricants – the solvents could damage the chain's sealing rings.* Apply the oil to the area where the sideplates overlap – not the middle of the rollers **(see illustration)**.

> **HAYNES HiNT** *Apply the lubricant to the top of the lower chain run, so centrifugal force will work the oil into the chain when the bike is moving. After applying the lubricant, let it soak in a few minutes before wiping off any excess.*

> ⚠ **Warning: Take care not to get any lubricant on the tyres or brake system components. If any of the lubricant inadvertently contacts them, clean it off thoroughly using a suitable solvent or dedicated brake cleaner before riding the machine.**

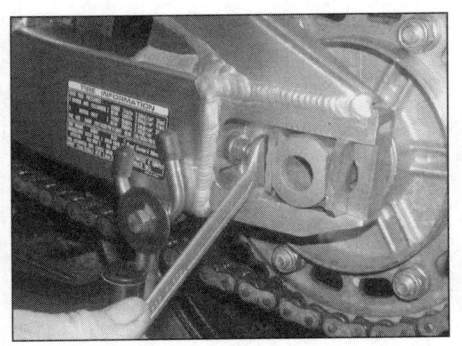

1.10a Turn each adjuster by an equal amount . . .

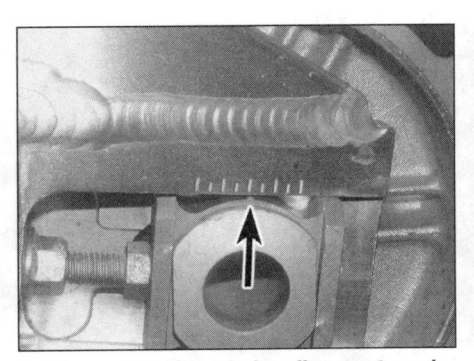

1.10b . . . then check the alignment marks as described – note the index pointer on the adjustment marker (arrowed)

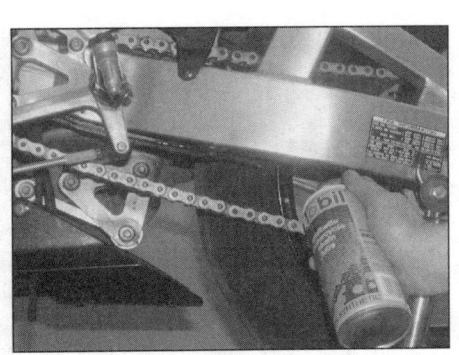

1.15 Apply the lubricant to the overlapping sections of the sideplates

Routine maintenance and servicing

2.3 Idle speed adjuster (arrowed)

3.1a Front brake pad wear indicator (arrowed)

3.1b Rear brake pad wear indicator (arrowed)

2 Idle speed – check and adjustment

Every 4000 miles (6000 km) or 6 months

1 The idle speed should be checked and adjusted after checking the valve clearances, and when it is obviously too high or too low. Before adjusting the idle speed, make sure the valve clearances and spark plug gaps are correct, and the air filter is clean. Also, turn the handlebars from side-to-side and check the idle speed does not change as you do. If it does, the throttle cables may not be adjusted or routed correctly, or may be worn out. This is a dangerous condition that can cause loss of control of the bike. Be sure to correct this problem before proceeding.

2 The engine should be at normal operating temperature, which is usually reached after 10 to 15 minutes of stop-and-go riding. Place the motorcycle on its sidestand, and make sure the transmission is in neutral.

3 The idle speed adjuster is a knurled knob located on the left-hand side of the machine between the frame and the fairing side panel **(see illustration)**. With the engine idling, adjust the speed by turning the knob until the speed listed in this Chapter's Specifications is obtained. Turn the screw clockwise to increase idle speed, and anti-clockwise to decrease it.

4 Snap the throttle open and shut a few times, then recheck the idle speed. If necessary, repeat the adjustment procedure.

5 If a smooth, steady idle can't be achieved check the starter valves (see Chapter 4).

3 Brake pads – wear check

Every 4000 miles (6000 km) or 6 months

1 Each brake pad has wear indicators in the form of cut-outs in the friction material. The wear indicators should be plainly visible by looking at the edges of the friction material from the best vantage point – on the front brake look at the bottom edge of the pad from below the caliper, and on the rear brake look at the rear edge of the pad from behind the caliper, but note that an accumulation of road dirt and brake dust could make them difficult to see **(see illustrations)**. If the indicators aren't visible, then the amount of friction material remaining should be, and it will be obvious when the pads need replacing.

Honda do not specify a minimum thickness for the friction material, but anything less than 1 mm should be considered worn. **Note:** *Some after-market pads may use different indicators to those on the original equipment.*

2 If the pads are worn to or beyond the wear indicator (i.e. the bottom of the cut-out on the front pads or the beginning of the cut-out on the rear pads) or there is little friction material remaining, they must be replaced with new ones, though it is advisable to renew the pads before they become this worn. If the pads are dirty or if you are in doubt as to the amount of friction material remaining, remove them for inspection (see Chapter 7). If the pads are excessively worn, check the brake discs (see Chapter 7).

3 Refer to Chapter 7 for details of pad replacement.

4 Clutch – check and adjustment

Every 4000 miles (6000 km) or 6 months

1 Check that the clutch lever operates smoothly and easily.

2 If the clutch lever operation is heavy or stiff, remove the cable (see Chapter 2) and lubricate it (see Section 7). If the cable is still stiff, replace it with a new one. Install the lubricated or new cable (see Chapter 2).

3 With the cable operating smoothly, check that the clutch lever is correctly adjusted. Periodic adjustment is necessary to compensate for wear in the clutch plates and stretch of the cable. Check that the amount of freeplay at the clutch lever end is within the specifications listed at the beginning of the Chapter **(see illustration)**.

4 If adjustment is required, this can be done first at the lever end of the cable. On RR-Y and RR-1 (2000 and 2001) models loosen the adjuster lockring **(see illustration)**. On all models turn the adjuster in or out until the required amount of freeplay is obtained **(see**

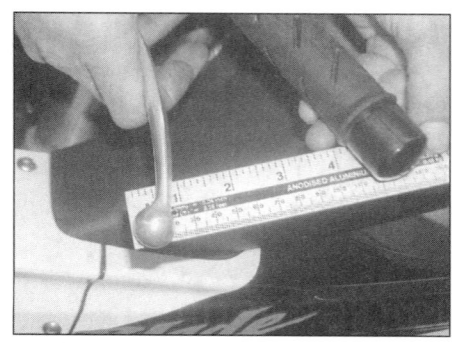

4.3 Measure the amount of freeplay at the clutch lever end as shown

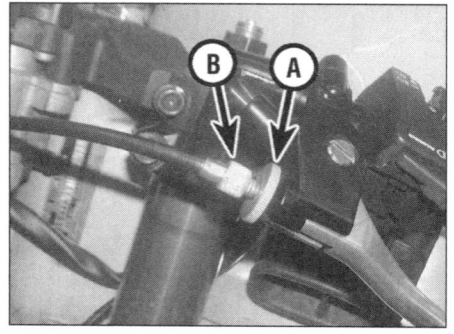

4.4a On RR-Y and RR-1 (2000 and 2001) models slacken the lockring (A) and turn the adjuster (B) in or out as required

Routine maintenance and servicing 1•9

4.4b On RR-2 and RR-3 (2002 and 2003) models turn the adjuster in or out as required

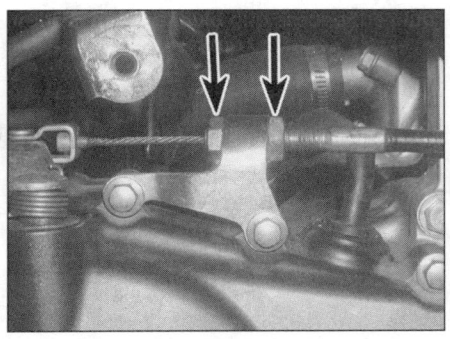

4.6 Slacken and adjust the nuts (arrowed) as described

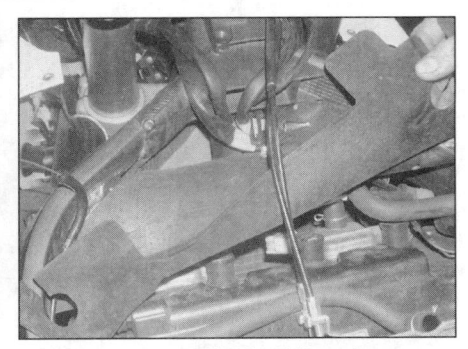

5.2 Lift the heat shield off the valve cover, noting how it fits round the hoses

illustration). To increase freeplay, thread the adjuster into the lever bracket. To reduce freeplay, thread the adjuster out of the bracket. On RR-Y and RR-1 (2000 and 2001) models tighten the lockring. When adjusting the cable make sure that the slot in the adjuster, and on RR-Y and RR-1 (2000 and 2001) models in the lockring, are not aligned with the slot in the lever bracket – these slots are to allow removal of the cable, and if they are all aligned while the bike is in use the cable could jump out. Also make sure the adjuster is not threaded too far out of the bracket so that it is only held by a few threads – this will leave it unstable and the threads could be damaged.

5 If all the adjustment has been taken up at the lever, thread the adjuster all the way into the bracket to give the maximum amount of freeplay, then back it out one turn – this resets the adjuster to its start point. Now set the correct amount of freeplay using the adjuster on the clutch end of cable. The adjuster is set in a bracket on the clutch cover on the right-hand side of the engine. Remove the right-hand fairing side panel for access (see Chapter 8).

6 Use the nuts on each end of the threaded section in the cable to adjust freeplay **(see illustration)**. To increase freeplay, slacken the front nut and tighten the rear nut until the freeplay is as specified, then tighten the front nut. To reduce freeplay, slacken the rear nut and tighten the front nut until the freeplay is as specified, then tighten the rear nut. Subsequent adjustments can now be made using the lever adjuster only.

5 Spark plugs – check

Every 4000 miles (6000 km) – US and Canada RR-Y and RR-1 models

Every 8000 miles (12,000 km) – European RR-Y and RR-1 models

Every 16,000 miles (24,000 km) – All RR-2 and RR-3 models

Note: *All models are equipped with plugs that have an iridium coated centre electrode. The plugs must be treated differently to conventional plugs, so be sure to follow the Steps as described – do not treat them in the same way as conventional plugs.*

1 Make sure your spark plug socket is the correct size before attempting to remove the plugs – a suitable one is supplied in the motorcycle's tool kit which is stored under the seat.

2 Remove the air filter housing (see Chapter 4). Lift the rubber heat shield **(see illustration)**.

3 Clean the area around the plug caps to prevent any dirt falling into the spark plug channels.

4 Check that the cylinder location is marked on each combined ignition coil/spark plug cap, and on RR-Y and RR-1 (2000 and 2001) models note the blue colour code on the No. 2 cylinder wiring, then disconnect the coil/cap wiring connectors **(see illustration)**. Pull the coil/cap off each spark plug **(see illustration)**.

5 Using either the plug removing tool supplied in the bike's toolkit or a deep socket type wrench, unscrew and remove the plugs from the cylinder head **(see illustrations)**. Lay each plug out in relation to its cylinder; if any plug shows up a problem it will then be easy to identify the troublesome cylinder.

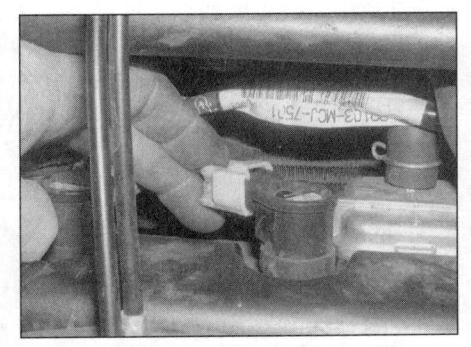

5.4a Disconnect the coil/cap wiring connector . . .

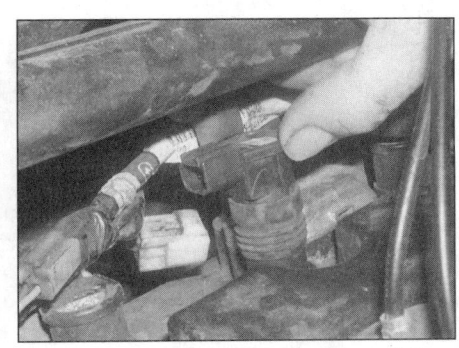

5.4b . . . then pull the coil/cap up off the plug

5.5a If you are using the Honda tool, locate it onto the plug and use a ring spanner to turn it

5.5b Lift the plug out with the tool – the rubber insert should grip around the plug top

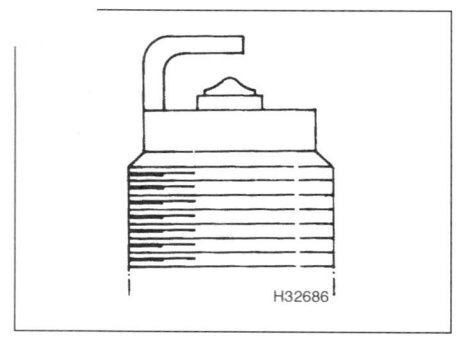

Routine maintenance and servicing

5.7a If the centre electrode has rounded off the plug is worn

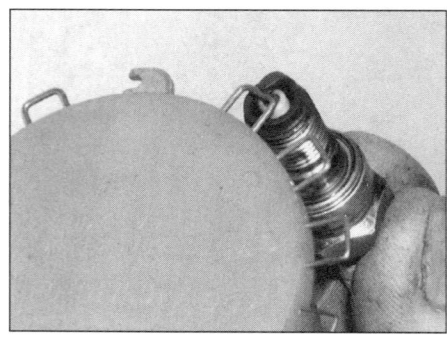

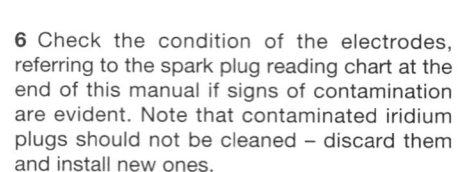

5.7b Using a wire type gauge ...

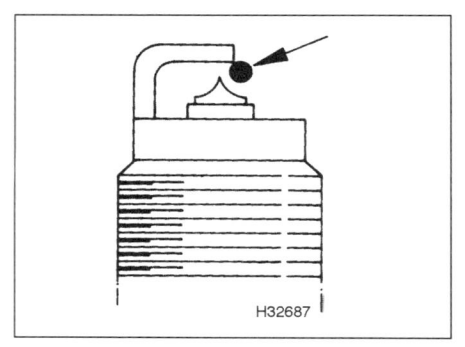

5.7c ... to measure the spark plug electrode gap

6 Check the condition of the electrodes, referring to the spark plug reading chart at the end of this manual if signs of contamination are evident. Note that contaminated iridium plugs should not be cleaned – discard them and install new ones.

7 Examine the pointed iridium-tipped centre electrode; if the tip has rounded off, the plug is worn **(see illustration)**. Measure the gap between the two electrodes with a wire type gauge only **(see illustrations)** – do not use blade type feeler gauges because the iridium tip might be damaged. The gap should be as given in the Specifications at the beginning of this chapter; if the electrodes have worn and the gap is wider than it should be, or for some reason the gap is narrower than it should be (if the plug has been dropped for instance) a new plug must be installed. Do not bend the outer electrode to adjust the gap.

8 Check the threads, the washer and the ceramic insulator body for cracks and other damage.

9 Note that Honda advise that the specified iridium plugs only must be fitted – do not substitute with conventional plugs.

10 Fit the plug into the end of the tool, then use the tool to insert the plug **(see illustration 5.5b)**. Since the cylinder head is made of aluminium, which is soft and easily damaged, thread the plugs as far as possible into the head turning the tool by hand. Once the plugs are finger-tight, the job can be finished with a spanner on the tool supplied or a socket drive **(see illustration 5.5a)**. If new plugs are being used, tighten them by 1/2 a turn after the washer has seated. If the old plugs are being reused, tighten them by 1/8 to 1/4 turn after they have seated, or if a torque wrench can be applied, tighten the spark plugs to the torque

> **HAYNES HiNT** As the plugs are quite recessed, slip a short length of hose over the end of the plug to use as a tool to thread it into place. The hose will grip the plug well enough to turn it, but will start to slip if the plug begins to cross-thread in the hole – this will prevent damaged threads.

setting specified at the beginning of the Chapter. Otherwise tighten them according to the instructions on the box. Do not over-tighten them.

11 Install the spark plug coils/caps, making sure they locate correctly onto the plugs **(see illustration 5.4b)**. Reconnect the coil/cap wiring connectors, making sure they are securely connected to the correct cylinder – on RR-Y and RR-1 (2000 and 2001) models the blue colour coded wiring must connect to the No. 2 cylinder coil/cap **(see illustration 5.4a)**. Install all other components previously removed.

> **HAYNES HiNT** Stripped plug threads in the cylinder head can be repaired with a Heli-Coil insert – see 'Tools and Workshop Tips' in the Reference section.

6 Spark plugs – renewal

Every 8000 miles (12,000 km) – US and Canada RR-Y and RR-1 models

Every 16,000 miles (24,000 km) – European RR-Y and RR-1 models

Every 32,000 miles (48,000 km) – All RR-2 and RR-3 models

1 Remove the old spark plugs as described in Section 5 and install new ones at the specified interval for your model. Note that Honda advise that the specified iridium plugs must be fitted – do not substitute with conventional plugs.

7 Sidestand, lever pivots and cables – lubrication

Every 8000 miles (12,000 km) or 12 months

1 Since the controls, cables and various other components of a motorcycle are exposed to the elements, they should be lubricated periodically to ensure safe and trouble-free operation.

2 Lubricate the footrests, clutch and brake levers, brake pedal, gearchange lever and linkage, and sidestand pivot. In order for the lubricant to be applied where it will do the most good, the component should be disassembled. The lubricant recommended by Honda for each application is listed at the beginning of the Chapter. If chain or cable lubricant is being used, it can be applied to the pivot joint gaps and will usually work its way into the areas where friction occurs, so less disassembly of the component is needed (however it is always better to do so and clean off all dirt and old lubricant first). If motor oil or light grease is being used, apply it sparingly as it may attract dirt (which could cause the controls to bind or wear at an accelerated rate). **Note:** *One of the best alternative lubricants for the control lever pivots is a dry-film lubricant (available from many sources by different names).*

3 To lubricate the cables, disconnect the relevant cable at its upper end, then lubricate it with a pressure adapter and aerosol lubricant, or if one is not available, using the set-up shown **(see illustrations)**. See Chapter 4 for the throttle cable removal procedures.

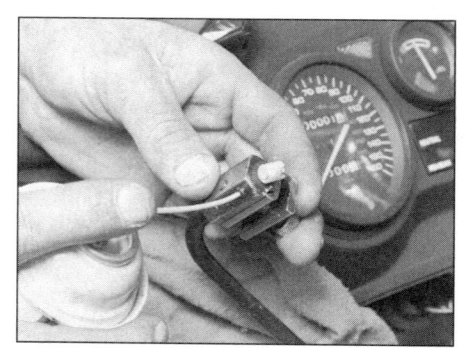

7.3a Lubricating a cable with a pressure lubricator. Make sure the tool seals around the inner cable

Routine maintenance and servicing 1•11

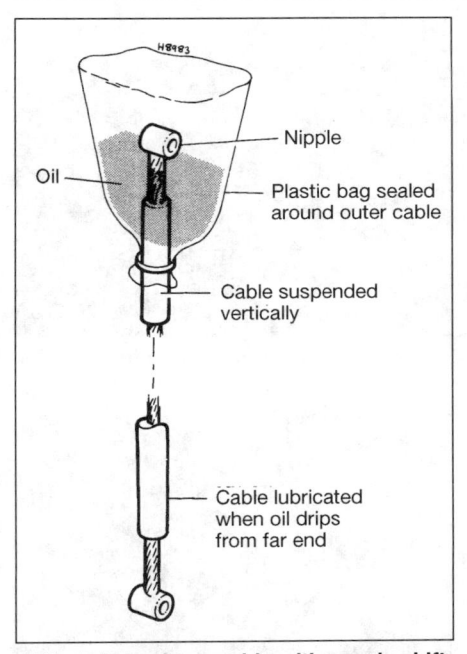

7.3b Lubricating a cable with a makeshift funnel and motor oil

8 Engine – oil and oil filter change

Every 8000 miles (12,000 km) or 12 months

⚠️ **Warning:** *Be careful when draining the oil, as the exhaust pipes, the engine, and the oil itself can cause severe burns.*

1 Consistent routine oil and filter changes are the single most important maintenance procedure you can perform on a motorcycle. The oil not only lubricates the internal parts of the engine, transmission and clutch, but it also acts as a coolant, a cleaner, a sealant, and a protectant. Because of these demands, the oil takes a terrific amount of abuse and should be replaced often with new oil of the recommended grade and type. The oil filter should be changed with every oil change.

> **HAYNES HiNT**
> *Saving a little money on the difference in cost between a good oil and a cheap oil won't pay off if the engine is damaged.*

2 Before changing the oil, warm up the engine so the oil will drain easily. Make sure the bike is on level ground, or if there is a slight slope make sure the front of the bike is pointing up the hill. Put the motorcycle on its sidestand rather than using an auxiliary stand. The oil drain plug is at the back of the engine on the left-hand side so the angles created by a slope and using the sidestand will help the oil to drain. On RR-Y and RR-1 (2000 and 2001) models remove the lower fairing (See Chapter 8). On RR-2 and RR-3 (2002 and 2003) models remove the fairing side panels (see Chapter 8).

3 Position a clean drain tray below the engine. Unscrew the oil filler cap from the clutch cover to vent the crankcase and to act as a reminder that there is no oil in the engine **(see illustration)**.

4 Unscrew the oil drain plug from the rear left-hand side of the engine and allow the oil to flow into the drain tray **(see illustrations)**. Check the condition of the sealing washer on the drain plug and replace it with a new one if it is damaged or worn – it is advisable to use a new one whatever the condition of the old one.

5 When the oil has completely drained, fit the plug into the sump, using a new sealing washer if necessary, and tighten it to the torque setting specified at the beginning of the Chapter **(see illustrations)**. Avoid overtightening, as it is quite easy to damage the threads in the sump.

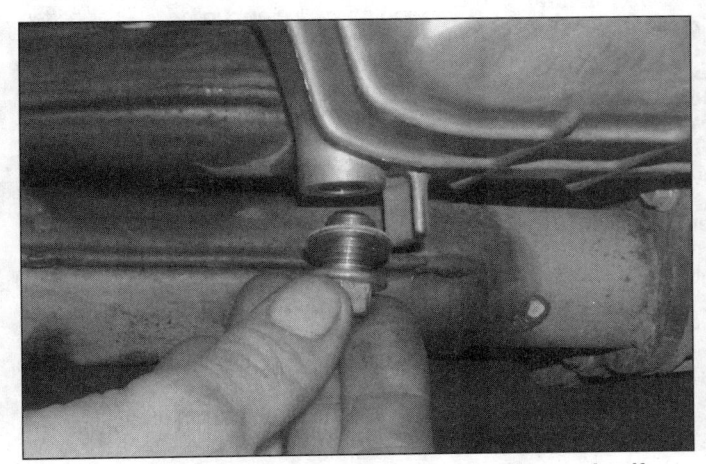

8.3 Unscrew the oil filler cap to act as a vent ...

8.4a ... then unscrew the oil drain plug (arrowed) ...

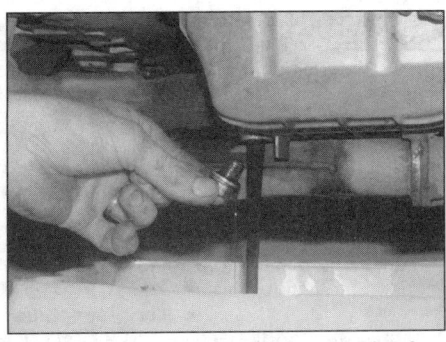

8.4b ... and allow the oil to completely drain

8.5a Install the drain plug, using a new sealing washer if necessary, ...

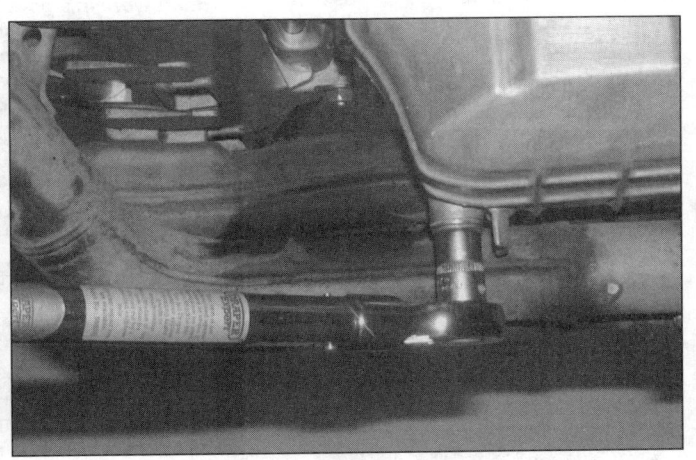

8.5b ... and tighten it to the specified torque setting

1•12 Routine maintenance and servicing

6 Now place the drain tray below the oil filter on the front of the engine. Unscrew the oil filter using a filter socket (one can be obtained as a kit with the new filter from Honda dealers, or separately under part No. 07HAA-PJ70101, or otherwise there are commercially available equivalents available from good accessory dealers), a filter removing strap or a chain-wrench, and tip any residual oil into the drain tray **(see illustrations)**. The filter socket is preferable because it allows a means of tightening the new filter to the correct torque.

7 Smear clean engine oil onto the rubber seal on the new filter and thread it onto the engine **(see illustration)**. Tighten it to the specified torque setting using the filter socket if available **(see illustration)**, or tighten the filter as tight as possible by hand, or by the number of turns specified on the filter itself or its packaging. **Note:** *Do not use a strap or chain filter removing tool to tighten the filter as you will damage it.*

8 Refill the engine to the proper level using the recommended type and amount of oil (see *Daily (pre-ride) checks*). With the motorcycle vertical, the oil level should lie between the maximum and minimum level lines on the inspection window (see *Daily (pre-ride) checks*). Install the filler cap **(see illustration 8.3)**. Start the engine and let it run for two or three minutes (make sure that the oil pressure light extinguishes after a few seconds). Shut it off, wait a few minutes, then check the oil level. If necessary, add more oil to bring the level close to the maximum line, but do not go above it. Check around the drain plug and the oil filter for leaks. A leak around the drain plug probably means a new washer is needed. A leak around the filter probably means it is not tight enough. Install the lower fairing or fairing side panels as required according to model (see Chapter 8).

9 The old oil drained from the engine cannot be re-used and should be disposed of properly. Check with your local refuse disposal company, disposal facility or environmental agency to see whether they will accept the used oil for recycling. Don't pour used oil into drains or onto the ground.

> **HAYNES HINT**
> *Check the old oil carefully – if it is very metallic coloured, then the engine is experiencing wear from break-in (new engine) or from insufficient lubrication. If there are flakes or chips of metal in the oil, then something is drastically wrong internally and the engine will have to be disassembled for inspection and repair. If there are pieces of fibre-like material in the oil, the clutch is experiencing excessive wear and should be checked.*

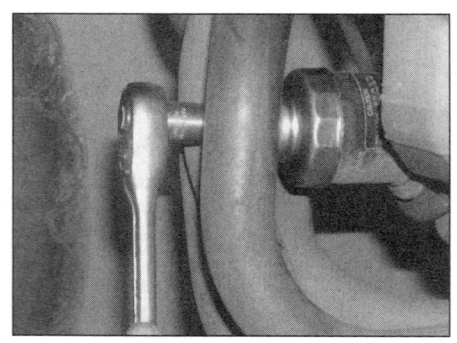

8.6a Unscrew the filter using a filter removing tool . . .

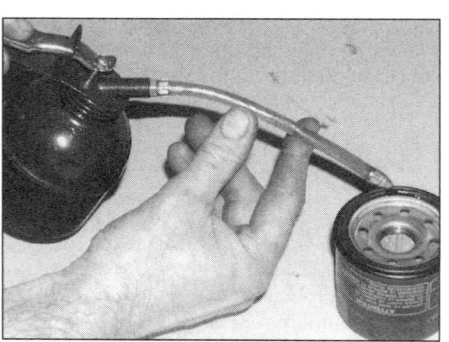

8.7a Smear clean oil onto the seal . . .

9 Fuel system – check

Every 8000 miles (12,000 km) or 12 months

⚠ *Warning: Petrol (gasoline) is extremely flammable, so take extra precautions when you work on any part of the fuel system. Don't smoke or allow open flames or bare light bulbs near the work area, and don't work in a garage where a natural gas-type appliance is present. If you spill any fuel on your skin, rinse it off immediately with soap and water. When you perform any kind of work on the fuel system, wear safety glasses and have a fire extinguisher suitable for a Class B type fire (flammable liquids) on hand.*

1 Raise the fuel tank (see Chapter 4). Check the tank and the fuel hoses for signs of leakage, deterioration and damage; in particular check that there is no leakage from the fuel hoses. Replace any hoses which are cracked or deteriorated with new ones (see Section 34).

2 Also check for signs of fuel leakage between the injectors, the fuel rail and the throttle bodies. If there are, remove the injectors and install new O-rings and seals (see Chapter 4).

3 If there are signs of fuel starvation or the

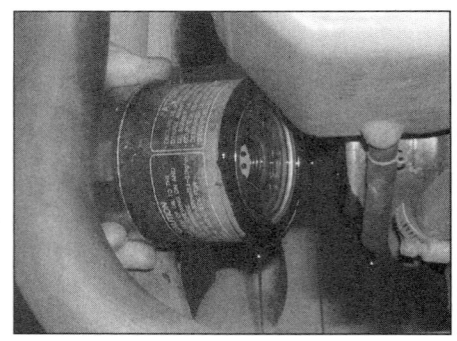

8.6b . . . and allow the oil to drain

8.7b . . . then install the filter and tighten it as described

machine has covered a high mileage, the fuel filters could be in need of cleaning or renewal (see Section 36).

10 Battery – check

Every 8000 miles (12,000 km) or 12 months

1 All models covered in this manual are fitted with a sealed MF (maintenance free) battery. **Note:** *Do not attempt to remove the battery caps to check the electrolyte level or battery specific gravity. Removal will damage the caps, resulting in electrolyte leakage and battery damage. All that should be done is to check that the terminals are clean and tight and that the casing is not damaged or leaking. See Chapter 9 for further details.*

2 If the machine is not in regular use, disconnect the battery and give it a refresher charge every month to six weeks (see Chapter 9).

11 Throttle cables – check

Every 8000 miles (12,000 km) or 12 months

1 Make sure the throttle grip rotates smoothly and freely from fully closed to fully open with

Routine maintenance and servicing 1•13

11.3 Throttle cable freeplay is measured in terms of twistgrip rotation

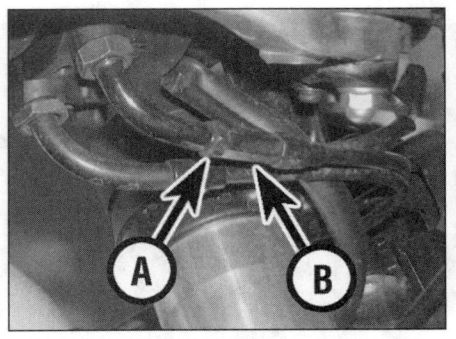

11.4 Throttle cable adjuster locknut (A) and adjuster (B) – throttle end

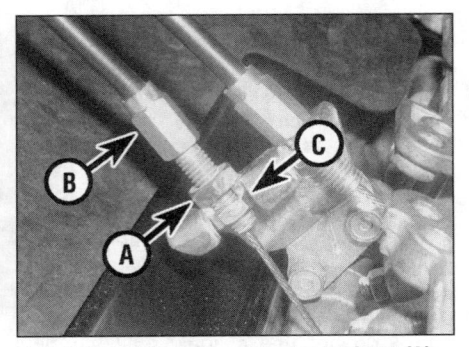

11.6 Throttle cable adjuster locknut (A), adjuster (B) and lower nut (C)

the front wheel turned at various angles. The grip should return automatically from fully open to fully closed when released.

2 If the throttle sticks, this is probably due to a cable fault. Remove the cables (see Chapter 4) and lubricate them (see Section 7). Check that the inner cables slide freely and easily in the outer cables. If not, replace the cables with new ones. With the cables removed, make sure the throttle twistgrip rotates freely on the handlebar. If necessary, unscrew the handlebar end-weight and slide the twistgrip off the handlebar. Clean any old grease from the bar and the inside of the tube. Smear some new grease of the specified type onto the bar, then refit the twistgrip. When fitting the end-weight, align the boss with the cut-out on the inner weight inside the handlebar. Clean the threads of the end-weight retaining screw, then apply a suitable non-permanent thread locking compound and tighten it to the torque setting specified at the beginning of the Chapter. Install the cables, making sure they are correctly routed (see Chapter 4). If this fails to improve the operation of the throttle, the cables must be replaced with new ones. Note that in very rare cases the fault could lie in the throttle bodies rather than the cables, necessitating their removal and inspection (see Chapter 4).

3 With the throttle operating smoothly, check for a small amount of freeplay in the cables, measured in terms of the amount of twistgrip rotation before the throttle opens, and compare the amount to that listed in this Chapter's Specifications **(see illustration)**. If it's incorrect, adjust the cables to correct it as follows.

4 Initially adjust freeplay using the adjuster in the throttle opening cable where it leaves the throttle/switch housing on the handlebar. Loosen the locknut and turn the adjuster in or out as required until the specified amount of freeplay is obtained (see this Chapter's Specifications), then retighten the locknut **(see illustration)**.

5 If the adjuster has reached its limit of adjustment, reset it to its start point by turning it fully in, so that freeplay is at a maximum, then remove the air filter housing (see Chapter 4), and adjust the cable at the throttle body end.

6 The adjuster is on the lower cable in the bracket. Slacken the adjuster locknut, then screw the adjuster in or out as required, making sure the lower nut remains captive in the bracket, thereby threading itself along the adjuster as you turn it, until the specified amount of freeplay is obtained, then tighten the locknut **(see illustration)**. Subsequent adjustments can be made at the throttle end when required. If the cable cannot be adjusted as specified, replace it with a new one (see Chapter 4). Check that the throttle twistgrip operates smoothly and snaps shut quickly when released.

⚠ **Warning: Turn the handlebars all the way through their travel with the engine idling. Idle speed should not change. If it does,** the cables may be routed incorrectly. Correct this condition before riding the bike.

12 Cooling system – check

Every 8000 miles (12,000 km) or 12 months

⚠ **Warning: The engine must be cool before beginning this procedure.**

1 Check the coolant level in the reservoir (see Daily (pre-ride) checks).

2 On RR-Y and RR-1 (2000 and 2001) models, remove the right-hand air duct cover, the lower fairing and the fairing side panels, and the right-hand heat guard (see Chapter 8). On RR-2 and RR-3 (2002 and 2003) models remove the right-hand air duct cover and the fairing side panels (see Chapter 8). Check the entire cooling system for evidence of leakage. Examine each rubber coolant hose along its entire length. Look for cracks, abrasions and other damage. Squeeze each hose at various points to see whether they are dried out or hard **(see illustration)**. They should feel firm, yet pliable, and return to their original shape when released. If necessary, replace them with new ones (see Chapter 3).

3 Check for evidence of leaks at each cooling system joint and around the pump on the left-hand side of the engine. Tighten the hose clips carefully to prevent future leaks. If the pump is leaking around the cover, check that the bolts are tight. If they are, remove the cover and replace the O-ring with a new one (see Chapter 3). If it is leaking around the crankcase, remove the pump and replace the body O-ring with a new one (see Chapter 3).

4 To prevent leakage of coolant from the cooling system to the lubrication system and vice versa, two seals are fitted on the pump shaft. On the bottom of the pump housing there is a drain hole **(see illustration)**. If either seal fails, the drain allows the coolant or oil to escape and prevents them mixing. The seal on the water pump side is of the mechanical type which bears on the rear face of the

12.2 Check all the coolant hoses as described

12.4 Check the pump drain hole (arrowed) for signs of leakage

1•14 Routine maintenance and servicing

impeller. The second seal, which is mounted behind the mechanical seal is of the normal feathered lip type. If on inspection the drain shows signs of leakage, remove the pump and replace it with a new one – it comes as an assembly, and the seals are not available separately (see Chapter 3).

5 Check the radiator for leaks and other damage. Leaks in the radiator leave tell-tale scale deposits or coolant stains on the outside of the core below the leak. If leaks are noted, remove the radiator (see Chapter 3) and have it repaired or replace it with a new one – do not use a liquid leak stopping compound to try to repair leaks.

6 Check the radiator fins for mud, dirt and insects, which may impede the flow of air through the radiator. If the fins are dirty, remove the radiator (see Chapter 3) and clean it using water or low pressure compressed air directed through the fins from the inner side of the radiator. If the fins are bent or distorted, straighten them carefully with a screwdriver. If the air flow is restricted by bent or damaged fins over more than 20% of the radiator's surface area, replace the radiator with a new one.

7 Remove the pressure cap from the radiator filler neck by holding it with a heavy cloth and turning it anti-clockwise until it reaches a stop **(see illustration)**. If you hear a hissing sound (indicating there is still pressure in the system), wait until it stops. Now press down on the cap and continue turning it until it can be removed. Check the condition of the coolant in the system. If it is rust-coloured or if accumulations of scale are visible, drain and flush the system and refill it with new coolant (See Section 26). Check the cap seal for cracks and other damage. If in doubt about the pressure cap's condition, have it tested by a Honda dealer or replace it with a new one.

8 Check the anti-freeze content of the coolant with an anti-freeze hydrometer. Sometimes coolant looks like it's in good condition, but might be too weak to offer adequate protection. If the hydrometer indicates a weak mixture, drain, flush and refill the system (see Section 26).

9 Install the cap by turning it clockwise until it reaches the first stop then push down on it and continue turning until it can turn no further. Start the engine and let it reach normal operating temperature, then check for leaks again. As the coolant temperature increases, the electric fan (mounted on the back of the radiator) should come on automatically and the temperature should begin to drop. If it does not, refer to Chapter 3 and check the fan and fan circuit carefully.

10 If the coolant level is consistently low, and no evidence of leaks can be found, have the entire system pressure checked by a Honda dealer.

11 Check the oil cooler on the front of the engine for any signs of oil leakage between it and the filter, and between it and the engine.

12.7 Remove the pressure cap (arrowed) as described

Tighten the filter or the cooler bolt as required (remove the filter to access it – see Section 8), but if the leakage persists you will have to fit a new filter or a new O-ring between the cooler and the engine, as required (see Chapter 2). Check that the coolant pipes to the cooler are tight, and that there is no evidence of coolant leakage from the body of the cooler. If there is, the cooler is damaged and must be replaced with a new one.

13 Brake system – check

Every 8000 miles (12,000 km) or 12 months

1 A routine general check of the brake system will ensure that any problems are discovered and remedied before the rider's safety is jeopardised.

2 Check the brake lever and pedal for loose connections, improper or rough action, excessive play, bends, and other damage. Replace any damaged parts with new ones (see Chapter 7).

3 Make sure all brake component fasteners are tight. Check the brake pads for wear (see Section 3) and make sure the fluid level in the reservoirs is correct (see *Daily (pre-ride) checks*). Look for leaks at the hose connections and check for cracks in the hoses and unions **(see illustration)**. If the

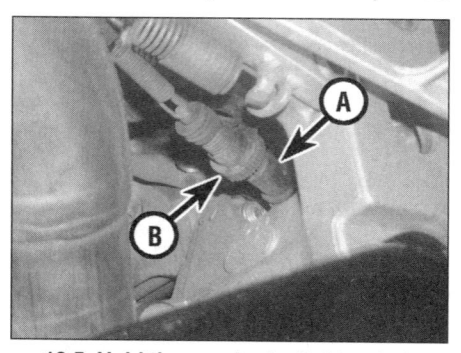

13.5 Hold the rear brake light switch body (A) and turn the adjuster ring (B) as required

13.3 Flex the hoses and check for cracks, bulges and leaking fluid

lever or pedal is spongy, bleed the brakes (see Chapter 7).

4 Make sure the brake light operates when the front brake lever is pulled in. The front brake light switch, mounted on the underside of the master cylinder, is not adjustable. If it fails to operate properly, check it (see Chapter 9).

5 Make sure the brake light is activated just before the rear brake takes effect. If adjustment is necessary, hold the switch and turn the adjuster ring on the switch body until the brake light is activated when required **(see illustration)**. The switch is mounted behind the brake pedal. If the brake light comes on too late or not at all, turn the ring clockwise (when looked at from the front) so the switch threads out of the bracket. If the brake light comes on too soon or is permanently on, turn the ring anti-clockwise so the switch threads into the bracket. If the switch doesn't operate the brake light, check it (see Chapter 9).

6 The front brake lever has a span adjuster which alters the distance of the lever from the handlebar **(see illustration)**. Each setting is identified by a number on the adjuster which aligns with the arrow on the lever. Turn the adjuster ring until the setting which best suits the rider is obtained. Do not set the adjuster between the defined settings.

7 The height of the rear brake pedal can be adjusted to suit the rider's preference. Slacken the clevis locknut on the master cylinder pushrod, then turn the pushrod using a spanner on the hex at the top of the rod until

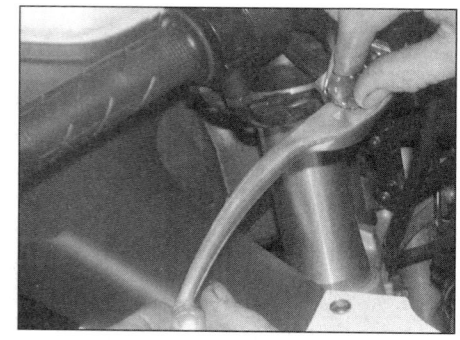

13.6 Adjusting the front brake lever span

Routine maintenance and servicing 1•15

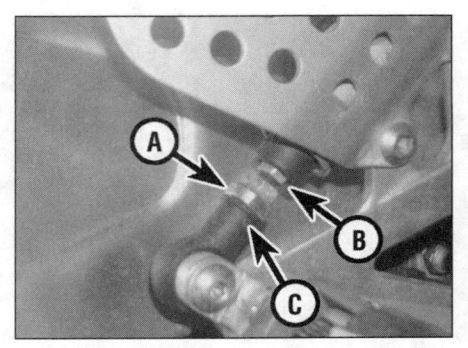

13.7 Slacken the locknut (A) and turn the pushrod using the hex (B) to adjust pedal height. Check that some of the pushrod is visible in the hole (C)

the pedal is at the desired height **(see illustration)**. Always make sure that some part of the pushrod is visible via the hole in the top of the clevis – if not there will not be enough of the rod threaded into the clevis to make it secure. On completion tighten the locknut securely. Adjust the rear brake light switch after adjusting the pedal height (see Step 5).

14 Headlight aim – check and adjustment

Every 8000 miles (12,000 km) or 12 months

Note: *An improperly adjusted headlight may cause problems for oncoming traffic or provide poor, unsafe illumination of the road ahead. Before adjusting the headlight aim, be sure to consult with local traffic laws and regulations – for UK models refer to MOT Test Checks in the Reference section.*

1 The headlight beam can adjusted both horizontally and vertically. Before making any adjustment, check that the tyre pressures are correct and the suspension is adjusted as required. Make any adjustments to the headlight aim with the machine on level ground, with the fuel tank half full and with an assistant sitting on the seat. If the bike is usually ridden with a passenger on the back, have a second assistant to do this.

15.1 Check the springs as described

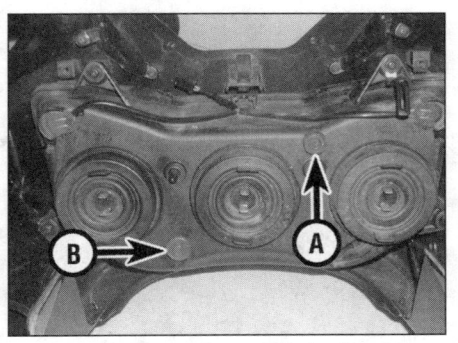

14.2 Vertical adjuster (A), horizontal adjuster (B)

2 Vertical adjustment is made by turning the adjuster screw on the top right corner of the centre headlight unit **(see illustration)**. Turn it clockwise to move the beam up, and anti-clockwise to move it down.
3 Horizontal adjustment is made by turning the adjuster bolt on the bottom left corner of the centre headlight unit **(see illustration 14.2)**. Turn it clockwise to move the beam to the right, and anti-clockwise to move it to the left.

15 Sidestand and starter interlock circuit – check

Every 8000 miles (12,000 km) or 12 months

1 Check the stand springs for damage and distortion **(see illustration)**. The springs must be capable of retracting the stand fully and holding it retracted when the motorcycle is in use. If a spring is sagged or broken it must be replaced with a new one.
2 Lubricate the stand pivot regularly (see Section 7).
3 Check the stand and its mount for bends and cracks. Stands can often be repaired by welding.
4 Check the operation of the starter interlock circuit as follows: Firstly, make sure the transmission is in neutral, then retract the stand and start the engine. Pull in the clutch lever and select a gear. Extend the sidestand. The engine should stop as the sidestand is extended.

16.7a Checking for play in the swingarm bearings

Secondly, make sure the engine is in neutral and the sidestand is down, the start the engine. Pull the clutch lever in and select a gear. The engine should cut out. Also check that when the sidestand is down the engine can only be started if the transmission is in neutral, and when the sidestand is up and the transmission is in gear the engine can only be started if the clutch lever is pulled in. If the circuit does not operate as described, check the sidestand switch, neutral switch and the clutch switch, and the circuit between them (see Chapter 9).

16 Suspension – check

Every 8000 miles (12,000 km) or 12 months

1 The suspension components must be maintained in top operating condition to ensure rider safety. Loose, worn or damaged suspension parts decrease the motorcycle's stability and control.

Front suspension

2 While standing alongside the motorcycle, apply the front brake and push on the handlebars to compress the forks several times. See if they move up-and-down smoothly without binding. If binding is felt, the forks should be disassembled and inspected (see Chapter 6).
3 Inspect the area around the dust seal on the bottom of the fork tube for signs of oil leakage, then carefully lever the seal down using a flat-bladed screwdriver and inspect the area around the fork seal. If leakage is evident, the seals must be replaced with new ones (see Chapter 6). If there is evidence of corrosion between the seal retaining ring and its groove in the fork slider spray the area with a penetrative lubricant, otherwise the ring will be difficult to remove if needed. Press the dust seal back into the bottom of the fork tube on completion.
4 Check the tightness of all suspension nuts and bolts to be sure none have worked loose.

Rear suspension

5 Inspect the rear shock absorber for fluid leakage and tightness of its mountings. If leakage is found, the shock must be replaced with a new one (see Chapter 6).
6 With the aid of an assistant to support the bike, compress the rear suspension several times. It should move up-and-down freely without binding. If any binding is felt, the worn or faulty component must be identified and checked (see Chapter 6). The problem could be due to either the shock absorber, the suspension linkage components or the swingarm components.
7 Support the motorcycle on an auxiliary stand so that the rear wheel is off the ground. Grab the swingarm and rock it from side-to-side – there should be no discernible movement at the rear **(see illustration)**. If there's a little

1•16 Routine maintenance and servicing

16.7b Checking for play in the rear shock mountings and suspension linkage bearings

17.4 Checking for play in the steering head bearings

17.6 Slacken the fork clamp bolt (arrowed) on each side

movement or a slight clicking can be heard, inspect the tightness of all the swingarm and rear suspension mounting bolts and nuts, referring to the torque settings specified at the beginning of Chapter 6, and re-check for movement. Next, grasp the top of the rear wheel and pull it upwards – there should be no discernible freeplay before the shock absorber begins to compress **(see illustration)**. Any freeplay felt in either check indicates worn bearings in the suspension linkage or swingarm, or worn shock absorber mountings. The worn components must be identified and replaced with new ones (see Chapter 6).

8 To make an accurate assessment of the swingarm bearings, remove the rear wheel (see Chapter 7) and the bolt securing the suspension linkage assembly to the swingarm (see Chapter 6). Grasp the rear of the swingarm with one hand and place your other hand at the junction of the swingarm and the frame. Try to move the rear of the swingarm from side-to-side. Any wear (play) in the bearings should be felt as movement between the swingarm and the frame at the front. If there is any play the swingarm will be felt to move forward and backward at the front (not from side-to-side). Next, move the swingarm up and down through its full travel. It should move freely, without any binding or rough spots. If there is any play in the swingarm or if it does not move freely, remove the bearings for inspection (see Chapter 6).

17 Steering head bearings – freeplay check and adjustment

Every 8000 miles (12,000 km) or 12 months

1 This motorcycle is equipped with caged ball steering head bearings which can become dented, rough or loose during normal use of the machine. In extreme cases, worn or loose steering head bearings can cause steering wobble – a condition that is potentially dangerous.

Check

2 On RR-Y and RR-1 (2000 and 2001) models remove the lower fairing and on RR-2 and RR-3 (2002 and 2003) models remove the fairing side panels (see Chapter 8). Raise the front wheel off the ground using an auxiliary stand placed under the engine. Always make sure that the bike is properly supported and secure. Due to the shape of the sump it is not advisable to use only one jack under the engine as the bike could easily unbalance – either use two jacks, or place blocks of wood or axle stands under the engine after raising it on the jack as extra support. Whenever a jack or axle stand is used, place a block wood between the jack or stand head and the engine to distribute the weight more evenly.

3 Point the front wheel straight-ahead and slowly move the handlebars from side-to-side. Any dents or roughness in the bearing races will be felt and the bars will not move smoothly and freely. Again point the wheel straight-ahead, and tap the front of the wheel to one side. The wheel should 'fall' under its own weight to the limit of its lock, indicating that the bearings are not too tight. Check for similar movement to the other side. If available, attach one end of a spring balance (graduated zero to 30 N) to the outer end of the rubber grip on the handlebars. With the steering straight-ahead, pull on the balance and check the reading at which the handlebars start to turn. If the reading is below the minimum value specified in the pre-load range given in the Specifications at the beginning of the Chapter, the steering head is too loose, if the reading is above the maximum value specified the steering head is too tight. If the steering doesn't perform as described, and it's not due to the resistance of cables or hoses, then the bearings should be adjusted as described below.

4 Next, grasp the bottom of the forks and gently pull and push them forward and backward **(see illustration)**. Any looseness or freeplay in the steering head bearings will be felt as front-to-rear movement of the forks. If play is felt, adjust the bearings as described below.

> **HAYNES HINT** *Make sure you are not mistaking any movement between the bike and stand, or between the stand and the ground, for freeplay in the bearings. Do not pull and push the forks too hard – a gentle movement is all that is needed. Freeplay between the fork slider and the fork tube due to worn bushes can also be misinterpreted as steering head bearing play – do not confuse the two.*

Adjustment

5 As a precaution, remove the fuel tank (see Chapter 4) and the fairing (see Chapter 8). Though not actually necessary, this will prevent the possibility of damage should a tool slip.

6 Slacken the fork clamp bolts in the top yoke **(see illustration)**.

7 Unscrew and remove the steering stem nut **(see illustration)**.

8 Gently ease the top yoke up off the fork tubes and position it clear of the head bearings, using a rag to protect other components **(see illustration)**.

17.7 Unscrew the steering stem nut (arrowed) . . .

17.8 . . . and gently ease the yoke up off the forks

Routine maintenance and servicing 1•17

17.9a Bend down the tabs securing the locknut . . .

17.9b . . . then unscrew the locknut . . .

17.9c . . . and remove the lockwasher

9 Bend the lockwasher tabs out of the notches in the locknut **(see illustration)**. Unscrew the locknut using either your fingers (it shouldn't be tight), a C-spanner or a suitable drift located in one of the notches **(see illustration)**. Remove the lockwasher, bending up the remaining tabs to release it from the adjuster nut if necessary **(see illustration)**. Inspect the tabs for cracks or signs of fatigue. If there is any sign of damage, discard the lockwasher and use a new one; otherwise the old one can be re-used, but note that Honda do recommend using a new one as a matter of course.

10 Using either the C-spanner or drift, slacken the adjuster nut slightly until pressure is just released, then tighten it until all freeplay is removed, yet the steering is able to move freely **(see illustration)**. The object is to set the adjuster nut so that the bearings are under a very light loading, just enough to remove any freeplay, but not so much that the steering does not move freely from side-to-side as described in the check procedure above. If the Honda service tool (part No. 07916-3710101) or a suitable peg spanner (which can be made by cutting castellations into an old socket) is available, tighten the adjuster nut first to the initial torque setting specified at the beginning of the Chapter, then turn the steering from lock-to-lock five times, then slacken the adjuster nut and retighten it to the final torque setting specified. On RR-2 (2002) models with caged ball bearings and all RR-3 (2003) models now tighten the nut by a further 90° (1/4 turn). However do not rely on the torque setting method alone and assume the loading to be correct – check the physical feel as described as well. If you have the spring balance (see Step 3), set the adjuster nut so that the steering starts to move at around the mid-point of the pre-load range given in the Specifications at the beginning of the Chapter.

Caution: Take great care not to apply excessive pressure because this will cause premature failure of the bearings.

11 If the bearings cannot be correctly adjusted, disassemble the steering head and check the bearings and races (see Chapter 6).

12 With the bearings correctly adjusted, install the lockwasher, using a new one if the tabs are weakened or cracked, onto the adjuster nut and fit the two short tabs into the slots in the adjuster nut **(see illustration 17.9c)**.

17.10 Adjust the bearings as described using either a C-spanner or a drift

13 Hold the adjuster nut to prevent it from moving, then install the locknut and tighten it finger-tight **(see illustration)**. Tighten the locknut further (but no more than 90°) until its notches align with the remaining lockwasher tabs, making sure the adjuster nut does not turn as well (though that is unlikely). Secure the locknut in position by bending up the long lock washer tabs into its notches **(see illustration)**.

14 Fit the top yoke onto the steering stem, locating the lug on the top of each handlebar clamp into its hole in the underside of the

17.13a Thread the locknut on and tighten it as described

17.13b Bend the tabs up into the notches in the locknut

1•18 Routine maintenance and servicing

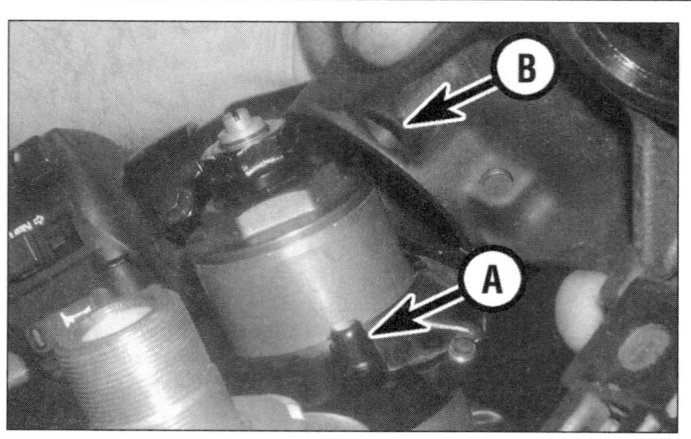

17.14a Locate each lug (A) in its hole (B)

17.14b Install the steering stem nut . . .

yoke **(see illustration)**. Install the steering stem nut and tighten it to the torque setting specified at the beginning of the Chapter **(see illustrations)**. Now tighten both the top yoke fork clamp bolts to the specified torque **(see illustration 17.6)**.

15 Check the bearing adjustment as described above and re-adjust if necessary.

16 Install the fuel tank (see Chapter 4) and the fairing (see Chapter 8).

18 Nuts and bolts – tightness check

Every 8000 miles (12,000 km) or 12 months

1 Since vibration of the machine tends to loosen fasteners, all nuts, bolts, screws, etc, should be periodically checked for proper tightness.

2 Pay particular attention to the following:
  Spark plugs
  Engine oil drain plug and coolant drain plugs
  Engine mounting bolts
  Lever and pedal bolts
  Footrest and stand bolts
  Shock absorber and suspension linkage bolts and swingarm pivot bolts
  Handlebar clamp bolts
  Front axle bolt and axle clamp bolts
  Steering stem nut
  Front fork clamp bolts (top and bottom yoke) and fork top bolts
  Rear axle nut
  Brake caliper and master cylinder mounting bolts
  Brake hose banjo bolts and caliper bleed valves
  Brake disc bolts
  Exhaust system bolts/nuts

3 If a torque wrench is available, use it along with the torque specifications at the beginning of this and other Chapters.

19 Wheels and tyres – general check

Every 8000 miles (12,000 km) or 12 months

Tyres

1 Check the tyre condition and tread depth thoroughly – see *Daily (pre-ride) checks*.

Wheels

2 Cast wheels are virtually maintenance free, but they should be kept clean and checked periodically for cracks and other damage. Also check the wheel runout and alignment (see Chapter 7). Never attempt to repair damaged cast wheels; they must be replaced with new ones. Check the valve rubber for signs of damage or deterioration and have it replaced if necessary. Also, make sure the valve stem cap is in place and tight.

20 Pulse secondary air injection (PAIR) system – check

Every 8000 miles (12,000 km) or 12 months

1 Remove the air filter housing (see Chapter 4). Lift the rubber heat shield **(see illustration 5.2)**.

2 Visually inspect the hoses between the reed valves on the valve cover and the PAIR control valve above it, and between the control valve and the air filter housing, for kinks and splits and any other damage or deterioration **(see illustration)**. Make sure

17.14c . . . and tighten it to the specified torque

20.2 Check the PAIR system hoses as described

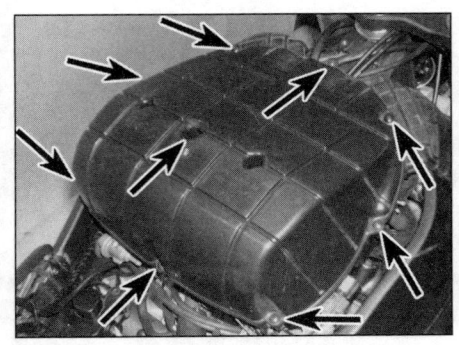

Routine maintenance and servicing 1•19

21.2a Undo the screws (arrowed) ...

21.2b ... remove the cover ...

21.2c ... and lift the element from the housing, noting how it fits

that all hoses are securely connected with a clamp on each end. Renew any hoses that are damaged or deteriorated.
3 See Chapter 4 for further information and tests on the system.

21 Air filter – renewal

Every 12,000 miles (18,000 km) or 18 months

Caution: If the machine is continually ridden in wet or dusty conditions, the filter should be replaced more frequently.

1 Raise the fuel tank (see Chapter 4). On RR-Y and RR-1 (2000 and 2001) models, disconnect the intake air temperature (IAT) sensor wiring connector.
2 Undo the screws securing the air filter housing cover and remove it **(see illustrations)**. Remove the element from the housing, noting how it fits, and discard it **(see illustration)**.
3 Fit the new filter element into the housing, making sure it is properly seated, then install the cover. On RR-Y and RR-1 (2000 and 2001) models, connect the IAT (intake air temperature) sensor wiring connector.
4 Install the fuel tank (see Chapter 4).
5 To clean the filter in between renewal intervals, tap it on a hard surface to dislodge any dirt and use compressed air to clear the element, directing the air in the opposite way to normal flow **(see illustration)**. Do not use any solvents or cleaning agents on the element as it is pre-treated with a dust adhesive.

22 Evaporative emission control (EVAP) system – check (California models)

Every 12,000 miles (18,000 km) or 18 months

1 Raise the fuel tank (see Chapter 4). Visually inspect all the system hoses between the fuel tank, the purge control solenoid valve, and the canister for kinks and splits and any other damage or deterioration. Make sure that the hoses are securely connected with a clamp on each end. Replace any hoses that are damaged or deteriorated.
2 Check the EVAP canister and the valve for cracks or other damage.
3 See Chapter 4 for further information and tests on the system. Note that there is an emission control system hose routing diagram on a label stuck to the top of the air filter housing, and an information label under the passenger seat.

23 Brake fluid – change

Every 12,000 miles (18,000 km) or two years

1 The brake fluid should be changed at the prescribed interval or whenever a master cylinder or caliper overhaul is carried out. Refer to the brake bleeding and fluid change section in Chapter 7.

HAYNES HiNT *Old brake fluid is invariably much darker in colour than new fluid, making it easy to see when all old fluid has been expelled from the system.*

21.5 Direct the air in the opposite direction of normal flow

24 Valve clearances – check and adjustment

Every 16,000 miles (24,000 km)

1 The engine must be completely cool for this maintenance procedure, so let the bike stand overnight before beginning.
2 Remove the spark plugs (see Section 5). Remove the valve cover (see Chapter 2).
3 Make a chart or sketch of all valve positions so that a note of each clearance can be made against the relevant valve.
4 Referring to Chapter 2, Section 9, remove the cap bolt and sealing washer from the end of the cam chain tensioner body and retract and hold the tensioner plunger as described. This releases all tension in the cam chain and ensures that the valve clearances are measured with the correct clearance between the camshafts and their holders. In case you are worried, Honda say that there is no chance of the chain jumping teeth on the sprockets with the tensioner removed, though it is worth keeping an eye on this throughout the procedure. If you are unable to fabricate the tool to hold the plunger retracted, remove the tensioner body as well to avoid having to hold the screwdriver in place while you check the clearances.
5 Unscrew the timing inspection cap from the clutch cover **(see illustration)**. To check the valve clearances the engine must be turned

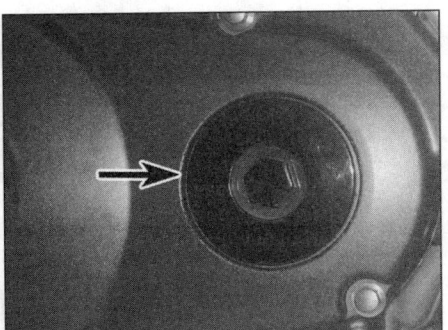

24.5 Remove the timing inspection cap (arrowed)

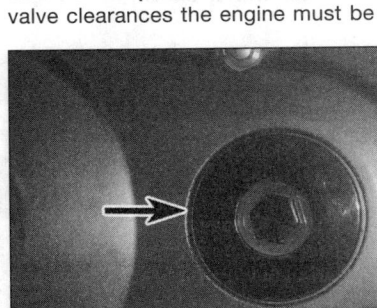

1•20 Routine maintenance and servicing

24.6a Turn the engine clockwise . . .

24.6b . . . until the line next to the T mark aligns with the notch (arrowed) . . .

24.6c . . . and the camshaft sprocket marks are as shown

so that the valve being checked is closed. The engine can be turned using a suitable spanner or a socket on the timing rotor bolt and turning it in a clockwise direction only **(see illustration 24.6a)**. Alternatively, place the motorcycle on an auxiliary stand so that the rear wheel is off the ground, select a high gear and rotate the rear wheel by hand in its normal direction of rotation.

6 Turn the engine clockwise until the line next to the T mark on the timing rotor aligns with the static timing mark, which is a notch in the inspection hole rim, and the IN and EX marks on the intake and exhaust camshaft sprockets respectively are facing away from each other and are flush with the cylinder head top surface **(see illustrations)**. If the sprocket

24.7 Insert the feeler gauge between the base of the cam lobe and the top of the follower as shown

marks are facing towards each other, rotate the engine clockwise one full turn (360°) until the line next to the T mark again aligns with the static timing mark. The sprocket marks will now be facing away.

7 With the engine in this position, check the clearances on the Nos. 1 and 3 cylinder intake valves. Insert a feeler gauge of the same thickness as the correct valve clearance (see Specifications) between the camshaft lobe and the follower of each valve and check that it is a firm sliding fit – you should feel a slight drag when the you pull the gauge out **(see illustration)**. If not, use the feeler gauges to obtain the exact clearance. Record the measured clearance on the chart.

8 Now rotate the engine 180° clockwise until the line next to the T mark on the timing rotor is diametrically opposite the static timing mark – the scribed line on the rotor will now be in the 12 o'clock position **(see illustration)**. With the engine in this position, check the clearances on the Nos. 2 and 4 cylinder exhaust valves using the method described in Step 7.

9 Now rotate the engine 180° clockwise until the line next to the T mark aligns with the static timing mark again **(see illustration 24.6b)**. With the engine in this position, check the clearances on the Nos. 2 and 4 cylinder intake valves using the method described in Step 7.

10 Now rotate the engine 180° clockwise until the line next to the T mark on the timing

rotor is once again diametrically opposite the static timing mark and the scribed line is at the 12 o'clock position **(see illustration 24.8)**. With the engine in this position, check the clearances on the Nos. 1 and 3 cylinder exhaust valves using the method described in Step 7.

11 When all clearances have been measured and charted, identify whether the clearance on any valve falls outside the specified range. If any do, the shim must be replaced with one of a thickness which will restore the correct clearance.

12 Shim replacement requires removal of the camshafts (see Chapter 2). There is no need to remove both camshafts if shims from only one side of the cylinder need replacing. Place rags over the spark plug holes and the cam chain tunnel to prevent a shim from dropping into the engine on removal. Work on one valve at a time to prevent the possibility of mixing up the followers, which must be returned to their original location. If you want to remove more than one shim and follower at a time, store them in a marked container or bag, denoting which cylinder and which valve the shim and follower are from, so that they do not get mixed up.

13 With the camshaft removed, remove the cam follower of the valve in question, then retrieve the shim from the inside of the follower **(see illustrations)**. The follower is best removed with a magnet or using the suction created by a valve lapping tool, but

24.8 Turn the engine 180° so the marks are aligned as shown

24.13a Carefully lift out the follower using grips, a lapping tool or a magnet . . .

24.13b . . . and retrieve the shim from inside it . . .

Routine maintenance and servicing 1•21

24.13c ... or from the top of the valve

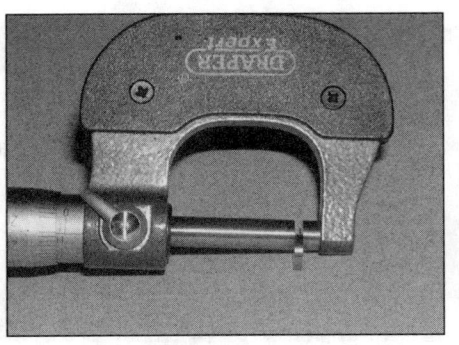

24.14 Check the thickness of the shim using a micrometer

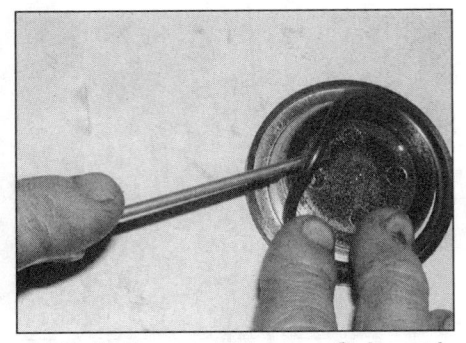

24.19 Fit the cap using a new O-ring and smear it and the threads with grease

long nosed pliers can be used with care. If the shim is not in the follower, pick it out of the top of the valve spring retainer using either a magnet, a screwdriver with a dab of grease on it (the shim will stick to the grease), or a very small screwdriver and a pair of pliers **(see illustration)**. Do not allow the shim to fall into the engine.

14 A size mark should be stamped on one face of the shim – a shim marked 175 is 1.75 mm thick. If the mark is not visible measure the shim thickness using a micrometer **(see illustration)**. It is recommended that the shim is measured anyway to check whether it has worn.

15 Calculate the required replacement shim by using the formula a = (b − c) + d, where a is the required shim size, b is the measured valve clearance, c is the specified valve clearance, and d is the existing shim thickness. For example:
The measured clearance of an inlet valve is 0.22 mm, so b = 0.22
The specified clearance range for an inlet valve is 0.13 to 0.19 mm, the mid-point being 0.16 mm, so c = 0.16
The thickness of the existing shim is 2.00 mm, so d = 2.0
Therefore, the required replacement shim a = 0.22 − 0.16 + 2.0 (a = 2.06 mm)

Note: *If the required replacement shim is greater than 2.800 mm (the largest available), the valve is probably not seating correctly due to a build-up of carbon deposits and should be checked and cleaned or resurfaced as required (see Chapter 2).*

16 Shims are available in 0.025 mm increments from 1.200 mm to 2.800 mm. Obtain the replacement shim, then lubricate it with molybdenum disulphide oil (a 50/50 mixture of molybdenum disulphide grease and engine oil) and fit it into the recess in the top of the valve spring retainer with the size mark facing up **(see illustration 24.13c)**.

17 Check that the shim is correctly seated, then lubricate the follower with molybdenum disulphide oil and install it on the valve, making sure it fits squarely in its bore **(see illustration 24.13a)**. Repeat the process for any other valves until the clearances are correct, then install the camshafts (see Chapter 2).

18 Rotate the crankshaft clockwise several turns to seat the new shim(s), then check the clearances again. Remove the cam chain tensioner locking key to allow the tensioner to take up cam chain slack, then install the end bolt with a new sealing washer. If the locking key was not used and the tensioner body was removed, install it (see Chapter 2). Rotate the engine again and check that the valve timing marks align correctly (see Step 6) before fitting the valve cover.

19 Install all disturbed components in a reverse of the removal sequence. Install the timing inspection cap using a new O-ring if required, and smear the O-ring and the cap threads with grease **(see illustration)**. Tighten the cap to the torque setting specified at the beginning of the Chapter.

20 On completion, check and adjust the idle speed (see Section 2).

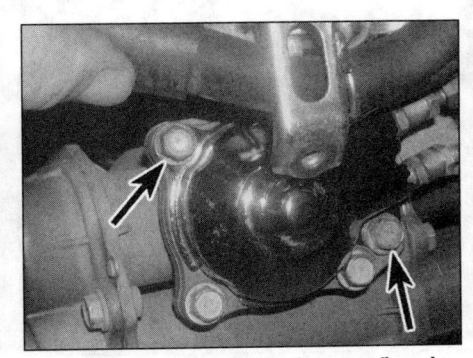

25 H-VIX system – check and cable adjustment

Every 16,000 miles (24,000 km)

1 Remove the rider's seat (see Chapter 8). On RR-Y and RR-1 (2000 and 2001) models remove the lower fairing (see Chapter 8). On RR-2 and RR-3 (2002 and 2003) models remove the left-hand fairing side panel (see Chapter 8).

2 Remove the air filter (see Section 21).

3 Unscrew the two exhaust control valve cover bolts and remove the cover – you will have to lift the coolant hose to access the top front bolt **(see illustration)**.

Functional check

4 Turn the ignition switch ON and make sure the kill switch is in the run position. Check the position of the index line on the control valve pulley – it should be pointing horizontally forward and align with the index line on the valve housing **(see illustration)**.

5 Start the engine and allow it to warm up. Increase the revs to approximately 3000 rpm – at this speed the servo behind the air filter housing should turn and move the control valve pulley clockwise 90° so that the index line now points vertically up.

6 Now increase the revs to approximately 8000 rpm – at this speed the servo should turn the pulley anti-clockwise 180° so that the index line now points vertically down and aligns with the index line on the valve housing. As it turns, simultaneously check that the pulley on the air intake control valve in the air filter housing also turns so that the intake flap opens.

7 As the servo turns in the above checks, make sure that the pulleys on the exhaust and intake control valves are turning smoothly and freely. If the servo tries to turn but a pulley doesn't, remove the cables and check them, and lubricate or replace them with new ones as necessary (see Chapter 4). If the pulley operates, but the movement of the valve is sticky or not complete, first check the cables, and if they are good, disassemble the valve

25.3 Unscrew the bolts (arrowed) and remove the cover

25.4 The index line (A) on the pulley should align with the line (B) on the housing

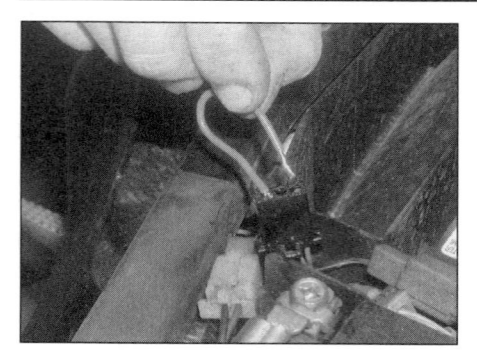

1•22 Routine maintenance and servicing

25.9a Locate the connector and short between its terminals

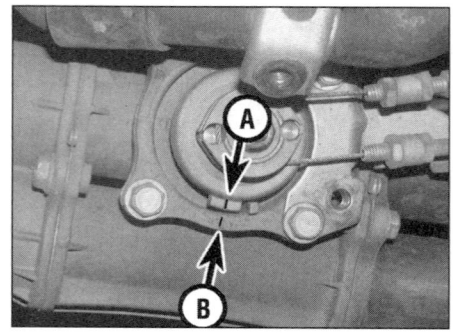

25.9b The index line (A) on the exhaust pulley should align with the line (B) on the housing...

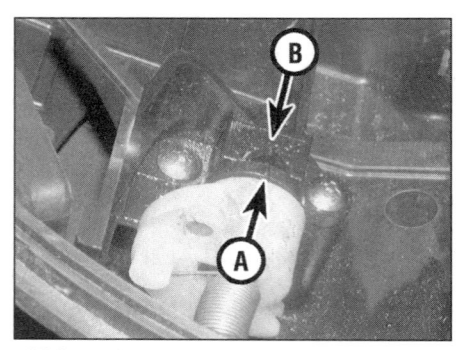

25.9c ...and the index line (A) on the air intake pulley should align with the line (B) on the holder

for cleaning and inspection (see Chapter 4). If the servo does not operate, refer to Chapter 4 for further tests.

8 Smear some grease over the cable ends on the pulley. Apply some copper grease to the cover bolt threads. Install the cover and tighten the bolts to the torque setting specified at the beginning of the Chapter.

Cable check and adjustment

9 Locate the service check connector – it is an open 3-pin connector located just behind the battery on its right-hand end. Connect between the two outer terminals using a short length of auxiliary wire with bared ends **(see illustration)**. Check that the servo turns as you connect the wire, and the index line on the exhaust control valve pulley points vertically down and aligns with the index line on the valve housing **(see illustration)**. Also check that the index mark on the air intake control valve pulley aligns with the index mark on the valve holder **(see illustration)**. If the index lines on either control valve pulley are out of alignment adjust the cable(s) as described below.

10 If no adjustment is required, or on completion of the adjustment procedure, disconnect the auxiliary wire from the service check connector **(see illustration 25.9a)**. Install the fairing panel(s) as required, and the seat.

Exhaust control valve

11 Slacken the locknut on the lower cable adjuster, then turn the adjuster nut as required until the index lines align, then tighten the locknut **(see illustration)**. Now slacken the locknuts on the upper cable and set the adjuster in the bracket so all freeplay in the cable is removed, but take care not to introduce any tension in the cable that will affect its position.

12 Smear some grease over the cable ends in the pulley. Apply some copper grease to the exhaust valve cover bolt threads. Install the cover and tighten the bolts to the torque setting specified at the beginning of the Chapter.

Air intake control valve

13 Slacken the locknut on the cable adjuster, then turn the adjuster as required until the index lines align, then turn the adjuster by 1/2 turn in (towards the locknut) to create some freeplay, then tighten the locknut **(see illustration)**.

14 Install the air filter (see Section 21).

26 Cooling system – draining, flushing and refilling

Every 24,000 miles (36,000 km) or two years

⚠️ **Warning:** *Allow the engine to cool completely before performing this maintenance operation. Also, don't allow anti-freeze to come into contact with your skin or the painted surfaces of the motorcycle. Rinse off spills immediately with plenty of water. Anti-freeze is highly toxic if ingested. Never leave anti-freeze lying around in an open container or in puddles on the floor; children and pets are attracted by its sweet smell and may drink it. Check with local authorities (councils) about disposing of anti-freeze. Many communities have collection centres which will see that anti-freeze is disposed of safely. Anti-freeze is also combustible, so don't store it near open flames.*

Draining

1 On RR-Y and RR-1 (2000 and 2001) models, remove the right-hand air duct cover,

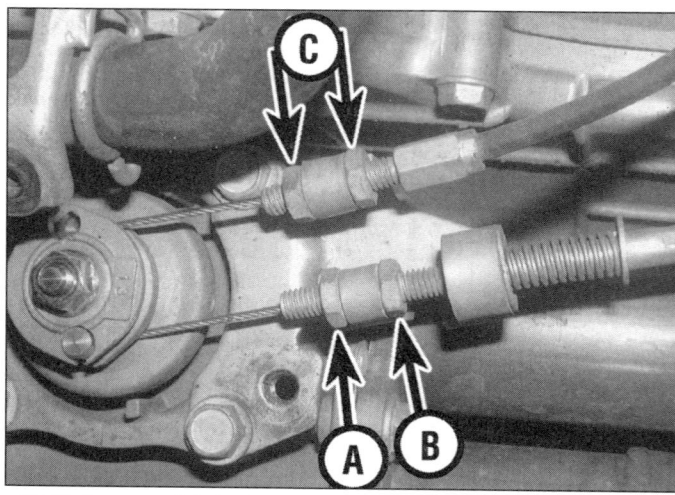

25.11 Lower cable locknut (A) and adjuster nut (B), upper cable locknuts (C)

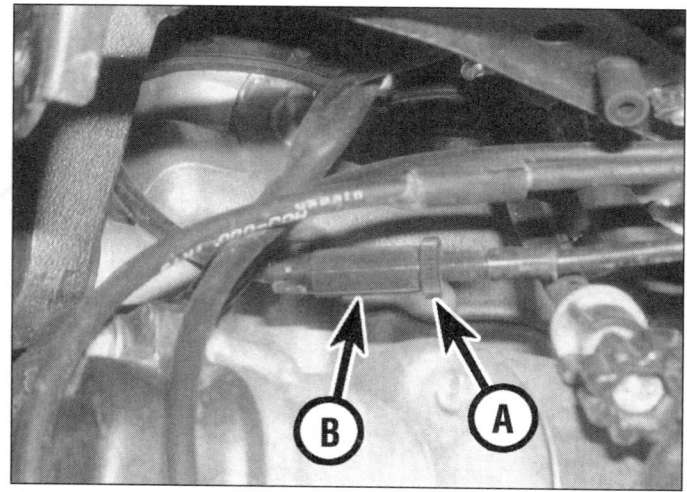

25.13 Slacken the locknut (A) and turn the adjuster (B) as required

Routine maintenance and servicing 1•23

26.3a Unscrew the drain plug (arrowed) . . .

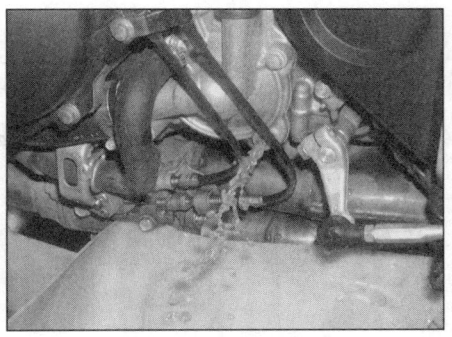

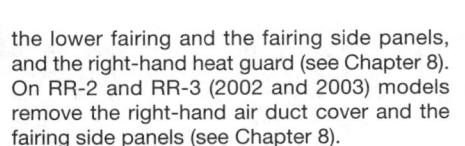

26.3b . . . and allow the coolant to drain

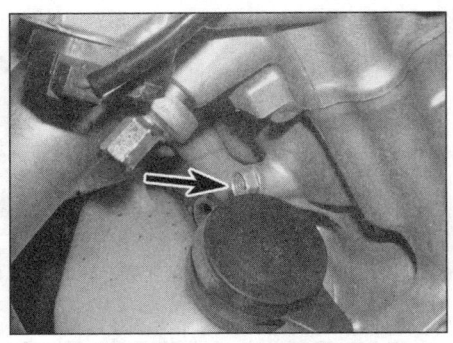

26.4 Unscrew the cylinder drain plug (arrowed) and drain the cylinder jacket

the lower fairing and the fairing side panels, and the right-hand heat guard (see Chapter 8). On RR-2 and RR-3 (2002 and 2003) models remove the right-hand air duct cover and the fairing side panels (see Chapter 8).

2 Remove the pressure cap from the top of the radiator by covering it with a heavy cloth and turning it anti-clockwise until it reaches a stop **(see illustration 12.7)**. If you hear a hissing sound (indicating there is still pressure in the system), wait until it stops. Now press down on the cap and continue turning the cap until it can be removed. Also remove the coolant reservoir cap.

3 Position a suitable container beneath the water pump on the left-hand side of the engine. Unscrew the drain bolt and allow the coolant to completely drain from the system **(see illustrations)**. Retain the old sealing washer for use during flushing.

4 Now position the container beneath the front of the engine on the left-hand side. Unscrew the cylinder drain plug and allow the coolant to completely drain from the cylinder jacket **(see illustration)**. Retain the old sealing washer for use during flushing.

5 Place the container on the right-hand side of the engine. Disconnect the radiator overflow hose (the top hose) from the radiator filler neck, then place it below the level of the reservoir and allow it to drain into the container **(see illustrations)**.

Flushing

6 Flush the system with clean tap water by inserting a garden hose in the radiator filler neck. Allow the water to run through the system until it is clear and flows out cleanly. If the radiator is extremely corroded, remove it (see Chapter 3) and have it cleaned by a specialist. Also flush the reservoir, then fit the radiator overflow hose back onto the radiator filler neck.

7 Clean the drain holes in the water pump and cylinder bock then install the drain bolts using the old sealing washers.

8 Fill the cooling system with clean water mixed with a flushing compound. Make sure the flushing compound is compatible with aluminium components, and follow the manufacturer's instructions carefully. Fit the radiator cap.

9 Start the engine and allow it to reach normal operating temperature. Let it run for about ten minutes.

10 Stop the engine. Let it cool for a while, then cover the pressure cap with a heavy rag and turn it anti-clockwise to the first stop, releasing any pressure that may be present in the system. Once the hissing stops, push down on the cap and remove it completely.

11 Drain the system once again.

12 Fill the system with clean water and repeat the procedure in Steps 6 to 11.

Refilling

13 Install the drain bolts using new sealing washers **(see illustrations)**.

14 Fill the system to the base of the radiator filler neck with the proper coolant mixture (see this Chapter's Specifications) **(see illustration)**.

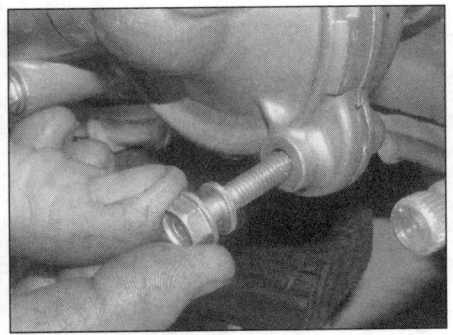

26.5a Detach the hose from the filler neck . . .

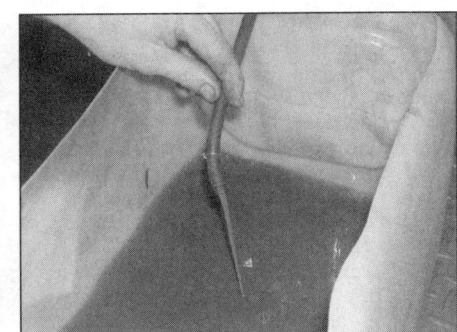

26.5b . . . and allow the reservoir to drain

26.13a Install the pump drain bolt . . .

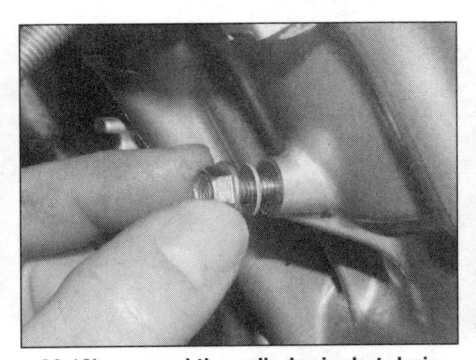

26.13b . . . and the cylinder jacket drain bolt using new sealing washers

26.14 Fill the system and bleed it as described

1•24 Routine maintenance and servicing

Note: *Pour the coolant in slowly to minimise the amount of air entering the system.* Fill the reservoir to the UPPER level line (see *Daily (pre-ride) checks*).

15 Start the engine and allow it to idle for 2 to 3 minutes. Flick the throttle twistgrip part open 3 or 4 times, so that the engine speed rises to approximately 4000 to 5000 rpm, then stop the engine. This process will bleed any trapped air bubbles from the system.

16 If necessary, top up the coolant level to the base of the radiator filler neck, then install the pressure cap. Also top up the coolant reservoir to the UPPER level line.

17 Start the engine and allow it to reach normal operating temperature, then shut it off. Let the engine cool then remove the pressure cap as described in Step 2. Check that the coolant level is still up to the base of the upper radiator filler neck. If it's low, add the specified mixture until it reaches the base of the filler neck. Refit the cap.

18 Check the coolant level in the reservoir and top up if necessary.

19 Check the system for leaks. Install the heat guard, air duct cover and fairing panels as required for your model (see Chapter 8).

20 Do not dispose of the old coolant by pouring it down the drain. Instead pour it into a heavy plastic container, cap it tightly and take it into an authorised disposal site or service station – see **Warning** at the beginning of this Section.

Non-scheduled maintenance

27 Cylinder compression – check

1 Among other things, poor engine performance may be caused by leaking valves, incorrect valve clearances, a leaking head gasket, or worn pistons, rings and/or cylinder walls. A cylinder compression check will help pinpoint these conditions and can also indicate the presence of excessive carbon deposits in the cylinder heads.

2 The only tools required are a compression gauge and a spark plug wrench. A compression gauge with a threaded end for the spark plug hole is preferable to the type which requires hand pressure to maintain a tight seal. Depending on the outcome of the initial test, a squirt-type oil can may also be needed.

3 Make sure the valve clearances are correctly set (see Section 24) and that the cylinder head nuts are tightened to the correct torque setting (see Chapter 2).

4 Refer to *Fault Finding Equipment* in the Reference section for details of the compression test. Refer to the specifications at the beginning of the Chapter for compression figures.

28 Engine oil pressure – check

1 The oil pressure warning light should come on when the ignition (main) switch is turned ON and extinguish a few seconds after the engine is started – this serves as a check that the warning light bulb is sound. If the oil pressure light comes on whilst the engine is running, low oil pressure is indicated – stop the engine immediately and carry out an oil level check *(see Daily (pre-ride) checks)*.

2 An oil pressure check must be carried out if the warning light comes on when the engine is running yet the oil level is good (Step 1). It can also provide useful information about the condition of the engine's lubrication system.

3 To check the oil pressure, a suitable gauge and adapter (which screws into the crankcase) will be needed. Honda provide a gauge and adapter (part Nos. 07506-3000000 and 07510 MA70000) for this purpose, or one can be obtained commercially. You will also need some rags to catch and mop up any residual oil that gets lost in between removing the oil gallery plug and installing the gauge – place the bike on its sidestand so that the oil gathers at the other end of the gallery to reduce spillage.

4 On RR-Y and RR-1 (2000 and 2001) models, remove the lower fairing (see Chapter 8). On RR-2 and RR-3 (2002 and 2003) models remove the right-hand fairing side panel (see Chapter 8). Check the oil level (see *Daily (pre-ride) checks*). Warm the engine up to normal operating temperature then stop it.

5 Unscrew the main oil gallery plug, which is threaded into the crankcase on the right-hand side of the engine (directly below the timing inspection cap in the front of the clutch cover), and screw the adapter in its place **(see illustration)**. Connect the oil pressure gauge to the adapter.

6 Start the engine and briefly increase the engine speed to 5400 rpm whilst watching the gauge reading. The oil pressure should be similar to that given in the Specifications at the start of this Chapter.

7 If the pressure is significantly lower than the standard, either the pressure relief valve is stuck open, the oil pump or its drive mechanism is faulty, the oil strainer or filter is blocked, or there is other engine damage. Also make sure the correct grade oil is being used. Begin diagnosis by checking the oil filter, strainer and relief valve, then the oil pump (see Chapter 2). If those items check out okay, chances are the bearing oil clearances are excessive and the engine needs to be overhauled.

8 If the pressure is too high, either an oil passage is clogged, the relief valve is stuck closed or the wrong grade of oil is being used.

9 Stop the engine and unscrew the gauge and adapter from the crankcase.

10 Install the main gallery plug, using a suitable non-permanent thread locking compound, and tighten it to the torque setting specified at the beginning of the Chapter. Check the oil level (see *Daily (pre-ride) checks*).

29 Wheel bearings – check

1 Wheel bearings will wear over a period of time and result in handling problems.

2 Support the motorcycle upright using an auxiliary stand, and support it so that the wheel being checked is off the ground (remove the fairing side panels and anything else that could be damaged before placing a support under the engine). Check for any play in the bearings by pushing and pulling the wheel against the axle **(see illustration)**. Also spin the wheel and check that it rotates smoothly.

3 If any play is detected in the hub, or if the wheel does not rotate smoothly (and this is not

28.5 Main oil gallery plug (arrowed)

29.2 Checking for play in the wheel bearings

Routine maintenance and servicing 1•25

due to brake or transmission drag), the wheel bearings must be removed and inspected for wear or damage (see Chapter 7).

30 Steering head bearings – re-greasing

1 Over a period of time the grease will harden or may be washed out of the bearings by incorrect use of jet washes.
2 Disassemble the steering head for re-greasing of the bearings. Refer to Chapter 6 for details.

31 Swingarm and suspension linkage bearings – re-greasing

1 Over a period of time the grease will harden or dirt will penetrate the bearings due to failed seals.
2 The suspension is not equipped with grease nipples. Remove the swingarm and suspension linkage as described in Chapter 6 for greasing of the bearings.

32 Brake caliper and master cylinder seals – renewal

1 Brake seals will deteriorate over a period of time and lose their effectiveness, leading to sticking operation or fluid loss, or allowing the ingress of air and dirt. Refer to Chapter 7 and dismantle the components for seal renewal.

33 Brake hoses – renewal

1 The hoses will in time deteriorate with age and should be renewed regardless of their apparent condition. Refer to Chapter 7 and disconnect the brake hoses from the master cylinders and calipers. Always replace the banjo union sealing washers with new ones.

34 Fuel system hoses – renewal

⚠️ **Warning: Petrol (gasoline) is extremely flammable, so take extra precautions when you work on any part of the fuel system. Don't smoke or allow open flames or bare light bulbs near the work area, and don't work in a garage where a natural gas-type appliance is present. If you spill any fuel on your skin, rinse it off immediately with soap and water. When you perform any kind of work on the fuel system, wear safety glasses and have a fire extinguisher suitable for a Class B type fire (flammable liquids) on hand.**

1 The fuel system hoses should be renewed at the first signs of cracking or hardening. This includes all the vent and drain hoses, and the vacuum hoses.
2 Remove the fuel tank (see Chapter 4). Disconnect the hoses, noting the routing of each hose and where it connects (see Chapter 4 if required). It is advisable to make a sketch of the various hoses before removing them to ensure they are correctly installed.
3 Secure each new hose to its unions using new clamps or sealing washers where fitted. Refer to Chapter 4 for torque settings for the fuel delivery and return hose fasteners. Run the engine and check for leaks before taking the machine out on the road.

35 Front forks – oil change

1 Fork oil degrades over a period of time and loses its damping qualities. Refer to the fork oil change procedure in Chapter 6, Section 7. The forks do not need to be completely disassembled.

36 Fuel filter – cleaning and renewal

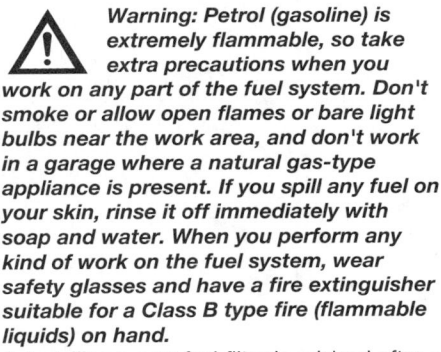

⚠️ **Warning: Petrol (gasoline) is extremely flammable, so take extra precautions when you work on any part of the fuel system. Don't smoke or allow open flames or bare light bulbs near the work area, and don't work in a garage where a natural gas-type appliance is present. If you spill any fuel on your skin, rinse it off immediately with soap and water. When you perform any kind of work on the fuel system, wear safety glasses and have a fire extinguisher suitable for a Class B type fire (flammable liquids) on hand.**

1 Installing a new fuel filter is advised after a particularly high mileage has been covered. It is also necessary if fuel starvation is suspected. Honda do not specify a replacement interval – fuel is so clean now that this may not always be necessary. Check the condition of the inside of your tank – if it is old and there is evidence of rust, remove, drain and clean the tank (see Chapter 4), and fit a new filter afterwards.
2 The filter is fitted alongside the fuel pump inside the tank. To change the filter, remove the fuel pump assembly (see Chapter 4). Undo the clamp screw and release the filter from its holder, noting which way round it fits **(see illustration)**.
3 Have a rag handy to soak up any residual fuel, then release the clamps and disconnect the hoses from the filter, noting which fits where. Discard the filter. Fit the new filter into the hoses and secure them with the clamps. Secure the filter with the clamp.
4 A fuel strainer is fitted in the base of the fuel pump assembly housing. It is basically a wad of coarse steel wool and strains the fuel returning to the tank **(see illustration)**. Check that there are no large particles of dirt caught in it and clean it through using petrol if necessary. If it is clogged with debris, replace it with a new wad. Install the fuel pump assembly (see Chapter 4). Start the engine and check that there are no leaks.

37 Throttle body starter valve – synchronisation

⚠️ **Warning: Take great care not to burn your hand on the hot engine unit when accessing the gauge take-off points on the intake manifolds. Do not allow exhaust gases to build up in the work area; either perform the check outside or use an exhaust gas extraction system.**

Honda do not specify this as a service item, and say that it need only be carried out if the starter valves have been removed from the throttle body assembly. The procedure does not alter the setting of the throttle valves themselves (these are pre-set at the factory and fixed), but only the starter valves, which control the idle speed when the engine is cold and warming up. However on high mileage machines the linkage could wear and produce uneven idling when cold and warming up. If this is the case, then the synchronisation procedure should be carried out (see Chapter 4).

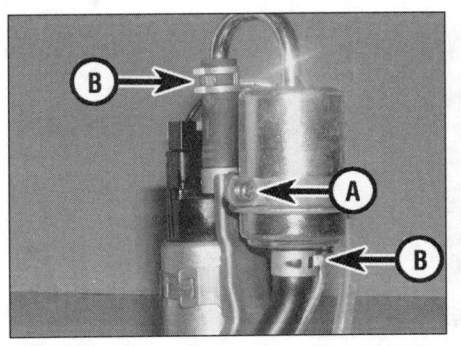

36.2 Fuel filter clamp screw (A) and hose clamps (B)

36.4 Check the strainer for debris

Notes

Chapter 2
Engine, clutch and transmission

Contents

Alternator – removal and installation	see Chapter 9
Camshafts and followers – removal, inspection and installation	10
Cam chain, tensioner blade and guide blades – removal, inspection and installation	11
Cam chain tensioner – removal, inspection and installation	9
Clutch – check	see Chapter 1
Clutch and cable – removal, inspection and installation	16
Connecting rods and bearings – removal, inspection and installation	25
Crankcase halves and cylinder bores – inspection and servicing	23
Crankcase halves – separation and reassembly	22
Crankshaft and main bearings – removal, inspection and installation	28
Cylinder head – removal and installation	12
Cylinder head and valves – disassembly, inspection and reassembly	14
Engine – compression check	see Chapter 1
Engine – removal and installation	5
Engine disassembly and reassembly – general information	6
Gearchange mechanism – removal, inspection and installation	19
General information	1
Idle speed – check and adjustment	see Chapter 1
Initial start-up after overhaul	31
Main and connecting rod bearings – general information	24
Major engine repair – general note	4
Neutral switch – check, removal and installation	see Chapter 9
Oil and filter – change	see Chapter 1
Oil cooler – removal and installation	7
Oil level – check	see *Daily (pre-ride) checks*
Oil pressure – check	see Chapter 1
Oil pressure switch – check, removal and installation	see Chapter 9
Oil pump – removal, inspection and installation	18
Oil sump, oil strainer and pressure relief valve – removal, inspection and installation	17
Operations possible with the engine in the frame	2
Operations requiring engine removal	3
Pistons – removal, inspection and installation	26
Piston rings – inspection and installation	27
Pulse generator coil assembly – removal and installation	see Chapter 5
Recommended running-in procedure	32
Selector drum and forks – removal, inspection and installation	20
Spark plugs – check and renewal	see Chapter 1
Starter clutch – check, removal and installation	15
Starter motor – removal and installation	see Chapter 9
Timing rotor – removal and installation	21
Transmission shafts and bearings – removal and installation	29
Transmission shafts – disassembly, inspection and reassembly	30
Valve clearances – check and adjustment	see Chapter 1
Valve cover – removal and installation	8
Valves/valve seats/valve guides – servicing	13

Degrees of difficulty

Easy, suitable for novice with little experience	Fairly easy, suitable for beginner with some experience	Fairly difficult, suitable for competent DIY mechanic	Difficult, suitable for experienced DIY mechanic	Very difficult, suitable for expert DIY or professional

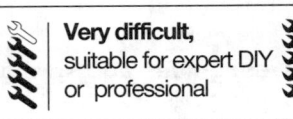

Specifications

General

Type	Four-stroke in-line four
Capacity	
RR-Y and RR-1 (2000 and 2001) models	929 cc
RR-2 and RR-3 (2002 and 2003) models	954 cc
Bore	
RR-Y and RR-1 (2000 and 2001) models	74.0 mm
RR-2 and RR-3 (2002 and 2003) models	75.0 mm
Stroke	54.0 mm
Compression ratio	
RR-Y and RR-1 (2000 and 2001) models	11.3 to 1
RR-2 and RR-3 (2002 and 2003) models	11.5 to 1
Cylinder numbering	1 to 4 from left to right
Firing order	1-2-4-3

General (continued)

Valve timing
  RR-Y and RR-1 (2000 and 2001) models
    Intake valve opens .... 25° BTDC
    Intake valve closes .... 35° ABDC
    Exhaust valve opens .... 40° BBDC
    Exhaust valve closes .... 20° ATDC
  RR-2 and RR-3 (2002 and 2003) models
    Intake valve opens .... 25° BTDC
    Intake valve closes .... 38° ABDC
    Exhaust valve opens .... 41° BBDC
    Exhaust valve closes .... 22° ATDC
Cooling system .... Liquid cooled
Clutch .... Wet multi-plate
Transmission .... Six-speed constant mesh
Final drive .... Chain

Camshafts and followers

Intake lobe height
  RR-Y and RR-1 (2000 and 2001) models
    Standard .... 36.48 to 36.72 mm
    Service limit (min) .... 36.45 mm
  RR-2 and RR-3 (2002 and 2003) models
    Standard .... 36.74 to 36.98 mm
    Service limit (min) .... 36.72 mm
Exhaust lobe height
  RR-Y and RR-1 (2000 and 2001) models
    Standard .... 36.08 to 36.32 mm
    Service limit (min) .... 36.50 mm
  RR-2 and RR-3 (2002 and 2003) models
    Standard .... 36.45 to 36.69 mm
    Service limit (min) .... 36.43 mm
Oil clearance
  Standard .... 0.020 to 0.062 mm
  Service limit (max) .... 0.10 mm
Runout (max)
  RR-Y and RR-1 (2000 and 2001) models .... 0.05 mm
  RR-2 and RR-3 (2002 and 2003) models .... 0.04 mm
Camshaft follower diameter
  Standard .... 25.978 to 25.993 mm
  Service limit (min) .... 25.97 mm
Camshaft follower bore diameter
  Standard .... 26.010 to 26.026 mm
  Service limit (min) .... 26.04 mm

Cylinder head

Warpage (max) .... 0.10 mm

Valves, guides and springs

Valve clearances .... see Chapter 1
Stem diameter
  Intake valve
    Standard .... 4.475 to 4.490 mm
    Service limit (min) .... 4.465 mm
  Exhaust valve
    Standard .... 4.465 to 4.480 mm
    Service limit (min) .... 4.455 mm
Guide bore diameter – intake and exhaust valves
  Standard .... 4.500 to 4.512 mm
  Service limit (max) .... 4.540 mm
Stem-to-guide clearance
  Intake valve .... 0.010 to 0.037 mm
  Exhaust valve .... 0.020 to 0.047 mm
Seat width – intake and exhaust valves
  Standard .... 0.90 to 1.10 mm
  Service limit (max) .... 1.50 mm
Valve guide height above cylinder head
  Intake valve .... 14.3 to 14.6 mm
  Exhaust valve .... 12.4 to 12.7 mm

Engine, clutch and transmission 2•3

Valves, guides and springs (continued)

Valve spring free length
  RR-Y and RR-1 (2000 and 2001) models
    Inner spring – intake and exhaust
      Standard .... 34.80 mm
      Service limit (min) .... 34.1 mm
    Outer spring – intake and exhaust
      Standard .... 37.97 mm
      Service limit (min) .... 37.2 mm
  RR-2 and RR-3 (2002 and 2003) models
    Intake – inner spring
      Standard .... 34.80 mm
      Service limit (min) .... 33.1 mm
    Intake – outer spring
      Standard .... 37.97 mm
      Service limit (min) .... 36.1 mm
    Exhaust
      Standard .... 39.60 mm
      Service limit (min) .... 37.6 mm

Starter clutch

Starter driven gear hub OD
  Standard .... 51.699 to 51.718 mm
  Service limit (min) .... 51.684 mm

Clutch

Friction plates 9
Plain plates 8
Friction plate thickness
  Standard .... 2.92 to 3.08 mm
  Service limit (min) .... 2.6 mm
Plain plate warpage (max) .... 0.3 mm
Spring free length
  Standard .... 48.8 mm
  Service limit (min) .... 47.4 mm
Clutch guide OD
  Standard .... 34.975 to 34.991 mm
  Service limit (min) 34.97mm
Clutch guide ID
  Standard .... 25.000 to 25.021 mm
  Service limit (max) .... 25.03 mm
Input shaft OD at clutch guide
  Standard .... 24.980 to 24.993 mm
  Service limit (max) .... 24.96 mm

Oil pump

Oil pressure .... see Chapter 1
Inner rotor tip-to-outer rotor clearance
  Standard .... 0.15 mm
  Service limit (max) .... 0.20 mm
Outer rotor-to-body clearance
  Standard .... 0.15 to 0.22 mm
  Service limit (max) .... 0.35 mm
Rotor endfloat
  Standard .... 0.02 to 0.07 mm
  Service limit (max) .... 0.10 mm

Selector drum and forks

Selector fork end thickness
  Standard .... 5.93 to 6.00 mm
  Service limit (min) .... 5.90 mm
Selector fork bore ID
  Standard .... 12.000 to 12.018 mm
  Service limit (max) .... 12.03 mm
Selector fork shaft OD
  Standard .... 11.957 to 11.968 mm
  Service limit (min) .... 11.95 mm

Cylinder bores
Bore
    RR-Y and RR-1 (2000 and 2001) models
        Standard .... 74.005 to 74.020 mm
        Service limit (max) .... 74.15 mm
    RR-2 and RR-3 (2002 and 2003) models
        Standard .... 75.005 to 75.015 mm
        Service limit (max) .... 75.15 mm
Warpage (max) .... 0.05 mm
Ovality (out-of-round) (max) .... 0.10 mm
Taper (max) .... 0.10 mm
Cylinder compression .... see Chapter 1

Connecting rods
Small-end internal diameter
    Standard .... 17.016 to 17.034 mm
    Service limit (max) .... 17.04 mm
Small-end-to-piston pin clearance
    Standard .... 0.016 to 0.040 mm
    Service limit .... 0.06 mm
Big-end side clearance
    Standard .... 0.05 to 0.20 mm
    Service limit (max) .... 0.3 mm
Big-end oil clearance
    Standard .... 0.030 to 0.052 mm
    Service limit (max) .... 0.062 mm

Pistons
RR-Y and RR-1 (2000 and 2001) models
    Piston diameter (measured 13 mm up from skirt, at 90° to piston pin axis)
        Standard .... 73.965 to 73.985 mm
        Service limit (min) .... 73.90 mm
RR-2 and RR-3 (2002 and 2003) models
    Piston diameter (measured 4 mm up from skirt, at 90° to piston pin axis)
        Standard .... 74.960 to 74.980 mm
        Service limit (min) .... 74.895 mm
Piston-to-bore clearance
    Standard pistons and rings .... 0.020 to 0.055 mm*
    Oversize (+0.25) pistons and rings .... 0.015 to 0.050 mm
Piston pin diameter
    Standard .... 16.994 to 17.000 mm
    Service limit (min) .... 16.98 mm
Piston pin bore diameter in piston
    Standard .... 17.002 to 17.008 mm
    Service limit (max) .... 17.03 mm
Piston pin-to-piston pin bore clearance .... 0.002 to 0.014 mm

If the piston-to-bore clearance exceeds the maximum, the cylinders can be rebored – Honda supply +0.25 oversize pistons and rings. Following rebore, the piston-to-bore clearance must be as specified for oversize pistons

Piston rings
Ring end gap (installed)
    Top ring
        Standard .... 0.28 to 0.38 mm
        Service limit (max) .... 0.50 mm
    Second ring
        Standard .... 0.40 to 0.55 mm
        Service limit (max) .... 0.70 mm
    Oil ring side-rail
        Standard .... 0.20 to 0.70 mm
        Service limit (max) .... 0.9 mm
Ring-to-groove clearance
    Top ring
        Standard .... 0.030 to 0.065 mm
        Service limit (max) .... 0.08 mm
    Second ring
        Standard .... 0.015 to 0.045 mm
        Service limit (max) .... 0.06 mm

Engine, clutch and transmission 2•5

Crankshaft and bearings
Main bearing oil clearance
  RR-Y and RR-1 (2000 and 2001) models
    Journals 1 and 5 (outer journals)
      Standard .... 0.017 to 0.035 mm
      Service limit (max) .... 0.045 mm
    Journals 2, 3 and 4 (inner journals)
      Standard .... 0.027 to 0.045 mm
      Service limit (max) .... 0.055 mm
  RR-2 and RR-3 (2002 and 2003) models
    Standard .... 0.017 to 0.035 mm
    Service limit (max) .... 0.045 mm
Runout (max) .... 0.03 mm

Transmission
Gear ratios (no. of teeth)
  Primary reduction .... 1.520 to 1 (73/48T)
  Final reduction
    European models .... 2.625 to 1 (42/16T)
    US and Canada models .... 2.687 to 1 (43/16T)
  1st gear .... 2.692 to 1 (35/13T)
  2nd gear .... 1.933 to 1 (29/15T)
  3rd gear .... 1.600 to 1 (32/20T)
  4th gear .... 1.400 to 1 (28/20T)
  5th gear .... 1.285 to 1 (27/21T)
  6th gear .... 1.190 to 1 (25/21T)
Input shaft 5th and 6th gears ID
  Standard .... 31.000 to 31.025 mm
  Service limit (max) .... 31.04 mm
Input shaft 5th and 6th gears bush OD
  Standard .... 30.950 to 30.975 mm
  Service limit (min) .... 30.93 mm
Input shaft 5th and 6th gears gear-to-bush clearance
  Standard .... 0.025 to 0.075 mm
  Service limit (max) .... 0.11 mm
Input shaft 5th gear bush ID
  Standard .... 27.985 to 28.006 mm
  Service limit (max) .... 28.02 mm
Input shaft OD at 5th gear bush point
  Standard .... 27.967 to 27.980 mm
  Service limit (min) .... 27.957 mm
Input shaft-to-bush clearance at 5th gear bush point
  Standard .... 0.005 to 0.039 mm
  Service limit (max) .... 0.08 mm
Output shaft 1st gear ID
  Standard .... 26.000 to 26.021 mm
  Service limit (max) .... 26.04 mm
Output shaft 2nd, 3rd and 4th gears ID
  Standard .... 33.000 to 33.025 mm
  Service limit (max) .... 33.04 mm
Output shaft 2nd gear bush ID
  Standard .... 29.985 to 30.006 mm
  Service limit (min) .... 30.02 mm
Output shaft 3rd and 4th gears bush OD
  Standard .... 32.950 to 32.975 mm
  Service limit (min) .... 32.93 mm
Output shaft 3rd and 4th gears gear-to-bush clearance
  Standard .... 0.025 to 0.075 mm
  Service limit (max) .... 0.11 mm
Output shaft OD at 2nd gear bush point
  Standard .... 29.967 to 29.980 mm
  Service limit (min) .... 29.96 mm
Output shaft-to-bushing clearance at 2nd gear bush point
  Standard .... 0.005 to 0.039 mm
  Service limit (max) .... 0.08 mm

Torque settings

Cam chain front guide blade bolt	12 Nm
Cam chain tensioner blade bolt	10 Nm
Camshaft holder bolts	12 Nm
Camshaft sprocket bolts	20 Nm
Cam pulse generator rotor bolts	12 Nm
Clutch nut	127 Nm
Clutch spring bolts	12 Nm
Connecting rod nuts	35 Nm
Crankcase breather separator bolts	12 Nm

Crankcase bolts
   RR-Y and RR-1 (2000 and 2001) models
      Lower crankcase 9 mm bolts . . . . 35 Nm
      Lower crankcase 10 mm bolt . . . . 39 Nm
      Lower crankcase 8 mm bolts . . . . 24 Nm
      Upper crankcase 8 mm bolts . . . . 24 Nm
   RR-2 and RR-3 (2002 and 2003) models
      Lower crankcase 9 mm bolts
         Initial setting . . . . 10 Nm
         Second setting . . . . 20 Nm
         Final setting . . . . + 150°
      Lower crankcase 10 mm bolt . . . . 39 Nm
      Lower crankcase 8 mm bolts . . . . 25 Nm
      Upper crankcase 8 mm bolts . . . . 25 Nm

Cylinder head
   9 mm bolts . . . . 51 Nm
   8 mm bolts . . . . 24 Nm

Engine mountings
   Rear mounting bolt nut . . . . 54 Nm
   Rear mounting pinch bolt . . . . 26 Nm
   Front mounting bolts . . . . 39 Nm
   Centre mounting bolts . . . . 54 Nm
   Swingarm pivot bolt nut . . . . 118 Nm
   Swingarm pivot pinch bolts . . . . 26 Nm
   Engine bracket bolt nuts . . . . 42 Nm
   Engine bracket pinch bolts . . . . 26 Nm
   Shock absorber lower mounting bolt nut . . . . 44 Nm

Oil cooler bolt	74 Nm
Oil pump assembly bolt	8 Nm
Oil pump driven sprocket bolt	15 Nm
Selector drum bearing/fork shaft retainer bolts	12 Nm
Selector drum cam bolt	23 Nm
Stopper arm bolt	12 Nm
Starter clutch bolts	16 Nm
Timing inspection cap	18 Nm
Timing rotor bolt	59 Nm
Transmission input shaft bearing retainer plate bolts	12 Nm
Valve cover bolts	10 Nm

1 General information

The engine/transmission unit is a liquid-cooled in-line four cylinder. The sixteen valves are operated by double overhead camshafts which are chain driven off the right-hand end of the crankshaft. The engine/transmission is a unit assembly constructed from aluminium alloy. The crankcase divides horizontally.

The crankcase incorporates a wet sump, pressure-fed lubrication system which uses a dual rotor trochoidal oil pump that is chain-driven off the back of the clutch. The system has an oil filter, an oil pressure switch, and a pressure relief valve which is in the feed from the pump to the filter.

The alternator is on the left-hand end of the crankshaft and has the starter clutch mounted behind it. The water pump is on the left-hand side of the engine, and its drive shaft is keyed to the oil pump drive shaft. The pulse generator coil and ignition timing rotor are on the right-hand end of the crankshaft.

Power from the crankshaft is routed to the transmission via the clutch. The clutch is of the wet, multi-plate type and is gear-driven off the crankshaft. The clutch is operated by cable. The transmission is a six-speed constant-mesh unit. Final drive to the rear wheel is by chain and sprockets.

2 Operations possible with the engine in the frame

The components and assemblies listed below can be removed without having to remove the engine/transmission assembly

Engine, clutch and transmission 2•7

from the frame. If, however, a number of areas require attention at the same time, removal of the engine is recommended.
  Valve cover
  Cam chain tensioner and blades
  Camshafts and cam chain
  Ignition timing rotor and pulse generator coil assembly
  Clutch
  Gearchange mechanism
  Alternator
  Oil filter and oil cooler
  Oil sump, oil pump, oil strainer and oil pressure relief valve
  Starter motor
  Starter clutch
  Water pump
  Selector drum and forks (though easier with engine removed)

3 Operations requiring engine removal

It is necessary to remove the engine/transmission assembly from the frame to gain access to the following components.
  Cylinder head
  Pistons, piston rings and cylinder bores
  Transmission shafts
  Crankshaft and bearings
  Connecting rods and bearings

4 Major engine repair – general note

1 It is not always easy to determine when or if an engine should be completely overhauled, as a number of factors must be considered.
2 High mileage is not necessarily an indication that an overhaul is needed, while low mileage, on the other hand, does not preclude the need for an overhaul. Frequency of servicing is probably the single most important consideration. An engine that has regular and frequent oil and filter changes, as well as other required maintenance, will most likely give many miles of reliable service. Conversely, a neglected engine, or one which has not been run in properly, may require an overhaul very early in its life.
3 Exhaust smoke and excessive oil consumption are both indications that piston rings and/or valve guides are in need of attention, although make sure that the fault is not due to oil leakage.
4 If the engine is making obvious knocking or rumbling noises, the connecting rods and/or main bearings are probably at fault.
5 Loss of power, rough running, excessive valve train noise and high fuel consumption may also point to the need for an overhaul, especially if they are all present at the same time. If a complete tune-up does not remedy the situation, major mechanical work is the only solution.

6 An engine overhaul generally involves restoring the internal parts to the specifications of a new engine. The piston rings and main and connecting rod bearings are usually replaced and the cylinder walls honed or, if necessary, re-bored (oversize pistons are available), during a major overhaul. Generally the valve seats are re-ground, since they are usually in less than perfect condition at this point. The end result should be a like new engine that will give as many trouble-free miles as the original.
7 Before beginning the engine overhaul, read through the related procedures to familiarise yourself with the scope and requirements of the job. Overhauling an engine is not all that difficult, but it is time consuming. Plan on the motorcycle being tied up for a minimum of two weeks. Check on the availability of parts and make sure that any necessary special tools, equipment and supplies are obtained in advance.
8 Most work can be done with typical workshop hand tools, although a number of precision measuring tools are required for inspecting parts to determine if they must be renewed. Often a dealer will handle the inspection of parts and offer advice concerning reconditioning and renewal. As a general rule, time is the primary cost of an overhaul so it does not pay to install worn or substandard parts.
9 As a final note, to ensure maximum life and minimum trouble from a rebuilt engine, everything must be assembled with care in a spotlessly clean environment.

5 Engine – removal and installation

Caution: The engine is very heavy. Engine removal and installation should be carried out with the aid of at least one assistant; personal injury or damage could occur if the engine falls or is dropped.

Note 1: *The sidestand, the swingarm and the suspension linkage are all bolted to a bracket, which itself bolts to the engine, and so they must be removed before removing the engine. Therefore it is necessary to support the bike using either a hoist or some form of auxiliary stand that does not attach to either the rear wheel or swingarm. We found the best method to be two axle stands placed under the pillion footrest brackets, with the front brake lever tied to the handlebar so that the brake is locking the front wheel – this will prevent the bike rolling off the axle stands should it be knocked or moved as the engine is manoeuvred out. If the rear wheel is off the ground place a support under it so that it does not drop when the suspension is detached – ideally the rear wheel should be supported, but without any weight on it or force through it so that the shock absorber is uncompressed.*
Note 2: *To unscrew the swingarm pivot bolt nut you need a 24 mm hex key, and to do it up you will need a second one to counter-hold the bolt head as the pinch bolts cannot be tightened before the pivot bolt. 24 mm hex keys are not readily available, but it is worth trying an automotive or commercial vehicle tool supplier – sump keys are sometimes that big. Alternatively Honda can supply the tools (part No. 07930-KA50100), or alternatively a tool can be fabricated using a 24 mm head bolt with two nuts threaded onto it and locked very tightly together. To counter-hold the pivot bolt head on installation a 24 mm nut inserted half-way into the head and held on the protruding section using a spanner or pipe wrench should suffice – it worked well for us. Alternatively weld a bar onto the nut to act as a handle.*

Removal

1 Support the bike as described in the **Note** above, making sure it is on level ground. Work can be made easier by raising the machine to a suitable working height on an hydraulic ramp or a suitable platform. Make sure the motorcycle is secure and will not topple over (also see *Tools and Workshop Tips* in the Reference section).
2 Remove the rider's seat, fairing and fairing side panels, and on RR-Y and RR-1 (2000 and 2001) models the lower fairing (see Chapter 8).
3 If the engine is dirty, particularly around its mountings, wash it thoroughly. This will make work much easier and rule out the possibility of caked on lumps of dirt falling into some vital component.
4 Drain the engine oil and the cooling system (see Chapter 1). Remove the oil filter (see Chapter 1).
5 Disconnect the negative (–ve) lead from the battery (see Chapter 9).
6 Remove the fuel tank, the air filter housing and the throttle bodies (see Chapter 4). Plug the engine inlet manifolds with clean rag.
7 Remove the radiator along with its hoses, noting their routing (see Chapter 3). Also remove the coolant reservoir along with its hoses and the radiator bracket (see Chapter 3). Note the routing of the overflow hose through the fairing side panel bracket on the right-hand side of the engine.
8 Remove the exhaust system (see Chapter 4). Free the exhaust control valve cables from the guide on the alternator cover **(see illustration)**.

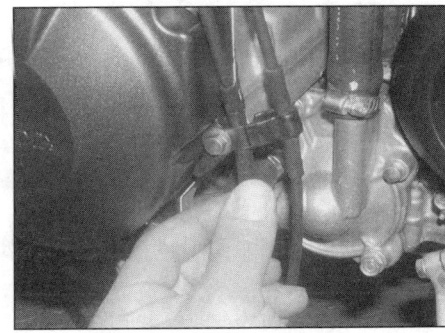

5.8 Free the cables from their guide

2•8 Engine, clutch and transmission

5.10a Unscrew the nut and detach the lead

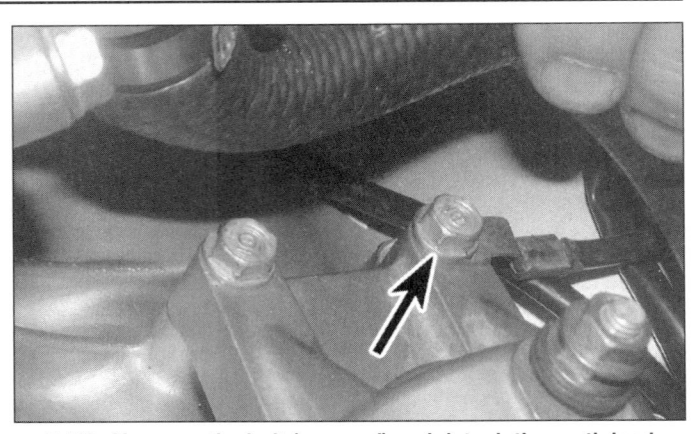

5.10b Unscrew the bolt (arrowed) and detach the earth lead

9 Remove the spark plug caps/ignition HT coils (see Chapter 5).

10 If required, remove the starter motor (see Chapter 9). If you want to leave the starter motor in situ, pull back the rubber cover on its terminal, then unscrew the nut and disconnect the lead **(see illustration)**. Also unscrew the mounting bolt securing the earth lead and detach the lead **(see illustration)**.

11 Unscrew the bolts securing the clutch cable bracket to the clutch cover **(see illustration)**. Displace the bracket, noting how it locates, and free the cable end from the release arm. Position the cable clear of the engine.

12 On RR-Y and RR-1 (2000 and 2001) models disconnect the sidestand switch (green 2-pin) and speed sensor (3-pin black) wiring connectors – they are housed inside the rubber boot above the crankcase **(see illustration)**. Also disconnect the oil pressure switch and neutral switch sub-harness (blue 2-pin) wiring connector and the ignition pulse generator coil (red 2-pin) wiring connector, both located near the loom by the right-hand frame beam. Release the wiring from any clips or ties, noting its routing, and feed it through to its source so that it does not impede engine removal. If required, disconnect the wires from the neutral switch and oil pressure switch (by removing the screw) and remove the sub-harness **(see illustrations 5.13b and 5.13c)**. Disconnect the coolant temperature sensor wiring connector **(see illustration 5.13d)**.

13 On RR-2 and RR-3 (2002 and 2003) models disconnect the sidestand switch (green 2-pin) and speed sensor (3-pin natural), and engine sub-harness (blue 2-pin) wiring connectors – they are housed inside the rubber boot above the crankcase. The sub-harness feeds the oil pressure switch and neutral switch. Also disconnect the ignition pulse generator coil (red 2-pin) wiring connector, located on the inside of the right-hand frame beam **(see illustration)**. Release the wiring from any clips or ties, noting its routing, and feed it through to its source so that it does not impede engine removal. If required, disconnect the wires from the neutral switch and oil pressure switch (by removing the screw) and remove the sub-harness **(see illustrations)**. Disconnect the coolant temperature sensor wiring connector **(see illustration)**.

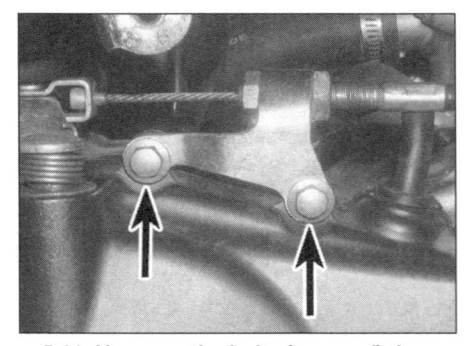

5.11 Unscrew the bolts (arrowed) then detach the cable end from the release arm

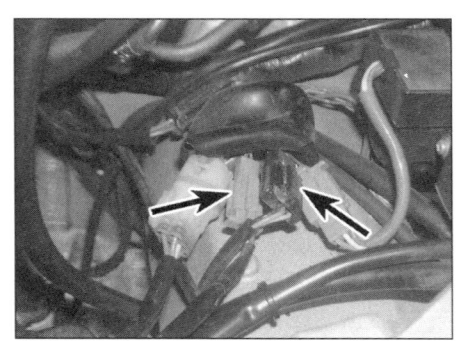

5.12 Disconnect the wiring connectors (arrowed) in the boot

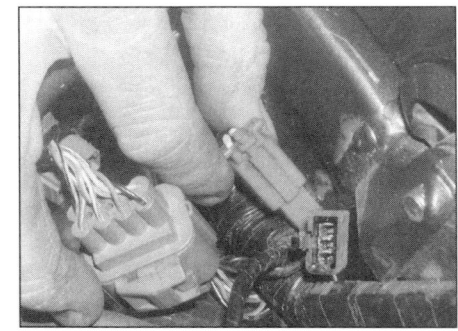

5.13a Disconnect the pulse generator coil wiring connector

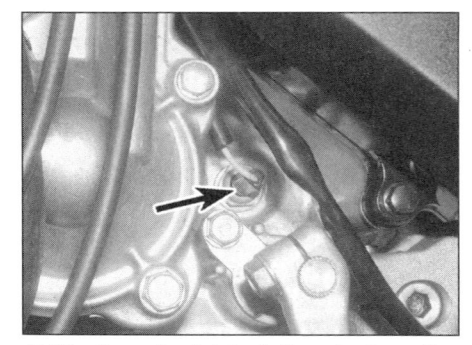

5.13b If required detach the wire from the neutral switch (arrowed) . . .

5.13c . . . and the oil pressure switch

5.13d Disconnect the ECT sensor wiring connector

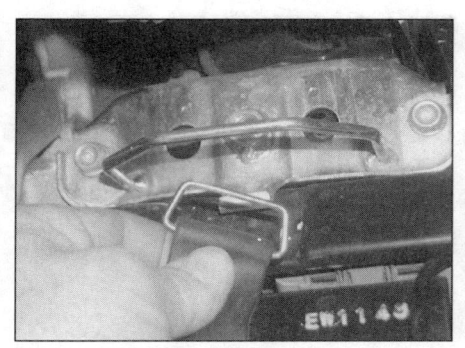

Engine, clutch and transmission 2•9

5.14a Release the strap and remove the ECM cover

5.14b Remove the cover, noting how it locates at the back

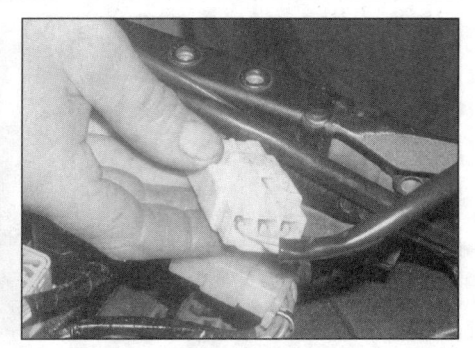

5.14c Disconnect the alternator wiring connector

5.15 Note the alignment of the punch mark, then remove the bolt and slide the arm off the shaft

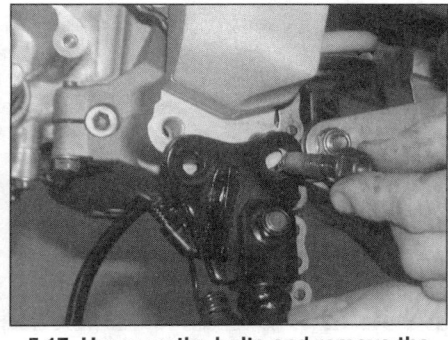

5.17 Unscrew the bolts and remove the stand assembly

5.18a Unscrew the bolt and detach the reservoir

14 On RR-Y and RR-1 (2000 and 2001) models release the battery strap and remove the ECM cover **(see illustration)**. On RR-2 and RR-3 (2002 and 2003) models remove the battery (see Chapter 9), then remove the ECM cover. Trace the wiring from the alternator and disconnect it at the white 3-pin wiring connector with the three yellow wires **(see illustrations)**. Feed the wiring down and coil it on the crankcase, releasing it from any ties and noting its routing.

15 Unscrew the gearchange linkage arm pinch bolt and slide the arm off the shaft, noting how the slot in the arm aligns with the punch mark on the shaft **(see illustration)**. Unscrew the footrest bracket mounting bolts and remove the footrest/gearchange lever and linkage assembly.

16 Remove the front sprocket (see Chapter 6).

17 Unscrew the sidestand bracket bolts and remove the stand assembly **(see illustration)**.

18 You must now separate the rear suspension assembly from the engine – there are two ways of doing this:

Method 1: *If you are experienced at working on bikes and have the physical strength and dexterity, and do not wish to remove the individual assemblies separately for cleaning and/or checking, you can remove the entire rear wheel, swingarm and suspension linkage assembly as one, bringing the complete rear brake system with it, as follows: Unscrew the bolt securing the rear master cylinder reservoir and lay it on some rag on the swingarm* **(see illustration 5.18a)**. *Trace the wiring from the brake light switch and disconnect it at the connector. Feed the wiring down to the switch, noting its routing. Unscrew the right-hand footrest mounting bolts and displace the footrest/brake pedal/master cylinder assembly* **(see illustration 5.18b)**. *Tie it to the swingarm with some rag between them. Pull the collar out of the exhaust system mounting rubber, then carefully press the rubber out of its lug* **(see illustrations 5.18c and 5.18d)**. *Unscrew the nuts and withdraw the bolts securing the suspension linkage arm to the engine bracket and the bottom of the shock absorber to the*

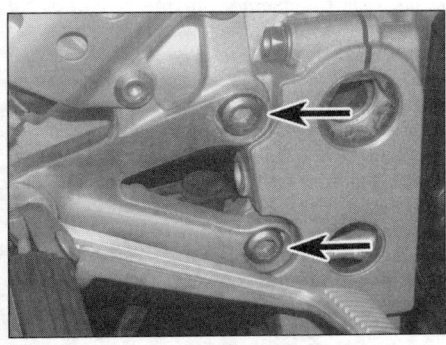

5.18b Unscrew the bolts (arrowed) and detach the bracket from the frame

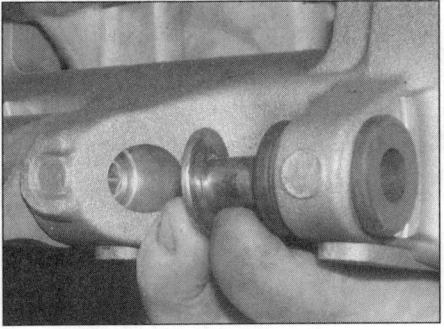

5.18c Withdraw the collar . . .

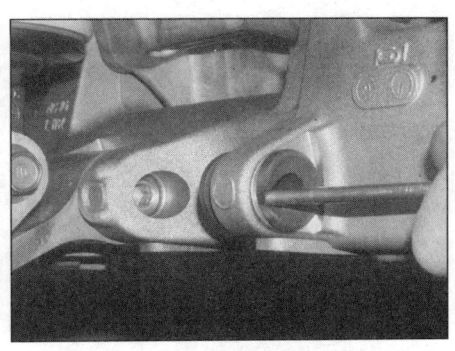

5.18d . . . and press the rubber out

2•10 Engine, clutch and transmission

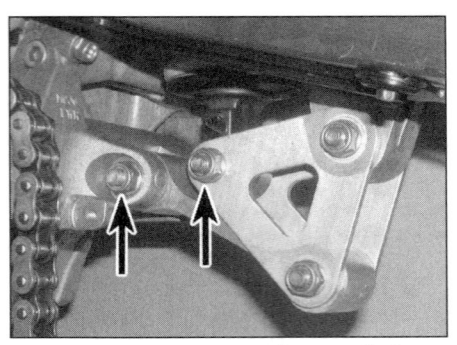

5.18e Unscrew the nuts (arrowed), then withdraw the bolts and detach the linkage assembly

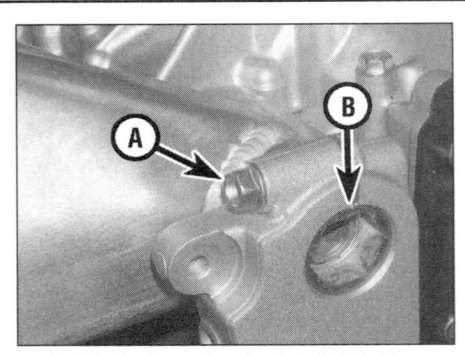

5.18f Slacken the pinch bolt (A), then unscrew the nut (B) as described

5.18g Slacken the pinch bolt (arrowed)

5.18h Withdraw the pivot bolt . . .

5.18i . . . and carefully manoeuvre the rear suspension and brake assembly out of the frame

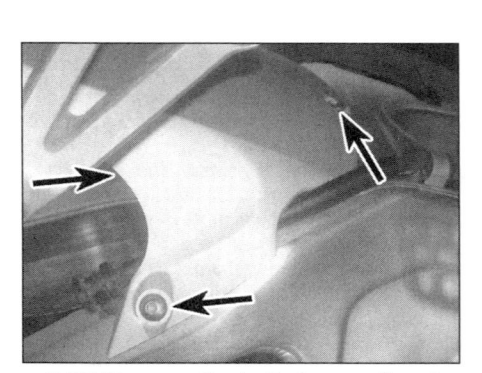

5.18j Unscrew the bolts (arrowed) and remove the hugger

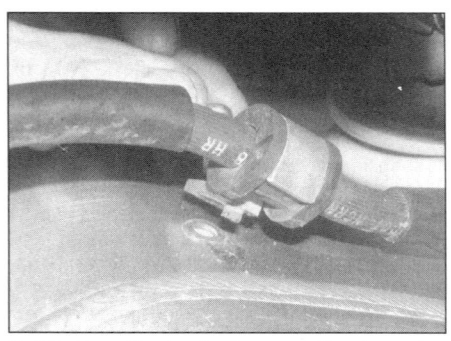

5.18k Free the brake hose, noting how the guide locates

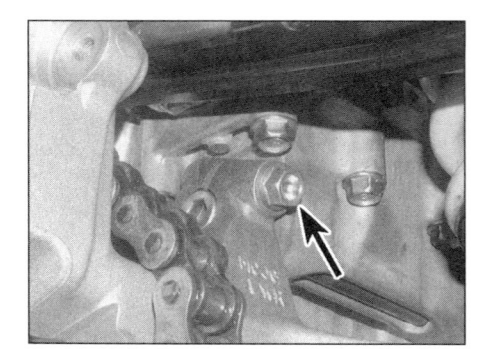

5.19a Slacken the upper pinch bolt (arrowed) . . .

5.19b . . . and the lower pinch bolt (arrowed) . . .

linkage plates *(see illustration 5.18e)*. Slacken the pinch bolt on the right-hand end of the swingarm pivot *(see illustration 5.18f)*. Unscrew the nut on the right-hand end of the pivot bolt using one of the tools described in the **Note 2** above. Slacken the pinch bolt on the left-hand end of the swingarm pivot *(see illustration 5.18g)*. Withdraw the pivot bolt and carefully manoeuvre the rear wheel/swingarm/ suspension linkage/brake assembly out, leaving the shock absorber attached to the frame *(see illustrations 5.18h and 5.18i)*.

Method 2: If you prefer to remove the individual assemblies separately, it is best to leave the brake system behind, as follows: On RR-Y and RR-1 (2000 and 2001) models undo the screw securing each rear brake hose guide to the swingarm. On RR-2 and RR-3 (2002 and 2003) models unscrew the bolts securing the rear hugger to the swingarm and remove it, noting how one bolt secures the brake hose rear guide, then undo the screw securing the brake hose front guide to the swingarm *(see illustrations 5.18j and 5.18k)*. On all models trace the wiring from the brake light switch and disconnect it at the connector. Feed the wiring down to the switch, noting its routing. Unscrew the right-hand footrest mounting bolts and displace the footrest/brake pedal/master cylinder assembly *(see illustration 5.18b)*. Tie it to the rear sub-frame with some rag between them (on RR-2 and RR-3 (2002 and 2003) models remove the seat cowling if required). Remove the rear wheel (see Chapter 7). Tie the brake caliper assembly to the rear footrest bracket. Remove the swingarm and rear suspension linkage (see Chapter 6). Note that you do not have to separate the linkage plates from the swingarm or separate the linkage components – just remove the linkage arm to engine bracket bolt and the linkage plates to shock absorber bolt, leaving the arm attached to the plates, and the plates attached to the swingarm *(see illustration 5.18e)*.

19 If required now (this can be done after the engine has been removed, and the bracket makes a good handle for manoeuvring the engine without adding much weight), slacken the pinch bolts on the upper and lower engine bracket bolts **(see illustrations)**. Unscrew the

Engine, clutch and transmission 2•11

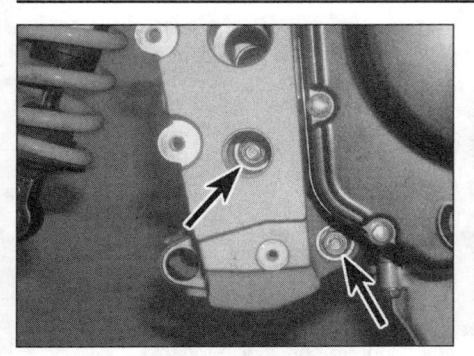

5.19c ... then unscrew the nuts (arrowed) ...

5.19d ... accessing the top one via the hole in the bracket as shown

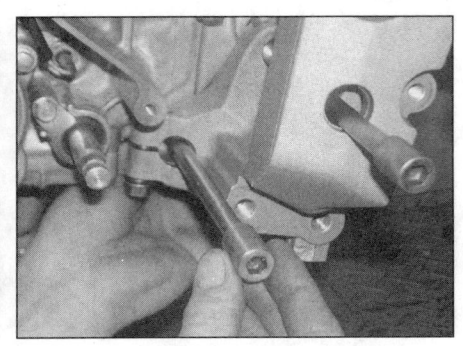

5.19e Withdraw the bolts ...

nuts on the bolts and withdraw the bolts from the engine and bracket **(see illustrations)**. Manoeuvre the bracket off the engine and remove it **(see illustration)**.

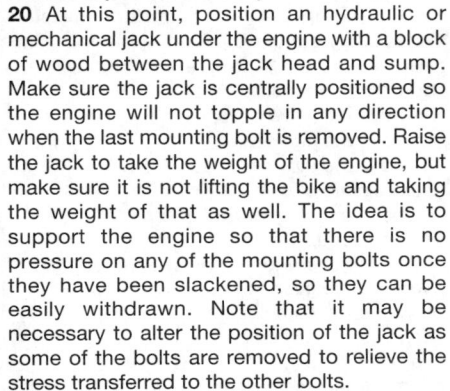

20 At this point, position an hydraulic or mechanical jack under the engine with a block of wood between the jack head and sump. Make sure the jack is centrally positioned so the engine will not topple in any direction when the last mounting bolt is removed. Raise the jack to take the weight of the engine, but make sure it is not lifting the bike and taking the weight of that as well. The idea is to support the engine so that there is no pressure on any of the mounting bolts once they have been slackened, so they can be easily withdrawn. Note that it may be necessary to alter the position of the jack as some of the bolts are removed to relieve the stress transferred to the other bolts.

21 Unscrew the front and centre engine mounting bolts on the left-hand side **(see illustration)**. Unscrew the front and centre engine mounting bolts on the right-hand side and remove the spacers from between the frame and engine **(see illustrations)**. Note the difference in size between the front and centre bolts.

22 Slacken the pinch bolt on the rear engine mounting bolt **(see illustration)**. Unscrew the nut on the left-hand end of the bolt **(see illustration)**. Check that the engine is properly supported by the jack. Withdraw the rear mounting bolt from the right-hand side **(see illustration)**. The engine can now be removed

5.19f ... and remove the bracket

5.21a Unscrew the front and centre bolts on the left-hand side

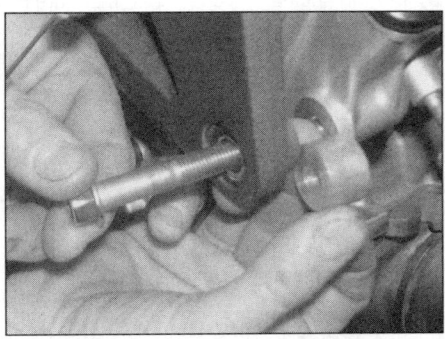

5.21b Unscrew the front bolt on the right-hand side and remove the spacer ...

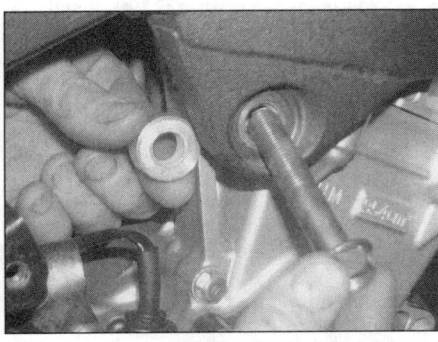

5.21c ... then unscrew the centre bolt and remove its spacer

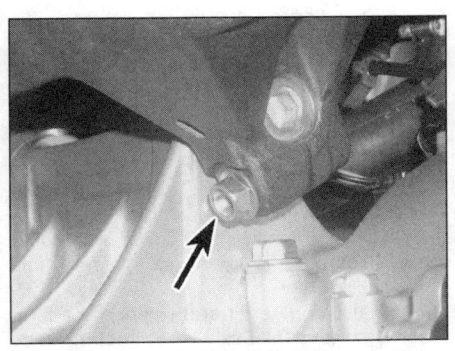

5.22a Slacken the pinch bolt (arrowed) ...

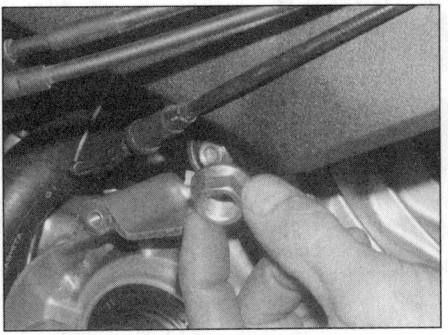

5.22b ... then unscrew the nut ...

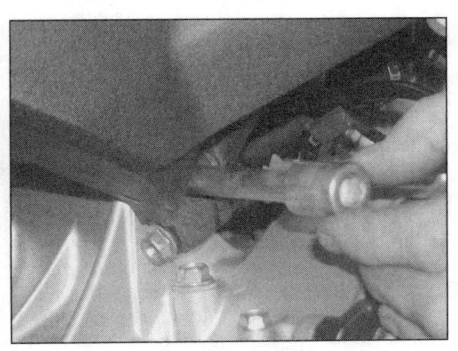

5.22c ... and withdraw the bolt

5.22d Remove the engine as described

from the frame (see *Caution* at beginning of this Section). Check that all wiring, cables and hoses are well clear, then carefully lower the jack a bit and manoeuvre the engine forward so that the cam chain tensioner (if not removed) clears the frame **(see illustration)**. Fully lower the jack, then with the aid of an assistant remove the jack from under the engine and remove the engine.

23 If required and not already done (this must be done if the crankcases are being separated), refer to Step 19 and remove the engine bracket.

Installation

24 If required now (this can be done after the engine has been installed), manoeuvre the engine bracket onto the engine, aligning the bolt holes **(see illustration 5.19f)**. Install the bracket bolts from the left-hand side and lightly tighten their nuts – do not fully tighten them until after the swingarm pivot bolt has been installed and tightened **(see illustrations 5.19e and 5.19c)**.

25 Manoeuvre the engine into position under the frame and lift it onto the jack **(see illustration 5.22d)**. Raise the engine to align all the mounting bolt holes, taking care not to catch the cam chain tensioner on the frame. Note that it may be necessary to adjust the jack as some of the bolts are installed and tightened to realign the other bolt holes.

26 Install the rear mounting bolt from the right-hand side and tighten its nut finger-tight **(see illustrations 5.22c and 5.22b)**.

27 Install the front (10 mm) and centre (12 mm) bolts on the right-hand side with their spacers and tighten them finger-tight **(see illustrations 5.21b and 5.21c)**.

28 Install the front (10 mm) and centre (12 mm) bolts on the left-hand side and tighten them finger-tight **(see illustration 5.21a)**.

29 Counter-hold the rear mounting bolt and tighten the nut to the torque setting specified at the beginning of the Chapter. Now tighten the rear mounting pinch bolt to the specified torque setting **(see illustration 5.22a)**.

30 Tighten the front and centre mounting bolts on the left-hand side to their specified torque settings.

31 Tighten the front and centre mounting bolts on the right-hand side to their specified torque settings.

32 If not already done, refer to Step 24 and install the engine bracket.

33 Install the swingarm, suspension linkage, rear wheel and brake assemblies according to your removal method, referring to Chapters 6 and 7, and to the torque settings given at the beginning of those Chapters – whatever your method, after installing the swingarm pivot bolt and tightening its nut, tighten the engine bracket bolt nuts and pinch bolts to the torque settings specified at the beginning of this Chapter. Thereafter install and tighten the suspension linkage and shock absorber bolts and nuts.

34 The remainder of the installation procedure is the reverse of removal, noting the following points:
- Use new gaskets on the exhaust pipe connections.
- When fitting the gearchange linkage arm onto the gearchange shaft, align the slit in the arm with punch mark on the shaft, and tighten the pinch bolt securely **(see illustration 5.15)**.
- Make sure all wires, cables and hoses are correctly routed and connected, and secured by any clips or ties.
- Refill the engine with oil and coolant (see Chapter 1).
- Adjust the throttle and clutch cable freeplay.
- Adjust the drive chain (see Chapter 1).
- Start the engine and check that there are no oil or coolant leaks. Adjust the idle speed (see Chapter 1).

6 Engine disassembly and reassembly – general information

Disassembly

1 Before disassembling the engine, thoroughly clean and degrease its external surfaces. This will prevent contamination of the engine internals, and will also make working a lot easier and cleaner. A high flash-point solvent, such as paraffin (kerosene) can be used, or better still, a proprietary engine degreaser. Use old paintbrushes and toothbrushes to work the solvent into the various recesses of the casings. Take care to exclude solvent or water from the electrical components and intake and exhaust ports.

⚠ **Warning: The use of petrol (gasoline) as a cleaning agent should be avoided because of the risk of fire.**

2 When clean and dry, position the engine on the workbench, leaving suitable clear area for working. Gather a selection of small containers, plastic bags and some labels so that parts can be grouped together in an easily identifiable manner. Also get some paper and a pen so that notes can be taken.

You will also need a supply of clean rag, which should be as absorbent as possible.

3 Before commencing work, read through the appropriate section so that some idea of the necessary procedure can be gained. When removing components note that great force is seldom required, unless specified (checking the specified torque setting of the particular bolt being removed will indicate how tight it is, and therefore how much force should be needed). In many cases, a component's reluctance to be removed is indicative of an incorrect approach or removal method – if in any doubt, re-check with the text.

4 When disassembling the engine, keep 'mated' parts together (including gears, cylinder bores, pistons, connecting rods, valves, etc, that have been in contact with each other during engine operation). These 'mated' parts must be reused or replaced as an assembly.

5 A complete engine/transmission disassembly should be done in the following general order with reference to the appropriate Sections.
  Remove the valve cover
  Remove the camshafts
  Remove the cylinder head
  Remove the clutch
  Remove the alternator/starter clutch (see Chapter 9)
  Remove the starter motor (see Chapter 9)
  Remove the gearchange mechanism
  Remove the cam chain and blades
  Remove the oil sump
  Remove the oil pump
  Remove the selector drum and forks
  Separate the crankcase halves
  Remove the connecting rods and pistons
  Remove the crankshaft
  Remove the transmission shafts

Reassembly

6 Reassembly is accomplished by reversing the general disassembly sequence.

7 Oil cooler – removal and installation

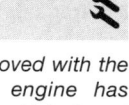

Note: *The oil cooler can be removed with the engine in the frame. If the engine has been removed, ignore the steps which do not apply.*

Removal

1 The cooler is located on the front of the engine. On RR-Y and RR-1 (2000 and 2001) models remove the lower fairing (see Chapter 8). On RR-2 and RR-3 (2002 and 2003) models remove the fairing side panels (see Chapter 8).

2 Drain the engine oil and remove the filter (see Chapter 1). Drain the coolant (see Chapter 1). Remove the reservoir for improved access if required (see Chapter 3).

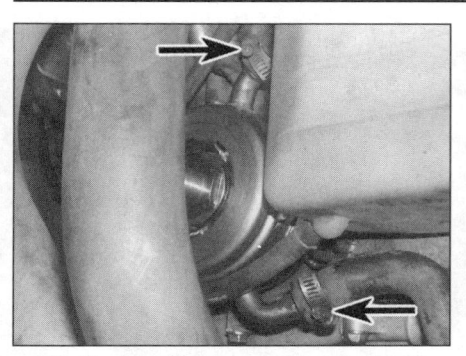

Engine, clutch and transmission 2•13

7.3 Slacken the clamps (arrowed) and detach the hoses

7.4 Unscrew the bolt and remove the cooler

7.6a Fit a new O-ring into the groove

7.6b Locate the slot in the bracket over the lug on the crankcase (arrow)

7.6c Make sure the dot on the washer faces towards the engine

3 Slacken the clamp securing each hose to the cooler and detach the hoses **(see illustration)**.
4 Unscrew the cooler bolt using a socket and remove the lockwasher **(see illustration)**. Remove the cooler, noting how the slotted bracket locates over the lug on the crankcase. Discard the O-ring as a new one must be used.
5 Check the cooler body for cracks and dents and any evidence of coolant leakage and replace it with a new one if necessary. Also check the hoses for splits, cracks, hardening and deterioration and fit new ones if required.

Installation

6 Installation is the reverse of removal, noting the following:
● Ensure the mating surfaces of the crankcase and the cooler are clean and dry.
● Use a new O-ring on the cooler body and smear it with clean engine oil. Make sure it seats in its groove **(see illustration 7.6a)**.
● Locate the slot in the bracket on the cooler over the lug on the crankcase **(see illustration 7.6b)**.
● Apply clean engine oil to the cooler bolt threads and seating surface. Fit the lockwasher with its concave side, marked with a dot, facing the cooler **(see illustration 7.6c)**. Tighten the bolt to the torque setting specified at the beginning of the Chapter.
● Make sure the coolant hoses are pressed fully onto their unions and are secured by the clamps **(see illustration 7.3)**.
● Fit a new oil filter and fill the engine with oil (see Chapter 1).
● Refill the cooling system (see Chapter 1).

8 Valve cover – removal and installation

Note: *The valve cover can be removed with the engine in the frame. If the engine has been removed, ignore the steps which do not apply.*

Removal

1 Remove the left-hand fairing side panel (see Chapter 8). Disconnect the radiator fan wiring connector and release the connector from its bracket **(see illustration)**.
2 Remove the air filter housing (see Chapter 4). Lift the rubber heat shield, noting how it fits **(see illustration)**.

8.1 Disconnect the cooling fan wiring connector

8.2 Lift the heat shield, noting how it locates round the cable and hoses

2•14 Engine, clutch and transmission

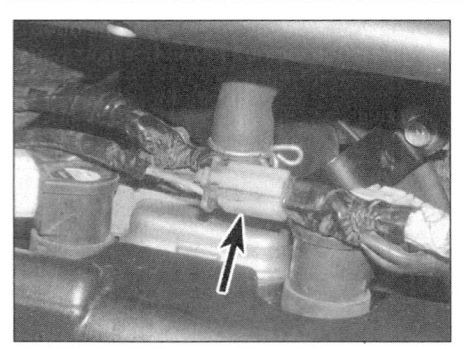

8.3 Disconnect the wiring connector (arrowed)

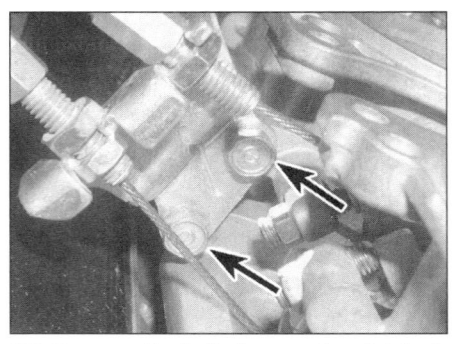

8.6 Unscrew the bolts (arrowed) and detach the bracket, then free the cable ends

8.8a Unscrew the bolts (arrowed) . . .

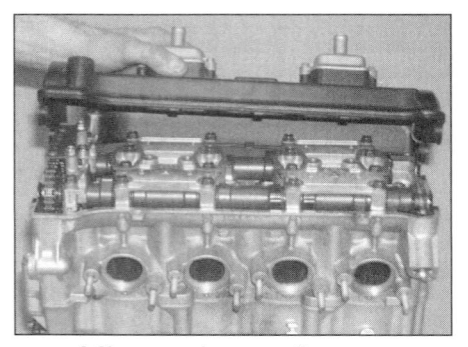

8.8b . . . and remove the cover

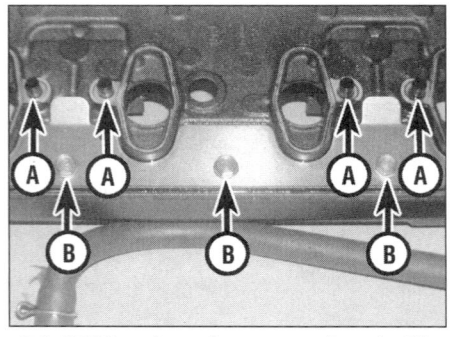

8.9 PAIR system air passage dowels (A), breather separator bolts (B)

3 Disconnect the cam pulse generator wiring connector (see illustration).
4 Remove the PAIR control valve along with its hoses (see Chapter 4).
5 Remove the spark plug caps/ignition HT coils (see Chapter 5).
6 Unscrew the two throttle cable bracket bolts and displace the bracket from the throttle body assembly – take care not to drop the bolts (see illustration). Detach the cable ends from the throttle cam.
7 Remove the PAIR system reed valve covers, and the valves and their base plates if loose (see Chapter 4).
8 Unscrew the four valve cover bolts and lift the cover off the cylinder head (see illustrations). If it is stuck, do not try to lever it off with a screwdriver. Tap it gently around the sides with a rubber hammer or block of wood to dislodge it. Note the rubber washers for the bolts and remove them if they are loose (see illustration 8.14a). The rubber gasket is normally glued into the groove in the cover, and is best left there if it is reusable. If the gasket is in any way damaged, deformed or deteriorated, remove it.
9 Note the four dowels that link the PAIR system air passages between the valve cover and cylinder head and remove them for safekeeping if they are loose (which is unlikely), taking care not to drop them if they are not in the valve cover (see illustration).
10 If required, unscrew the crankcase breather separator bolts and remove the separator (see illustration 8.9). Discard the gasket. Clean out the breather chamber.

Installation

11 If removed, fit the breather separator onto the valve cover using a new gasket. Apply a suitable non-permanent thread locking compound to the bolts and tighten them to the torque setting specified at the beginning of the Chapter (see illustration 8.9).
12 If removed, fit the PAIR system dowels into the valve cover (see illustration 8.9).
13 Examine the valve cover gasket for signs of damage or deterioration and install a new one if necessary. If a new one is used, clean all traces of the old glue from the groove in the cover and clean it and the cylinder head mating surface with solvent. Fit the new gasket into the groove, using a suitable glue, sealant or grease to hold it in place (see illustration). Also apply a suitable sealant to the cut-outs in the cylinder head (see illustration).
14 Position the valve cover on the cylinder

8.13a Make sure the gasket locates in the groove and stays there

8.13b Apply a sealant to the cutouts in the cylinder head

Engine, clutch and transmission 2•15

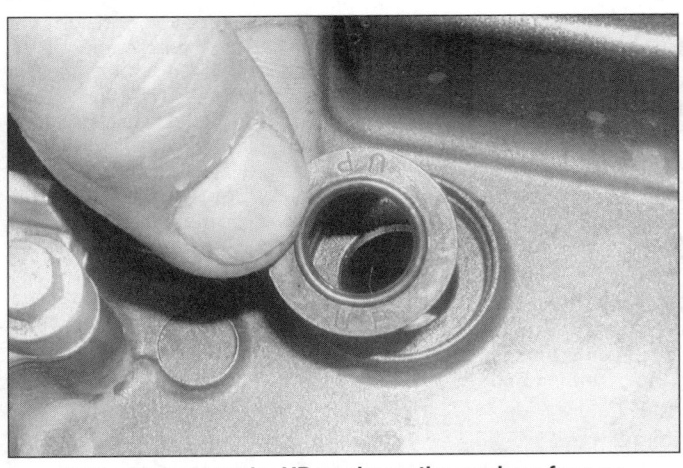

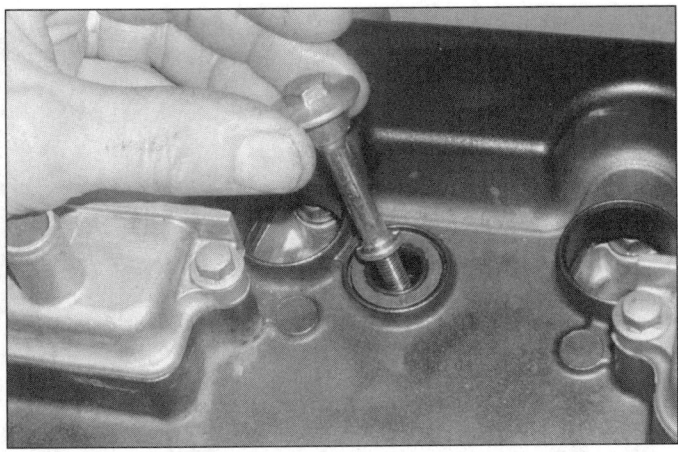

8.14a Make sure the UP marks on the washers face up

8.14b Install the bolts and tighten them to the specified torque

head, making sure the gasket stays in place **(see illustration 8.8b)**. If removed, fit the rubber washers into the cover, using new ones if required, and making sure they are installed with the UP mark facing up **(see illustration)**. Install the cover bolts and tighten them to the specified torque setting **(see illustration)**.

15 Install the remaining components in the reverse order of removal.

9 Cam chain tensioner – removal and installation

Note: *The cam chain tensioner can be removed with the engine in the frame, however there is very little clearance between it and the frame which could make the plunger retraction procedure tricky, depending on your tools and dexterity. The only alternative is to remove the engine. If the engine has been removed, ignore the steps which do not apply.*

Removal

1 Remove the air filter housing, and for best access the throttle bodies (see Chapter 4).
2 Unscrew the tensioner cap bolt and remove the sealing washer **(see illustration)**.
3 If a Honda tensioner locking key is available, insert it in the end of the tensioner so that it engages the slotted plunger and turn it clockwise until the plunger is fully retracted, then push the key into the slots in the end of the tensioner body to lock it **(see illustration)**. Unscrew the tensioner mounting bolts, then withdraw the tensioner from the engine. Note that a home-made equivalent of the Honda tool can be easily made out of a piece of 1 mm steel plate cut to the dimensions shown **(see illustration)**.
4 If a locking key is not available, first slacken the tensioner mounting bolts slightly. Insert a small flat-bladed screwdriver in the end of the tensioner so that it engages the slotted plunger. Turn the screwdriver clockwise until the plunger is fully retracted and hold it in this

position while unscrewing the tensioner mounting bolts **(see illustration)**. Remove the bolts and sealing washers, then withdraw the tensioner from the engine and release the screwdriver – the plunger will spring back out once the screwdriver is removed, but can be easily reset on installation.
5 Discard the gasket and sealing washer as new ones must be used on installation. Do not dismantle the tensioner.

Installation

6 Check that the plunger moves smoothly

9.2 Unscrew the cap bolt and remove the washer

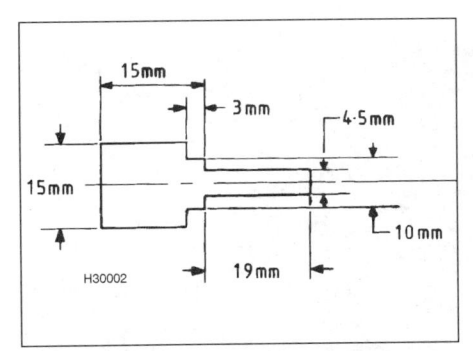

9.3b A copy of Honda's locking key can be made using 1 mm thick steel cut to the dimensions shown

when wound into the tensioner and springs back out freely when released. Ensure the tensioner and cylinder block surfaces are clean and dry.
7 If the locking key described above is being used, insert it in the end of the tensioner so that it engages the slotted plunger and turn it clockwise until the plunger is fully retracted, then push the key into the slotted end of the tensioner body to lock it in this position **(see illustrations 9.3a and 9.3b)**. Fit a new gasket onto the tensioner body, then install the tensioner with its mounting bolts and tighten

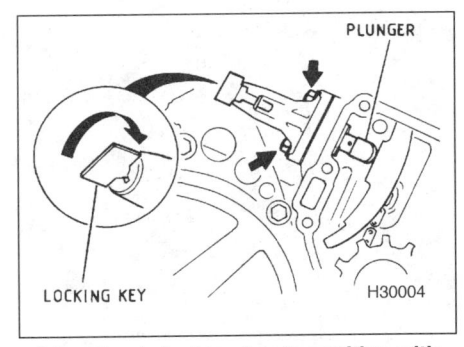

9.3a Honda locking key in position with tensioner plunger retracted. Tensioner mounting bolts arrowed

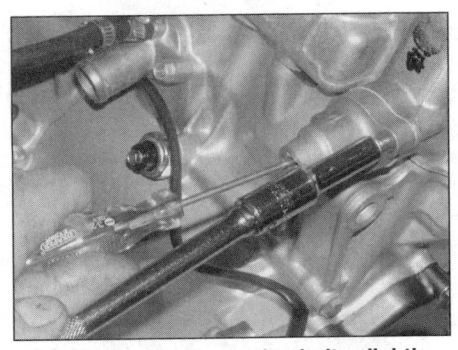

9.4 Slacken the mounting bolts slightly, then insert the screwdriver, retract the tensioner and unscrew the mounting bolts

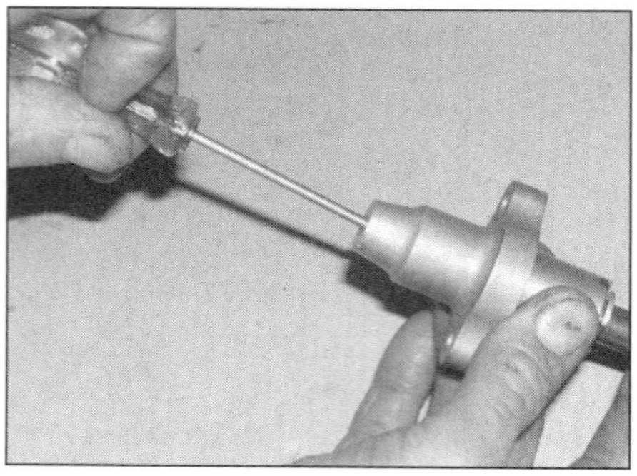

9.8a Insert the screwdriver and retract the plunger...

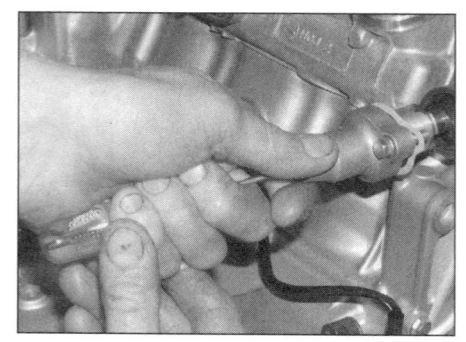

9.8b ...then install the tensioner using a new gasket and tighten the mounting bolts

them. Remove the key, then install the tensioner cap bolt with a new sealing washer and tighten it **(see illustration 9.2)**.

8 If the key is not available, insert a small flat-bladed screwdriver in the end of the tensioner so that it engages the slotted plunger **(see illustration)**. Turn the screwdriver clockwise until the plunger is fully retracted and hold it in this position whilst the tensioner is installed. Fit a new gasket onto the tensioner body, then install the tensioner with its mounting bolts and tighten them **(see illustration)**. Release and remove the screwdriver, then install the tensioner end bolt with a new sealing washer and tighten it **(see illustration 9.2)**.

9 Install the throttle bodies if removed, and the air filter housing.

10 Camshafts and followers – removal, inspection and installation

Note: *The camshafts can be removed with the engine in the frame. Place clean rags over the spark plug holes and the cam chain tunnel to prevent any component from dropping into the engine.*

Removal

1 Remove the valve cover (see Section 8). Remove the cam pulse generator (see Chapter 4).
2 Unscrew the timing inspection cap from the clutch cover **(see illustration)**. The engine must be turned so that the No. 1 piston is at TDC (top dead centre) on its compression stroke. The engine can be turned using a suitable spanner or socket on the timing rotor bolt and turning it in a clockwise direction only. Alternatively, place the motorcycle on an auxiliary stand so that the rear wheel is off the ground, select a high gear and rotate the rear wheel by hand in its normal direction of rotation (removing the spark plugs first will make this a lot easier – see Chapter 1).
3 Turn the engine until the line next to the T mark on the timing rotor aligns with the static timing mark, which is a notch in the inspection hole rim, and the IN and EX marks on the intake and exhaust camshaft sprockets respectively are facing away from each other and are flush with the cylinder head top surface **(see illustrations)**. If the marks are facing towards each other, rotate the engine clockwise one full turn (360°) until the line next to the T mark again aligns with the static timing mark. The sprocket marks will now be facing away.
4 Either remove the cam chain tensioner (see Section 9), or if you prefer, obtain or fabricate the tensioner locking tool, then retract and lock the tensioner plunger using the tool as described in Section 9, Step 3. Also unscrew the bolts securing the top cam chain guide and remove it **(see illustration)**.
5 There are three camshaft holders, each bridging both camshafts **(see illustration)**. Of

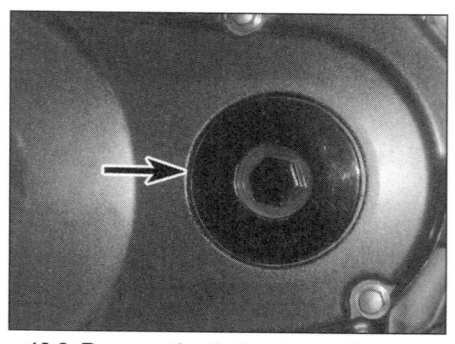

10.2 Remove the timing inspection cap (arrowed)

10.3a Turn the engine clockwise using a socket on the timing rotor bolt...

10.3b ...until the line next to the T mark aligns with the notch (arrowed)...

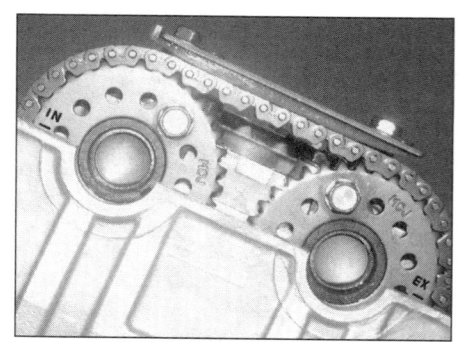

10.3c ...and the camshaft sprocket marks are as shown

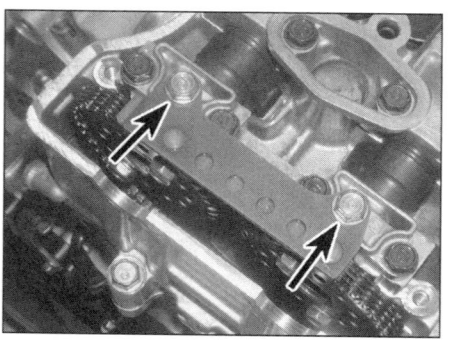

10.4 Unscrew the bolts (arrowed) and remove the guide

10.5 Camshaft holders (arrowed)

Engine, clutch and transmission 2•17

10.6 Unscrew the bolts as described and remove the holders

10.7 Note the identity mark on each camshaft

10.9a Carefully lift out the follower using a lapping tool, grips or a magnet . . .

10.9b . . . and retrieve the shim from inside it . . .

10.9c . . . or from the top of the valve

the two larger holders, the one on the right-hand end is marked R and the one on the left-hand end L, and these letters are at the front. Mark an arrow on the small holder adjacent to the cam chain to denote which side points to the front of the engine. Note the numbers marked on the holders, adjacent to each bolt. These numbers denote the **tightening** sequence for the holder bolts.

6 Unscrew the camshaft holder bolts, slackening them evenly and a very little at a time in a **reverse** of the tightening sequence marked on the holders. Remove the bolts, noting which fits where as some are different, and lift off the holders, noting how they fit **(see illustration)**. Note the sealing washers fitted with the eight bolts around the spark plug bores. Remove the sealing rings from their grooves around the spark plug holes on the underside of the holders, noting how they also locate around the PAIR system air passage dowels **(see illustration 10.30a)**. Discard them as new ones must be used. Do not remove the holder bolt or air passage dowels unless they are loose and liable to drop out.

Caution: Make sure the holders lift up squarely and evenly and do not stick on a dowel or distort from some of the bolts being slackened more than the others as they or a camshaft could easily break.

7 If both camshafts are being removed, remove the intake camshaft first. Carefully lift each camshaft off the head and disengage the sprocket from the chain **(see illustrations 10.29 and 10.28)**. The camshafts are marked on the sprocket end for identification. The intake camshaft is marked I and the exhaust camshaft is marked E **(see illustration)**. If the marks aren't clear make your own as the camshafts must be installed in their original location.

8 While the camshafts are out do not rotate the crankshaft – the chain may drop down and bind between the crankshaft and case, which could damage these components. Wire the chain to another component or secure it using a rod of some sort to prevent it from dropping. Place rags over the spark plug holes and the cam chain tunnel to prevent anything from dropping into the engine on removal.

9 If the followers and shims are being removed from the cylinder head, obtain a container which is divided into sixteen compartments, and label each compartment with the location of a valve, i.e. intake or exhaust camshaft, left or right valve. If a container is not available, use labelled plastic bags (egg cartons also do very well!). Remove the cam follower of the valve in question, then retrieve the shim from the inside of the follower **(see illustrations)**. The follower is best removed with a magnet or using the suction created by a valve lapping tool, but long nosed pliers can be used with care. If the shim is not in the follower, pick it out of the top of the valve spring retainer using either a magnet, a screwdriver with a dab of grease on it (the shim will stick to the grease), or a very small screwdriver and a pair of pliers **(see illustration)**. Do not allow the shim to fall into the engine.

10 The sprockets can be separated from the camshafts if required by removing the two bolts that hold them **(see illustration)**. Both sprockets are identical and are therefore interchangeable, but mark them according to their camshaft so they can be installed in their original position. Also make alignment marks between the sprocket and the camshaft so that the sprocket can be installed the correct way round to avoid confusion when setting up the timing.

11 Note the alignment and fitting of the cam pulse generator rotor on the left-hand end of the exhaust camshaft and remove it if required – it is secured by two bolts **(see illustration)**.

10.10 Camshaft sprocket bolts (arrowed)

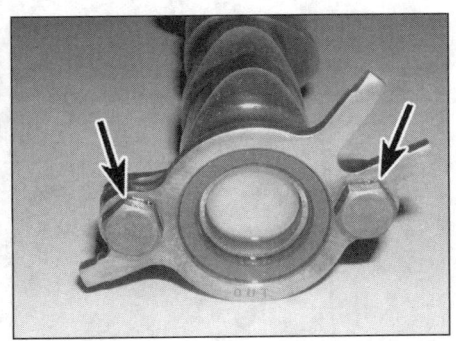

10.11 Cam pulse generator bolts (arrowed)

10.13 Measure the height of the camshaft lobes with a micrometer

Inspection

12 Inspect the bearing surfaces of the camshaft holders and cylinder head and the corresponding journals on the camshafts. Look for score marks, deep scratches and evidence of spalling (a pitted appearance). Check the oil passages for clogging.

13 Check the camshaft lobes for heat discoloration (blue appearance), score marks, chipped areas, flat spots and spalling. Measure the height of each lobe with a micrometer **(see illustration)** and compare the results to the minimum height listed in this Chapter's Specifications. If damage is noted or wear is excessive, the camshaft must be replaced with a new one.

14 Check the amount of camshaft runout by supporting each end on V-blocks, and measuring any runout using a dial gauge. If the runout exceeds the specified limit the camshaft must be replaced with a new one.

HAYNES HiNT *Refer to Tools and Workshop Tips in the Reference section for details of how to read a micrometer and dial gauge.*

15 Next, check the camshaft journal oil clearances. In order to negate the probability of the camshafts rotating (due to the fact that some of the lobes will be depressing their valves) as the holder bolts are tightened down, which will disturb the Plastigauge and lead to a false measurement, the valves should be removed from the cylinder head. To do this the head must be removed from the engine, which means the engine must be removed from the frame (see Sections 5, 12 and 14). Clean the camshafts and the bearing surfaces in the cylinder head and camshaft holder with a clean lint-free cloth, then lay each camshaft in its correct location in the cylinder head (see Step 7) **(see illustrations 10.29 and 10.28).**

16 Cut some strips of Plastigauge and lay one piece on each journal, parallel with the camshaft centreline. Make sure the camshaft holder dowels are installed **(see illustration 10.30a).** If the valves are installed, install the holders and tighten the bolts as described in Step 30. If the valves have been removed, install the holders as described in Step 30, noting that you can do each holder separately rather than all three at the same time, and tighten the bolts evenly and a little at a time in a criss-cross sequence, making sure the holders are pulled down squarely onto the dowels. While doing this, don't let the camshafts rotate, or the Plastigauge will be disturbed and you will have to start again.

17 Now unscrew the camshaft holder bolts as described in Step 6 (valves installed) or evenly and a little at a time in a criss-cross sequence (valves removed), and lift off the holder.

18 To determine the oil clearance, compare the crushed Plastigauge (at its widest point) on each journal to the scale printed on the Plastigauge container. Compare the results to this Chapter's Specifications. If the oil clearance is greater than specified, replace the camshaft with a new one and recheck the clearance. If the clearance is still too great, also replace the cylinder head and holder with new ones.

HAYNES HiNT *Before replacing the camshafts, cylinder head or holders because of damage, check with local machine shops specialising in motorcycle engine work. In the case of the camshafts, it may be possible for cam lobes to be welded, reground and hardened, at a cost far lower than that of a new camshaft. If the bearing surfaces in the case or holders are damaged, it may be possible for them to be bored out to accept bearing inserts. Due to the cost of new components it is recommended that all options be explored before condemning them as trash!*

19 Except in cases of oil starvation, the cam chain should wear very little. If the chain has stretched excessively, which makes it difficult to maintain proper tension, or if it is stiff or the links are binding or kinking, replace it with a new one. Refer to Section 11 for replacement.

20 Check the sprockets for wear, cracks and other damage, and replace them with new ones if necessary (see Steps 10 and 24). If the sprockets are worn, the cam chain is also worn, and so probably is the sprocket on the crankshaft. If severe wear is apparent, the entire engine should be disassembled for inspection.

21 Inspect the cam chain guides and tensioner blade (see Section 11).

22 Inspect the outer surface of each cam follower for evidence of scoring or other damage. If a follower is in poor condition, it is probable that the bore in the cylinder head in which it works is also damaged. Check for clearance between each follower and its bore. Measure the outer diameter of each follower and the inner diameter of its bore and compare the results to the Specifications **(see illustrations).** If any follower is worn beyond its service limit replace it with a new one. If any bore is worn beyond its limit, is seriously

10.22a Measure the external diameter of each follower...

10.22b ... and the internal diameter of each bore

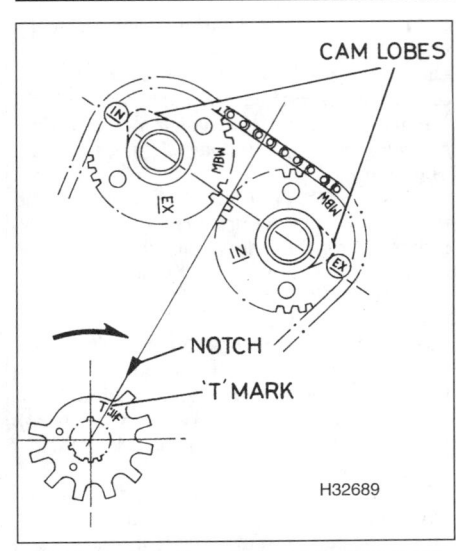

Engine, clutch and transmission 2•19

10.24 Camshaft and sprocket alignment, and timing mark alignment

10.28 Install the exhaust camshaft as described...

10.29 ...then install the intake camshaft

out-of-round or tapered, replace the cylinder head with a new one.

Installation

23 If removed, fit the cam pulse generator rotor onto the left-hand end of the exhaust camshaft – align it so that when the No. 1 cylinder (left-hand) lobes are facing up, the OUT mark on the rotor is at the bottom and faces out **(see illustration 10.11)**. Apply a suitable non-permanent thread locking compound to the bolts and tighten them to the specified torque setting. If necessary, tighten the bolts after the camshafts are installed so the shaft can be counter-held easily.

24 If separated, fit the sprockets onto the camshafts. Make sure they are installed the correct way round and in their original location as identified by the marks made on removal (Step 10) **(see illustration and 10.10)**. Apply a suitable non-permanent thread locking compound to the sprocket bolts and tighten them to the torque setting specified at the beginning of the Chapter. If necessary, tighten the bolts after the camshafts are installed so the shaft can be counter-held easily.

25 If removed, lubricate each shim and its follower with molybdenum disulphide oil (a 50/50 mixture of molybdenum disulphide grease and engine oil). Fit each shim into its recess in the top of the valve spring retainer with the size mark facing up, making sure it is correctly seated **(see illustration 10.9c)**.
Note: *It is most important that the shims and followers are returned to their original valves otherwise the valve clearances will be inaccurate.* Install each follower, making sure it fits squarely in its bore **(see illustration 10.9a)**.

26 Make sure the bearing surfaces on the camshafts and in the cylinder head are clean, then apply molybdenum disulphide oil (a 50/50 mixture of molybdenum disulphide grease and engine oil) to each of them. Also apply it to the camshaft journals and lobes. Make sure that none gets on the mating surfaces between the holder and the head, or in the bolt holes.

27 Check that the line next to the T mark on the timing rotor aligns with the notch in the inspection hole rim **(see illustration 10.3b and 10.24)**. If both camshafts have been removed, install the exhaust camshaft first, then the intake.

28 Lay the exhaust camshaft (marked E) onto the head with the EX mark on the sprocket facing forward and level with the cylinder head top mating surface, and the No. 1 cylinder lobes facing forward **(see illustrations 10.3c and 10.24)**. Fit the cam chain around the sprocket as you install the camshaft, pulling up on the chain to remove all slack in the front run between the crankshaft and the camshaft **(see illustration)**.

29 Lay the intake camshaft (marked I) onto the head with the IN mark on the sprocket facing back and level with the cylinder head top mating surface, and the No. 1 cylinder lobes facing diagonally back and up **(see illustration 10.3c and 10.24)**. Fit the cam chain around the sprocket as you install the camshaft, pulling on it to remove all slack from between the two camshaft sprockets **(see illustration)**. Any slack in the chain must lie in the rear run of the chain between the intake camshaft and the crankshaft so that it is later taken up by the tensioner.

30 Make sure the bearing surfaces in the camshaft holders are clean, then apply molybdenum disulphide oil (a 50/50 mixture of molybdenum disulphide grease and engine oil) to each of them. Make sure the camshaft holder bolt dowels are installed. Fit new sealing rings into the grooves around the spark plug holes on the underside of the main holders, making sure they also locate around the PAIR system air passage dowels **(see illustration)**. Lay the holders in the head **(see illustration)**. Install all the holder bolts, not forgetting new sealing washers with the eight bolts around the spark plug bores, and fitting the six longer bolts into the holes where dowels are fitted **(see illustration)**. Tighten the bolts evenly and a little at a time in the correct numerical sequence (i.e. 1 to 20) to the torque setting specified at the

10.30a Fit new sealing rings over the air passage dowels and into the grooves around the plug holes

10.30b Install all three holders...

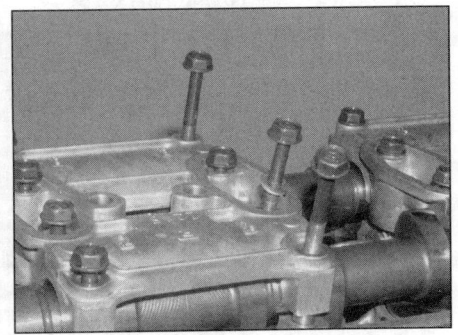

10.30c ...making sure the bolts are correctly located and tightened as described

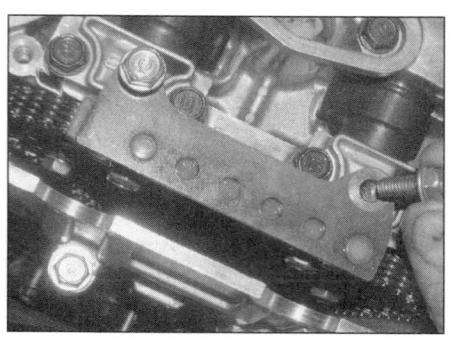

2•20 Engine, clutch and transmission

10.32 Install the top cam chain guide

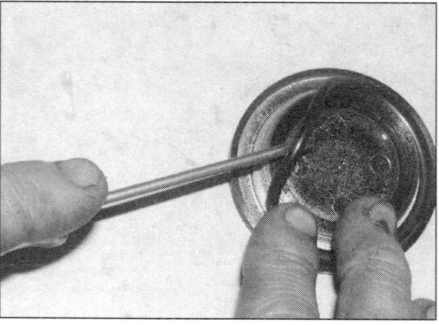

10.34 Fit the cap using a new O-ring and smear it and the threads with grease

beginning of the Chapter, making sure the dowels in the holders enter their bores in the head, and the holders are being pulled down evenly and squarely.

Caution: Whilst tightening the bolts, make sure the holders are being pulled evenly and squarely down and are not binding on the dowels or tilting to one side – if they do, adjust the relevant bolts until the holders are again square to the head. A holder or camshaft is likely to break if they are not tightened down evenly and squarely.

31 If the tensioner locking tool was used to retract and hold the tensioner plunger, remove it to release the plunger. If the tensioner was removed, use a piece of wooden dowel to press on the back of the cam chain tensioner blade via the tensioner bore in the cylinder block to ensure that any slack in the cam chain is taken up and transferred to the rear run of the chain (where it will later be taken up by the tensioner). At this point check that all the timing marks are still in **exact** alignment as described in Step 3 **(see illustrations 10.3b, 10.3c and 10.24)**. Note that it is easy to be slightly out (one tooth on the sprocket) without the marks appearing drastically out of alignment. If the marks are out, verify which sprocket is misaligned, then either reinstall the tensioner locking tool and retract the plunger, or remove the wooden dowel. Unscrew the sprocket's bolts and slide it off the camshaft, then disengage it from the chain. Move the camshaft round as required, then fit the sprocket back into the chain and onto the camshaft, and check the marks again. With everything correctly aligned, apply a suitable non-permanent thread locking compound to the sprocket bolts and tighten them to the torque setting specified at the beginning of the Chapter.

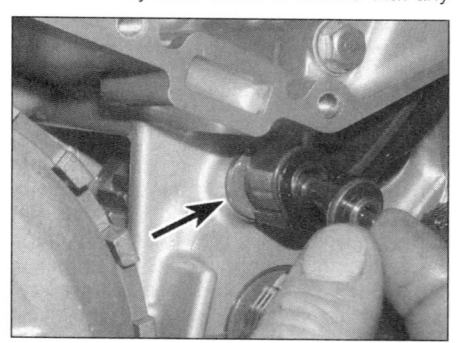

Caution: If the marks are not aligned exactly as described, the valve timing will be incorrect and the valves may strike the pistons, causing extensive damage to the engine.

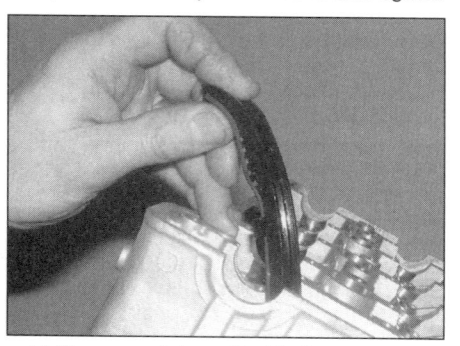

32 Install the cam chain top guide and tighten its bolts securely **(see illustration)**. Either install the cam chain tensioner (see Section 9), or remove the tensioner locking tool, according to the method you used earlier.

33 Turn the engine clockwise through two full turns and check again that all the timing marks still align (see Step 3) **(see illustrations 10.3a, 10.3b, 10.3c and 10.24)**. Check the valve clearances and adjust them if necessary (see Chapter 1).

34 Install the timing inspection cap using a new O-ring if required, and smear the O-ring and the cap threads with grease **(see illustration)**. Tighten the cap to the torque setting specified at the beginning of the Chapter.

35 Install the valve cover (see Section 8). Install the cam pulse generator (see Chapter 4).

11 Cam chain, tensioner blade and guide blades – removal, inspection and installation

Note: The cam chain and its blades can be removed with the engine in the frame. If the engine has been removed, ignore the steps which do not apply.

Removal

1 Remove the camshafts – this procedure involves removing the top guide blade (see Section 10).
2 Remove the timing rotor (see Section 21).
3 Unscrew the tensioner blade pivot bolt and draw the blade out of the top of the engine, noting the washer that fits behind it **(see illustrations)**.
4 Unscrew the guide blade pivot bolt and draw the blade out of the top of the engine **(see illustrations)**. Note the washer with the bolt and the collar that fits into the back of the pivot bore **(see illustration)**.

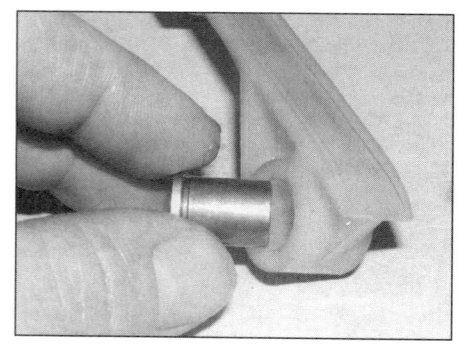

11.3a Unscrew the pivot bolt, noting the washer (arrowed) behind the blade . . .

11.3b . . . and draw the blade out of the engine

11.4a Remove the guide blade pivot bolt . . .

11.4b . . . and lift the blade out of the cylinder head

11.4c Note the collar in the end of the blade

Engine, clutch and transmission 2•21

5 Draw the cam chain off the crankshaft sprocket and out of the engine **(see illustration)**. If required, slide the sprocket off the end of the crankshaft, noting the wide spline that means it can only be installed in one position **(see illustration)**.

Inspection

Cam chain

6 Check the chain for binding, kinks and any obvious damage and replace it with a new one if necessary. Check the camshaft and crankshaft sprocket teeth for wear and renew the cam chain, camshaft sprockets and crankshaft as a set if necessary.

Tensioner and guide blades

7 Check the sliding surface and edges of the blades for excessive wear, deep grooves, cracking and other obvious damage, and replace them with new ones if necessary.

Installation

8 Installation of the chain and blades is the reverse of removal. Do not omit the washer that fits behind the tensioner blade **(see illustration 11.3a)**, or the collar that fits into the back of the front guide blade and the washer with its bolt **(see illustrations 11.4c and 11.4a)**.

9 Apply a suitable non-permanent thread locking compound to the pivot bolt threads and tighten them to the torque settings specified at the beginning of the Chapter.

12 Cylinder head – removal and installation

Note: *To remove the cylinder head the engine must be removed from the frame.*

Removal

1 Remove the engine from the frame (see Section 5).
2 Remove the camshafts, followers and shims (see Section 10).
3 If required remove the thermostat housing (see Chapter 3).
4 The cylinder head is secured by two 8 mm bolts and ten 9 mm bolts with fitted washers (i.e. they can't be separated from the bolts) **(see illustrations)**. First unscrew and remove the 8 mm bolts. Now unscrew and remove the 9 mm bolts, slackening them evenly and a little at a time in a criss-cross pattern working from the outside to the middle until they are all loose.
5 Hold the cam chain up and pull the cylinder head up off the block, then pass the cam chain down through the tunnel **(see illustration)**. Do not let the chain fall into the crankcase – secure it with a piece of wire or metal bar to prevent it from doing so. If the head is stuck, tap around the joint faces with a soft-faced mallet. Do not attempt to free the head by inserting a screwdriver between the head and block mating surfaces – you'll damage them.

6 Remove the cylinder head gasket and discard it as a new one must be used. If they are loose, remove the dowels from the crankcase or the underside of the cylinder head **(see illustration 12.10)**.

7 Check the cylinder head gasket and the mating surfaces on the cylinder head and crankcase for signs of leakage, which could indicate warpage. Refer to Section 14 and check the cylinder head gasket surface for warpage.

8 Clean all traces of old gasket material from the cylinder head and crankcase. If a scraper is used, take care not to scratch or gouge the soft aluminium. Be careful not to let any of the gasket material fall into the crankcase, the cylinder bore or the oil and coolant passages.

11.5a Disengage the cam chain from the sprocket . . .

11.5b . . . and slide the sprocket off the crankshaft

Installation

9 Lubricate the cylinder bores with engine oil. If removed, fit the dowels into the crankcase **(see illustration 12.10)**.

10 Ensure both cylinder head and crankcase mating surfaces are clean. Lay the new head gasket over the cam chain and blades and onto the crankcase, locating it over the dowels and making sure all the holes are correctly aligned **(see illustration)**. Never reuse the old gasket.

11 Carefully fit the cylinder head onto the block, making sure it locates correctly onto the dowels **(see illustration 12.5)**. Feed the cam chain up through the tunnel as you install the head, then secure it in place with a piece of wire to prevent it from falling back down.

12 Apply some molybdenum disulphide oil

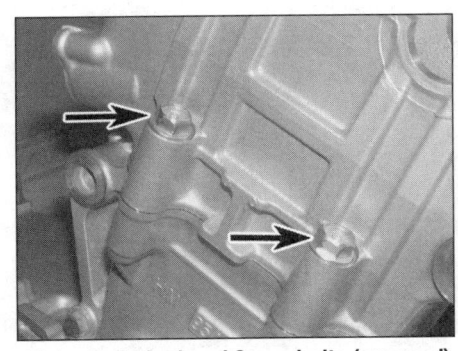

12.4a Cylinder head 8 mm bolts (arrowed)

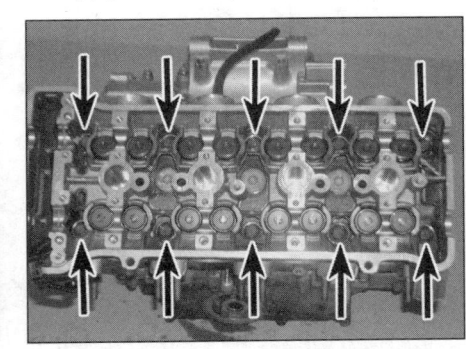

12.4b Cylinder head 9 mm bolts (arrowed)

12.5 Carefully lift the head up off the block

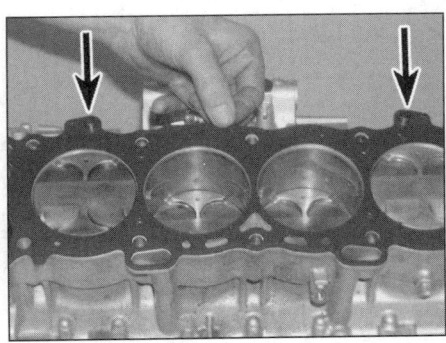

12.10 Install the dowels (arrowed) then lay the new gasket on the block

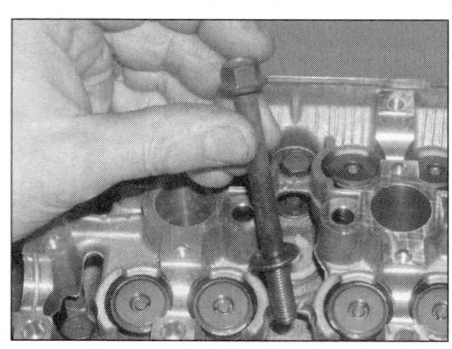

12.12a Lubricate the 9 mm bolts as described and fit them . . .

12.12b . . . then lubricate the 8 mm bolts and fit them . . .

12.12c . . . and tighten them as described to specified torque setting

(a 50/50 mixture of molybdenum disulphide grease and engine oil) to the threads and the underside of the heads of all the bolts. Note that if new bolts are being installed, first remove the anti-rust coating by cleaning them with solvent. Install the bolts and tighten them all finger-tight **(see illustrations)**. First tighten the 9 mm bolts evenly and a little at a time in a criss-cross pattern working from the middle to the outside to the torque setting specified at the beginning of the Chapter **(see illustration)**. Now tighten the 8 mm bolts to the specified torque.

13 Install the remaining components in a reverse of their removal sequence, referring to the relevant Sections or Chapters (see Steps 1 to 3).

13 Valves/valve seats/valve guides – servicing

1 Because of the complex nature of this job and the special tools and equipment required, most owners leave servicing of the valves, valve seats and valve guides to a professional. However, you can make an initial assessment of whether the valves are seating correctly, and therefore sealing, by pouring a small amount of solvent into each of the valve ports. If the solvent leaks past any valve into the combustion chamber area the valve is not seating correctly and sealing.

2 You can also remove the valves from the cylinder head, clean the components, check them for wear to assess the extent of the work needed, and, unless a valve service is required, grind in the valves (see Section 14). The head can then be reassembled.

3 A dealer service department will remove the valves and springs, replace the valves and guides, recut the valve seats, check and replace the valve springs, spring retainers and collets (as necessary), replace the valve stem seals with new ones and reassemble the valve components.

4 After the valve service has been performed, the head will be in like-new condition. When the head is returned, be sure to clean it again very thoroughly before installation on the engine to remove any metal particles or abrasive grit that may still be present from the valve service operations. Use compressed air, if available, to blow out all the holes and passages.

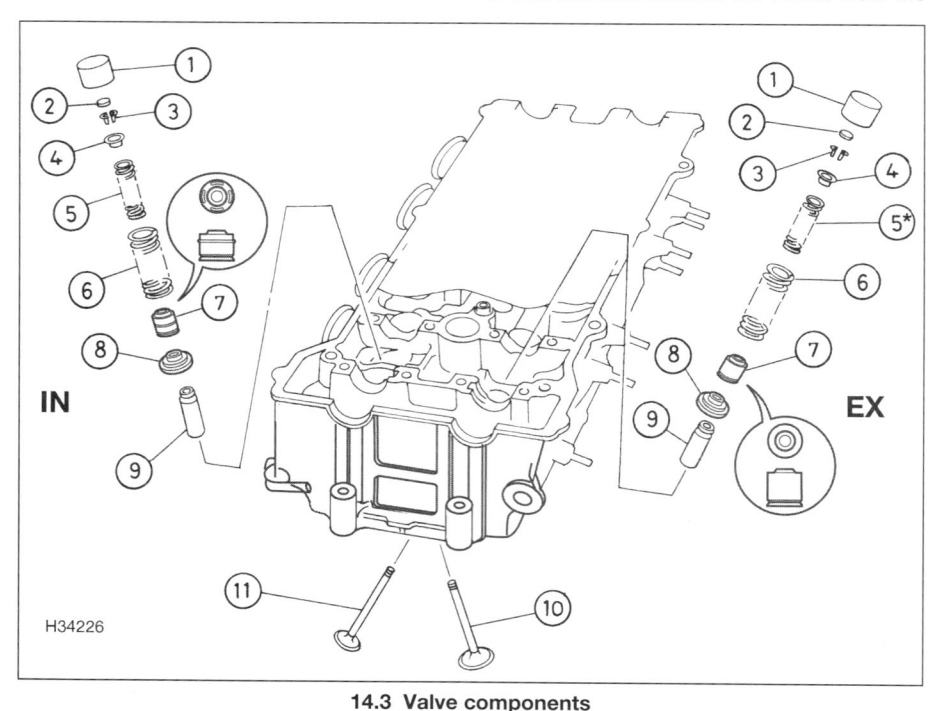

14 Cylinder head and valves – disassembly, inspection and reassembly

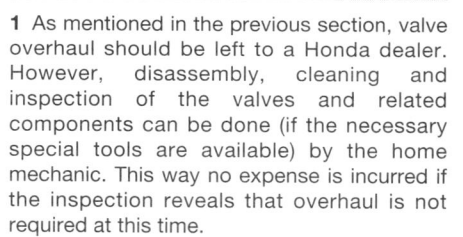

1 As mentioned in the previous section, valve overhaul should be left to a Honda dealer. However, disassembly, cleaning and inspection of the valves and related components can be done (if the necessary special tools are available) by the home mechanic. This way no expense is incurred if the inspection reveals that overhaul is not required at this time.

2 To disassemble the valve components without the risk of damaging them, a valve spring compressor is absolutely essential. Make sure it is suitable for motorcycle work.

Disassembly

3 Before proceeding, arrange to label and store the valves along with their related components in such a way that they can be returned to their original locations without getting mixed up **(see illustration)**. A good way to do this is to use the same container as the followers and shims are stored in (see Section 10), or to obtain a separate container which is divided into sixteen compartments,

14.3 Valve components

1 Follower
2 Shim
3 Collets
4 Spring retainer
5 Inner spring*
6 Outer spring
7 Valve stem seal
8 Spring seat
9 Valve guide
10 Intake valve
11 Exhaust valve
*Inner exhaust valve spring fitted to RR-Y and RR-1 (2000 and 2001) models only

Engine, clutch and transmission 2•23

TOOL TiP

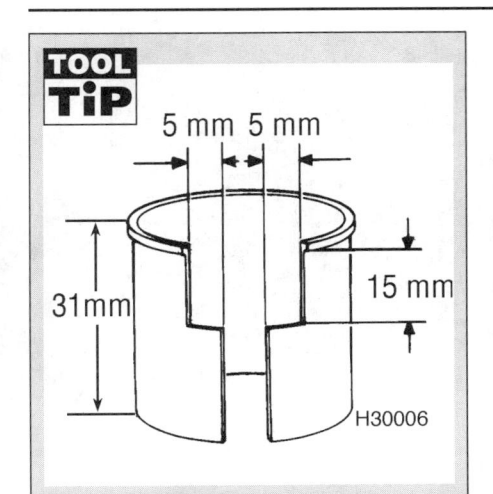

Protect the follower bore in the cylinder head from scratches by the valve spring compressor using either the Honda tool (Part No. 07HMG-MR70002) or by fabricating a shield from a 35 mm film canister. Cut the canister to the dimensions shown.

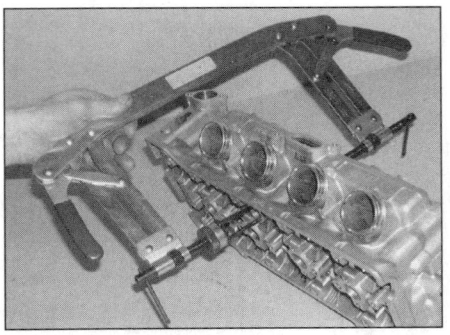

and label each compartment with the location of a valve, i.e. intake or exhaust camshaft, left or right valve. If a container is not available, use labelled plastic bags (egg cartons also do very well).

4 Clean all traces of old gasket material from the cylinder head. If a scraper is used, take care not to scratch or gouge the soft aluminium; refer to *Tools and Workshop Tips* for details of gasket removal methods.

5 Compress the valve spring(s) on the first valve with a spring compressor, making sure it is correctly located onto each end of the valve assembly **(see illustrations and Tool Tip)**. On the underside of the head make sure the plate on the compressor only contacts the valve and not the soft aluminium of the head – if the plate is too big for the valve, use a spacer between them. Do not compress the springs any more than is absolutely necessary. Remove the collets, using either needle-nose pliers, tweezers, a magnet or a screwdriver with a dab of grease on it **(see illustration)**. Carefully release the valve spring compressor and remove it. Remove the spring retainer, noting which way up it fits **(see illustration 14.28c)**. Remove the spring(s), noting that the closer wound coils are at the bottom – all models have an inner and outer spring with each intake valve; RR-Y and RR-1 (2000 and 2001) models have an inner and outer spring fitted with each exhaust valve, while RR-2 and RR-3 (2002 and 2003) models have only one spring with each exhaust valve **(see illustrations 14.28b and 14.28a and 14.3)**. Press down on the top of the valve stem and draw the valve out from the underside of the head **(see illustration 14.27)**. If the valve binds in the guide (won't pull through), push it back into the head and deburr the area around the collet groove with a very fine file or whetstone **(see illustration)**.

6 Once the valve has been removed and labelled, pull the valve stem seal off the top of the valve guide with pliers and discard it (the old seals should never be reused) **(see illustration)**. Now remove the spring seat **(see illustration)**. The seat is difficult to get hold of, so either use a small magnet or turn the head upside down and tip it out, taking care not to lose it. Note the difference between the intake and exhaust valve stem seals **(see illustration 14.3)**.

7 Repeat the procedure for the remaining valves. Remember to keep the parts for each valve together and in order so they can be reinstalled in the same location.

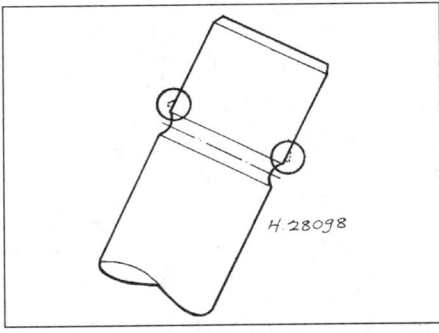

14.5a Compressing the valve springs using a valve spring compressor

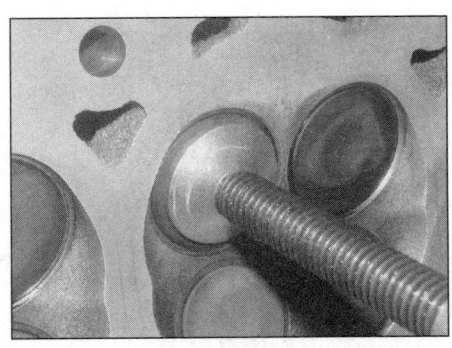

14.5b Make sure the compressor locates correctly both on the bottom of the valve . . .

14.5c . . . and on the top of the spring retainer

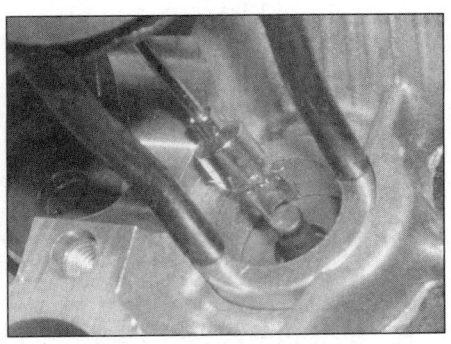

14.5d Remove the collets with needle-nose pliers, tweezers, a magnet or a screwdriver with a dab of grease on it

14.5e If the valve stem won't pull through the guide, deburr the area above the collet groove

14.6a Pull the seal off the valve stem . . .

14.6b . . . then remove the spring seat

14.13 Measure the valve seat width with a ruler (or for greater precision use a vernier caliper)

14.14a Measure the valve stem diameter with a micrometer

14.14b Insert a small bore gauge into the valve guide (arrowed), then expand and lock it and pull it out out, then measure the bore gauge with a micrometer

8 Next, clean the cylinder head with solvent and dry it thoroughly. Compressed air will speed the drying process and ensure that all holes and recessed areas are reached.

9 Clean all of the valve springs, collets, retainers and spring seats with solvent and dry them thoroughly. Do the parts from one valve at a time so they don't get mixed up.

10 Scrape off any deposits that may have formed on the valve, then use a motorised wire brush to remove deposits from the valve heads and stems. Again, make sure the valves do not get mixed up.

Inspection

11 Inspect the head very carefully for cracks and other damage. If cracks are found, a new head is required. Check the camshaft bearing surfaces for wear and evidence of seizure. Check the camshafts and holders for wear as well (see Section 10).

12 Using a precision straight-edge and a feeler gauge set to the warpage limit listed in the specifications at the beginning of the Chapter, check the head gasket mating surface for warpage. Refer to *Tools and Workshop Tips* in the Reference section for details of how to use the straight-edge.

13 Examine the valve seats in the combustion chamber. If they are pitted, cracked or burned, the head will require work beyond the scope of the home mechanic. Measure the valve seat width and compare it to this Chapter's Specifications (see illustration). If it exceeds the service limit, or if it varies around its circumference, overhaul is required.

14 Working on one valve and guide at a time, measure the valve stem diameter (see illustration). Clean the valve's guide using a guide reamer to remove any carbon build-up – insert the reamer from the underside of the head and turn it clockwise only. Now measure the inside diameter of the guide (at both ends and in the centre of the guide) with a small bore gauge, then measure the gauge with a micrometer (see illustration). Measure the guide at the ends and at the centre to determine if they are worn in a bell-mouth pattern (more wear at the ends). Subtract the stem diameter from the valve guide diameter to obtain the valve stem-to-guide clearance. If the stem-to-guide clearance is greater than listed in this Chapter's Specifications, renew whichever component is beyond its specification limits. If the valve guide is within specifications, but is worn unevenly, it should be renewed. Repeat for the other valves.

15 Carefully inspect each valve face, stem and collet groove area for cracks, pits and burned spots.

16 Rotate the valve and check for any obvious indication that it is bent, in which case it must be replaced with a new one. Check the end of the stem for pitting and excessive wear. The presence of any of the above conditions indicates the need for valve servicing.

17 Check the end of each valve spring for wear and pitting. Measure the spring free lengths and compare them to the specifications (see illustration). If any spring is shorter than specified it has sagged and must be replaced with a new one. Also place the spring upright on a flat surface and check it for bend by placing a ruler against it, or alternatively lay it against a set square (see illustration). If the bend in any spring is excessive, it must be replaced with a new one.

18 Check the spring seats, retainers and collets for obvious wear and cracks. Any questionable parts should not be reused, as extensive damage will occur in the event of failure during engine operation.

19 If the inspection indicates that no overhaul work is required, the valve components can be reinstalled in the head.

Reassembly

20 Unless a valve service has been performed, before installing the valves in the head they should be ground in (lapped) to ensure a positive seal between the valves and seats. This procedure requires coarse and fine valve grinding compound and a valve grinding tool (either hand-held or drill driven). If a grinding tool is not available, a piece of rubber or plastic hose can be slipped over the valve stem (after the valve has been installed in the guide) and used to turn the valve.

21 Apply a small amount of coarse grinding compound to the valve face and some molybdenum disulphide oil (a 50/50 mixture of molybdenum disulphide grease and engine oil) to the valve stem, then slip the valve into the guide (see illustration 14.27). Note: *Make sure each valve is installed in its correct guide and be careful not to get any grinding compound on the valve stem.*

22 Attach the grinding tool to the valve and rotate the tool between the palms of your hands. Use a back-and-forth motion (as though rubbing your hands together) rather than a circular motion (i.e. so that the valve

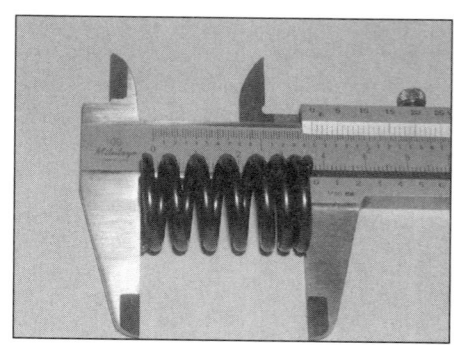

14.17a Measure the free length of the valve springs ...

14.17b ... and check them for squareness

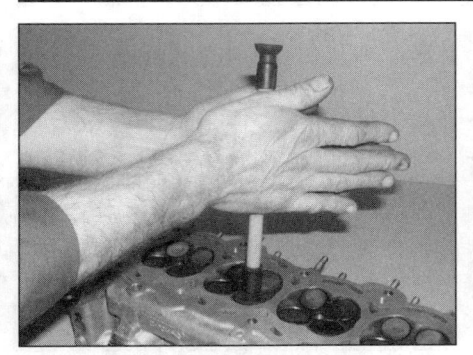

Engine, clutch and transmission 2•25

14.22 Rotate the valve grinding tool back-and-forth between the palms of your hands

14.25 Fit the spring seat using a rod to guide it if necessary

14.26a Fit a new valve stem seal ...

rotates alternately clockwise and anti-clockwise rather than in one direction only) **(see illustration)**. If a motorised tool is being used, take note of the correct drive speed for it – if your drill runs too fast and is not variable, use a hand tool instead. Lift the valve off the seat and turn it at regular intervals to distribute the grinding compound properly. Continue the grinding procedure until the valve face and seat contact area is of uniform width, and unbroken around the entire circumference.

23 Carefully remove the valve and wipe off all traces of grinding compound, making sure none gets in the guide. Use solvent to clean the valve and wipe the seat area thoroughly with a solvent soaked cloth.

24 Repeat the procedure with fine valve grinding compound, then use solvent to clean the valve and flush the guide, and wipe the seat area thoroughly with a solvent soaked cloth. Repeat the entire procedure for the remaining valves. On completion thoroughly clean the entire head again, then blow through all passages with compressed air. Make sure all traces of the grinding compound have been removed before assembling the head.

25 Working on one valve at a time, lay the spring seat in place in the cylinder head, making sure it is the correct way round **(see illustration)**. As it is easy to cock the seat on the top of the valve guide, and then tricky to get it to sit properly, install it using a rod as a guide for it to slide down.

14.26b ... and use your fingers, a deep socket or special tool to press it squarely into place

26 Separate your new valve stem seals between the intake type and the exhaust type so as to not mix them up **(see illustration 14.3)**. Fit a new valve stem seal onto the guide, using finger pressure, a stem seal fitting tool or an appropriate size deep socket, to push the seal squarely onto the end of the valve guide until it is felt to clip into place **(see illustrations)**. Make sure the seal does not get cocked sideways as it could be damaged.

27 Coat the valve stem with molybdenum disulphide oil (a 50/50 mixture of molybdenum disulphide grease and engine oil), then install it into its guide, rotating it slowly to avoid damaging the seal **(see illustration)**. Check that the valve moves up-and-down freely in the guide.

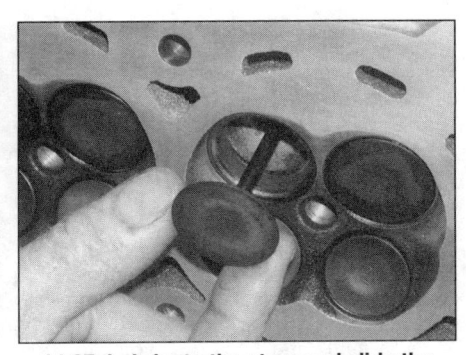

14.27 Lubricate the stem and slide the valve into its correct location

28 Next, install the spring(s), with their closer-wound coils facing down into the cylinder head **(see illustrations)**. All models have an inner and outer spring with each intake valve; RR-Y and RR-1 (2000 and 2001) models have an inner and outer spring fitted with each exhaust valve, while RR-2 and RR-3 (2002 and 2003) models have only one spring with each exhaust valve. Fit the spring retainer, with its shouldered side facing down so that it fits into the top of the spring(s) **(see illustration)**.

29 Compress the valve spring(s) with a spring compressor, making sure it is correctly located onto each end of the valve assembly **(see illustrations 14.5a, 14.5b and 14.5c)**. On the underside of the head make sure the

14.28a Fit the inner valve spring ...

14.28b ... and the outer valve spring ...

14.28c ... then fit the spring retainer

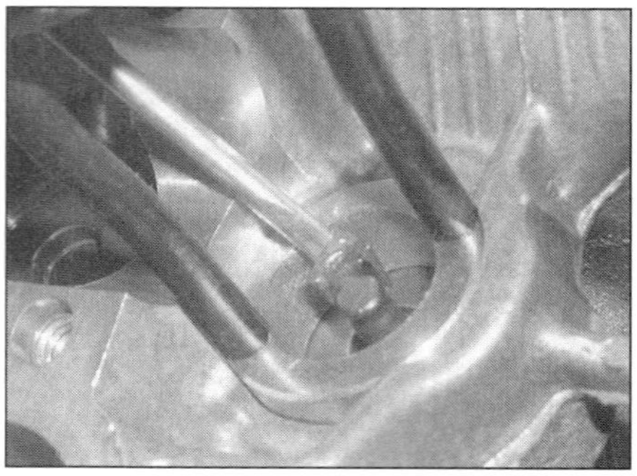

2•26 Engine, clutch and transmission

14.29a Locate each collet in its groove in the top of the valve stem

14.29b Make sure both collets remain in place and lock into the groove as the compressor is released

15.3 Check the operation of the clutch as described

plate on the compressor only contacts the valve and not the soft aluminium of the head – if the plate is too big for the valve, use a spacer between them. Do not compress the springs any more than is necessary to slip the collets into place. Apply a small amount of grease to the collets to help hold them in place. Locate each collet in turn into the groove in the valve stem using either needle-nose pliers, tweezers or a screwdriver with a dab of grease on it **(see illustration)**. Carefully release the compressor, making sure the collets seat and lock as you do **(see illustration)**. Check that the collets are securely locked in the retaining groove.

30 Support the cylinder head on blocks so the valves can't contact the workbench top, then very gently tap the top of the valve stem with a brass drift. This will help seat the collets in the groove. If you don't have a brass drift, fit the shim into its recess in the top of the valve spring retainer and use a soft-faced hammer and a piece of wood as an interface.

> **HAYNES HiNT** *Check for proper sealing of the valves by pouring a small amount of solvent into each of the valve ports. If the solvent leaks past any valve into the combustion chamber area the valve grinding operation on that valve should be repeated.*

31 Repeat the procedure for the remaining valves. Remember to keep the parts for each valve together, and separate from the other valves, so they can be reinstalled in the same location. After the cylinder head and camshafts have been installed, check and adjust the valve clearances as required (see Chapter 1).

15 Starter clutch – check, removal, inspection and installation

Note: *The starter clutch can be removed with the engine in the frame. If the engine has been removed, ignore the steps which do not apply.*

Check

1 The operation of the starter clutch can be checked while it is in situ. Remove the starter motor (see Chapter 9). Check that the idle/reduction gear is able to rotate freely clockwise as you look at it via the starter motor aperture, but locks when rotated anti-clockwise. If not, the starter clutch is faulty and should be removed for inspection.

Removal

2 Remove the alternator rotor – the starter clutch is mounted on the back of it (see Chapter 9).

Inspection

3 With the alternator rotor face down on a workbench, check that the starter driven gear rotates freely in an anti-clockwise direction and locks against the rotor in a clockwise direction **(see illustration)**. If it doesn't, the starter clutch should be dismantled for further investigation.

4 Withdraw the starter driven gear from the starter clutch. If the gear appears stuck, rotate it anti-clockwise as you withdraw it to free it from the starter clutch.

5 Check the condition of the sprags inside the clutch body – if they are damaged, marked or flattened at any point, they should be replaced with new ones **(see illustration)**. Also check the corresponding surface on the driven gear hub. Measure the outside diameter of the hub and check that it has not worn beyond the service limit specified **(see illustration)**. To remove the sprag assembly, hold the rotor using a holding strap and unscrew the six bolts inside the rotor **(see illustration)**. Separate the sprag assembly from the rotor, noting how it fits. Install the new assembly in a reverse sequence. Apply clean engine oil to the sprags. Apply a suitable non-permanent thread locking compound to the bolts and tighten them to the torque setting specified at the beginning of the Chapter.

6 Check the needle roller bearing on the

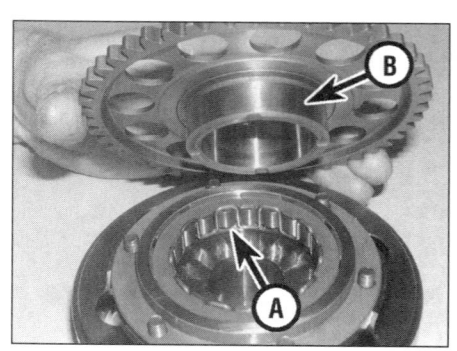

15.5a Check the sprags (A) and the hub (B)

15.5b Measure the external diameter of the hub

15.5c Unscrew the bolts and remove the sprag assembly

Engine, clutch and transmission 2•27

15.6 Check the needle bearing as described

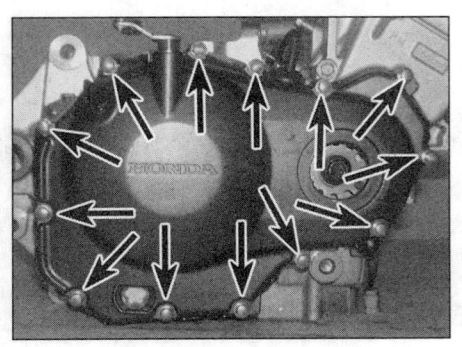

16.4 Unscrew the bolts (arrowed) and remove the cover

crankshaft and the bearing surface in the starter driven gear hub **(see illustration)**. Check for play between the bearing and its surface on the crankshaft. If the bearing surfaces show signs of excessive wear or the bearing itself is worn or damaged, they should be replaced with new ones. Refer to *Tools and Workshop Tips* in the Reference Section at the end of the book for bearing removal and installation methods – the bearing must be drawn off the shaft using a puller, and installed using an hydraulic or mechanical press. Note that a home-made press can be assembled using a suitable piece of metal tubing, suitable washers and the alternator rotor bolt – by tightening the bolt with the tubing between it and the bearing the bearing will be pushed onto the shaft. Place a washer on each end of the tubing to protect the bearing and the bolt.

7 Check the teeth of the starter idle/reduction gear and the corresponding teeth of the starter driven gear and starter motor drive shaft. Replace the gears and/or starter motor if worn or chipped teeth are discovered on related gears. Also check the idle/reduction gear shaft for damage, and check that the gear is not a loose fit on the shaft. Replace the shaft with a new one if necessary.

Installation

8 Lubricate the needle roller bearing on the crankshaft with clean engine oil **(see illustration 15.6)**. Lubricate the outside of the starter driven gear hub with clean engine oil, then fit the gear into the clutch, rotating it anti-clockwise as you do so to spread the sprags and allow the hub to enter.

9 Install the alternator rotor (see Chapter 9).

16 Clutch and cable – removal, inspection and installation

Note 1: *The clutch can be removed with the engine in the frame. If the engine has been removed, ignore the steps which don't apply.*
Note 2: *The clutch nut must be discarded and a new one used on installation – it is best to obtain the new nut in advance.*

Clutch removal

1 On RR-Y and RR-1 (2000 and 2001) models remove the lower fairing and the right-hand fairing side panel (see Chapter 8). On RR-2 and RR-3 (2002 and 2003) models remove both fairing side panels (see Chapter 8). Drain the engine oil (see Chapter 1).

2 Unscrew the bolts securing the clutch cable bracket to the clutch cover **(see illustration 5.11)**. Displace the bracket, noting how it locates, and free the cable end from the release arm **(see illustration 16.34c)**.

3 Raise the fuel tank (see Chapter 4). Trace the ignition pulse generator coil wiring from the top of the clutch cover and disconnect it at the red 2-pin wiring connector **(see illustration 5.13a)**.

4 Working evenly in a criss-cross pattern, unscrew the clutch cover bolts **(see illustration)**.

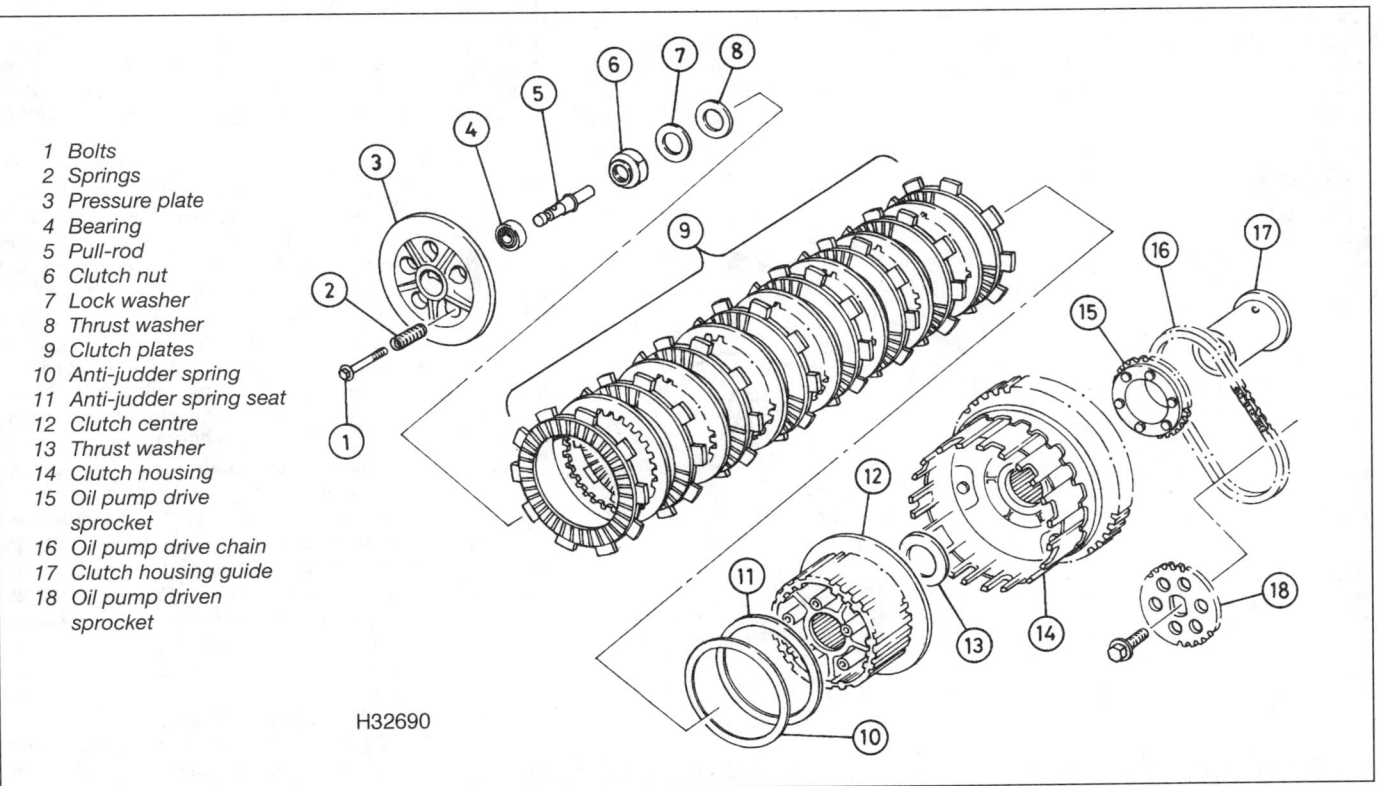

1 Bolts
2 Springs
3 Pressure plate
4 Bearing
5 Pull-rod
6 Clutch nut
7 Lock washer
8 Thrust washer
9 Clutch plates
10 Anti-judder spring
11 Anti-judder spring seat
12 Clutch centre
13 Thrust washer
14 Clutch housing
15 Oil pump drive sprocket
16 Oil pump drive chain
17 Clutch housing guide
18 Oil pump driven sprocket

16.5a Clutch components

2•28 Engine, clutch and transmission

16.5b Unscrew the bolts (arrowed) and remove the springs . . .

16.5c . . . then remove the pressure plate

16.6 Remove the clutch plates, keeping them in order

16.7a Unstake the nut . . .

16.7b . . . then unscrew it as described and remove the washers

Remove the cover, turning the release lever arm back (anti-clockwise) as you do to disengage the shaft from the pull-rod, and being prepared to catch any residual oil. Remove the two dowels from either the cover or the crankcase if they are loose.

5 Working in a criss-cross pattern, gradually slacken the clutch spring bolts until spring pressure is released (see illustrations). To prevent the assembly from turning, cover it with a rag and hold it securely – the bolts are not very tight. If available, have an assistant to hold the clutch while you unscrew the bolts. Remove the bolts and springs, then remove the pressure plate (see illustration). Remove the pull-rod from either the back of the pressure plate or the end of the shaft (see illustration 16.28a).

6 Remove the clutch friction and plain plates, noting how they fit and keeping them in the same order as you remove them (see illustration). Note how the tabs on the outer friction plate locate in the shallow slots in the housing, while the rest sit in the deep slots. The outer and inner friction plates are different to the rest – both have one tab end coloured green, and the inner friction plate has a larger internal diameter so that it fits over the anti-judder spring and spring seat. If the colour codes are difficult to identify, make sure you keep the plates in the correct order. Remove the anti-judder spring and spring seat, noting which way round they fit (see illustrations 16.26b and 16.26a).

7 The clutch nut is staked against the input shaft. Unstake the nut using a hammer and punch – take care not to damage the threads on the end of the shaft (see illustration). To remove the clutch nut, the input shaft must be locked. This can be done in several ways. If the engine is in the frame, engage 6th gear and have an assistant hold the rear brake on hard with the rear tyre in firm contact with the ground. Alternatively, the Honda service tool (Pt. No. 07724-0050002), or a similar commercially available or home-made tool (see *Tool tip*), can be used to stop the clutch centre from turning whilst the nut is slackened (see illustration). Unscrew the nut and remove the lock washer and the thrust washer (see illustrations 16.25d, 16.25c and 16.25b). Discard the nut as a new one must be used on installation.

8 Remove the clutch centre and the thrust washer from the shaft (see illustrations 16.24b and 16.24a).

9 Slide the clutch housing off the shaft, noting that you may have to prevent the guide in the centre of the housing from sliding with it by pressing on its rim using a very small

> **TOOL TIP** 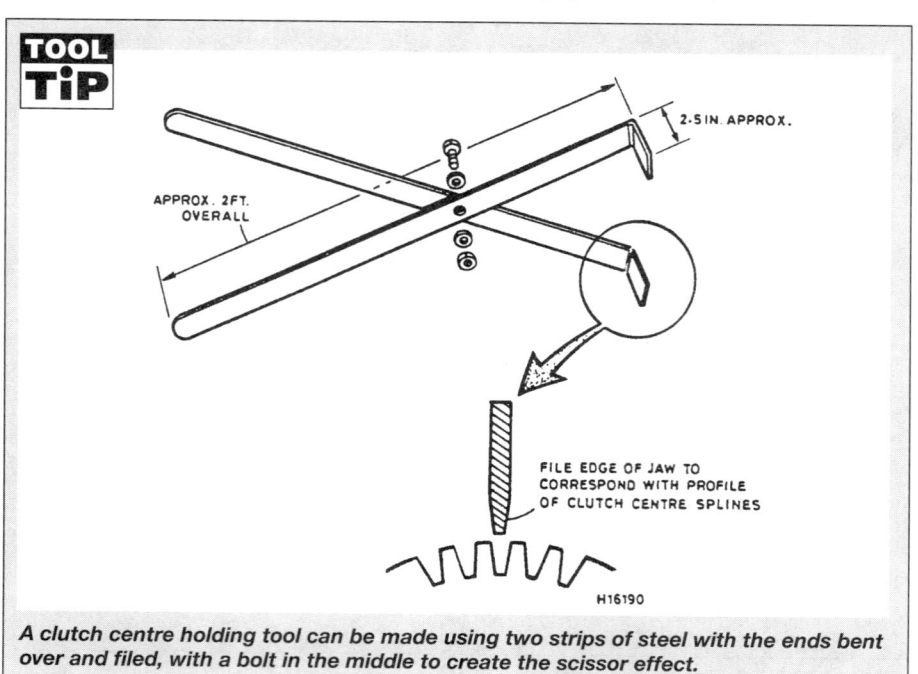
>
> *A clutch centre holding tool can be made using two strips of steel with the ends bent over and filed, with a bolt in the middle to create the scissor effect.*

Engine, clutch and transmission 2•29

16.9 Draw the housing off the shaft

16.10 Lock the sprocket as shown while unscrewing the bolt

16.11 Measuring clutch friction plate thickness

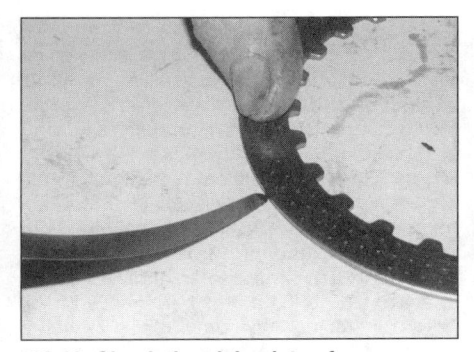

16.12 Check the plain plates for warpage

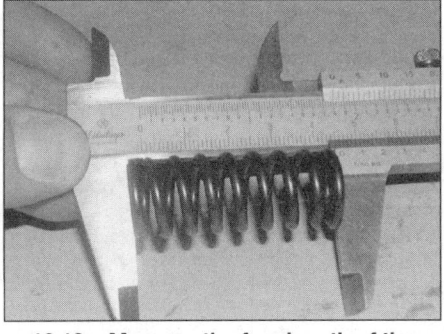

16.13a Measure the free length of the clutch springs . . .

16.13b . . . and check them for squareness

screwdriver **(see illustration)**. If the guide slides with the housing, it brings the oil pump drive chain with it which could distort the chain. Note how the holes in the back of the housing engage with the pins on the oil pump drive sprocket.

10 If required, lock the oil pump driven sprocket bolt to prevent it from turning and unscrew the bolt **(see illustration)**. Remove the driven sprocket, the chain and the drive sprocket **(see illustrations 16.22d, 16.22c, 16.22b and 16.22a)**. Also remove the clutch housing guide from the input shaft **(see illustration 16.21)**.

Clutch inspection

11 After an extended period of service the clutch friction plates will wear and promote clutch slip. Measure the thickness of each friction plate using a vernier caliper **(see illustration)**. If any plate has worn to or beyond the service limits given in the Specifications at the beginning of the Chapter, the friction plates must be renewed as a set. Also, if any of the plates smell burnt or are glazed, they must be renewed as a set.

12 The plain plates should not show any signs of excess heating (bluing). Check for warpage using a flat surface and feeler gauges **(see illustration)**. If any plate exceeds the maximum permissible amount of warpage, or shows signs of bluing, all plain plates must be renewed as a set.

13 Measure the free length of each clutch spring using a vernier caliper **(see illustration)**. If any spring is below the service limit specified, renew all the springs as a set. Place each spring upright on a flat surface and check it for bend by placing a ruler against it, or alternatively lay it against a set square **(see illustration)**. If the bend in any spring is excessive, renew all the springs as a set. Also check the anti-judder spring and spring seat for damage or distortion and replace them with new ones if necessary.

14 Inspect the friction plates and the clutch housing for burrs and indentations on the edges of the protruding tabs on the plates and/or the slots in the housing **(see illustration)**. Similarly

16.14a Check the friction plate tabs and housing slots . . .

16.14b . . . and the plain plate teeth and centre slots as described

16.15a Check the bearing surface on the guide . . .

16.15b . . . and the bearing itself in the housing

16.17 Check the pressure plate and its bearing

check for wear between the inner teeth of the plain plates and the slots in the clutch centre **(see illustration)**. Wear of this nature will cause clutch drag and slow disengagement during gear changes as the plates will snag when the pressure plate is lifted. With care a small amount of wear can be corrected by dressing with a fine file, but if this is excessive the worn components should be renewed.

15 Inspect the clutch housing guide bearing surface and the needle roller bearing in the clutch housing **(see illustrations)**. If there are any signs of wear, pitting or other damage the affected parts must be renewed. The bearing is a press fit in the housing – refer to *Tools and Workshop Tips* in the Reference Section for details on bearing removal and installation. When removing the old bearing, note carefully at what depth it sits in the centre. Install the new bearing so that it sits in exactly the same place (which should be flush with the outer surface of the centre), and so the marked end will be facing out when the housing is installed.

16 Using a vernier caliper, measure the internal and external diameter of the clutch housing guide, and the external diameter of the input shaft where the guide sits **(see illustration 16.15a)**. Compare the measurements to the specifications at the beginning of the Chapter and renew the guide or shaft if they are worn beyond their service limit. Also check the mating surfaces for signs of damage or scoring.

17 Check the pressure plate and its bearing for signs of wear or damage and roughness **(see illustration)**. Check that the bearing outer race is a good fit in the centre of the lifter, and that the inner race rotates freely without any rough spots. Check the pull-rod end and the corresponding cut-out in the release lever shaft for signs of wear or damage. Replace any parts necessary with new ones.

18 Check the release mechanism in the clutch cover for a smooth action. If the action is stiff or rough, withdraw the shaft, noting how the return spring ends locate, and remove the washer. Clean and check the oil seal and the two needle bearings in the cover **(see illustrations)**. The seal is available separately and can be renewed by levering the old one out with a screwdriver and pressing the new one in. The needle bearings for the shaft are not listed as being available separately from the cover, and so it may be necessary to fit a new cover if they are worn or damaged, though it is always worth contacting a Honda dealer and/or bearing specialist. Lubricate the bearings with oil before installing the shaft. Make sure the return spring ends locate correctly **(see illustration)**.

19 Check the teeth of the primary driven gear on the back of the clutch housing and the corresponding teeth of the primary drive gear on the crankshaft. Renew the clutch housing and/or crankshaft if worn or chipped teeth are discovered.

Clutch installation

20 Remove all traces of old sealant from the crankcase and clutch cover surfaces.

21 If removed, smear the inside and outside of the clutch housing guide with molybdenum disulphide oil (a 50/50 mixture of molybdenum disulphide grease and engine oil), then slide the guide onto the input shaft with the flanged end inwards **(see illustration)**.

22 Slide the oil pump drive sprocket onto the shaft, making sure the pins face out, and slip the chain around the sprocket **(see**

16.18a Withdraw the shaft . . .

16.18b . . . and check the oil seal and bearings

16.18c Make sure the spring ends (arrowed) locate correctly

16.21 Slide the guide onto the shaft

Engine, clutch and transmission 2•31

16.22a Slide the drive sprocket and chain onto the guide

16.22b Fit the driven sprocket into the chain with the OUT mark facing out . . .

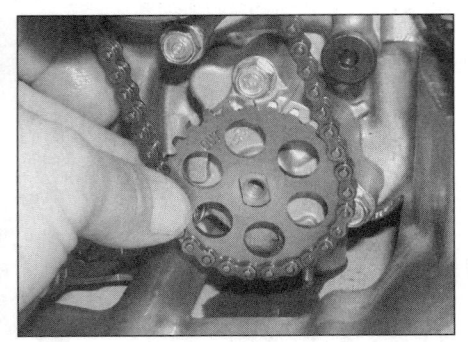

16.22c . . . and locate it onto the flats on the end of the shaft

16.22d Install the bolt with its washer . . .

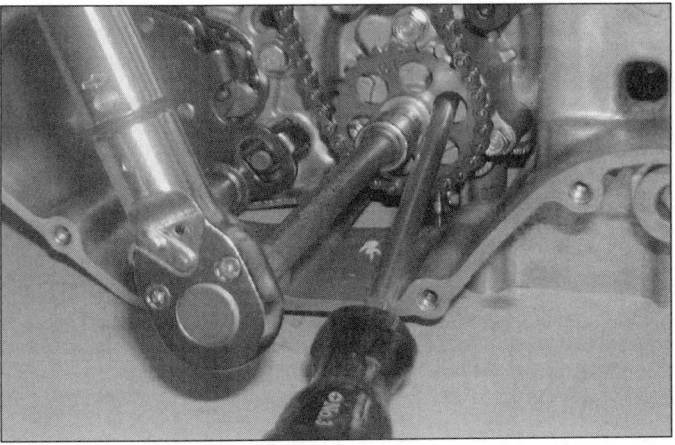

16.22e . . . and tighten it to the specified torque, locking the shaft as shown

illustration). Engage the driven sprocket with the chain, making sure the OUT mark faces out, then locate the sprocket on the oil pump **(see illustrations)**. Apply a suitable non-permanent thread locking compound to the sprocket bolt and tighten it to the torque setting specified at the beginning of the chapter (see **Tool Tip**), not forgetting its washer **(see illustrations)**.

23 To install the clutch housing it is

> **TOOL TiP** *Insert a screwdriver or drift through one of the holes in the sprocket and lock it against the crankcase to prevent the sprocket from turning whilst tightening the bolt.*

necessary to align the primary drive sub-gear teeth with the main gear **(see illustration)**. To do this locate a suitable screwdriver or rod into the hole in the crankcase and into one of the holes in the sub-gear – turn the engine clockwise using a spanner on the timing rotor bolt to align the holes **(see illustration)**. With the sub-gear locked in position, turn the crankshaft against the spring pressure between the sub-gear and main gear until the teeth align as shown **(see illustration)**. With the crankshaft held in this position slide the clutch housing over the housing guide on the

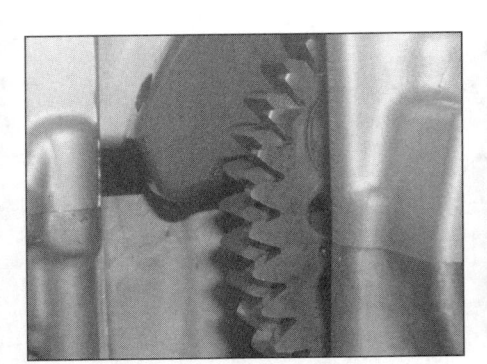

16.23a The offset teeth between the gears need to be aligned to install the clutch housing

16.23b To do this insert a screwdriver into the hole in the crankcase and locate it in one of the holes in the sub-gear . . .

16.23c . . . then turn the crankshaft until the teeth align as shown, and hold it there . . .

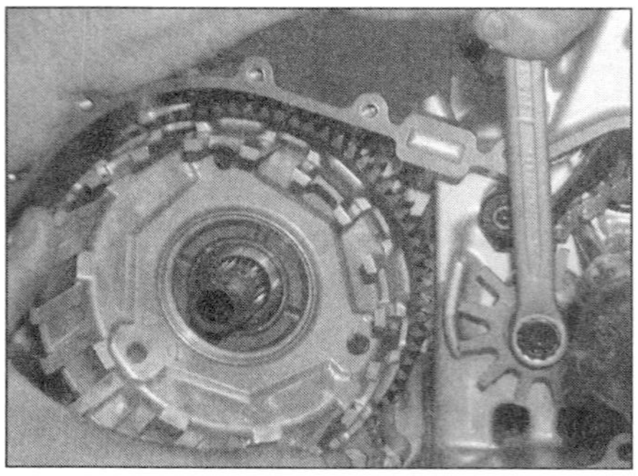

16.23d . . . while installing the clutch housing

16.23e The pins on the oil pump drive sprocket must locate in the holes in the back of the housing . . .

16.23f . . . you may need to turn the driven sprocket by hand to align the pins as you install the housing

input shaft, making sure that the primary drive and driven gear teeth engage (once they have you can let go of the spanner), and the pins on the oil pump drive sprocket locate in the holes in the rear of the housing – turn the sprocket with your finger while pressing on the housing until the pins are felt to locate and the housing moves in a bit further, then double-check by making sure the sprocket can't turn independently of the housing **(see illustrations)**.

24 Slide the thrust washer onto the shaft **(see illustration)**. Slide the clutch centre onto the shaft splines **(see illustration)**.

25 Install the thrust washer **(see illustration)**. Install the lock washer with its OUT mark facing out **(see illustration)**. Smear the new clutch nut threads with oil, then thread it onto the input shaft and, using the method employed on removal to lock the shaft (see Step 7), tighten the nut to the torque setting specified at the beginning of the Chapter **(see illustrations)**. Stake the collar of the nut into the indent on the end of the shaft **(see illustration)**.

26 Fit the anti-judder spring seat into the clutch centre, then fit the spring so that its outer edge is raised off the seat (concave face) and facing outwards **(see illustrations)**.

27 Coat each clutch plate with engine oil prior to installation, then build up the plates as

16.24a Fit the thrust washer . . .

16.24b . . . and the clutch centre

16.25a Fit the thrust washer . . .

16.25b . . . and the lock washer, with the OUT mark facing out

16.25c Fit a new clutch nut . . .

16.25d . . . and tighten it to the specified torque

16.25e Stake the nut against the detent in the shaft end

Engine, clutch and transmission 2•33

16.26a Fit the anti-judder spring seat ...

16.26b ... and the spring. With its outer edge raised off the seat

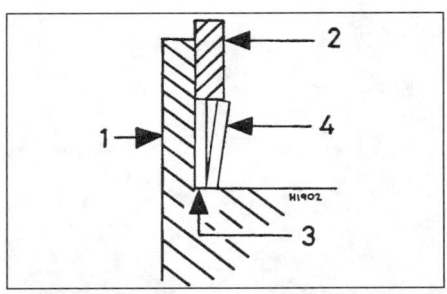

16.26c Anti-judder spring installation details
1. Clutch centre
2. Friction plate with green tab and larger internal diameter
3. Spring seat
4. Anti-judder spring

follows. Fit the friction plate with the green coloured tab and the larger internal diameter over the spring and spring seat **(see illustration 16.26c)**, then fit a plain plate, then alternate friction plates and plain plates until all except the outer green-coded friction plate are installed, then fit that with its tabs fitting into the shallow slots in the housing **(see illustrations)**.

28 Lubricate the bearing in the pressure plate. Fit the pull-rod into the bearing **(see illustration)**. Fit the pressure plate onto the clutch, engaging the protrusions on its inner rim in the slots in the clutch centre **(see illustration)**. Install the springs and the bolts, not forgetting their washers, and tighten them evenly in a criss-cross sequence to the specified torque setting **(see illustration)**. Counter-hold the clutch housing to prevent it turning when tightening the spring bolts.

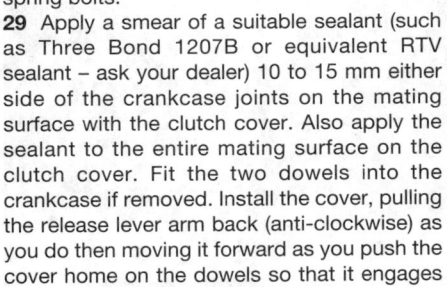

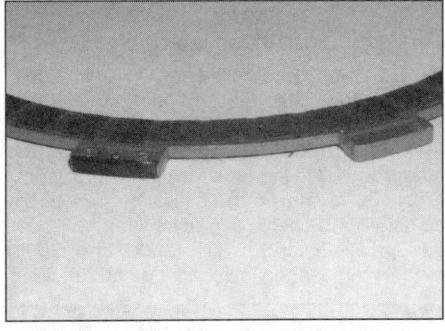

29 Apply a smear of a suitable sealant (such as Three Bond 1207B or equivalent RTV sealant – ask your dealer) 10 to 15 mm either side of the crankcase joints on the mating surface with the clutch cover. Also apply the sealant to the entire mating surface on the clutch cover. Fit the two dowels into the crankcase if removed. Install the cover, pulling the release lever arm back (anti-clockwise) as you do then moving it forward as you push the cover home on the dowels so that it engages

16.27a Identify the green-tabbed friction plate with the larger internal diameter ...

16.27b ... and fit that first, locating it over the anti-judder assembly ...

16.27c ... followed by a plain plate, then alternate between friction plates and plain plates

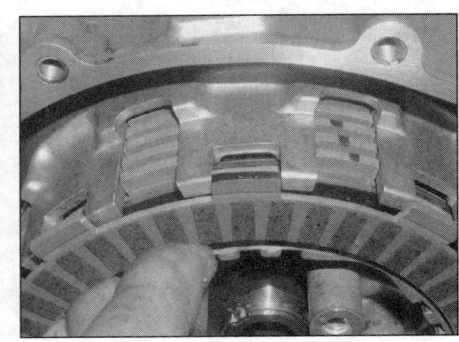

16.27d Locate the tabs on the second green-tabbed friction plate into the shallow slots in the housing

16.28a Fit the pull-rod into the bearing

16.28b Fit the pressure plate, making sure it locates in the slots ...

16.28c ... then install the springs and bolts and tighten them as described

2•34 Engine, clutch and transmission

16.29 Install the cover as described

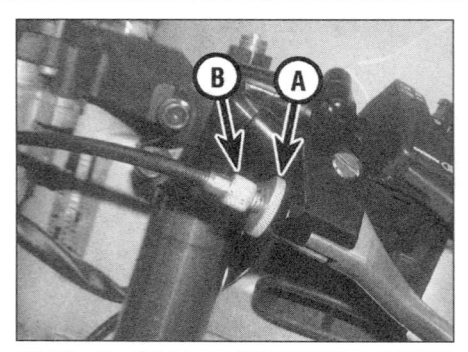

16.32a On RR-Y and RR-1 (2000 and 2001) models slacken the lockring (A) and turn the adjuster (B) in

16.32b On RR-2 and RR-3 (2002 and 2003) models turn the adjuster in

behind the pull-rod end **(see illustration)**. Install all the bolts (except the clutch cable bracket bolts) finger tight, then tighten them evenly and a little at a time in a criss-cross pattern.

30 Reconnect the pulse generator coil wiring connector **(see illustration 5.13a)**. Engage the clutch cable end in the release lever arm, then locate the bracket on the cover and tighten the bolts **(see illustration 5.11)**.

31 Fill the engine with oil (see Chapter 1). Lower the fuel tank (see Chapter 4). Install the fairing panels as required according to model (See Chapter 8).

Clutch cable renewal

32 On RR-Y and RR-1 (2000 and 2001) models fully slacken the lockring on the adjuster at the handlebar end of the cable **(see illustration)**. On all models thread the adjuster fully in **(see illustration)**. This provides freeplay in the cable and resets the adjuster to the beginning of its span.

33 On RR-Y and RR-1 (2000 and 2001) models remove the lower fairing and the right-hand fairing side panel (see Chapter 8). On RR-2 and RR-3 (2002 and 2003) models remove the right-hand fairing side panel (see Chapter 8).

34 Slacken the nuts on the threaded section of the cable in the bracket on the clutch cover **(see illustration)**. Thread the front nut as far up as it will go, then thread the rear nut off the cable **(see illustration)**. Slide the cable into the bracket to get some freeplay and free the cable end from the release lever, noting how it fits **(see illustration)**. Draw the cable out of the bracket **(see illustration)**.

16.34a Thread the front nut up ...

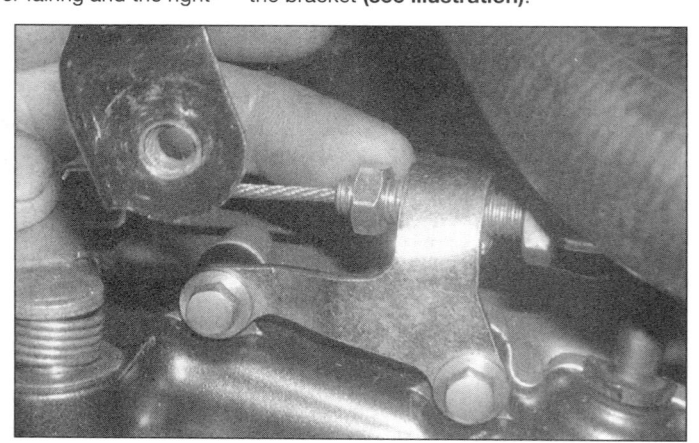

16.34b ... and the rear nut off ...

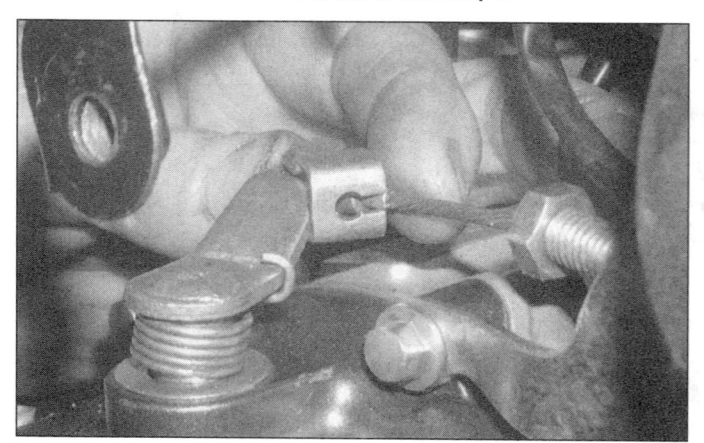

16.34c ... then free the cable end from the release lever ...

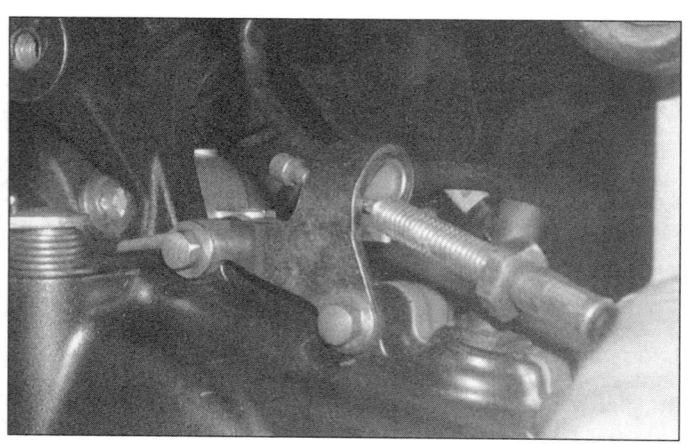

16.34d ... and draw the cable out of the bracket

16.35a Align the slot(s) in the adjuster (and lockring) with that in the bracket . . .

16.35b . . . then free the cable from the adjuster . . .

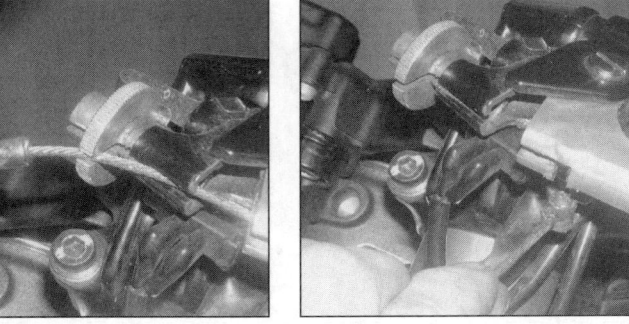

16.35c . . . and from the lever

35 Align the slots in the adjuster and lockring (RR-Y and RR-1 (2000 and 2001) models) at the handlebar end of the cable with that in the lever bracket, then pull the outer cable end from the socket in the adjuster and release the inner cable from the lever **(see illustrations)**. Remove the cable from the machine, noting its routing.

> **HAYNES HiNT** Before removing the cable from the bike, tape the lower end of the new cable to the upper end of the old cable. Slowly pull the lower end of the old cable out, guiding the new cable down into position. Using this method will ensure the cable is routed correctly.

36 Installation is the reverse of removal. Apply grease to the cable ends. Make sure the cable is correctly routed. Adjust the amount of clutch lever freeplay (see Chapter 1).

17 Oil sump, oil strainer and pressure relief valve – removal, inspection and installation

Note: *The oil sump, strainer and pressure relief valve can be removed with the engine in the frame. If the engine has been removed, ignore the steps which don't apply.*

Removal

1 On RR-Y and RR-1 (2000 and 2001) models remove the lower fairing and the fairing side panels (see Chapter 8). On RR-2 and RR-3 (2002 and 2003) models remove the fairing side panels (see Chapter 8). Drain the engine oil (see Chapter 1).
2 While the oil is draining, remove the exhaust system (see Chapter 4).
3 Unscrew the sump bolts, slackening them evenly in a criss-cross sequence to prevent distortion, and remove the sump **(see illustration)**.
4 Pull the strainer out, noting how it locates **(see illustration)**. Remove the rubber seal and discard it as a new one must be used.
5 To remove the pressure relief valve, pull it out of its socket in the crankcase – it is a push-fit **(see illustration)**. Discard the O-ring as a new one must be used.

Inspection

6 Remove all traces of sealant from the sump and crankcase mating surfaces, and clean the inside of the sump with solvent. Blow the sump dry with compressed air if available.
7 Clean the oil strainer in solvent and remove any debris caught in the mesh. If the strainer gauze is damaged, replace the strainer with a new one.
8 Push the relief valve plunger into the valve body and check that it moves smoothly and freely against spring pressure **(see illustration)**. If not, remove the circlip, noting that it is under spring pressure, then remove the spring seat, spring and plunger **(see illustrations)**. Clean all components in solvent, then check the plunger and the valve body for evidence of scoring, wear and any other damage. If any is found, replace the

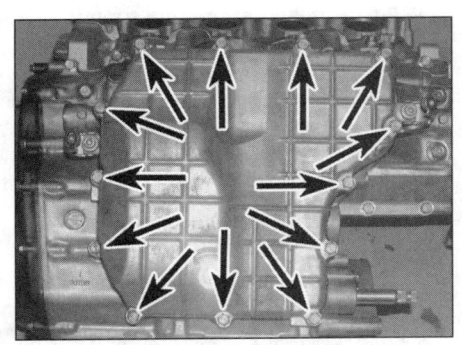

17.3 Unscrew the bolts (arrowed) and remove the sump

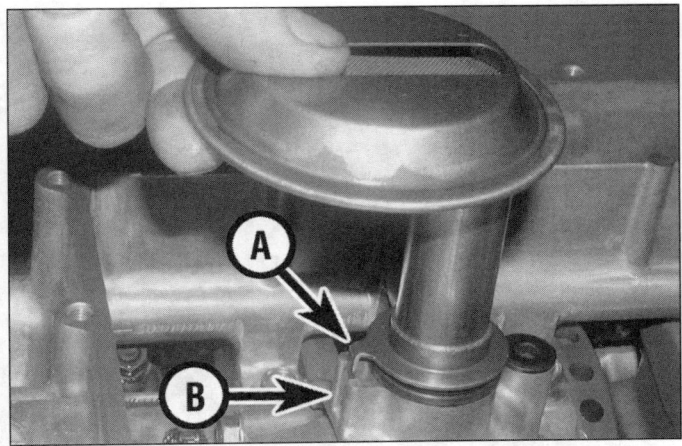

17.4 Remove the strainer, noting how the tab (A) locates in the groove (B)

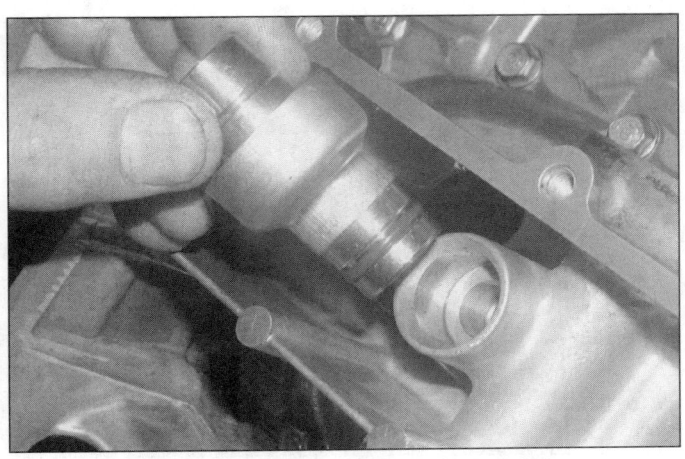

17.5 Pull the relief valve out of its socket

2•36 Engine, clutch and transmission

17.8a Push the plunger into the body and check that it moves smoothly

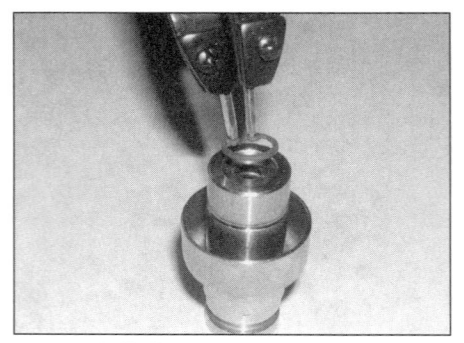

17.8b Remove the circlip . . .

17.8c . . . and the spring seat . . .

17.8d . . . then draw out the spring . . .

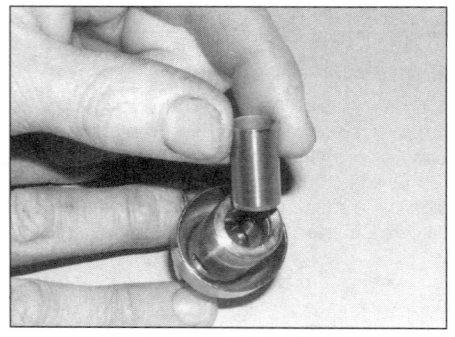

17.8e . . . and the plunger

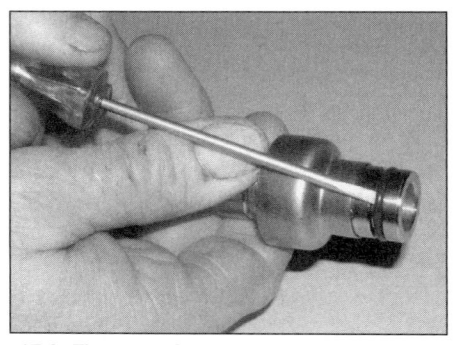

17.9 Fit a new O-ring onto the relief valve body

relief valve with a new one – individual components are not available. Otherwise, coat the plunger with oil and fit it closed end first back into the valve and recheck the movement. If it is good, install the spring and spring seat and secure them with the circlip.

Installation

9 Fit a new O-ring onto the relief valve and smear it with clean oil, then push the valve into its socket in the sump **(see illustration and 17.5)**.
10 Fit a new rubber seal smeared with clean oil into the strainer orifice in the crankcase **(see illustration)**. Do not fit it onto the strainer as it will distort when the strainer is fitted onto the pump. Fit the strainer, locating the tab in the cut-out in the crankcase **(see illustration 17.4)**.
11 Clean the mating surfaces of the sump and crankcase with solvent. Apply a suitable sealant (such as Three Bond 1207B or equivalent RTV sealant – ask your dealer) to the sump mating surface. Position the sump onto the crankcase and install the bolts finger-tight **(see illustration and 17.3)**. Tighten the bolts evenly and a little at a time in a criss-cross pattern.
12 Install the exhaust system, but do not yet fit the fairing panels (see Chapter 4).
13 Fill the engine with the correct type and quantity of oil as described in Chapter 1. Start the engine and check that there are no leaks around the sump.
14 Install the fairing side panels, and on RR-Y and RR-1 (2000 and 2001) models the lower fairing (see Chapter 8).

17.10 Lubricate the rubber seal and fit it into the crankcase

17.11 Apply the sealant then install the sump

Engine, clutch and transmission 2•37

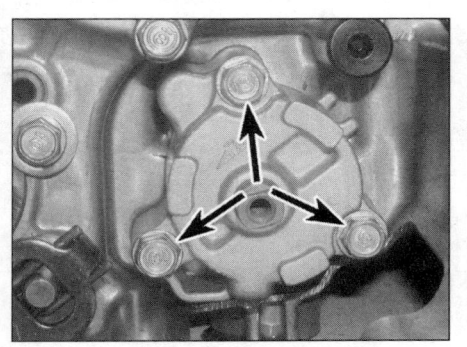

18.3a Unscrew the bolts (arrowed) . . .

18.3b . . . and remove the pump

18.4a Unscrew the bolt . . .

18 Oil pump – removal, inspection and installation

Note: *The oil pump can be removed with the engine in the frame. If the engine has been removed, ignore the steps which don't apply.*

Removal

1 Remove the sump and the oil strainer (see Section 17).
2 Remove the clutch, along with the oil pump drive and driven sprockets (see Section 16).
3 Unscrew the three bolts securing the pump to the crankcase, then remove the pump, noting how it fits **(see illustrations)**. Remove the dowels from either the crankcase or the pump if they are loose.

Inspection

Note: *When removing the rotors from the oil pump, note the punch mark in the outer rotor and which way it faces as it must be installed the same way round so that mated surfaces continue to run together.*

4 Unscrew the single bolt securing the cover to the pump body, then remove the cover, and remove the dowels (if in the pump and not already done) **(see illustrations)**.
5 Remove the thrust washer from the drive shaft **(see illustration)**. Withdraw the shaft from the pump and remove the drive pin, noting how it fits **(see illustration)**. Remove the inner and outer rotors, noting which way round they fit **(see illustrations)**.
6 Clean all the components in solvent.
7 Inspect the pump body and rotors for scoring and wear. If any damage, scoring or uneven or excessive wear is evident, replace the pump with a new one (individual components are not available).
8 Fit the inner and outer rotors into the pump body. Measure the clearance between the inner rotor tip and the outer rotor with a feeler gauge and compare it to the service limit listed in the specifications at the beginning of

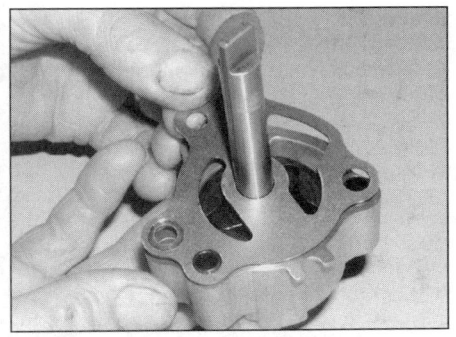

18.4b . . . then remove the cover . . .

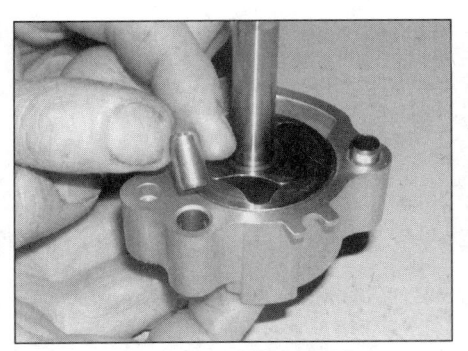

18.4c . . . and the dowels

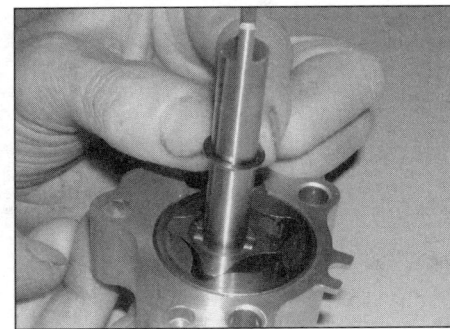

18.5a Remove the thrust washer . . .

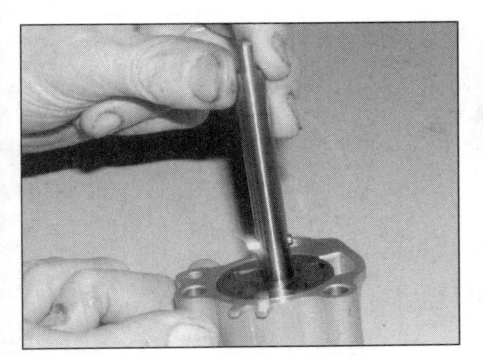

18.5b . . . then withdraw the shaft and remove the drive pin

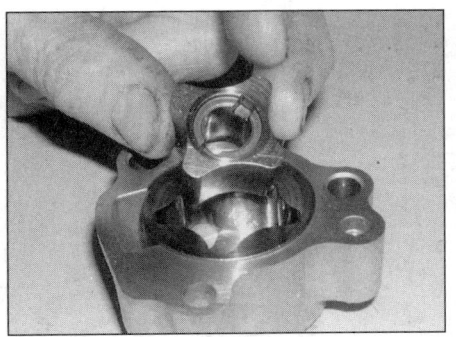

18.5c Remove the inner rotor . . .

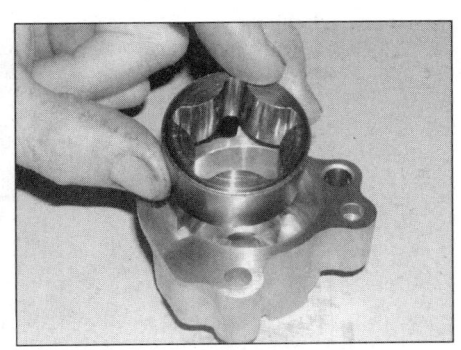

18.5d . . . and the outer rotor

2•38 Engine, clutch and transmission

18.8 Measure the inner rotor tip-to-outer rotor clearance as shown (inner cover shown removed for clarity)

18.9 Measure the outer rotor-to-body clearance as shown

18.10 Measure the rotor endfloat as shown

the Chapter **(see illustration)**. If the clearance measured is greater than the maximum listed, replace the pump with a new one.

9 Measure the clearance between the outer rotor and the pump body with a feeler gauge and compare it to the maximum clearance listed in the specifications at the beginning of the Chapter **(see illustration)**. If the clearance measured is greater than the maximum listed, replace the pump with a new one.

10 Lay a straight-edge across the rotors and the pump body and, using a feeler gauge, measure the rotor endfloat (the gap between the rotors and the straight-edge **(see illustration)**. If the clearance measured is greater than the maximum listed, replace the pump with a new one.

11 Check the pump drive chain and drive and driven sprockets for wear or damage, and renew them as a set if necessary.

12 If the pump is good, make sure all the components are clean, then lubricate them with new engine oil.

13 Fit the outer rotor into the pump body with the punch mark facing the same way as noted on removal **(see illustration 18.5d)**. Fit the inner rotor into the outer rotor with the cut-outs in the inner rotor facing out **(see illustration 18.5c)**. Slide the drive shaft through the inner rotor and pump body, making sure the end with the driven sprocket bolt hole goes through to the outer side and the tabbed end remains on the inner side **(see illustration)**. Slide the drive pin into its hole in the driveshaft and locate it into the cut-outs in the inner rotor **(see illustration)**. Slide the thrust washer onto the shaft so it covers the drive pin **(see illustration 18.5a)**. Fit the dowels into the pump **(see illustration 18.4c)**. Slide the cover onto the pump, then install the single bolt and tighten it to the torque setting specified at the beginning of the Chapter **(see illustrations 18.4b and 18.4a)**.

14 Rotate the pump shaft by hand and check it turns the rotors smoothly and freely.

Installation

15 Pour some clean engine oil into the pump and rotate the shaft to prime the pump.

16 Manoeuvre the pump into position, aligning it so the tabs are in the 2 o'clock position **(see illustration 18.3b)**. Rotate the drive shaft to align the tab on its end with the slot in the water pump shaft. Install the pump, making sure the shaft and dowels locate correctly. Fit the mounting bolts and tighten them **(see illustration 18.3a)**.

17 Install the pump drive and driven sprockets and chain, and the clutch (see Section 16).

18 Install the oil, strainer and sump (see Section 17).

19 Gearchange mechanism – removal, inspection and installation

Note: *The gearchange mechanism can be removed with the engine in the frame. If the engine has been removed, ignore the steps which don't apply.*

Removal

1 Make sure the transmission is in neutral. Remove the clutch (see Section 16). There is no need to remove the oil pump drive and driven sprockets and chain. Block the holes into the sump with clean rag to prevent anything falling in.

2 Unscrew the gearchange linkage arm pinch bolt and slide the arm off the shaft, noting how the slit in the arm aligns with the punch mark on the shaft **(see illustration 5.15)**.

3 Wrap a single layer of thin insulating tape around the gearchange shaft splines to protect the oil seal lips as the shaft is removed.

4 Note how the gearchange shaft centralising spring ends fit on each side of the locating pin in the casing, and how the pawls on the selector arm locate onto the pins on the end of the selector drum **(see illustration 19.10b)**. Grasp the end of the shaft and withdraw the shaft/arm assembly **(see illustration)**. Retrieve the washer from the crankcase if it didn't come with the shaft.

5 If required, note how the stopper arm spring ends locate and how the roller on the arm locates in the neutral detent on the selector drum cam, then unscrew the stopper arm bolt and remove the arm, the washer and

18.13a Slide the shaft through the pump as described

18.13b Locate the drive pin ends into the cut-outs in the inner rotor

19.4 Withdraw the shaft/arm assembly, noting how it fits

Engine, clutch and transmission 2•39

19.5 Note how the spring ends locate, and how the roller sits in the neutral detent, then unscrew the bolt (arrowed) and remove the arm

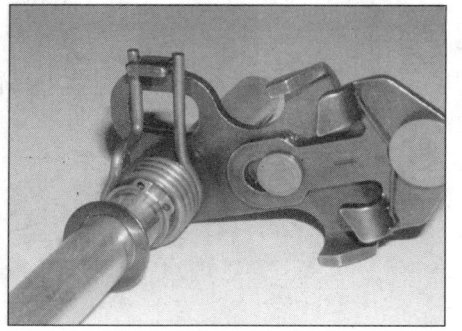

19.7a Check the selector arm assembly . . .

19.7b . . . and the stopper arm assembly as described

the spring, noting how they fit **(see illustration)**.

Inspection

6 Check the selector arm for cracks, distortion and wear of its pawls, and check for any corresponding wear on the pins on the selector drum cam. Also check the stopper arm roller and the detents in the selector drum cam for any wear or damage, and make sure the roller turns freely. Replace any components that are worn or damaged with new ones. Remove the selector drum cam by unscrewing the bolt in its centre **(see illustration 20.12e)**. Note the locating pin in the end of the drum and remove it for safekeeping if required. On installation, locate the pin in the cut-out in the back of the cam. Apply a suitable non-permanent thread locking compound to the cam bolt and tighten it to the torque setting specified at the beginning of the Chapter.

7 Inspect the shaft centralising spring and the stopper arm return spring for fatigue, wear or damage **(see illustrations)**. If any is found, they must be replaced with new ones. To replace the shaft spring, remove the circlip then slide the spring off the shaft, noting how its ends locate. Fit the new spring, locating the ends on each side of the tab, and secure it with the circlip, making sure it locates in its groove. Also check that the centralising spring locating pin in the crankcase is securely tightened. If it is loose, remove it and apply a non-permanent thread locking compound to its threads, then tighten it.

8 Check the gearchange shaft for straightness and damage to the splines. If the shaft is bent you can attempt to straighten it, but if the splines are damaged the shaft must be replaced with a new one. Also check the condition of the shaft oil seal in the left-hand side of the crankcase **(see illustration)**. If it is damaged, deteriorated or shows signs of leakage it must be replaced with a new one. Unscrew the bolt securing the seal retainer plate and remove the plate. Lever out the old seal with a screwdriver **(see illustration)**. With the seal removed, check the condition of the needle bearing, and replace that with a new one as well if necessary **(see illustration)**. Fit the new bearing and press or drive the new seal squarely into place using your fingers, a seal driver or suitable socket **(see illustration)**. Fit the retainer plate and tighten its bolt securely.

Installation

9 If removed, fit the bolt through the stopper arm, then fit the washer and the stopper arm return spring onto the bolt **(see illustration 19.7b)**. Apply a suitable non-permanent thread locking compound to the bolt. Install the arm, locating the roller onto the neutral detent on the selector drum and making sure the spring ends are positioned correctly **(see illustration and 19.5)**. Tighten the bolt to the torque setting specified at the

19.8a Check the oil seal and renew it if required – the retainer plate is secured by the bolt

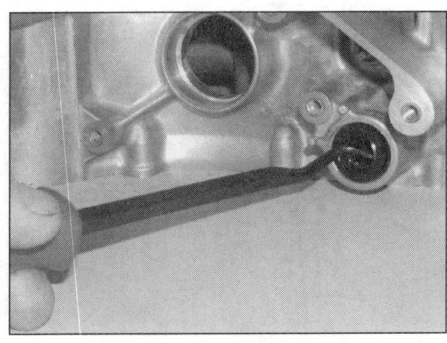

19.8b Lever out the old seal

19.8c Check the needle bearing (arrowed) . . .

19.8d . . . before pressing in a new seal

19.9 Locate the spring ends and stopper arm roller correctly

2•40 Engine, clutch and transmission

19.10a Slide the shaft into the engine...

19.10b ...and locate the selector arm and spring as shown

beginning of the Chapter. Check that the arm and spring ends are correctly positioned.

10 Check that the shaft centralising spring is properly positioned and slide the washer onto the shaft if removed **(see illustration 19.7a)**. Apply some grease to the lips of the gearchange shaft oil seal in the left-hand side of the crankcase. Slide the shaft into place and push it all the way through the case until the splined end comes out the other side **(see illustration)**. Locate the selector arm pawls onto the pins on the selector drum and the centralising spring ends onto each side of the locating pin in the crankcase **(see illustration)**.

11 Remove the rag that was blocking the sump, then install the clutch (see Section 16).

12 Remove the insulating tape from around the gearchange shaft splines. Slide the gearchange linkage arm onto the shaft, aligning its slit with the punch mark on the shaft **(see illustration 5.15)**. Install the pinch bolt and tighten it securely.

20 Selector drum and forks – removal, inspection and installation

Note: *The selector drum and forks can be removed and installed with the crankcases joined and the engine in the frame. Having* said that, it is much easier if the engine is removed so that you don't have to work upside down. If the crankcases are being separated anyway, remove the selector drum and forks afterwards as it is slightly easier.

Removal

1 The selector drum and forks are located in the lower crankcase half. If required, remove the engine (see Section 5) and, again if required, separate the crankcase halves (see Section 22). Otherwise, just remove the sump, and to improve access the oil strainer (see Section 17).

2 If not already done, remove the gearchange mechanism (see Section 19).

3 Before removing the selector forks, note that each fork carries an identification letter **(see illustration 20.6)**. The right-hand fork has R, the centre fork C, and the left-hand fork L. These letters all face the right-hand (clutch) side of the engine. If no letters are visible, mark them yourself using a felt pen. The R and L forks fit into the output shaft and the C fork fits into the input shaft.

4 Unscrew the selector drum bearing/fork shaft retainer bolts and remove them, noting how they fit **(see illustration)**.

5 Support the selector forks and withdraw the shaft from the casing, then remove the forks **(see illustrations)**. Withdraw the selector drum from the right-hand side of the engine **(see illustration)**.

6 Once removed from the case, slide the forks back onto the shaft to keep them in the correct order **(see illustration)**.

Inspection

7 Inspect the selector forks for any signs of wear or damage, especially around the fork ends where they engage with the groove in the pinion. Check that each fork fits correctly in its pinion groove. Check closely to see if the forks are bent. If the forks are in any way damaged they must be replaced with new ones.

8 Measure the thickness of the fork ends and

20.4 Unscrew the bolts (arrowed) and remove the retainers

20.5a Withdraw the shaft...

20.5b ...then remove the selector forks, noting how and which way round they fit...

20.5c ...and withdraw the selector drum

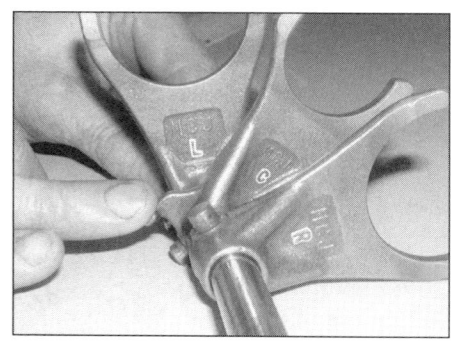

20.6 Keeps the forks in the correct order and way round

Engine, clutch and transmission 2•41

20.8 Measure the fork end thickness . . .

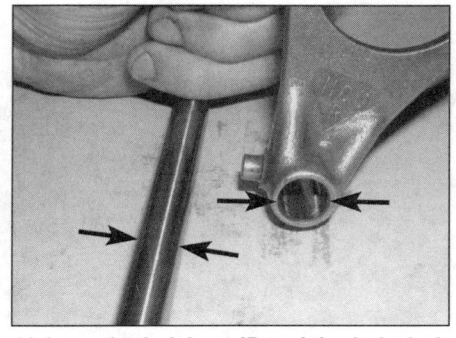

20.9 . . . the fork bore ID and the fork shaft OD as shown

20.11 Check the guide pins and their grooves in the drum

compare the readings to the specifications **(see illustration)**. Replace the forks with new ones if they are worn beyond their specifications.

9 Check that the forks fit correctly on their shaft. They should move freely with a light fit but no appreciable freeplay. Measure the internal diameter of the fork bores and the corresponding diameter of the fork shaft **(see illustration)**. Replace the forks and/or shaft with new ones if they are worn beyond their specifications. Check that the fork shaft holes in the casing are neither worn nor damaged.

10 Check the selector fork shaft for trueness by rolling it along a flat surface. A bent rod will cause difficulty in selecting gears and make the gearshift action heavy. Replace the shaft with a new one if it is bent.

11 Inspect the selector drum grooves and selector fork guide pins for signs of wear or damage **(see illustration)**. If either component shows signs of wear or damage the fork(s) and drum must be replaced with new ones.

12 Check that the selector drum bearing rotates freely and has no sign of freeplay between it and the casing. To fit a new bearing, remove the selector drum cam by unscrewing the bolt in its centre – pass a rod through the drum so as to counter-hold it **(see illustration)**. Note the locating pin in the end of the drum and remove it for safekeeping if required **(see illustration)**. Remove the old bearing and fit a new one (see *Tools and Workshop Tips* in the Reference Section if necessary) **(see illustration)**. Install the

selector drum cam, locating the pin in the cut-out in the back of the cam **(see illustration)**. Apply a suitable non-permanent thread locking compound to the cam bolt and tighten it to the torque setting specified at the beginning of the Chapter **(see illustration)**.

Installation

13 Slide the selector drum into position in the crankcase **(see illustration 20.5c)**. Make sure the drum end locates into its bore in the casing, and position it so that the neutral contact is against the neutral switch **(see illustration)**.

14 Lubricate the selector fork shaft with clean engine oil and slide it into the crankcase **(see illustration 20.5a)**, locating each fork's ends in turn in its pinion groove and its guide

20.12a Counter-hold the drum as shown and unscrew the bolt

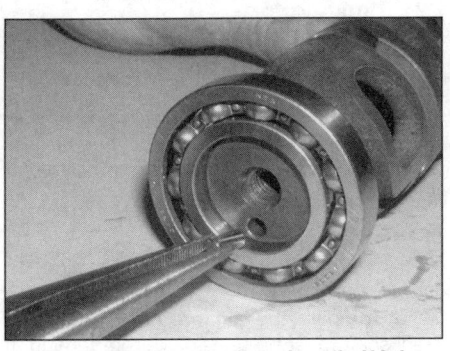

20.12b Remove the locating pin if it is loose

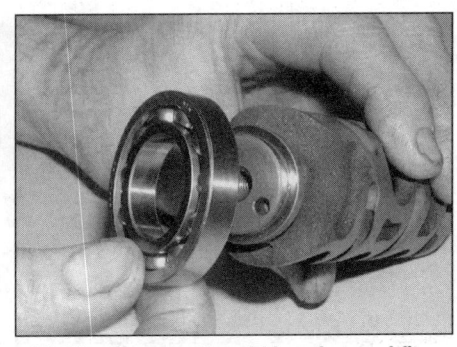

20.12c Remove the old bearing and fit a new one

20.12d Locate the cut-out (arrowed) over the pin

20.12e Apply a threadlock to the bolt

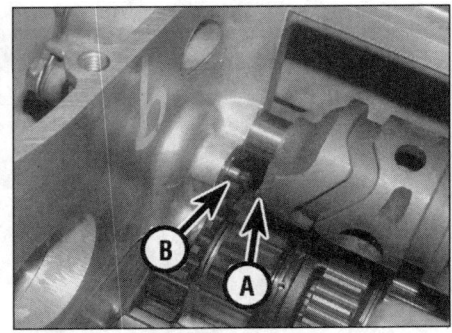

20.13 Position the drum so the raised neutral contact (A) is against the neutral switch contact (B)

20.14a Locate each fork in turn in its pinion ...

20.14b ... then pivot it so the guide pin locates in its groove in the drum and slide the shaft through ...

20.14c ... then slide the shaft end into its bore in the crankcase

pin in its groove in the selector drum, and sliding the shaft through each fork (making sure they are correctly positioned – see Step 3), and into its bore in the crankcase **(see illustrations)**.

15 Clean the threads of the selector drum retainer bolts, then apply a suitable non-permanent thread locking compound **(see illustration)**. Install the bolts and tighten them to the torque setting specified at the beginning of the Chapter.

16 Install any remaining components according to your removal procedure.

20.15 Clean the bolts and apply fresh threadlock

21 Timing rotor – removal and installation

Note: *The timing rotor can be removed with the engine in the frame. If the engine has been removed, ignore the steps which don't apply.*

Removal

1 Remove the clutch cover (see Section 16, Steps 1 to 4).

2 To remove the timing rotor bolt, the crankshaft must be prevented from turning. If the engine is in the frame, engage 6th gear and have an assistant hold the rear brake on hard with the rear tyre in firm contact with the ground. If the engine has been removed, use a rotor holding strap to counter-hold the clutch housing while slackening the timing rotor bolt **(see illustration)**. Alternatively, remove the alternator cover (see Chapter 9) and counter-hold the alternator rotor bolt using a spanner or socket – the alternator bolt is tighter than the timing rotor bolt so there is no danger of the wrong one coming undone.

3 Remove the bolt and its washer, then slide the timing rotor off the end of the crankshaft, noting how it locates over a wide spline so that it can only be fitted in one position **(see illustration)**.

Installation

4 Slide the timing rotor onto the end of the crankshaft, aligning the wide splines **(see illustration)**.

5 Apply some engine oil to the threads on the bolt. Install the bolt with its washer **(see illustration 21.3)** and tighten it to the torque setting specified at the beginning of the Chapter, using the same method as on removal to prevent the crankshaft turning **(see illustration)**.

6 Install the clutch cover (see Section 16, Steps 29 to 31).

21.2 Timing rotor bolt (arrowed)

21.3 Unscrew the bolt and remove the washer

21.4 Align the wide spline on the crankshaft with that in the rotor

21.5 Tighten the rotor bolt to the specified torque

Engine, clutch and transmission 2•43

22.3 Unscrew the three bolts and remove the plate

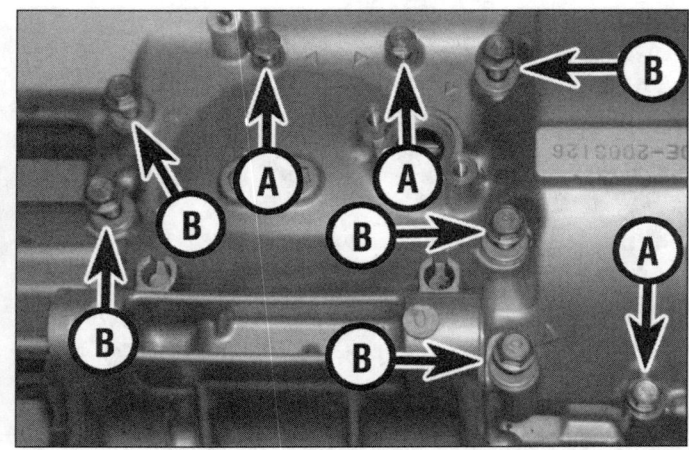

22.4 Upper crankcase 6 mm bolts (A) and 8 mm bolts (B)

22 Crankcase halves – separation and reassembly

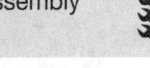

Note 1: *To separate the crankcase halves, the engine must be removed from the frame.*
Note 2: *On RR-2 and RR-3 (2002 and 2003) models, the 9 mm main journal crankcase bolts are of the stretch type, which can only be used in a running engine once, though they can be used when performing the oil clearance check detailed in Section 28 to prevent having to buy two sets of new bolts. The new bolts come pre-coated with an oil additive which must not be cleaned off.*

Separation

1 To access the pistons, connecting rods, crankshaft, bearings and transmission shafts, the crankcase must be split into its two halves. The selector drum and forks can be removed and installed with the crankcases joined.
2 Before the crankcases can be separated the following components must be removed:
Valve cover (Section 8)
Camshafts (Section 10) – see Note
Cylinder head (Section 12) – see Note
Alternator/starter clutch (Chapter 9)
Clutch, oil pump drive chain and sprockets (Section 16)
Cam chain and blades (Section 11) – see Note
Water pump (Chapter 3)
Gearchange mechanism (Section 19) – see Note
Oil cooler (Section 7)
Starter motor (Chapter 9) – see Note
Oil sump, strainer and pressure relief valve (Section 17).
Oil pump if required (Section 18)
Speed sensor (Chapter 9)

Note: *If the crankcases are being separated to inspect the crankshaft without removing it, the camshafts and cylinder head can remain in situ. To remove the crankshaft without removing the connecting rods and pistons, the camshafts must be removed but the head can stay. To inspect or remove the transmission shafts, the camshafts and cylinder head can remain in situ. However, if removal of the connecting rod assemblies is intended, full disassembly of the top-end is necessary. The cam chain can remain loose on the crankshaft after the camshafts have been removed, and can be taken off after the crankshaft has been removed. The gearchange mechanism can remain in situ unless the selector drum and forks are being removed.*

3 Unscrew the bolts securing the transmission input shaft bearing retainer plate and remove the plate, noting how it fits **(see illustration)**.
4 Unscrew the three 6 mm upper crankcase bolts, noting the two fitted with sealing washers, and the five 8 mm upper crankcase bolts, all fitted with sealing washers **(see illustration)**. Unscrew the bolts evenly, a little at a time and in a criss-cross sequence until they are finger-tight, then remove them. **Note:** *As each bolt is removed, store it in its relative position in a cardboard template of the crankcase halves. This will ensure all bolts and washers are installed in the correct location on reassembly. Note that new sealing washers should be used on assembly, though it is wise to keep the old ones with the bolts for the time being as a pattern for the new ones.*
5 Turn the engine upside down.
6 Unscrew the ten 6 mm, seven 8 mm and one 10 mm lower crankcase bolts **(see illustrations)**. Now unscrew the ten 9 mm

22.6a Lower crankcase 6 mm bolts (arrowed)

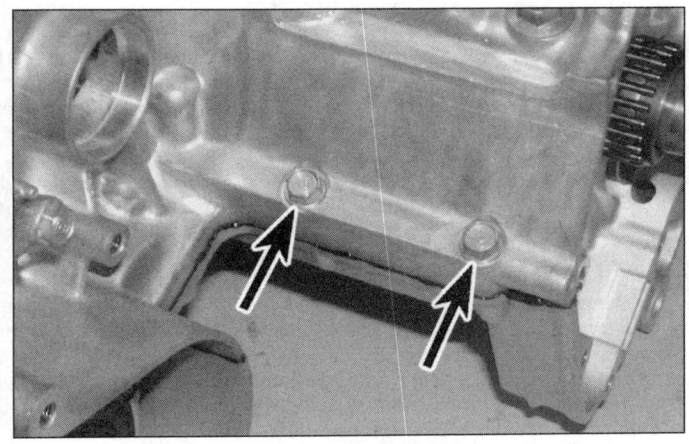

22.6b Lower crankcase 6 mm bolts (arrowed)

2•44 Engine, clutch and transmission

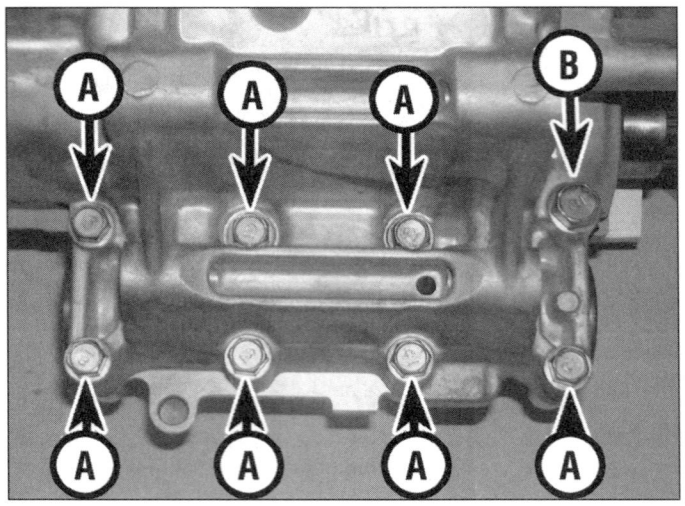

22.6c Lower crankcase 8 mm bolts (A) and 10 mm bolt (B)

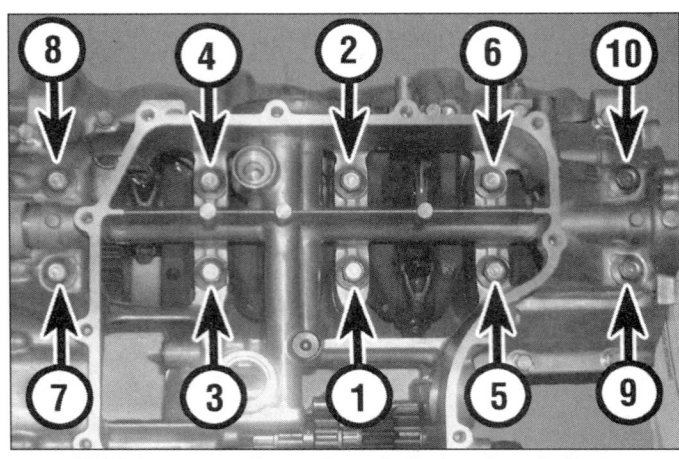

22.6d Lower crankcase 9 mm bolts (numbers indicate tightening sequence)

Bolts 2, 4, 6, 7 and 8 – shorter bolts
Bolts 1, 3, 5 and 9 – longer bolts   Bolt 10 – gold coloured

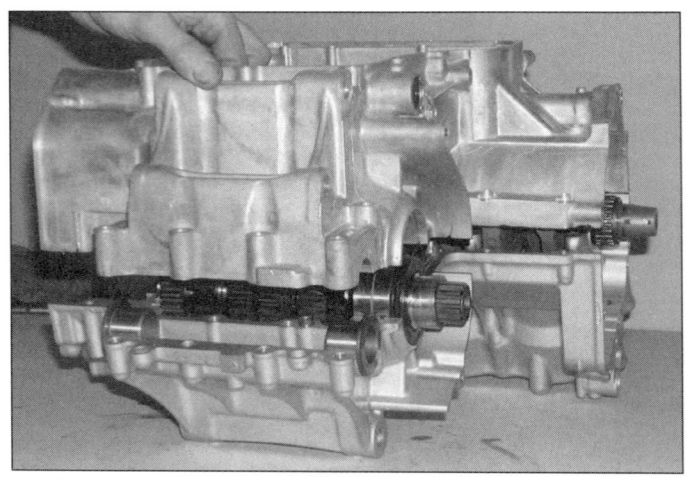

22.7 Carefully separate the crankcase halves

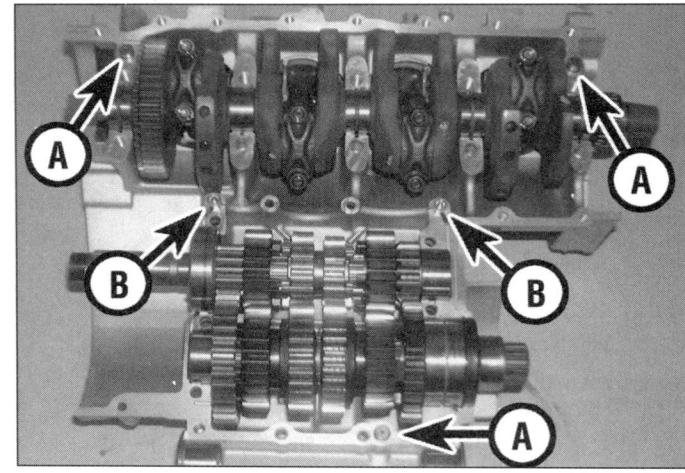

22.8a Remove the dowels (A) if they are loose, and the oil orifices (B) ...

bolts evenly, a little at a time and in a reverse of the tightening sequence, i.e. starting from the outside and working to the centre, until they are finger-tight, then remove them **(see illustration)**. Note: *As each bolt is removed, store it in its relative position in a cardboard template of the crankcase halves. This will ensure all bolts are installed in the correct location on reassembly – there are three different type bolts.* Refer to **Note 2** at the head of this Section for information regarding the 9 mm bolts fitted on RR-2 and RR-3 (2002 and 2003) models.

7 Carefully lift the lower crankcase half off the upper half, using a soft-faced hammer to tap around the joint to initially separate the halves if necessary **(see illustration)**. Note: *If the halves do not separate easily, make sure all fasteners have been removed. Do not try and separate the halves by levering against the crankcase mating surfaces as they are easily scored and will leak oil in the future if damaged.* The lower crankcase half will come away with the gearchange mechanism and the selector drum and forks (if not already removed), leaving the crankshaft and transmission shafts in the upper crankcase half.

8 Remove the three locating dowels from the crankcase if they are loose (they could be in either crankcase half), and the two oil orifices, noting how they fit **(see illustration)**. Remove the two collars from the swingarm pivot bore **(see illustration)**.

9 Refer to Sections 23 to 30 for the removal and installation of the components housed within the crankcases.

Reassembly

10 Remove all traces of sealant from the crankcase mating surfaces.

11 Ensure that all components and their bearings are in place in the upper and lower crankcase halves. If the transmission shafts have not been removed, check the condition of the oil seal on the left-hand end of the output shaft and replace it with a new one if it is damaged or deteriorated – it is advisable to fit a new one as a matter of course **(see illustration 29.5)**.

12 Generously lubricate the crankshaft and transmission shafts, particularly around the bearings, with clean engine oil, then use a rag

22.8b ... and the swingarm collars

Engine, clutch and transmission 2•45

soaked in high flash-point solvent to wipe over the mating surfaces of both crankcase halves to remove all traces of oil.

13 If removed, install the three locating dowels and the two oil orifices in the upper crankcase half **(see illustration 22.8a)**. Make sure the orifices are installed the correct way round **(see illustration)**. Also fit the swingarm pivot collars into the ends of the bore in the upper crankcase half **(see illustration 22.8b)**.

14 Apply a small amount of suitable sealant (Three-Bond 1207B or equivalent RTV sealant – ask your dealer) to the outer mating surface of the lower crankcase half as shown **(see illustrations)**.

Caution: Apply the sealant only to the shaded areas. Do not apply an excessive amount as it will ooze out when the case halves are assembled and may obstruct oil passages. Do not apply the sealant close to any of the bearing inserts or surfaces, or oil passages.

15 Check again that all components are in position, particularly that the bearing shells are still correctly located in the lower crankcase half. Carefully fit the lower crankcase half down onto the upper crankcase half, making sure the dowels all locate correctly **(see illustration 22.7)**. If the selector drum and forks are installed, make sure each selector fork locates correctly in the groove in its pinion.

16 Check that the lower crankcase half is correctly seated. Make sure the swingarm pivot collars are fully pushed in so the flanges meet the crankcases and are square in the bore.

Caution: The crankcase halves should fit together without being forced. If the casings are not correctly seated, remove the lower crankcase half and investigate the problem. Do not attempt to pull them together using the crankcase bolts as the casing will crack and be ruined.

17 On RR-Y and RR-1 (2000 and 2001) models, clean the threads of the ten 9 mm lower crankcase bolts. Apply some oil to the threads and the underside of the heads of the bolts and insert them in their original locations **(see illustration 22.6d)**. Secure all bolts finger-tight at first, then tighten them evenly and a little at a time in the numerical sequence

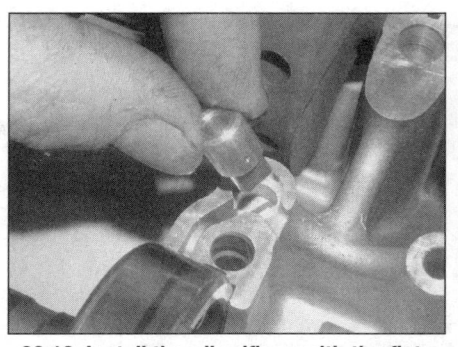

22.13 Install the oil orifices with the flats locating against the corresponding flats in their bores

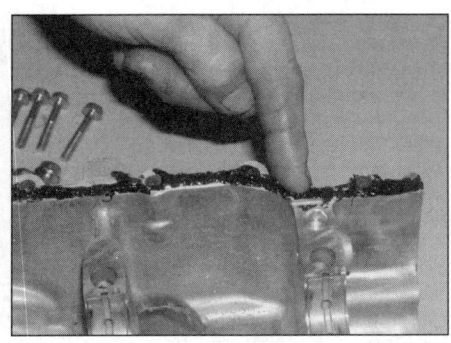

22.14a Apply the sealant . . .

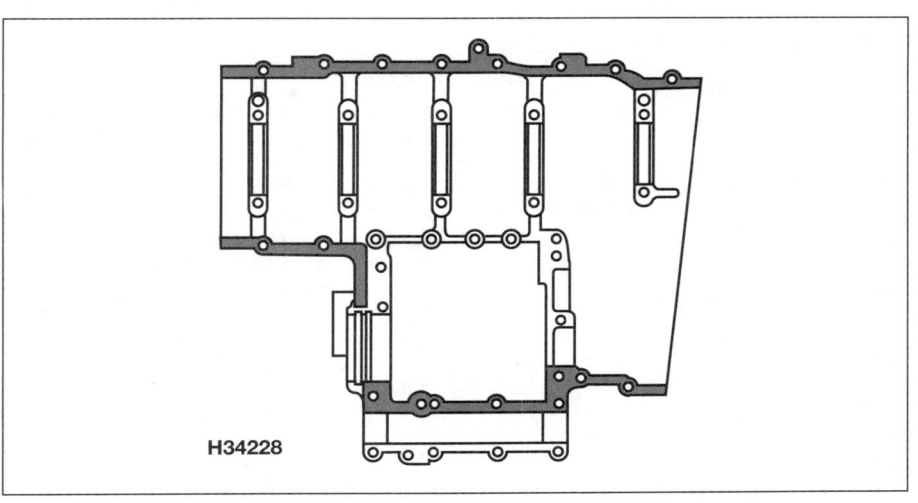

22.14b . . . to the shaded area shown

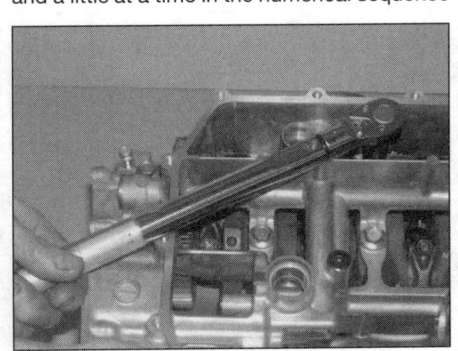

shown to the torque setting specified at the beginning of the Chapter **(see illustration)**.

18 On RR-2 and RR-3 (2002 and 2003) models, install the ten NEW 9 mm lower crankcase bolts, following the pattern of the template-stored old bolts to ensure the various different lengths are all in their correct location **(see illustration 22.6d)**. Secure all bolts finger-tight at first, then tighten them evenly in the numerical sequence shown to the initial torque setting specified at the beginning of the Chapter **(see illustration**

22.17 Tighten the bolts as described to the specified torque setting

22.18 Use a degree disc to finally tighten the bolts

22.17). Now tighten them in the same sequence to the second torque setting specified. Finally, using a degree disc, tighten each bolt in turn by a further 150°, again following the numerical sequence **(see illustration)**.

19 On all models clean the threads of the ten 6 mm, seven 8 mm and one 10 mm lower crankcase bolts and insert them in their original locations **(see illustrations 22.6a, 22.6b and 22.6c)**. Secure all bolts finger-tight at first, then tighten the 10 mm bolt followed by the 8 mm bolts evenly and a little at a time in a criss-cross sequence to the specified torque settings. Now tighten the 6 mm bolts, starting in the centre and working outwards.

20 Turn the engine over. Clean the threads of the three 6 mm and five 8 mm upper crankcase bolts and insert them in their original locations, not forgetting to fit new sealing washers with all the 8 mm bolts and the two forward 6 mm bolts – all bolt holes that require sealing washers are marked by a triangle **(see illustration 22.4)**. Secure all bolts finger-tight at first, then tighten them evenly and a little at a time in a criss-cross sequence, tightening the 8 mm bolts to the specified torque setting.

21 With all crankcase fasteners tightened, check that the crankshaft and transmission shafts rotate smoothly and easily. Check that the transmission shafts rotate freely and independently in neutral, then, if it is installed, rotate the selector drum by hand and select each gear in turn whilst rotating the input shaft. Check that all gears can be selected and that the shafts rotate freely in every gear. If there are any signs of undue stiffness, tight or rough spots, or of any other problem, the fault must be rectified before proceeding further.

22 Apply a suitable non-permanent thread locking compound to the transmission input shaft bearing retainer plate bolts, then install the plate with the OUTSIDE mark facing out and tighten the bolts to the specified torque setting **(see illustration 22.3)**.

23 Install all other removed assemblies in a reverse of the sequence given in Step 2.

23 Crankcase halves and cylinder bores – inspection and servicing

Crankcase halves

1 After the crankcases have been separated, remove the crankshaft, connecting rods and pistons, transmission shafts, selector drum and forks (if not already done), speed sensor, neutral switch and oil pressure switch, referring to the relevant Sections of this Chapter, and to Chapter 9 for the speed sensor and oil pressure and neutral switches. If there are any other components or assemblies that have not been removed as part of your strip-down procedure, for example the starter motor or the coolant inlet union, remove these as well, referring to the relevant Chapter.

2 Clean the crankcases thoroughly with new solvent and dry them with compressed air. Blow out all oil passages with compressed air.

3 Remove all traces of old gasket sealant from the mating surfaces. Clean up minor damage to the surfaces with a fine sharpening stone or grindstone.

Caution: Be very careful not to nick or gouge the crankcase mating surfaces or oil leaks will result. Check both crankcase halves very carefully for cracks and other damage.

4 Small cracks or holes in aluminium castings can be repaired with an epoxy resin adhesive as a temporary measure. Permanent repairs can only be done by argon-arc welding, and only a specialist in this process is in a position to advise on the economy or practical aspect of such a repair. If any damage is found that can't be repaired, replace the crankcase halves as a set.

5 Damaged threads can be economically reclaimed using a diamond section wire insert, for example of the Heli-Coil type (though there are other makes), which are easily fitted after drilling and re-tapping the affected thread.

6 Sheared studs or screws can usually be removed with extractors, which consist of a tapered, left-hand thread screw of very hard steel. These are inserted into a pre-drilled hole in the stud, and usually succeed in dislodging the most stubborn stud or screw. If a stud has sheared above its bore line, it can be removed using a conventional stud extractor which avoids the need for drilling.

HAYNES HiNT *Refer to Tools and Workshop Tips for details of installing a thread insert and using screw extractors.*

7 Install all components and assemblies, referring to the relevant Sections of this and the other Chapters, before reassembling the crankcase halves.

Cylinder bores

Note: *Do not attempt to separate the cylinder liners from the cylinder block.*

8 Check the cylinder walls carefully for scratches and score marks.

9 Using a precision straight-edge and a feeler gauge set to the warpage limit listed in the specifications at the beginning of the Chapter, check the block gasket mating surface for warpage. Refer to *Tools and Workshop Tips* in the Reference section for details of how to use the straight-edge. If warpage is excessive the crankcases must be replaced with new ones.

10 Using telescoping gauges and a micrometer (see *Tools and Workshop Tips*), check the dimensions of each cylinder to assess the amount of wear, taper and ovality. Measure near the top (but below the level of the top piston ring at TDC), centre and bottom (but above the level of the oil ring at BDC) of the bore, both parallel to and across the crankshaft axis **(see illustrations)**. Compare the results to the specifications at the beginning of the Chapter. If the cylinders are worn, oval or tapered beyond the service limit they can be rebored – an oversize (+ 0.25) set of pistons and rings and available. Note that the person carrying out the rebore must be aware of the piston-to-bore clearance for the oversize pistons and rings (see Specifications).

11 If the precision measuring tools are not available, take the upper crankcase to a Honda dealer or specialist motorcycle repair shop for assessment and advice.

12 If the cylinder bores are in good condition and the piston-to-bore clearance is within specifications (see Section 26), the cylinders should be honed (de-glazed). To perform this operation you will need the proper size flexible hone with fine stones, or a bottle-brush type hone, plenty of light oil or honing oil, some clean rags and an electric drill motor.

13 Hold the block sideways (so that the bores are horizontal rather than vertical) in a vice with soft jaws or cushioned with wooden blocks. Mount the hone in the drill motor, compress the stones and insert the hone into the cylinder. Thoroughly lubricate the cylinder, then turn on the drill and move the hone up and down in the cylinder at a pace which produces a fine cross-hatch pattern on the cylinder wall with the lines intersecting at an angle of approximately 60°. Be sure to use plenty of lubricant and do not take off any more material than is necessary to produce the desired effect. Do not withdraw the hone from the cylinder while it is still turning. Switch off the drill and continue to move it up and down in the cylinder until it has stopped turning, then compress the stones and withdraw the hone. Wipe the oil from the cylinder and repeat the procedure on the other cylinders. Remember, do not take too much material from the cylinder wall.

14 Wash the bores thoroughly with warm soapy water to remove all traces of the abrasive grit produced during the honing operation. Be sure to run a brush through the stud holes and flush them with running water. After rinsing, dry the cylinders thoroughly and apply a thin coat of light, rust-preventative oil to all machined surfaces.

15 If you do not have the equipment or desire to perform the honing operation, take the crankcase to a Honda dealer or specialist motorcycle repair shop.

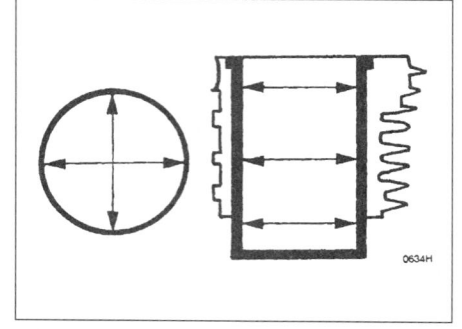

23.10a Measure the cylinder bore in the directions shown . . .

23.10b . . . using a telescoping gauge, then measure the gauge with a micrometer

Engine, clutch and transmission 2•47

24 Main and connecting rod bearings – general information

1 Even though main and connecting rod bearings are generally replaced with new ones during the engine overhaul, the old bearings should be retained for close examination as they may reveal valuable information about the condition of the engine.
2 Bearing failure can occur for many reasons: a of lack of lubrication, the presence of dirt or other foreign particles, overloading the engine, and corrosion. Regardless of the cause of bearing failure, it must be corrected before the engine is reassembled to prevent it from happening again.
3 When examining the main or connecting rod bearings, remove them from their journal and lay them out on a clean surface in the same general position as their location in the crankcase or on the crankshaft. This will enable you to match any noted bearing problems with the corresponding journal.
4 Dirt and other foreign particles get into the engine in a variety of ways. It may be left in the engine during assembly or it may pass through filters or breathers. It may get into the oil and from there into the bearings. Metal chips from machining operations and normal engine wear are often present. Abrasives are sometimes left in engine components after reconditioning operations, especially when parts are not thoroughly cleaned using the proper methods. Whatever the source, these foreign objects often end up embedded in the soft bearing material and are easily recognised. Large particles will not embed in the bearing and will score or gouge the bearing and journal. The best prevention for this is to clean all parts thoroughly and keep everything spotlessly clean during engine reassembly. Frequent and regular oil and filter changes are also recommended.
5 Lack of lubrication or lubrication breakdown has a number of interrelated causes. Excessive heat (which thins the oil), overloading (which squeezes the oil from the bearing face) and oil leakage or throw off (from excessive bearing clearances, worn oil pump or high engine speeds) all contribute to lubrication breakdown. Blocked oil passages will also starve a bearing and destroy it. When lack of lubrication is the cause of bearing failure, the bearing material is wiped or extruded from the steel backing of the bearing. Temperatures may increase to the point where the steel backing and the journal turn blue from overheating.

HAYNES HiNT *Refer to Tools and Workshop Tips for bearing fault finding.*

6 Riding habits can have a definite effect on bearing life. Full throttle low speed operation, or labouring the engine, puts very high loads on bearings, which tend to squeeze out the oil film. These loads cause the bearings to flex, which produces fine cracks in the bearing face (fatigue failure). Eventually the bearing material will loosen in pieces and tear away from the steel backing. Short trip riding leads to corrosion of bearings, as insufficient engine heat is produced to drive off the condensed water and corrosive gases produced. These products collect in the engine oil, forming acid and sludge. As the oil is carried to the engine bearings, the acid attacks and corrodes the bearing material.
7 Incorrect bearing installation during engine assembly will lead to bearing failure as well. Tight fitting bearings which leave insufficient bearing oil clearances result in oil starvation. Dirt or foreign particles trapped behind a bearing insert result in high spots on the bearing which lead to failure.
8 To avoid bearing problems, clean all parts thoroughly before reassembly, double check all bearing clearance measurements and lubricate the new bearings with clean engine oil during installation.

25 Connecting rods and bearings – removal, inspection and installation

Note: *To remove the connecting rods the engine must be removed from the frame and the crankcases separated.*

Removal

1 Remove the engine from the frame (see Section 5) and separate the crankcase halves (see Section 22).
2 Before removing the rods from the crankshaft, measure the side clearance (the gap between the connecting rod big-end and the crankshaft web) with a feeler gauge. If the clearance is greater than the service limit listed in this Chapter's Specifications, replace the rods with new ones. If the clearance is still excessive, replace the crankshaft with a new one.
3 Using paint or a felt marker pen, mark the relevant cylinder identity on each connecting rod and cap. Mark across the cap-to-connecting rod join and note which side of the rod faces the front of the engine to ensure that the cap and rod are fitted the correct way around on reassembly. Note that the number already across the rod and cap indicates rod size grade **(see illustration 25.21b)**. On RR-Y and RR-1 (2000 and 2001) models the piston crown is marked IN and this mark faces the intake side of the cylinder, and the oil hole in the big-end of the connecting rod should face the same way, to the back of the engine (intake side). On RR-2 and RR-3 (2002 and 2003) models the piston crown is marked O and this mark faces the intake side of the cylinder, and the bearing shell notch in the big-end of the connecting rod should face the same way, to the back of the engine (intake side) **(see illustration 26.2)**.
4 Remove the crankshaft (see Section 28).
5 Raise the crankcase onto wooden blocks to provide room for the connecting rod/piston assemblies to be removed from the tops of the bores. Alternatively turn the crankcase on its side, but remove the transmission shafts first otherwise they will fall out (see Section 29). Push each piston/connecting rod assembly up the bore and remove it from the top of its bore making sure the connecting rod does not mark the cylinder walls **(see illustrations)**. Keep the rod, cap, nuts/bolts, and (if they are to be reused) the bearing shells together in their correct positions to ensure correct installation.

HAYNES HiNT *To ease removal of the pistons, carefully remove any ridge of carbon built up on the top of each cylinder bore using a scraper. If there is a pronounced wear ridge, remove it using a ridge reamer.*

Caution: *Do not try to remove the piston/connecting rod from the bottom of the cylinder bore. The piston will not pass the crankcase main bearing webs. If the piston is pulled right to the bottom of the bore the oil control ring will expand and lock the piston in position. If this happens it is likely the ring will break.*

6 Immediately install the relevant bearing shells (if removed), bearing cap, and nuts on

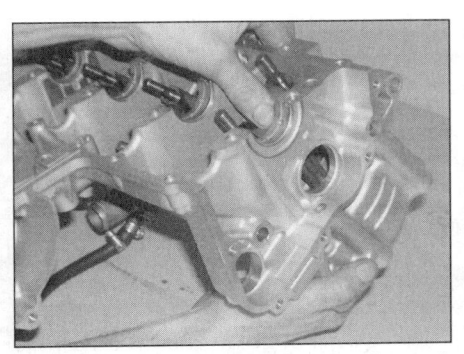

25.5a Push the piston and connecting rod assembly up ...

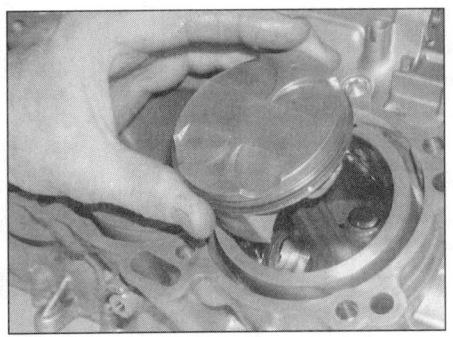

25.5b ... and withdraw it from the top of the cylinder

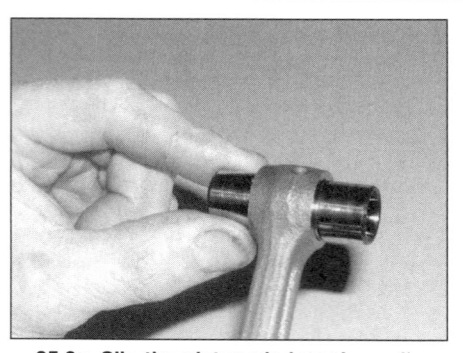

25.9a Slip the piston pin into the rod's small-end and rock it back-and-forth to check for looseness

25.9b Measure the external diameter of the pin . . .

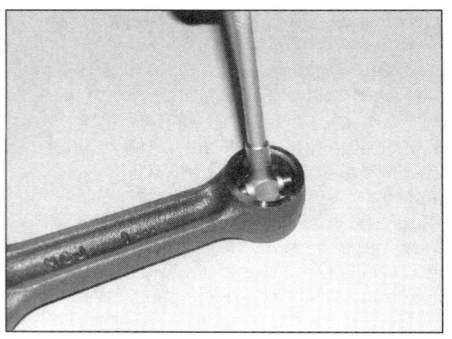

25.9c . . . and the internal diameter of the connecting rod small-end

each piston/connecting rod assembly so that they are all kept together as a matched set.

7 Remove the pistons from the connecting rods if required (see Section 26), but note that if you are doing a big-end oil clearance check they are best left installed as they will prevent the rod rotating on the crankpin and upsetting the Plastigauge.

Inspection

8 Check the connecting rods for cracks and other obvious damage.

9 Apply clean engine oil to the piston pin, insert it into the connecting rod small-end and check for any freeplay between the two **(see illustration)**. Measure the pin external diameter and the small-end bore diameter, then calculate the difference to obtain the small-end-to-piston pin clearance **(see illustrations)**. Compare the result to the specifications at the beginning of the Chapter. If the clearance is greater than specified, renew the components that are worn beyond their specified limits.

10 Refer to Section 24 and examine the connecting rod bearing shells. If they are scored, badly scuffed, corroded, or appear to have seized, new shells must be installed. Remove them by pushing their centres out to the side then lifting them out **(see illustration)**. Always replace the shells in the connecting rods as a set. If they are badly damaged, check the corresponding crankpin. Evidence of extreme heat, such as discoloration, indicates that lubrication failure has occurred.

Be sure to thoroughly check the oil pump and pressure regulator as well as all oil holes and passages before reassembling the engine.

11 Have the rods checked for twist and bend by a Honda dealer if you are in doubt about their straightness.

Oil clearance check

12 Whether new bearing shells are being fitted or the original ones are being reused, the connecting rod bearing oil clearance should be checked prior to reassembly. Check the clearance on one rod at a time.

13 Clean the backs of the bearing shells and the bearing housings in both the connecting rod and cap.

14 Press the bearing shells into their housings, making sure the tab on each shell engages the notch in the connecting rod/cap **(see illustration)**. Make sure the bearings are fitted in the correct location and take care not to touch any shell's bearing surface with your fingers. Lay the crankshaft in the upper crankcase half, making sure it is the correct way round and that all the main bearing shells are installed (see Section 28).

15 Cut a length of the appropriate size Plastigauge (it should be slightly shorter than the width of the crankpin). Place a strand of Plastigauge on the (cleaned) crankpin journal, making sure it is not over the oil hole. Refer to Step 25 and pull the connecting rod onto the crankpin, then fit the cap onto the rod **(see illustrations 28.26 and 28.2b)**. Make sure the cap is fitted the correct way around so the previously made markings align, and that the rod is facing the right way (see Step 3). Apply some clean oil to the threads and under the heads of the connecting rod nuts. Install the nuts and tighten them evenly and alternately, in two or three stages, to the torque setting specified at the beginning of the Chapter, all the time ensuring that the crankshaft does not rotate **(see illustrations 28.2a and 28.27)**. It is highly advisable to have an assistant to hold the crankshaft down in the crankcase while tightening the nuts as it could jump out.

16 Slacken the nuts and remove the connecting rod cap, again taking great care not to rotate the rod or crankshaft. Compare the width of the crushed Plastigauge on the crankpin to the scale printed on the Plastigauge envelope to obtain the connecting rod bearing oil clearance. Compare the reading to the specifications at the beginning of the Chapter.

17 On completion carefully scrape away all traces of the Plastigauge material from the crankpin and bearing shells using a fingernail or other object which is unlikely to score the shells.

18 If the clearance is within the range listed in this Chapter's Specifications and the bearings are in perfect condition, they can be reused. If the clearance is beyond the service limit, replace the bearing shells with new ones (see Steps 21 and 22). Check the oil clearance once again (the new shells may be thick enough to bring bearing clearance within the specified range). Always replace all of the shells at the same time.

19 If the clearance is still greater than the service limit listed in this Chapter's Specifications, the crankpin is worn and the crankshaft should be replaced with a new one.

20 Repeat the oil clearance check for the other connecting rods.

Bearing shell selection

21 Replacement bearing shells for the big-end bearings are supplied on a selected fit basis. Code letters and numbers stamped on the crankshaft and connecting rod are used to identify the correct replacement bearings. The crankpin journal size letters are stamped on the outside of the left-hand crankshaft web, and

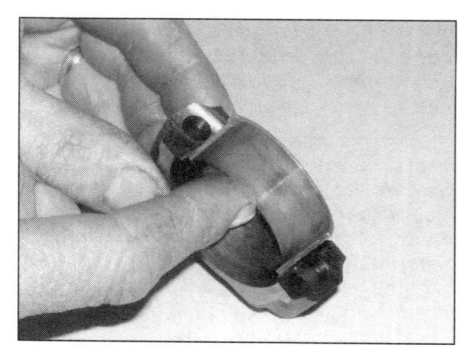

25.10 Remove the shells as shown

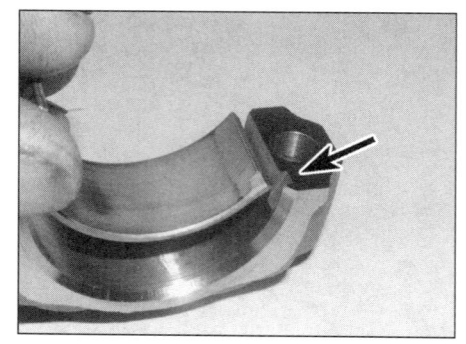

25.14 Fit each shell into its housing making sure each tab locates in its notch (arrow)

Engine, clutch and transmission 2•49

will be either an A, a B or a C **(see illustration)**. The first letter after the L is for the No. 1 cylinder connecting rod (left-hand journal), and the letters correspond consecutively for each cylinder. The connecting rod size code number is marked across the flat face of the connecting rod and cap and will be either a 1, a 2 or a 3 **(see illustration)**.

22 A range of bearing shells is available. To select the correct bearing shell colour code for a particular big-end, use the table below and cross-refer the crankpin journal size letter (stamped on the web) with the connecting rod size number (stamped on the rod). For example, if the crankpin size is B, and the connecting rod size is 1, then the bearing required is green. The colour is marked on the side of the shell **(see illustration)**.

	Connecting rod code		
Crankpin journal code	1 (39.000 to 39.006 mm)	2 (39.006 to 39.012 mm)	3 (39.012 to 39.018 mm)
A (35.997 to 36.003 mm)	E – Yellow	D – Green	C – Brown
B (35.991 to 35.997 mm)	D – Green	C – Brown	B – Black
C (35.985 to 35.991 mm)	C – Brown	B – Black	A – Blue

Installation

23 Fit the pistons onto the connecting rods (see Section 26).
24 Clean the backs of the bearing shells and the bearing housings in both cap and rod. If new shells are being fitted, ensure that all traces of the protective grease are cleaned off using paraffin (kerosene). Wipe the shells, cap and rod dry with a clean lint free cloth. Install the bearing shells in the connecting rods and caps, making sure the tab on each shell engages the notch in the connecting rod/cap **(see illustration 25.14)**. On RR-Y and RR-1 (2000 and 2001) models make sure the oil holes in the rod, cap and shells align. Lubricate the shells with molybdenum disulphide oil (a 50/50 mixture of molybdenum disulphide grease and clean engine oil).
25 Lubricate the pistons, rings and cylinder bore with clean engine oil **(see illustrations)**. Insert the piston/connecting rod assembly into the top of its bore, taking care not to allow the connecting rod to mark the bore. Make sure the IN or O mark (according to model – see Step 3) on the piston crown, and the connecting rod oil hole or bearing shell notch are on the intake side of the bore, then carefully compress and feed each piston ring into the bore until the piston crown is flush with the top of the bore **(see illustration 25.5b)**. If available, a piston ring compressor makes installation a lot easier **(see illustrations)**.
26 Install the crankshaft (see Section 28).
27 Reassemble the crankcase halves (see Section 22).

25.21a Crankpin journal size letters

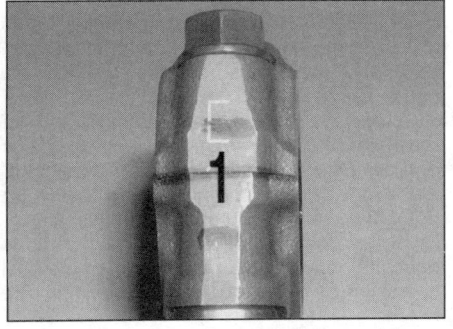

25.21b Connecting rod size number

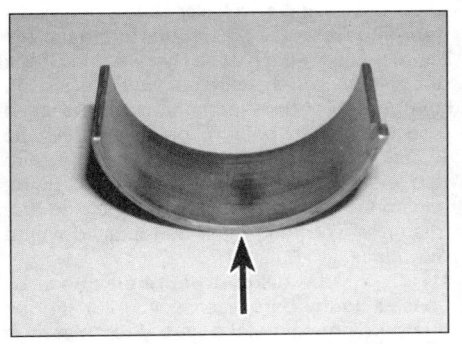

25.22 The colour code is on the edge of the shell (arrow)

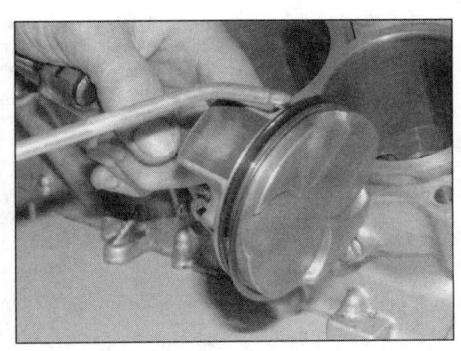

25.25a Lubricate the pistons and rings . . .

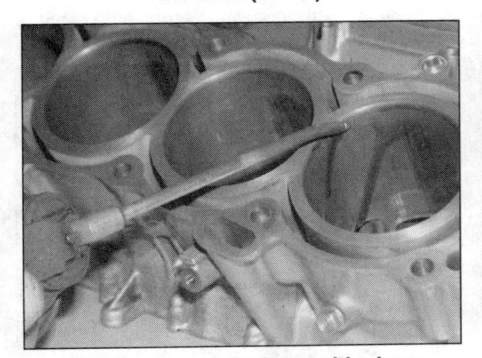

25.25b . . . and the bore with clean engine oil

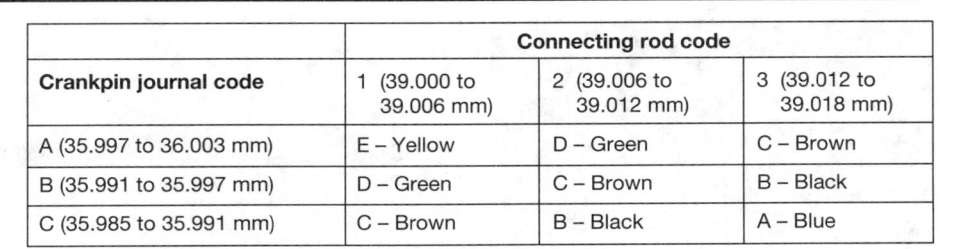

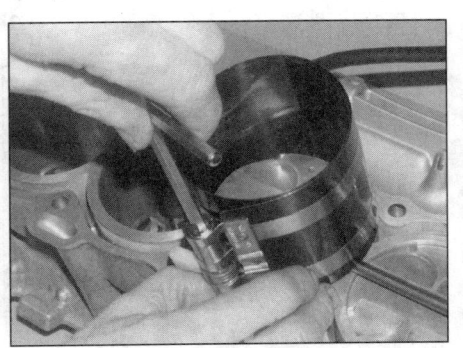

25.25c Fit the compressor over the piston and rings and compress the rings by tightening the bands on the compressor using an Allen key

25.25d Tap the top of the piston with a soft tool so that it enters the bore

26 Pistons – removal, inspection and installation

Note: *To remove the pistons the engine must be removed from the frame and the crankcase halves separated.*

Removal

1 Remove the connecting rods (see Section 25).

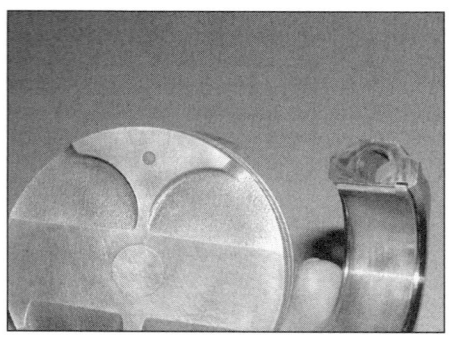

2•50 Engine, clutch and transmission

26.2 Note the IN or O mark on the piston which faces the intake side

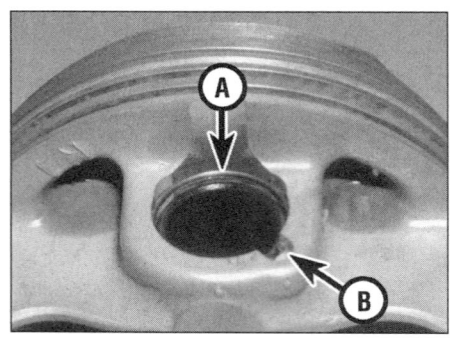

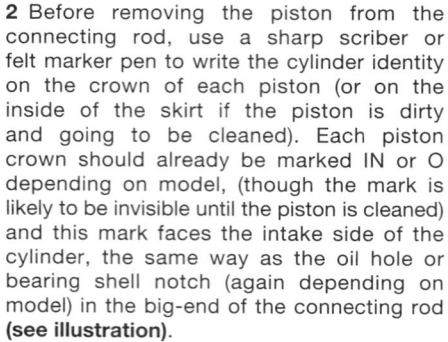

26.3a Prise out the circlip (A) using a suitable tool in the notch (B) . . .

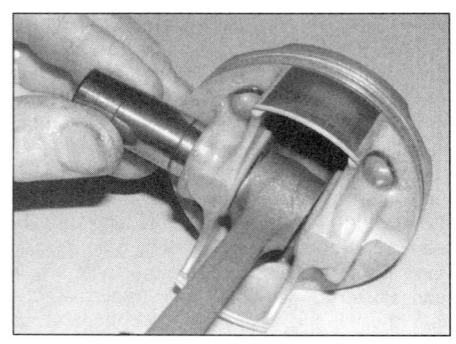

26.3b . . . then push out the pin and separate the piston from the rod

2 Before removing the piston from the connecting rod, use a sharp scriber or felt marker pen to write the cylinder identity on the crown of each piston (or on the inside of the skirt if the piston is dirty and going to be cleaned). Each piston crown should already be marked IN or O depending on model, (though the mark is likely to be invisible until the piston is cleaned) and this mark faces the intake side of the cylinder, the same way as the oil hole or bearing shell notch (again depending on model) in the big-end of the connecting rod **(see illustration)**.

3 Carefully prise out the circlip on one side of the piston using needle-nose pliers or a small flat-bladed screwdriver inserted into the notch **(see illustration)**. Push the piston pin out from the other side to free the piston from the connecting rod **(see illustration)**. Remove the other circlip and discard them as new ones must be used **(see illustration)**. When the piston has been removed, install its pin back into its bore so that related parts do not get mixed up.

> **HAYNES HiNT**
> If a piston pin is a tight fit in the piston bosses, soak a rag in boiling water then wring it out and wrap it around the piston – this will expand the alloy piston sufficiently to release its grip on the pin. If the piston pin is particularly stubborn, extract it using a drawbolt tool, but be careful to protect the piston's working surfaces.

Inspection

4 Using your thumbs or a piston ring removal and installation tool, carefully remove the rings from the pistons **(see illustrations 27.10, 27.9, 27.7c, 27.7b and 27.7a)**. Do not nick or gouge the pistons in the process. Carefully note which way up each ring fits and in which groove as they must be installed in their original positions if being reused. The upper surface of the top ring should be marked with the letter R at one end, and the second (middle) ring marked RN. The top and middle rings can also be identified by the fact that the top ring is narrower in width than the second (middle) ring, and their cross-section profiles are different.

5 Scrape all traces of carbon from the tops of the pistons. A hand-held wire brush or a piece of fine emery cloth can be used once most of the deposits have been scraped away. Do not, under any circumstances, use a wire brush mounted in a drill motor to remove deposits from the pistons; the piston material is soft and will be eroded away by the wire brush.

6 Use a piston ring groove cleaning tool to remove any carbon deposits from the ring grooves. If a tool is not available, a piece broken off an old ring will do the job. Be very careful to remove only the carbon deposits. Do not remove any metal and do not nick or gouge the sides of the ring grooves.

7 Once the deposits have been removed, clean the pistons with solvent and dry them thoroughly. If the identification mark previously made on the piston is cleaned off, be sure to re-mark it with the correct identity. Make sure the oil return holes below the oil ring groove are clear.

8 Carefully inspect each piston for cracks around the skirt, at the pin bosses and at the ring lands. Normal piston wear appears as even, vertical wear on the thrust surfaces of the piston. If the skirt is scored or scuffed, the engine may have been suffering from overheating and/or abnormal combustion, which caused excessively high operating temperatures. The oil pump should be checked thoroughly. Also check that the circlip grooves are not damaged.

9 A hole in the piston crown, an extreme to be sure, is an indication that abnormal combustion (pre-ignition) was occurring. Burned areas at the edge of the piston crown are usually evidence of spark knock (detonation). If any of the above problems exist, the causes must be corrected or the damage will occur again.

10 Measure the piston ring-to-groove clearance by laying each piston ring in its groove and slipping a feeler gauge in beside it **(see illustration)**. Make sure you have the correct ring for the groove (see Step 4). Check the clearance at three or four locations around the groove. If the clearance is greater than specified, renew both the piston and rings as a set. If new rings are being used, measure the clearance using the new rings. If the clearance is greater than that specified, the piston is worn and must be replaced with a new one.

11 Check the piston-to-bore clearance by measuring the bore (see Section 23) and the piston diameter. Make sure each piston is matched to its correct cylinder. Refer to the Specifications at the beginning of the Chapter and measure the piston the correct distance

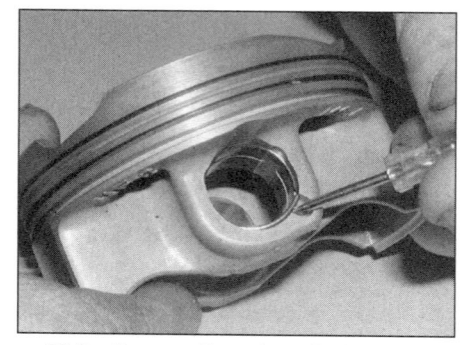

26.3c Remove the other clip from the other side

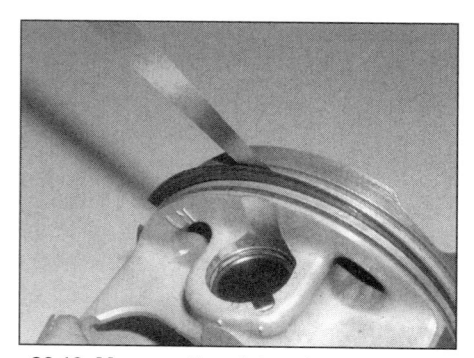

26.10 Measure the piston ring-to-groove clearance with a feeler gauge

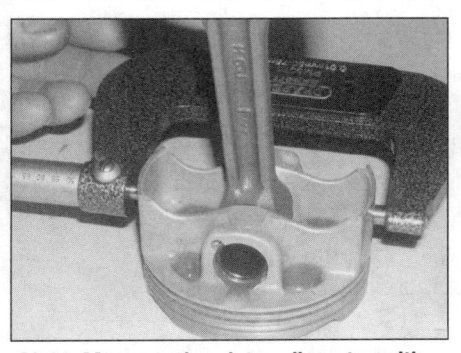

Engine, clutch and transmission 2•51

26.11 Measure the piston diameter with a micrometer at the specified distance from the bottom of the skirt

26.12a Slip the pin into the piston and try to rock it back-and-forth. If it's loose, replace the piston and pin

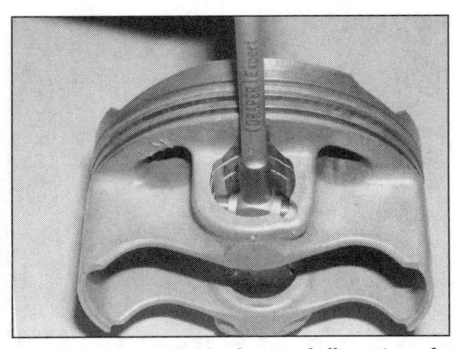

26.12b Measure the internal diameter of the bore in the piston

up from the bottom of the skirt and at 90° to the piston pin axis **(see illustration)**. Subtract the piston diameter from the bore diameter to obtain the clearance. If it is greater than the specified figure, the piston must be replaced with a new one (assuming the bore itself is within limits).

12 Apply clean engine oil to the piston pin, insert it into the piston and check for any freeplay between the two **(see illustration)**. Measure the pin external diameter **(see illustration 25.9b)**, and the pin bore in the piston **(see illustration)**. Calculate the difference to obtain the piston pin-to-piston pin bore clearance. Compare the result to the specifications at the beginning of the Chapter. If the clearance is greater than specified, replace the components that are worn beyond their specified limits. If not already done (see Section 25), repeat the measurements between the pin and the connecting rod small-end **(see illustration 25.9c)**.

Installation

13 Inspect and install the piston rings (see Section 27).
14 Lubricate the piston pin, the piston pin bore and the connecting rod small-end bore with molybdenum disulphide oil (a 50/50 mixture of molybdenum disulphide grease and clean engine oil).

15 When installing the pistons onto the connecting rods, on RR-Y and RR-1 (2000 and 2001) models make sure the IN mark on the piston crown faces the same way as the oil hole in the big-end of the connecting rod, and on RR-2 and RR-3 (2002 and 2003) models make sure the O mark faces the same way as the bearing shell notch in the big-end **(see illustration 26.2)**.
16 Install a *new* circlip in one side of the piston (do not reuse old circlips) **(see illustration 26.3c)**. Line up the piston on its correct connecting rod, and insert the piston pin from the other side **(see illustration 26.3b)**. Secure the pin with the other *new* circlip **(see illustration)**. When installing the circlips, compress them only just enough to fit them in the piston, and make sure they are properly seated in their grooves with the open end away from the removal notch.
17 Install the connecting rods (see Section 25) and reassemble the crankcase halves (see Section 22).

27 Piston rings – inspection and installation

1 It is good practice to renew the piston rings when an engine is being overhauled. Before installing the new rings, check the end gaps with the rings installed in the bore, as follows.

2 If new rings are being used, lay out each piston with a new ring set and keep them together so the rings will be matched with the same piston and bore during the end gap measurement procedure and engine assembly. If the old rings are being reused, make sure they are matched with their correct piston and cylinder.

3 To measure the installed ring end gap, insert the top ring into the top of the bore and square it up with the bore walls by pushing it in with the top of the piston. The ring should be about 20 mm below the top edge of the bore. Slip a feeler gauge between the ends of the ring and compare the measurement to the specifications at the beginning of the Chapter **(see illustration)**.

4 If the gap is larger or smaller than specified, double check to make sure that you have the correct rings before proceeding; excess end gap is not critical unless it exceeds the service limit. If the service limit is exceeded with new rings, check the bore for wear (see Section 23). If the gap is too small, the ring ends may come in contact with each other during engine operation, which can cause serious damage.

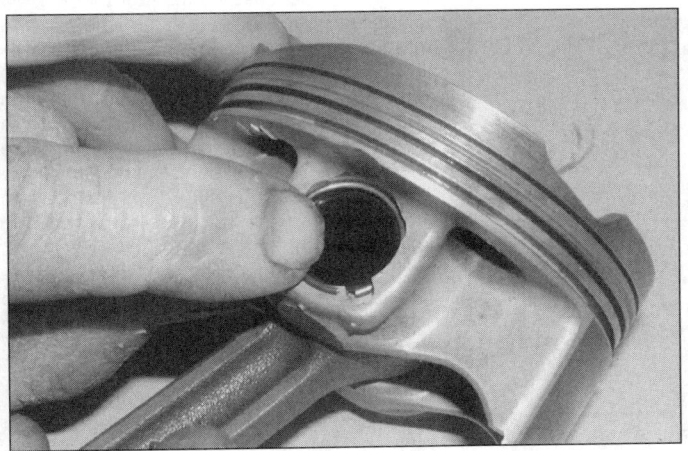

26.16 Make sure the clips locate in their groove

27.3 Measuring piston ring installed end gap

2•52 Engine, clutch and transmission

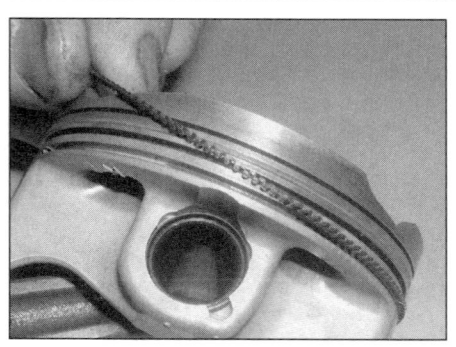

27.7a Install the oil ring expander in its groove . . .

27.7b . . . then fit the lower side-rail . . .

27.7c . . . and the upper side-rail on each side of it

27.9 Install the middle ring . . .

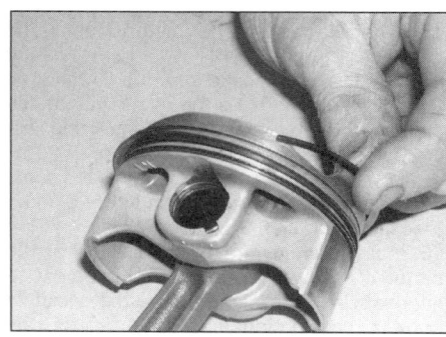

27.10 . . . and the top ring as described

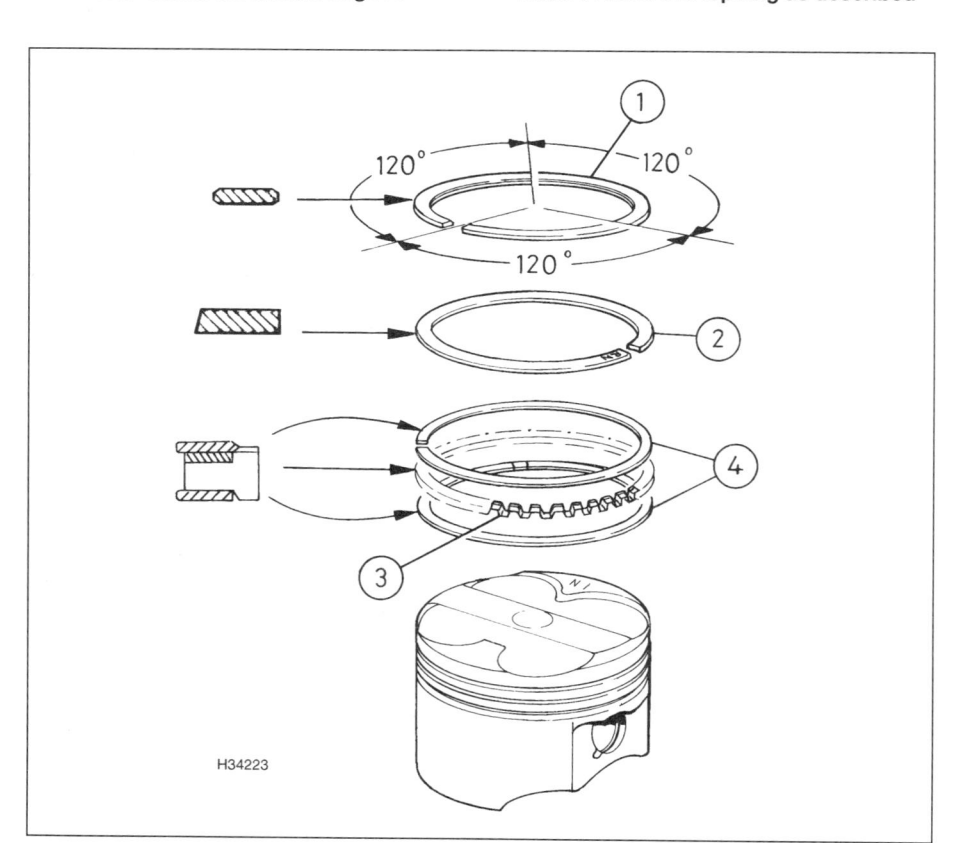

27.11 Piston ring installation details – stagger the ring end gaps as shown
1 Top ring   2 Middle ring   3 Oil ring expander   4 Oil ring side rails

5 Repeat the procedure for the middle ring and the oil control ring side-rails, but not the expander ring. Remember to keep the rings, pistons and bores matched up.

6 Once the ring end gaps have been checked/corrected, the rings can be installed on the pistons **(see illustration 27.11)**.

7 Install the oil control ring (lowest on the piston) first. It is composed of three separate components, namely the expander and the upper and lower side-rails. Slip the expander into the groove, making sure the ends don't overlap, then install the lower side-rail **(see illustrations)**. Do not use a piston ring installation tool on the side-rails as they may be damaged. Instead, place one end of the side-rail into the groove between the expander and the ring land. Hold it firmly in place and slide a finger around the piston while pushing the rail into the groove. Next, install the upper side-rail in the same manner **(see illustration)**. Check that the ends of the expander have not overlapped.

8 After the three oil ring components have been installed, check to make sure that both the upper and lower side-rails can be turned smoothly in the ring groove.

9 The upper surface of the top ring should be marked with the letter R at one end, and the second (middle) ring marked RN **(see illustration 27.11)**. The top and middle rings can also be identified by the fact that the top ring is narrower in width than the second (middle) ring, and by their different cross-section profiles. Install the second (middle) ring next. Make sure that the identification letter near the end gap is facing up. Fit the ring into the middle groove in the piston **(see illustration)**. Do not expand the ring any more than is necessary to slide it into place. To avoid breaking the ring, use a piston ring installation tool.

10 Finally, install the top ring in the same manner into the top groove in the piston **(see illustration)**. Make sure the identification letter near the end gap is facing up.

11 Once the rings are correctly installed, check they move freely without snagging and stagger their end gaps as shown **(see illustration)**.

Engine, clutch and transmission 2•53

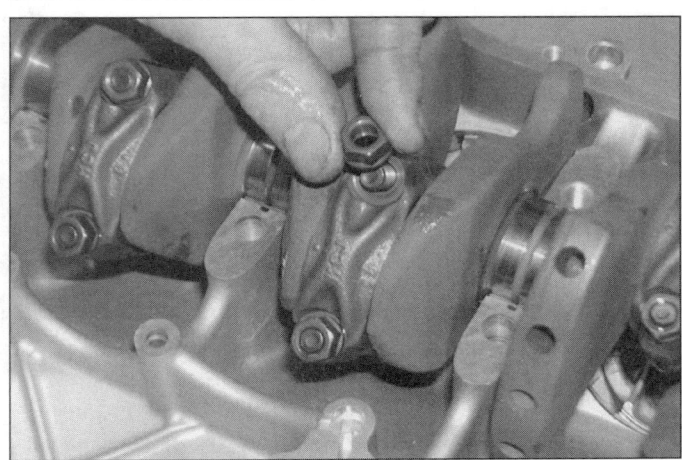

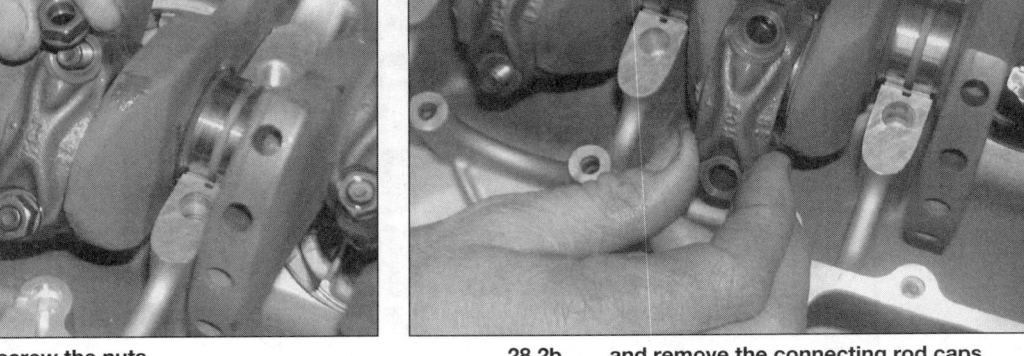

28.2a Unscrew the nuts . . .

28.2b . . . and remove the connecting rod caps

28 Crankshaft and main bearings – removal, inspection and installation

Note: *To remove the crankshaft the engine must be removed from the frame and the crankcase halves separated.*

Removal

1 Remove the engine from the frame (see Section 5) and separate the crankcase halves (see Section 22).

2 Using paint or a felt marker pen, mark the relevant cylinder identity on each connecting rod and cap. Mark across the cap-to-connecting rod join and note which side of the cap faces the front of the engine to ensure that they are fitted the correct way around onto the correct rod on reassembly. Note that the number already across the rod and cap indicates rod size grade **(see illustration 25.21b)**. Unscrew the connecting rod cap nuts and separate the caps from the crankpin (see illustrations). Push the rods and pistons up to the tops of the bores so that the bottom ends are clear of the crankshaft **(see illustration)**. **Note:** *If no work is to be carried out on the piston/connecting rod assemblies there is no need to remove them from the bores. If you do remove them, refer to Section 25.*

3 Lift the crankshaft out of the upper crankcase half, bringing the cam chain with it if it hasn't been removed, and taking care not to dislodge the main bearing shells **(see illustration)**.

4 The main bearing shells can be removed from the crankcase halves by pushing their centres to the side, then lifting them out **(see illustration 25.10)**. Keep the shells in order. On RR-2 and RR-3 (2002 and 2003) models remove the oil jets from the bearing housings in the upper crankcase half **(see illustration)**. Discard the O-rings as new ones must be used.

Inspection

5 Clean the crankshaft with solvent, squirting

28.2c Push the rods off the crankpins so they are clear

it under pressure through all the oil passages. On RR-2 and RR-3 (2002 and 2003) models also clean through the oil jets. If available, blow the crank dry with compressed air, and also blow through the oil passages. Check the primary drive gear and its sub-gear for wear or damage. If any of the main gear teeth are excessively worn, chipped or broken, the crankshaft must be replaced with a new one.

28.3 Lift the crankshaft out of the crankcase

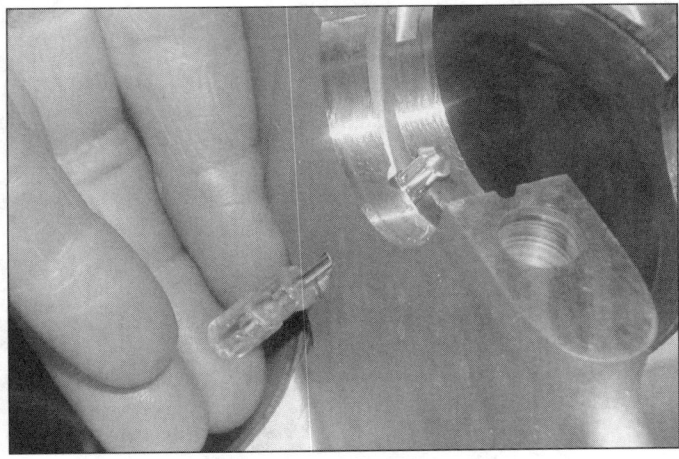

28.4 Push the oil jets up from the bottom

2•54 Engine, clutch and transmission

28.6a Remove the circlip . . .

28.6b . . . and the washer . . .

28.6c . . . then lever the sub-gear up, having noted its alignment, and remove it, noting how it locates

The sub-gear is available separately (see Step 6). If wear or damage is found, also inspect the primary driven gear on the back of the clutch housing (see Section 16).

6 To separate the sub-gear from the main gear, remove the large circlip and the sprung washer, noting which way round they fit **(see illustrations)**. Discard the circlip as a new one must be used. Remove the sub-gear, noting how one of its holes aligns with the hole in the main gear – mark this alignment as an aid to installation **(see illustration)**. Remove the springs, and if they are loose the stopper pins, noting how they fit **(see illustration)**. Check the springs for sag and distortion. When fitting the sub-gear back on, apply some molybdenum disulphide oil (a 50/50 mixture of molybdenum disulphide grease and clean engine oil) to the contact areas between the main gear, the sub gear and the sprung washer. Align the sub-gear as noted on removal, then butt the lugs on its inner face against the spring ends and twist it to compress the springs while pushing down until the gear is felt to locate **(see illustration)**. Fit the sprung washer so that its inner rim is raised off the sub-gear and facing out **(see illustration)**. Fit the new circlip with its large tab on the right as you look at it with the open ends at the top, and positioned so that the ends are at a right-angle to the cut-outs in the end of the crankshaft **(see illustrations)**.

7 Refer to Section 24 and examine the main bearing shells. If they are scored, badly scuffed or appear to have been seized, new bearings must be installed. Always replace the main bearings as a set. If they are badly damaged, check the corresponding crankshaft journals. Evidence of extreme heat, such as discoloration, indicates that lubrication failure has occurred. Be sure to thoroughly check the oil pump and pressure regulator as well as all oil holes and passages before reassembling the engine.

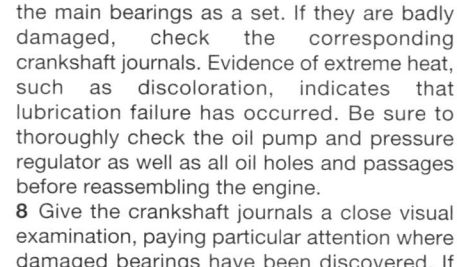

8 Give the crankshaft journals a close visual examination, paying particular attention where damaged bearings have been discovered. If the journals are scored or pitted in any way a new crankshaft will be required. Note that undersizes are not available, precluding the option of regrinding the crankshaft.

9 Place the crankshaft on V-blocks and check the runout at the main bearing journals using a dial gauge. Compare the reading to the maximum specified at the beginning of the Chapter. If the runout exceeds the limit, the crankshaft must be replaced with a new one.

Oil clearance check

10 Whether new bearing shells are being fitted or the original ones are being reused, the main bearing oil clearance should be checked before the engine is reassembled. Main bearing oil clearance is measured with a product known as Plastigauge.

11 Clean the backs of the bearing shells and the bearing housings in both crankcase halves.

12 Press the bearing shells into their cut-outs, ensuring that the tab on each shell

28.6d Note the layout of the springs

28.6e Align the gear as noted on removal, butting the lugs against the spring ends

28.6f Fit the washer with its inner rim raised up

28.6g Press the circlip down against the sprung washer so it locates in its groove . . .

28.6h . . . and position the open ends perpendicular to the flats on its groove rim

Engine, clutch and transmission 2•55

28.21a Main bearing journal size numbers

28.21b Main bearing housing size letters

engages in the notch in the crankcase (see illustration 28.24a). Make sure the bearings are fitted in the correct locations and take care not to touch any shell's bearing surface with your fingers.

13 Ensure the shells and crankshaft are clean and dry. Lay the crankshaft in position in the upper crankcase (see illustration 28.3). Install the three crankcase dowels if removed (see illustration 22.8a).

14 Cut five lengths of the appropriate size Plastigauge (they should be slightly shorter than the width of the crankshaft journals). Place a strand of Plastigauge on each (cleaned) journal, avoiding the oil hole. Make sure the crankshaft is not rotated.

15 Carefully fit the lower crankcase half onto the upper half (see illustration 22.7). Check that the lower half is correctly seated. **Note:** *Do not tighten the crankcase bolts if the casing is not correctly seated.* Clean the threads of the ten 9 mm lower crankcase bolts. Insert the bolts in their original locations (see illustration 22.6d). Secure all bolts finger-tight at first. On RR-Y and RR-1 (2000 and 2001) models, tighten the bolts evenly and a little at a time in the numerical sequence shown to the torque setting specified at the beginning of the Chapter (see illustration 22.17). On RR-2 and RR-3 (2002 and 2003) models, tighten the bolts evenly in the numerical sequence shown to the initial torque setting specified at the beginning of the Chapter. Now tighten them in the same sequence to the second torque setting specified. Finally, using a degree disc, tighten each bolt in turn by a further 150°, again following the numerical sequence (see illustration 22.18). Make sure that the crankshaft is not rotated as the bolts are tightened.

16 Slacken each bolt evenly and a little at a time in a reverse of the tightening sequence, i.e. starting from the outside and working to the centre, until they are all finger-tight, then remove the bolts. Carefully lift off the lower crankcase half, making sure the Plastigauge is not disturbed.

17 Compare the width of the crushed Plastigauge on each crankshaft journal to the scale printed on the Plastigauge envelope to obtain the main bearing oil clearance. Compare the reading to the specifications at the beginning of the Chapter.

18 On completion carefully scrape away all traces of the Plastigauge material from the crankshaft journal and bearing shells; use a fingernail or other object which is unlikely to score them.

19 If the oil clearance falls into the specified range, no bearing shell replacement is required (provided they are in good condition). If the clearance is beyond the service limit, refer to the marks on the case and the marks on the crankshaft and select new bearing shells (see Steps 21 and 22). Install the new shells and check the oil clearance once again (the new shells may bring the bearing clearance within the specified range). Always renew all of the shells at the same time.

20 If the clearance is still greater than the service limit listed in this Chapter's Specifications (even with replacement shells), the crankshaft journals are worn and the crankshaft should be replaced with a new one.

Main bearing shell selection

21 Replacement bearing shells for the main bearings are supplied on a selected fit basis. Code letters and numbers stamped on the crankshaft and crankcase are used to identify the correct replacement bearings. The crankshaft main bearing journal size numbers are stamped on the outside of the left-hand crankshaft web and will be either a 0, a 1, a 2 or a 3 (see illustration). The first letter, after the L, is for the left-hand journal, and the numbers correspond consecutively for each journal. The corresponding main bearing housing size letters are stamped into the left-hand side of the upper crankcase half and will be either an A, a B or a C (see illustration). The left-hand letter corresponds to the left-hand journal, and the letters correspond consecutively from left to right.

22 A range of bearing shells is available. To select the correct bearing for a particular journal, use the table below and cross-refer the main bearing journal size number (stamped on the crank web) with the main bearing housing size letter (stamped on the crankcase) to determine the colour code of the bearing required. For example, if the journal code is 3, and the housing code is A, then the bearing required is yellow. The colour is marked on the side of the shell (see illustration 25.22).

	Main bearing housing code		
Main bearing journal code	A	B	C
0	G – White	F – Red	E – Pink
1	F – Red	E – Pink	D – Yellow
2	E – Pink	D – Yellow	C – Green
3	D – Yellow	C – Green	B – Brown

2•56 Engine, clutch and transmission

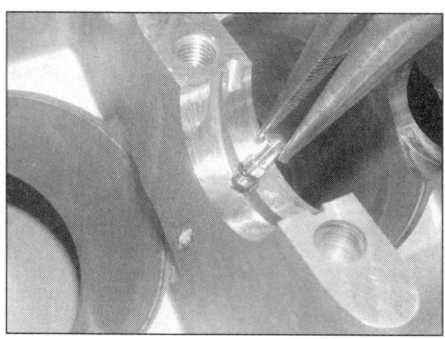

28.23 Fit the oil jets using new O-rings

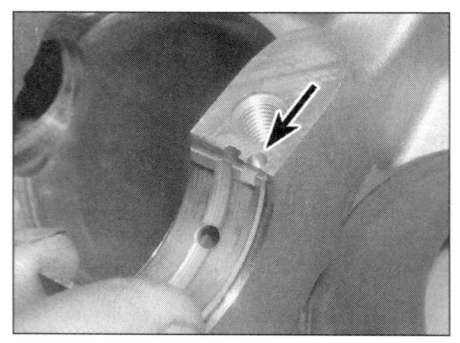

28.24a Fit the shells, locating the tabs in the notches (arrow) . . .

28.24b . . . and lubricate them with clean oil

Installation

23 Clean the backs of the bearing shells and the bearing cut-outs in both crankcase halves. If new shells are being fitted, ensure that all traces of the protective grease are cleaned off using paraffin (kerosene). Wipe the shells and crankcase halves dry with a lint-free cloth. Make sure all the oil passages and holes are clear, and blow them through with compressed air if it is available. On RR-2 and RR-3 (2002 and 2003) models install the oil jets with new O-rings smeared with oil in the bearing housings in the upper crankcase half **(see illustration)**.

24 Press the bearing shells into their locations. Make sure the tab on each shell engages in the notch in the casing **(see illustration)**. Make sure the bearings are fitted in the correct locations and take care not to touch any shell's bearing surface with your fingers. Lubricate each shell with molybdenum disulphide oil (a 50/50 mixture of molybdenum disulphide grease and clean engine oil) **(see illustration)**.

25 Lower the crankshaft into position in the upper crankcase, making sure all bearings remain in place **(see illustration 28.3)**.

26 Lubricate the crankpin with molybdenum disulphide oil (a 50/50 mixture of molybdenum disulphide grease and clean engine oil). Pull the connecting rod onto the crankpin and fit the cap onto the rod **(see illustration and 28.2b)**. Make sure the cap is fitted the correct way around so the previously made markings align, and that the rod is facing the right way (see Step 2 and Section 25).

27 Apply some clean oil to the threads and under the heads of the connecting rod nuts, then fit them and tighten them finger-tight **(see illustration 28.2a)**. Now tighten them evenly and alternately and a little at a time to the torque setting specified at the beginning of the Chapter **(see illustration)**. It is highly advisable to have an assistant to hold the crankshaft down in the crankcase while tightening the bolts as it could jump out.

28 Check that the crankshaft is free to rotate easily, then install the other connecting rods in the same way. Check to make sure that all components have been returned to their original locations using the marks made on disassembly.

29 Check that the rods rotate smoothly and freely on the crankpins. If there are any signs of roughness or tightness, remove the rods and recheck the bearing clearance. Sometimes tapping the bottom of the connecting rod cap will relieve tightness, but if in doubt, recheck the clearances.

30 Reassemble the crankcase halves (see Section 22).

29 Transmission shafts and bearings – removal and installation

Note: *To remove the transmission shafts the engine must be removed from the frame and the crankcases separated.*

Removal

1 Remove the engine from the frame and separate the crankcase halves (see Section 22).

2 Lift the output shaft and input shaft out of the casing, noting their relative positions, how they fit together, and how the hole in each needle bearing housing locates onto the

28.26 Pull the connecting rod onto the crankpin and fit the cap

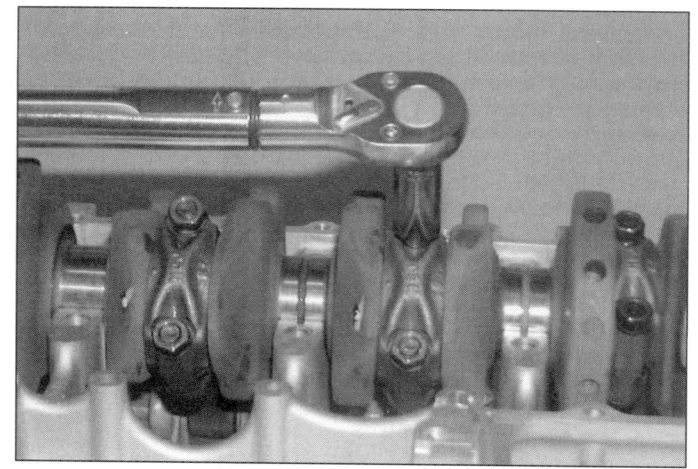

28.27 Tighten the nuts as described to the specified torque

Engine, clutch and transmission 2•57

29.2a Remove the output shaft . . .

29.2b . . . and the input shaft, noting how the bearings locate (arrows)

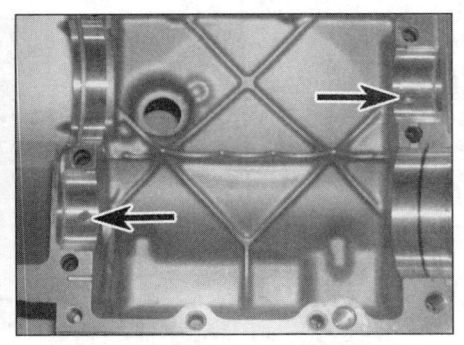

29.2c Remove the bearing dowels (arrowed) from their holes . . .

dowel **(see illustrations)**. If the shafts are stuck, use a soft-faced hammer and gently tap on the ends of the shafts to free them. Remove the needle bearing dowels – if they are not in their holes in the crankcase, remove them from the bearings on the shafts **(see illustration)**. Also remove the output shaft ball bearing half-ring retainer from either the crankcase or the bearing **(see illustration)**. If necessary, the input shaft and output shaft can be disassembled and inspected for wear or damage (see Section 30).

3 Referring to *Tools and Workshop Tips* (Section 5) in the Reference Section, check the bearings on the transmission shafts. Replace the bearings with new ones if necessary, noting that the left-hand bearing on the output shaft is not available separately from the shaft. Also check the condition of the output shaft oil seal and discard it if worn or damaged – note that a new one should be fitted as a matter of course **(see illustration 29.5)**.

Installation

4 Fit the needle bearing dowels into their holes in the crankcase, and the half-ring retainer into its slot **(see illustrations 29.2c and 29.2d)**.

5 Lubricate the left-hand end of the output shaft with clean oil and slide the output shaft oil seal on, preferably using a new one whatever the condition of the old one **(see illustration)**. Smear the seal outer lips with oil.

6 Lower the input shaft into position in the upper crankcase, making sure the hole in the needle bearing engages correctly with the dowel **(see illustration 29.2b)**.

7 Lower the output shaft into position in the upper crankcase **(see illustration 29.2a)**, making sure the hole in the needle bearing engages correctly with the dowel, the groove in the ball bearing engages correctly with the retainer in the bearing housing, and the locating pin sits in the cut-out in the crankcase **(see illustration)**. Also ensure that the oil seal lip locates in the crankcase groove.

Caution: *If the ring retainer or dowel do not locate correctly, the crankcase halves will not seat properly.*

8 Make sure both transmission shafts are correctly seated and their related pinions are correctly engaged.

9 Position the gears in the neutral position and check the shafts are free to rotate easily and independently (i.e. the input shaft can turn whilst the output shaft is held stationary) before proceeding further.

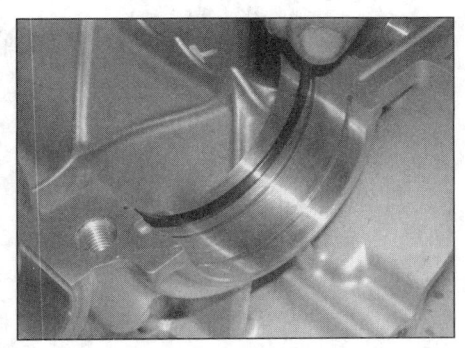

29.2d . . . and the bearing retainer from its groove

10 Reassemble the crankcase halves (see Section 22).

30 Transmission shafts – disassembly, inspection and reassembly

1 Remove the transmission shafts from the crankcase (see Section 29). Always disassemble the transmission shafts separately to avoid mixing up the components.

29.5 Lubricate the end of the shaft and fit the oil seal

29.7 Make sure the ring retainer (A), the bearing pin (B) and the oil seal lip (C) locate correctly as shown

2•58 Engine, clutch and transmission

30.2 Transmission input shaft components

1. Input shaft and bearing
2. Needle roller bearing outer race
3. Needle roller bearing
4. Thrust washer
5. 2nd gear pinion
6. Tabbed lockwasher
7. Slotted splined washer
8. 6th gear pinion splined bush
9. 6th gear pinion
10. Splined washer
11. Circlip
12. Combined 3rd/4th gear pinion
13. Circlip
14. Splined washer
15. 5th gear pinion bush
16. 5th gear pinion
17. Thrust washer

HAYNES HINT: When disassembling the transmission shafts, place the parts on a long rod or thread a wire through them to keep them in order and facing the proper direction.

Input shaft

Disassembly

2 Slide the outer race and needle bearing off the left-hand end of the shaft, followed by the thrust washer and the 2nd gear pinion – note which way around the pinion is fitted **(see illustration and 30.20d, 30.20c, 30.20b and 30.20a)**.

3 Slide the tabbed lockwasher off the shaft, then turn the slotted splined washer to offset the splines and slide it off the shaft, noting how they fit together **(see illustrations 30.19c, 30.19b and 30.19a)**. Slide the 6th gear pinion and its splined bush off the shaft, followed by the splined washer **(see illustrations 30.18c, 30.18b and 30.18a)**.

4 Remove the circlip securing the combined 3rd/4th gear pinion, then slide the pinion off the shaft **(see illustrations 30.17b and 30.17a)**.

5 Remove the circlip securing the 5th gear pinion, then slide the splined washer, the pinion and its bush, and the thrust washer off the shaft **(see illustrations 30.16e, 30.16d, 30.16c, 30.16b and 30.16a)**. The 1st gear pinion is integral with the shaft **(see illustration 30.11a)**.

6 If required, remove the caged ball bearing from the right-hand end of the shaft, referring to *Tools and Workshop Tips* in the Reference Section **(see illustration 30.15)**.

Inspection

7 Wash all of the components in clean solvent and dry them off.

8 Check the gear teeth for cracking, chipping, pitting and other obvious wear or damage. Any pinion that is damaged as such must be replaced.

9 Inspect the dogs and the dog holes in the gears for cracks, chips, and excessive wear especially in the form of rounded edges. Make sure mating gears engage properly. Replace the paired gears as a set if necessary.

10 Check for signs of scoring or bluing on the pinions, bushes and shaft. This could be caused by overheating due to inadequate lubrication. Check that all the oil holes and passages are clear. Replace any damaged pinions or bushes.

11 Check that each pinion moves freely on the shaft or its bush but without undue freeplay. Check that each bush moves freely on the shaft but without undue freeplay. If the necessary equipment is available the individual components listed can be measured and the results compared with the specifications at the beginning of this Chapter **(see illustrations)**.

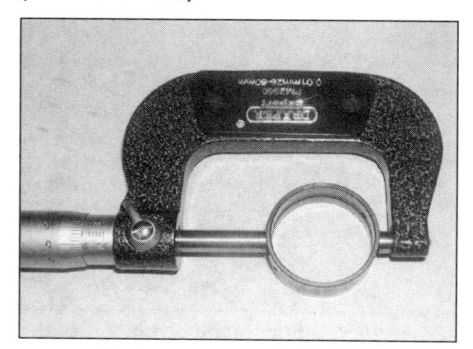

30.11a Measure the shaft OD . . .

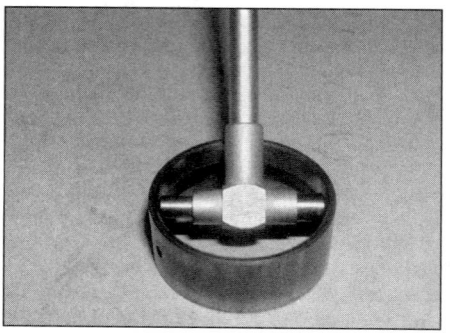

30.11b . . . the bush ID . . .

30.11c . . . the bush OD . . .

Engine, clutch and transmission 2•59

30.11d ... and the gear ID on the relevant components as listed in the Specifications

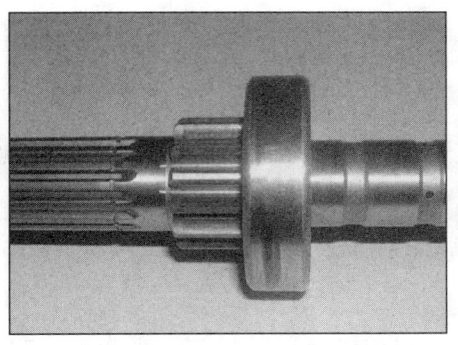

30.15 Fit the bearing onto the shaft as shown

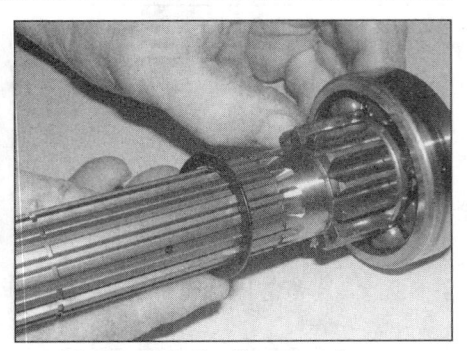

30.16a Slide the thrust washer ...

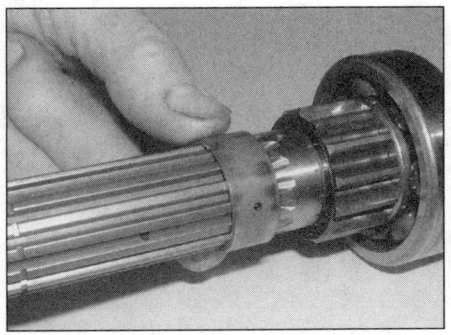

30.16b ... the 5th gear pinion bush ...

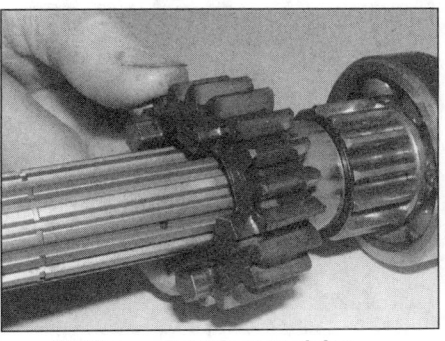

30.16c ... the 5th gear pinion ...

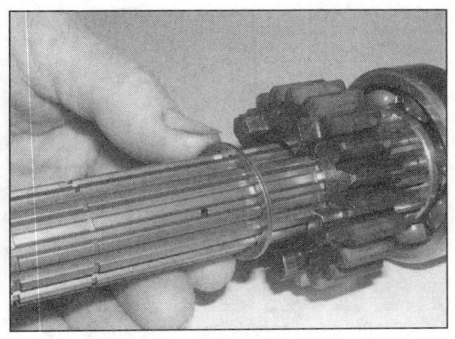

30.16d ... and the splined washer onto the shaft ...

12 The shaft is unlikely to sustain damage unless the engine has seized, placing an unusually high loading on the transmission, or the machine has covered a very high mileage. Check the surface of the shaft, especially where a pinion turns on it, and replace the shaft if it has scored or picked up, or if there are any cracks. Damage of any kind can only be cured by replacement.

13 Check the washers and circlips and replace any that are bent or appear weakened or worn. Use new ones if in any doubt. Note that it is good practice to renew all circlips when overhauling gearshafts.

Reassembly

14 During reassembly, apply molybdenum disulphide oil (a 50/50 mixture of molybdenum disulphide grease and clean engine oil) to the mating surfaces of the shaft, pinions and bushes. When installing the circlips, do not expand their ends any further than is necessary. Install the stamped circlips and washers so that their chamfered side faces the pinion it secures.

15 If removed, fit the caged ball bearing onto the right-hand end of the shaft, referring to *Tools and Workshop Tips* in the Reference Section **(see illustration)**.

16 Slide the thrust washer onto the left-hand end of the shaft, followed by the 5th gear pinion bush, aligning the oil hole in the bush with the hole in the shaft **(see illustrations)**. Fit the 5th gear pinion with its dogs facing away from the integral 1st gear **(see illustration)**. Slide the splined washer onto the shaft, then fit the circlip, making sure that it locates correctly in the groove in the shaft **(see illustrations)**.

17 Slide the combined 3rd/4th gear pinion onto the shaft with the larger 4th gear pinion facing the 5th gear pinion **(see illustration)**. Fit the circlip, making sure it is locates

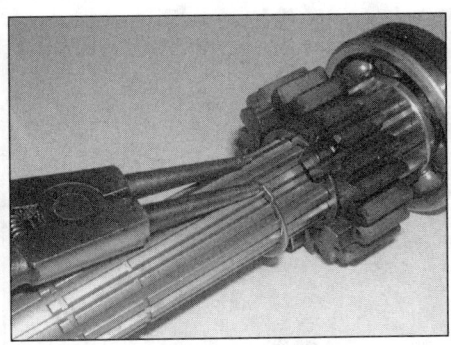

30.16e ... and secure them with the circlip ...

30.16f ... making sure it locates properly in its groove

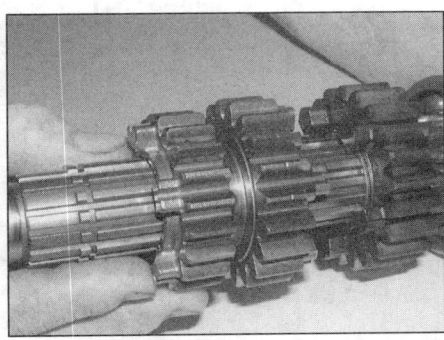

30.17a Slide the combined 3rd/4th gear pinion onto the shaft ...

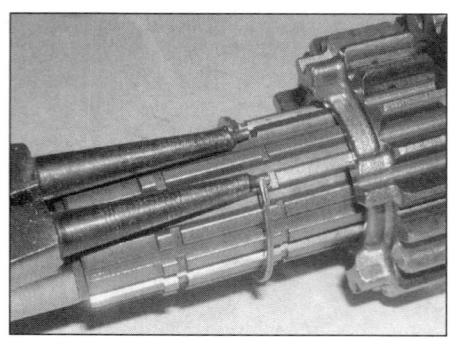

2•60 Engine, clutch and transmission

30.17b . . . and secure it with the circlip . . .

30.17c . . . making sure it locates properly in its groove

30.18a Slide the splined washer . . .

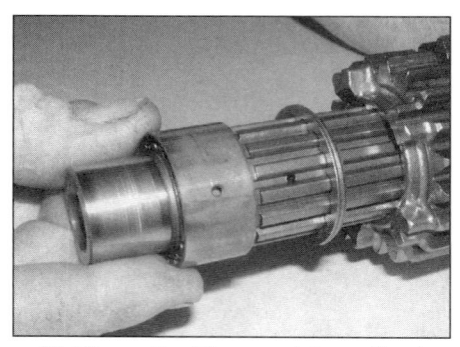

30.18b . . . the 6th gear pinion splined bush . . .

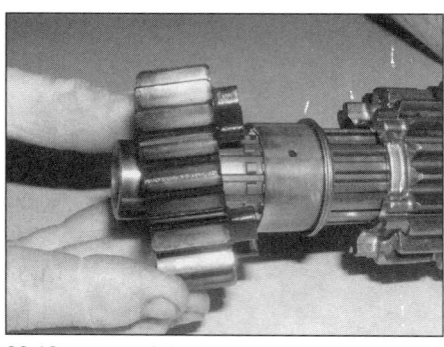

30.18c . . . and the 6th gear pinion onto the shaft

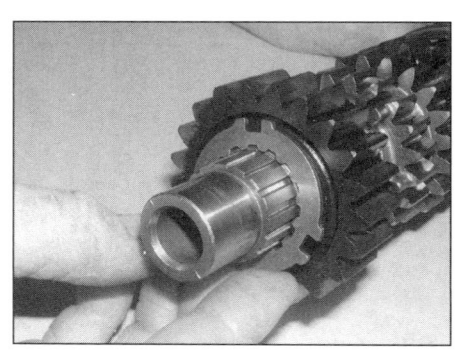

30.19a Slide on the slotted splined washer . . .

correctly in its groove in the shaft **(see illustrations)**.

18 Slide the splined washer onto the shaft, followed by the 6th gear pinion splined bush, aligning the oil hole in the bush with the hole in the shaft, and the 6th gear pinion, making sure its dogs face the 3rd/4th gear pinion **(see illustrations)**.

19 Slide the slotted splined washer onto the shaft and locate it in its groove, then turn it in the groove so that the splines on the washer align with the splines on the shaft and secure the washer in the groove **(see illustrations)**.

Slide the tabbed lockwasher onto the shaft, so that the tabs locate into the slots in the outer rim of the splined washer **(see illustrations)**.

20 Slide the 2nd gear pinion onto the end of the shaft with the chamfered side of the teeth facing the 6th gear pinion. Fit the thrust washer, followed by the needle roller bearing and its outer race over the end of the shaft **(see illustrations)**.

21 Check that all components have been correctly installed **(see illustration)**.

30.19b . . . and locate it as shown . . .

30.19c . . . then slide on the tabbed lockwasher . . .

30.19d . . . and locate it as shown

Engine, clutch and transmission 2•61

30.20a Slide the 2nd gear pinion . . .

30.20b . . . the thrust washer . . .

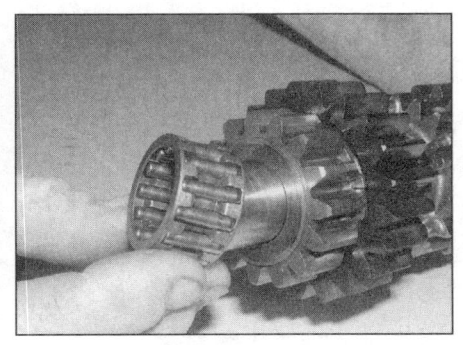

30.20c . . . the needle bearing . . .

30.20d . . . and the outer race onto the shaft

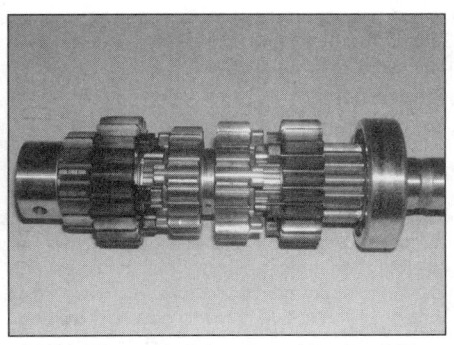

30.21 The complete assembly should be as shown

Output shaft

Disassembly

22 Slide the needle bearing off the right-hand end of the shaft (see accompanying illustration and illustration 30.37d).

23 Slide the thrust washer off the shaft, followed by the 1st gear pinion and its needle roller bearing, the thrust washer and the 5th gear pinion (see illustrations 30.37c, 30.37b, 30.37a, 30.36b and 30.36a).

24 Remove the circlip securing the 4th gear pinion, then slide the splined washer, the pinion and its splined bush off the shaft (see illustrations 30.35d, 30.35c, 30.35b and 30.35a).

1 Output shaft and bearing
2 Needle roller bearing outer race
3 Needle roller bearing
4 Thrust washer
5 1st gear pinion
6 Needle roller bearing
7 Thrust washer
8 5th gear pinion
9 Circlip
10 Splined washer
11 4th gear pinion
12 4th gear pinion splined bush
13 Tabbed lockwasher
14 Slotted splined washer
15 3rd gear pinion splined bush
16 3rd gear pinion
17 Splined washer
18 Circlip
19 6th gear pinion
20 Circlip
21 Splined washer
22 2nd gear pinion
23 2nd gear pinion bush

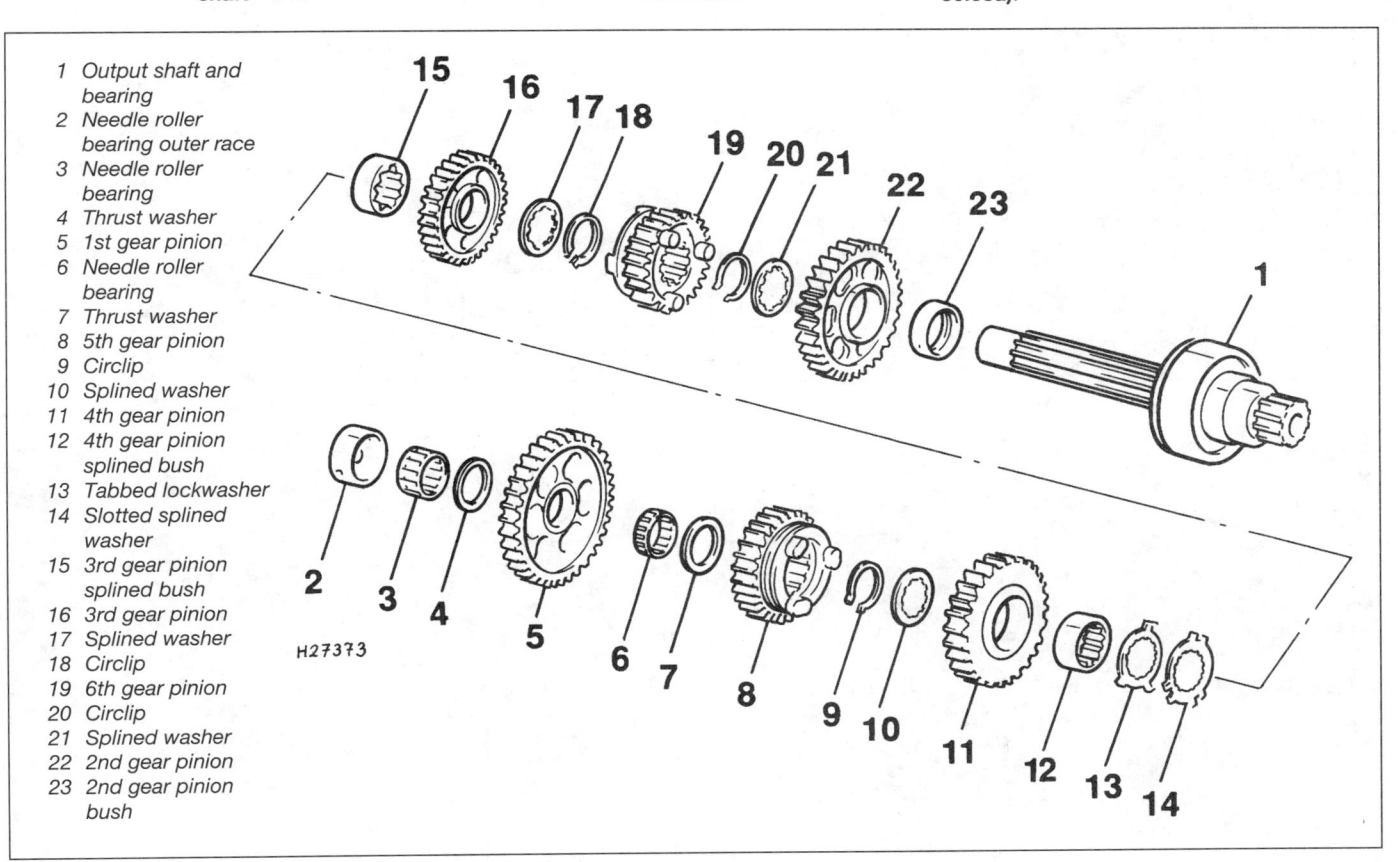

30.22 Transmission output shaft components

2•62 Engine, clutch and transmission

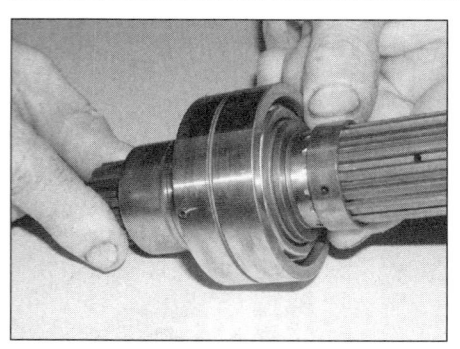

30.31a Slide the 2nd gear pinion bush . . .

30.31b . . . the 2nd gear pinion . . .

30.31c . . . and the splined washer onto the shaft . . .

25 Slide the tabbed lockwasher off the shaft, then turn the slotted splined washer to offset the splines and slide it off the shaft, noting how they fit together (see illustrations 30.34c, 30.34b and 30.34a).

26 Slide the 3rd gear pinion and its splined bush, followed by the splined washer, off the shaft (see illustrations 30.33c, 30.33b and 30.33a).

27 Remove the circlip securing the 6th gear pinion, then slide the pinion off the shaft (see illustrations 30.32b and 30.32a).

28 Remove the circlip securing the 2nd gear pinion, then slide the splined washer, the pinion and its bush off the shaft (see illustrations 30.31d, 30.31c, 30.31b and 30.31a).

Inspection

29 Refer to Steps 7 to 13 above.

Reassembly

30 During reassembly, apply engine oil to the mating surfaces of the shaft, pinions and bushes. When installing the circlips, do not expand the ends any further than is necessary. Install the stamped circlips and washers so that their chamfered side faces the pinion it secures.

31 Slide the 2nd gear pinion bush onto the shaft, aligning the oil hole in the bush with the hole in the shaft, then slide on the 2nd gear pinion with its dog holes facing away from the bearing, followed by the splined washer (see illustrations). Fit the circlip, making sure it is locates correctly in its groove in the shaft (see illustrations).

32 Slide the 6th gear pinion with its selector fork groove facing away from the 2nd gear pinion, then fit the circlip, making sure it is locates correctly in its groove in the shaft (see illustrations).

30.31d . . . and secure them with the circlip . . .

30.31e . . . making sure it locates in the groove

30.32a Slide the 6th gear pinion onto the shaft . . .

30.32b . . . and secure it with the circlip

Engine, clutch and transmission 2•63

30.33a Slide the splined washer . . .

30.33b . . . the 3rd gear pinion splined bush . . .

30.33c . . . and the 3rd gear pinion onto the shaft

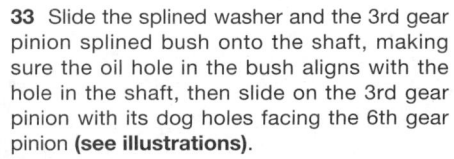

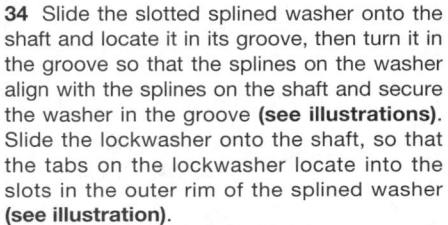

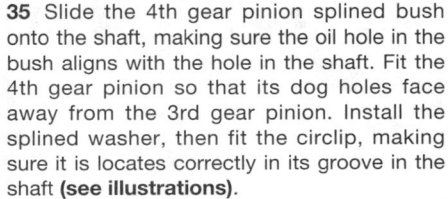

33 Slide the splined washer and the 3rd gear pinion splined bush onto the shaft, making sure the oil hole in the bush aligns with the hole in the shaft, then slide on the 3rd gear pinion with its dog holes facing the 6th gear pinion **(see illustrations)**.

34 Slide the slotted splined washer onto the shaft and locate it in its groove, then turn it in the groove so that the splines on the washer align with the splines on the shaft and secure the washer in the groove **(see illustrations)**. Slide the lockwasher onto the shaft, so that the tabs on the lockwasher locate into the slots in the outer rim of the splined washer **(see illustration)**.

35 Slide the 4th gear pinion splined bush onto the shaft, making sure the oil hole in the bush aligns with the hole in the shaft. Fit the 4th gear pinion so that its dog holes face away from the 3rd gear pinion. Install the splined washer, then fit the circlip, making sure it is locates correctly in its groove in the shaft **(see illustrations)**.

36 Slide the 5th gear pinion onto the shaft with its selector fork groove facing the 4th gear pinion, followed by the thrust washer **(see illustrations overleaf)**.

37 Slide the 1st gear pinion needle roller bearing onto the shaft, followed by the 1st gear pinion with its dog holes facing the 5th gear pinion. Fit the thrust washer then the

30.34a Slide the slotted splined washer onto the shaft . . .

30.34b . . . and locate it as shown

30.34c Slide the lockwasher onto the shaft and engage it with the slotted washer

30.35a Slide the 4th gear pinion splined bush . . .

30.35b . . . the 4th gear pinion . . .

30.35c . . . and the splined washer onto the shaft . . .

30.35d . . . and secure them with the circlip

2•64 Engine, clutch and transmission

30.36a Slide the 5th gear pinion . . .

30.36b . . . and the thrust washer onto the shaft

30.37a Slide the needle bearing . . .

needle roller bearing over the end of the shaft **(see illustrations)**.

38 Check that all components have been correctly installed **(see illustration)**.

31 Initial start-up after overhaul

1 Make sure the engine oil and coolant levels are correct (see *Daily (pre-ride) checks*). Make sure there is fuel in the tank.
2 Turn the engine kill switch to the ON position and shift the gearbox into neutral. Turn the ignition ON.
3 Start the engine and allow it to run at a moderately fast idle until it reaches operating temperature.

⚠️ **Warning: If the oil pressure warning light doesn't go off, or it comes on while the engine is running, stop the engine immediately.**

4 Check carefully for oil and coolant leaks and make sure the transmission and controls, especially the brakes, function properly before road testing the machine. Refer to Section 32 for the recommended running-in procedure.
5 Upon completion of the road test, and after the engine has cooled down completely, recheck the valve clearances (see Chapter 1) and check the engine oil and coolant levels (see *Daily (pre-ride) checks*).

32 Recommended running-in procedure

1 Treat the machine gently for the first few miles to make sure oil has circulated throughout the engine and any new parts installed have started to seat.
2 Even greater care is necessary if new pistons/rings or a new crankcase/bores have been fitted, and the bike will have to be run in as when new. This means greater use of the transmission and a restraining hand on the throttle until at least 300 miles (500 km) have been covered. There's no point in keeping to any

30.37b . . . the 1st gear pinion . . .

30.37d . . . then fit the bearing onto the end of the shaft

set speed limit – the main idea is to keep from labouring the engine and to gradually increase performance up to the 300 miles (500 km) mark. Experience is the best guide, since it's easy to tell when an engine is running freely.
3 If a lubrication failure is suspected, stop the engine immediately and try to find the cause. If an engine is run without oil, even for a short period of time, severe damage will occur.

30.37c . . . and the thrust washer onto the shaft . . .

30.38 The assembled shaft should be as shown

Chapter 3
Cooling system

Contents

Coolant hoses and unions – removal and installation 9	General information 1
Coolant level check ....see *Daily (pre-ride) checks*	Radiator – removal and installation 6
Coolant reservoir – removal and installation 8	Radiator pressure cap – check 2
Cooling fan and fan switch/relay – check and replacement 3	Temperature display and sensor – check and replacement 4
Cooling system checks ....see Chapter 1	Thermostat and housing – removal, check and installation 5
Cooling system draining, flushing and refilling ....see Chapter 1	Water pump – check, removal and installation 7

Degrees of difficulty

Easy, suitable for novice with little experience	Fairly easy, suitable for beginner with some experience	Fairly difficult, suitable for competent DIY mechanic	Difficult, suitable for experienced DIY mechanic	Very difficult, suitable for expert DIY or professional

Specifications

Coolant
Mixture type and capacity .... see Chapter 1

Cooling fan switch – RR-Y and RR-1 (2000 and 2001)
Switch closes (fan ON) .... 98 to 102°C
Switch opens (fan OFF) .... 93 to 97°C

Temperature gauge sensor
Resistance @ 80°C .... 2.1 to 2.6 K-ohms
Resistance @ 120°C .... 650 to 730 ohms

Thermostat
Opening temperature .... 80.5 to 83.5°C
Fully open .... 95°C
Valve lift .... 8 mm (min)

Radiator
Cap valve opening pressure .... 16 to 20 psi (1.1 to 1.4 Bar)

Torque settings
Cooling fan blade nut .... 3 Nm
Cooling fan motor nuts .... 5 Nm
Fan switch .... 18 Nm
Temperature sensor .... 23 Nm

3•2 Cooling system

1 General information

The cooling system uses a water/anti-freeze coolant to carry away excess heat from the engine and maintain as constant a temperature as possible. The cylinders are surrounded by a water jacket from which the heated coolant is circulated by thermo-syphonic action in conjunction with a water pump, which is driven by the oil pump. The hot coolant passes upwards to the thermostat and through to the radiator. The coolant then flows across the core of the radiator, then to the water pump and back to the engine where the cycle is repeated.

A thermostat is fitted in the system to prevent the coolant flowing through the radiator when the engine is cold, therefore accelerating the speed at which the engine reaches normal operating temperature. A dual circuit sensor (containing the temperature gauge sensor and the ECT (engine coolant temperature) sensor) mounted in the thermostat housing transmits information to the temperature gauge on the instrument panel, and to the ECM (electronic control module). A cooling fan fitted to the back of the radiator aids cooling in extreme conditions by drawing extra air through. On RR-Y and RR-1 (2000 and 2001) models the fan motor is controlled by a thermostatic switch fitted in the left-hand side of the radiator. On RR-2 and RR-3 (2002 and 2003) models the fan motor is controlled by a relay which receives a signal from the ECM which in turn receives information from the ECT sensor.

The complete cooling system is partially sealed and pressurised, the pressure being controlled by a valve contained in the spring-loaded radiator cap. By pressurising the coolant the boiling point is raised, preventing premature boiling in adverse conditions. The overflow pipe from the system is connected to a reservoir into which excess coolant is expelled under pressure. The discharged coolant automatically returns to the radiator by the vacuum created when the engine cools.

⚠️ *Warning: Do not remove the pressure cap from the radiator when the engine is hot. Scalding hot coolant and steam may be blown out under pressure, which could cause serious injury. When the engine has cooled, place a thick rag, like a towel, over the pressure cap; slowly rotate the cap anti-clockwise to the first stop. This procedure allows any residual pressure to escape. When the steam has stopped escaping, press down on the cap while turning it anti-clockwise and remove it. Caution: Do not allow anti-freeze to come in contact with your skin or painted surfaces of the motorcycle. Rinse off any spills immediately with plenty of water.*

Anti-freeze is highly toxic if ingested. Never leave anti-freeze lying around in an open container or in puddles on the floor; children and pets are attracted by its sweet smell and may drink it. Check with the local authorities about disposing of used anti-freeze. Many communities will have collection centres which will see that anti-freeze is disposed of safely.
Caution: At all times use the specified type of anti-freeze, and always mix it with distilled water in the correct proportion. The anti-freeze contains corrosion inhibitors which are essential to avoid damage to the cooling system. A lack of these inhibitors could lead to a build-up of corrosion which would block the coolant passages, resulting in overheating and severe engine damage. Distilled water must be used as opposed to tap water to avoid a build-up of scale which would also block the passages.

2 Radiator pressure cap – check

1 If problems such as overheating or loss of coolant occur, check the entire system as described in Chapter 1. The radiator cap opening pressure should be checked by a Honda dealer with the special tester required to do the job. If the cap is defective, replace it with a new one.

3 Cooling fan and fan switch/relay – check and replacement

Cooling fan

Check

1 If the engine is overheating and the cooling fan isn't coming on, first check the cooling fan circuit fuse (see Chapter 9). If the fuse is good, check the fan switch (RR-Y and RR-1 (2000 and 2001) models) or relay (RR-2 and RR-3 (2002 and 2003) models) as described below.
2 To test the cooling fan motor on RR-Y and RR-1 (2000 and 2001) models, remove the left-hand fairing side panel (see Chapter 8). Disconnect the fan wiring connector and the fan switch wiring connector **(see illustrations)**. Using a 12 volt battery and two jumper wires with suitable connectors, connect the battery positive (+ve) lead to the black/blue wire terminal on the fan side of the wiring connector, and the battery negative (–ve) lead to the fan switch wiring connector. Once connected the fan should operate. If it does not, and the wiring is all good, then the fan motor is faulty. Individual components are available for the fan assembly.
3 To test the cooling fan motor on RR-2 and RR-3 (2002 and 2003) models, remove the left-hand fairing side panel (see Chapter 8). Disconnect the fan wiring connector **(see illustration 3.2b)**. Using a 12 volt battery and two jumper wires with suitable connectors, connect the battery positive (+ve) lead to the black/blue wire terminal on the fan side of the wiring connector, and the battery negative (–ve) lead to the green wire terminal on the connector. Once connected the fan should operate. If it does not, and the wiring is all good, then the fan motor is faulty. Individual components are available for the fan assembly.

Replacement

Warning: The engine must be completely cool before carrying out this procedure.
4 Remove the radiator (see Section 6).
5 On RR-Y and RR-1 (2000 and 2001) models disconnect the wiring connector from the fan switch **(see illustration 3.2b)**.
6 Unscrew the bolts securing the fan assembly to the radiator, noting that on RR-Y and RR-1 (2000 and 2001) models one bolt also secures the earth (ground) wire **(see illustration)**. Free the fan wiring connector from its lug on the radiator and all the wiring from any clamps.
7 Unscrew the fan blade nut and remove the blade. Remove the spacer from the fan motor spindle. Undo the three nuts on the front of the fan motor and separate the motor from its bracket.
8 Installation is the reverse of removal. Apply a suitable non-permanent thread locking compound to the fan blade nut and tighten it

3.2a Disconnect the fan wiring connector (arrowed) . . .

3.2b . . . and the fan switch wiring connector (arrowed)

Cooling system 3•3

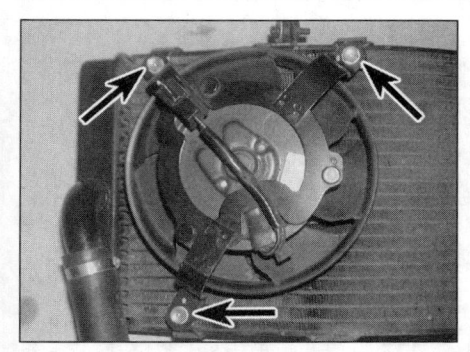

3.6 Undo the bolts (arrowed) and remove the fan assembly

to the torque setting specified at the beginning of the Chapter. Also tighten the fan motor nuts to the specified torque. On RR-Y and RR-1 (2000 and 2001) models do not forget to attach the earth (ground) cable to the radiator.
9 Install the radiator (see Section 7).

Cooling fan switch – RR-Y and RR-1 (2000 and 2001)

Check

10 If the engine is overheating and the cooling fan isn't coming on, first check the cooling fan circuit fuse (see Chapter 9). If the fuse is blown, check the fan circuit for a short to earth (see the wiring diagrams at the end of this book).
11 If the fuse is good, remove the left-hand fairing side panel (see Chapter 8). Disconnect the fan wiring connector **(see illustration 3.2a)**. Check that there is battery voltage at the black/blue wire terminal on the loom side of the connector with the ignition ON. If not, check the wiring and connections as described in Chapter 9, following the relevant wiring diagram. Turn the ignition OFF, then reconnect the fan wiring connector.
12 If the voltage is good, disconnect the fan switch wire **(see illustration 3.2b)**, then connect it to earth (ground) using a jumper wire. Turn the ignition switch ON. The fan should come on. If it does, the fan switch is defective and must be replaced with a new one. If it does not come on, check for battery voltage at the switch wiring connector with the ignition ON. If voltage is present, test the fan motor itself (see above). If there is no voltage, there is a break in the circuit between the fan connector and the switch connector.
13 If the fan is on the whole time, disconnect the wiring connector. The fan should stop. If it does, the switch is defective and must be replaced with a new one. If it doesn't, check the wiring between the switch and the fan for a short to earth, and the fan itself.
14 If the fan works but is suspected of cutting in at the wrong temperature, a more comprehensive test of the switch can be made as follows.
15 Remove the switch (see Steps 17 and 18). Fill a small heatproof container with coolant and place it on a stove. Connect the positive (+ve) probe of an ohmmeter to the terminal of the switch and the negative (–ve) probe to the switch body, and using some wire or other support suspend the switch in the coolant so that just the sensing portion and the threads are submerged **(see illustration)**. Also place a thermometer capable of reading temperatures up to 110°C in the coolant so that its bulb is close to the switch. **Note:** *None of the components should be allowed to directly touch the container.*

⚠ **Warning: This must be done very carefully to avoid the risk of personal injury.**

16 Initially the ohmmeter reading should be very high indicating that the switch is open (OFF). Heat the coolant, stirring it gently. When the temperature reaches around 98 to 102°C the meter reading should drop to around zero ohms, indicating that the switch has closed (ON). Now turn the heat off. As the temperature falls below 93 to 97°C the meter reading should show infinite (very high) resistance, indicating that the switch has opened (OFF). If the meter readings obtained are different, or they are obtained at different temperatures, then the switch is faulty and must be replaced with a new one. On completion, fit the switch as described in Steps 19 and 20.

Replacement

⚠ **Warning: The engine must be completely cool before carrying out this procedure.**

17 Drain the cooling system (see Chapter 1).
18 Disconnect the wiring connector from the fan switch on the left-hand side of the radiator **(see illustration 3.2b)**. Unscrew the switch and withdraw it from the radiator. Discard the O-ring as a new one must be used.
19 Install the switch using a new O-ring and some suitable sealant on the upper portion of the threads, and tighten it to the torque setting specified at the beginning of the Chapter. Take care not to overtighten the switch as the radiator could be damaged.
20 Reconnect the switch wiring and refill the cooling system (see Chapter 1).

Cooling fan relay – RR-2 and RR-3 (2002 and 2003)

Check

21 If the engine is overheating and the cooling fan isn't coming on, first check the cooling fan circuit fuse (see Chapter 9). If the fuse is blown, check the fan circuit for a short to earth (see the wiring diagrams at the end of this book).
22 If the fuse is good, remove the seat cowling (see Chapter 8). Displace the cooling fan relay and disconnect its wiring connector – the relay is the front one of the row of three on the right-hand side of the rear sub-frame **(see illustration)**.
23 Set a multimeter to the ohms x 1 scale and connect it across the relay's green/yellow and black/white wire terminals. There should be no continuity (infinite resistance). Using a fully-charged 12 volt battery and two insulated jumper wires, connect the positive (+ve) terminal of the battery to the red/green wire terminal on the relay, and the negative (–ve) terminal to the black/blue wire terminal on the relay. At this point the relay should be heard to click and the multimeter read 0 ohms (continuity). If this is the case the relay is proved good. If the relay does not click when

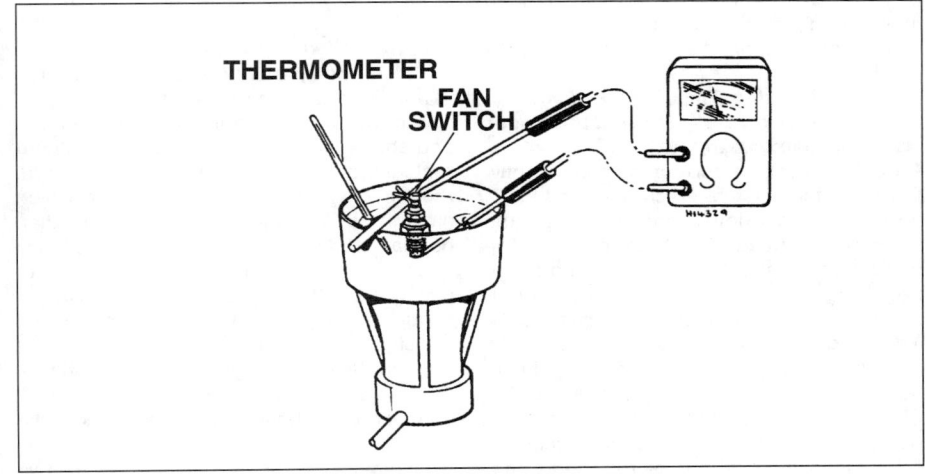

3.15 Cooling fan switch testing set-up

3.22 Cooling fan relay (arrowed)

3•4 Cooling system

battery voltage is applied and still indicates no continuity (infinite resistance) across its terminals, it is faulty and must be replaced with a new one.

24 If the relay is good, check for battery voltage at the red/green wire of the wiring connector with the ignition switch ON. If there is no voltage, check the wiring, referring to the relevant wiring diagram at the end of Chapter 9. If voltage is present, check that there is continuity to earth in the black/blue wire with the ignition switch OFF. If there is no continuity, check the wiring. Also check the wiring between the relay and the ECM (electronic control module), and between the relay and the fan motor wiring connector on the radiator (see the wiring diagrams at the end of this book) – there should be continuity in all wires.

25 If the fan is on the whole time, disconnect the relay wiring connector. The fan should stop. If it does, the relay is defective and must be replaced with a new one. If it doesn't, check the wiring between the relay and the fan for a short to earth, and the fan itself.

26 If the fan works but is suspected of cutting in at the wrong temperature, check the ECT sensor (see Section 4).

Replacement

27 Remove the seat cowling (see Chapter 8). The relay is the rearmost of the row of three on the right-hand side of the rear sub-frame **(see illustration 3.22)**.

28 Displace the cooling fan relay and disconnect its wiring connector. Remove the relay from its rubber sleeve.

29 Installation is the reverse of removal.

4 Temperature display and sensor – check and replacement

Temperature & warning display

Check

1 The circuit consists of the sensor mounted in the thermostat housing and the display which is part of the instrument cluster LCD unit. If the system malfunctions check the fuse. When the ignition is first switched on all the digital display segments and modes should come on temporarily – this serves as an indication that the LCD is functioning correctly.

2 Under normal operating conditions, when the coolant temperature is below 34°C the display will show '- -'. When the temperature is between 35°C and 132°C the display will show the actual temperature. Once the temperature reaches 122°C the display will start to flash, and the red malfunction indicator light (MIL) and the temperature warning symbol will come on. If this occurs stop the engine and check the coolant level in the reservoir (see *Daily (pre-ride) checks*). If the temperature goes above 132°C the display will continue to show that temperature.

3 If the display is not working at all, check the instrument cluster power input (see Chapter 9). If the power lines are good, then either the printed circuit board (PCB) or the LCD display unit could be faulty.

4 If the display as a whole works but the coolant function doesn't or is thought to be inaccurate, check the sensor (see below). If the sensor is good, remove the fairing (see Chapter 8) and check the wiring between the sensor and the instrument cluster for continuity. If the wiring is good the display is faulty.

Caution: Do not leave the ignition switched on for any longer than is necessary to take the reading, or the gauge may be damaged.

Replacement

5 The temperature display is part of the LCD unit in the instrument cluster. No individual components are available for the instrument cluster. If it is faulty, replace it with a new one (see Chapter 9).

Temperature sensor

Check

6 Drain the cooling system (see Chapter 1). The sensor is mounted in the thermostat housing.

7 Remove the sensor (see Steps 10 and 11 below).

8 Fill a small heatproof container with coolant and place it on a stove. Using an ohmmeter, connect the positive (+ve) probe of the meter to the green/blue wire terminal on the sensor, and the negative (–ve) probe to the body of the sensor. Using some wire or other support suspend the sensor in the coolant so that just the sensing head up to the threads is submerged, and with the head a minimum of 40 mm above the bottom of the container. Also place a thermometer capable of reading temperatures up to 130°C in the coolant so that its bulb is close to the sensor **(see illustration 3.15)**. Note: *None of the components should be allowed to directly touch the container.*

> **Warning:** *This must be done very carefully to avoid the risk of personal injury.*

9 Begin to heat the coolant, stirring it gently. When the temperature reaches around 80°C, turn the heat down and maintain the temperature steady for three minutes. The meter reading should be as specified at the beginning of the Chapter. Turn the heat on again. When the temperature reaches around 120°C, again turn the heat down and maintain it for three minutes. The meter reading should again be as specified at the beginning of the Chapter. If the meter readings obtained are different by a margin of 10% or more, then the sensor is faulty and must be replaced with a new one.

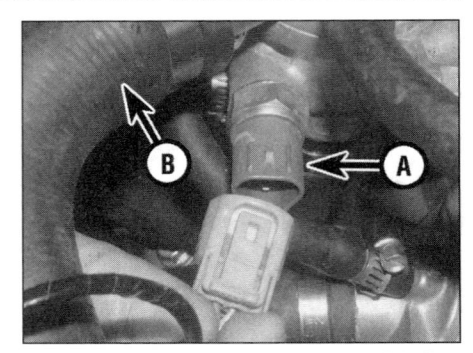

4.11 Coolant temperature sensor (A). To improve access detach the large bore hose (B) from the cover

Replacement

> **Warning:** *The engine must be completely cool before carrying out this procedure.*

10 Drain the cooling system (see Chapter 1). Remove the fuel tank (see Chapter 4). The sensor is mounted in the thermostat housing. To improve access slacken the clamp securing the large bore hose to the thermostat cover and detach it **(see illustration 4.11)**.

11 Disconnect the sensor wiring connector **(see illustration)**. Unscrew the sensor and remove it from the thermostat housing.

12 Fit a new sealing washer onto the sensor. Install the sensor and tighten it to the torque setting specified at the beginning of the Chapter. Connect the wiring.

13 Install the fuel tank (see Chapter 4). Refill the cooling system (see Chapter 1).

5 Thermostat and housing – removal, check and installation

Removal

Note: The complete thermostat housing can be removed without removing the thermostat itself.

> **Warning:** *The engine must be completely cool before carrying out this procedure.*

1 The thermostat is automatic in operation and should give many years service without requiring attention. In the event of a failure, the valve will probably jam open, in which case the engine will take much longer than normal to warm up. Conversely, if the valve jams shut, the coolant will be unable to circulate and the engine will overheat. Neither condition is acceptable, and the fault must be investigated promptly.

2 Drain the cooling system (see Chapter 1). Remove the throttle bodies (see Chapter 4). The thermostat housing is on the back of the engine in the middle.

3 To remove the thermostat, unscrew the two bolts securing the cover and detach it from

Cooling system 3•5

5.3a Unscrew the bolts (arrowed) and detach the cover . . .

5.3b . . . then withdraw the thermostat from the housing

5.4a Detach the hoses (arrowed) from the cover . . .

the housing **(see illustration)**. Withdraw the thermostat, noting how it fits **(see illustration)**.

4 To remove the thermostat housing, disconnect the coolant temperature sensor wiring connector **(see illustration 4.11)**. Slacken the clamps securing all the hoses to the cover and housing and detach them, noting which fits where **(see illustrations)**. Depending on your tools, you may need to remove the cover to access the right-hand bolt (see Step 3) – if you do, then remove the thermostat as well to prevent the possibility of damaging it (see Step 3). Unscrew the bolts securing the housing to the engine and remove the housing **(see illustration)**. Discard the O-ring.

Check

5 Examine the thermostat visually before carrying out the test. If it remains in the open position at room temperature, it should be replaced with a new one. Check the condition of the rubber seal around the thermostat and replace it with a new one if it is damaged, deformed or deteriorated.

6 Suspend the thermostat by a piece of wire in a container of cold water. Place a thermometer capable of reading temperatures up to 110°C in the water so that the bulb is close to the thermostat **(see illustration)**.

5.4b . . . and from the housing

Heat the water, noting the temperature when the thermostat opens, and compare the result with the specifications given at the beginning of the Chapter. Also check the amount the valve opens after it has been heated for a few minutes and compare the measurement to the specifications. If the readings obtained differ from those given, the thermostat is faulty and must be replaced with a new one.

7 In the event of thermostat failure, as an emergency measure only, it can be removed and the machine used without it (this is better than leaving a permanently closed thermostat in, but if it is permanently open, you might as well leave it in). **Note:** *Take care when starting*

5.4c Unscrew the bolts (arrowed) and remove the housing

the engine from cold as it will take much longer than usual to warm up. Ensure that a new unit is installed as soon as possible.

Installation

8 To install the thermostat, first make sure the seal is fitted around it and that it is in good condition, otherwise use a new one **(see illustration)**. Smear some clean coolant over the seal. Install the thermostat with the hole facing back and make sure it locates correctly **(see illustration 5.3b)**. Fit the cover onto the housing, then install the bolts and tighten them **(see illustration)**.

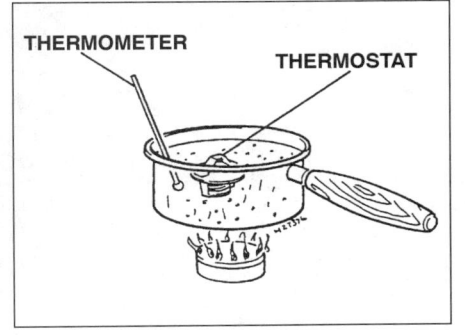

5.6 Thermostat testing set-up

5.8a Fit a new thermostat seal if necessary

5.8b Install the cover

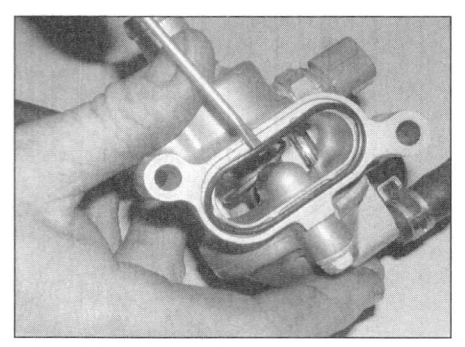

5.9a Fit a new O-ring into the groove . . .

5.9b . . . then install the housing

6.3a Detach the three hoses (arrowed) from the right-hand side of the radiator . . .

6.3b . . . and the hose (arrowed) from the left-hand side – RR-2 and RR-3 (2002 and 2003) model shown

RR-Y and RR-1 (2000 and 2001) models have two hoses on the left

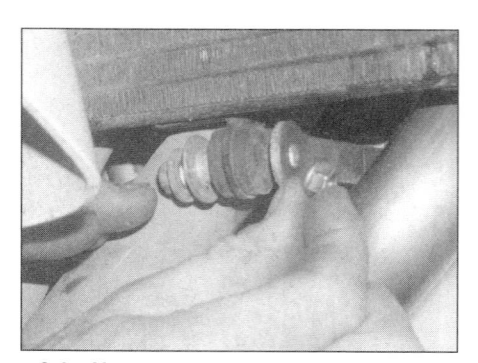

6.4a Unscrew the nut and withdraw the lower bolt

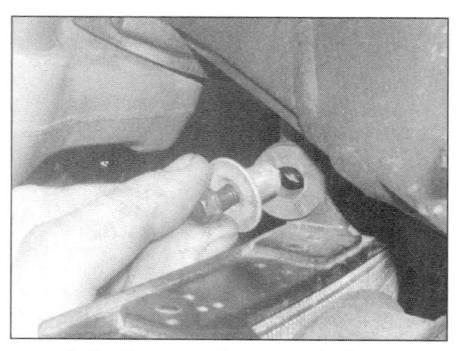

6.4b Unscrew the upper bolt . . .

6.4c . . . then draw the radiator to the right to free the grommet from the lug

9 To install the thermostat housing, fit a new O-ring into the groove, using a dab of grease to keep it in place if required **(see illustration)**. Fit the housing and tighten the bolts **(see illustration)**. Fit the thermostat and cover if removed (see above). Attach the hoses to their unions on the cover and housing and tighten the clamps **(see illustration 5.3a)**. Connect the temperature sensor wiring connector **(see illustration 4.11)**.

10 Install the throttle bodies (see Chapter 4). Refill the cooling system (see Chapter 1).

6 Radiator – removal and installation

Note: *If the radiator is being removed as part of the engine removal procedure, detach the hoses from their unions on the engine rather than on the radiator and remove the radiator with the hoses attached to it. Note the routing of the hoses.*

Removal

⚠️ **Warning: The engine must be completely cool before carrying out this procedure.**

1 Drain the cooling system (see Chapter 1). Remove the fairing and the fairing side panels (see Chapter 8). Where fitted, release the tie securing the clutch cable to the radiator lug on the right-hand side of the frame.

2 Disconnect the fan wiring connector **(see illustration 3.2a)**.

3 Slacken the clamps securing the hoses to the radiator and detach them, noting which fits where **(see illustrations)**.

4 Unscrew the nut on the radiator lower mounting bolt and remove the bolt **(see illustration)**. Unscrew the radiator upper mounting bolt, then draw the radiator to the right to free the top grommet from the lug **(see illustrations)**. Note the arrangement of the collars and rubber grommets in the radiator mounts.

5 If necessary, remove the cooling fan, and on RR-Y and RR-1 (2000 and 2001) models its switch, from the radiator (see Section 3). Check the radiator for signs of damage and clear any dirt or debris that might obstruct air flow and inhibit cooling. If the radiator fins are badly damaged or broken the radiator must be replaced with a new one. Also check the rubber mounting grommets, and renew them if necessary.

Installation

6 Installation is the reverse of removal, noting the following.
● Make sure the rubber grommet fits correctly onto the locating lug.
● Make sure the collars are correctly installed with the mounting bolts.
● Make sure that the fan wiring is correctly connected.
● Ensure the coolant hoses are in good condition (see Chapter 1), and are securely retained by their clamps, using new ones if necessary.
● On completion refill the cooling system as described in Chapter 1.

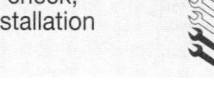

7 Water pump – check, removal and installation

Check

1 The water pump is located on the lower left-hand side of the engine. On RR-Y and RR-1 (2000 and 2001) models remove the lower fairing (see Chapter 8). On RR-2 and RR-3 (2002 and 2003) models remove the left-hand fairing side panel (see Chapter 8). Visually check the area around the pump for signs of leakage.

Cooling system 3•7

7.2 Check the drain hole (arrowed) for signs of leakage

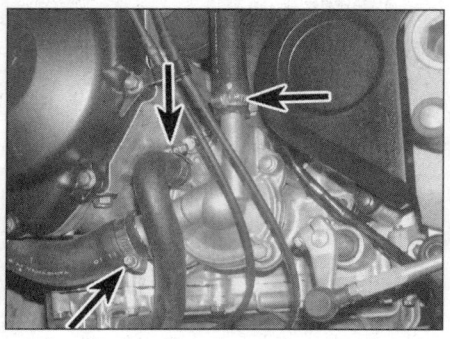

7.5a Slacken the clamps (arrowed) and detach the cover hoses

7.5b Water pump bolts (arrowed)

7.5c Remove the cover and discard the O-ring

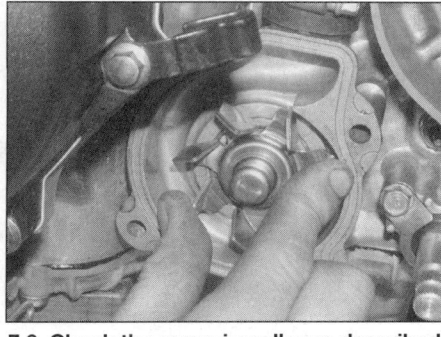

7.6 Check the pump impeller as described

2 To prevent leakage of water from the cooling system to the lubrication system and *vice versa*, two seals are fitted on the pump shaft. On the bottom of the pump housing there is also a drain hole **(see illustration)**. If either seal fails, the drain allows the coolant or oil to escape and prevents them mixing.

3 The seal on the water pump side is of the mechanical type which bears on the rear face of the impeller. The second seal, which is mounted behind the mechanical seal, is of the normal feathered lip type. If on inspection the drain shows signs of leakage, remove the pump and replace it with a new one – it comes as an assembly.

Removal

4 Drain the coolant (see Chapter 1).

5 To remove the cover only, leaving the pump body in place, slacken the clamps securing the coolant hoses to the pump cover and detach the hoses, noting which fits where **(see illustration)**. Unscrew the three remaining bolts (the bottom right bolt will have already been removed when draining the system) and remove the cover **(see illustrations)**. Remove the O-ring from the cover or pump and discard it as a new one must be used **(see illustration 7.9a)**.

6 Wiggle the water pump impeller back-and-forth and in-and-out **(see illustration)**. If there is excessive movement, replace the pump with a new one – see Step 7 to remove the body. Also check for corrosion or a build-up of scale in the pump body and clean or replace the pump as necessary.

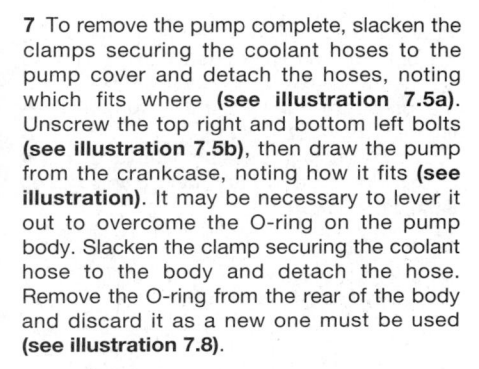

7 To remove the pump complete, slacken the clamps securing the coolant hoses to the pump cover and detach the hoses, noting which fits where **(see illustration 7.5a)**. Unscrew the top right and bottom left bolts **(see illustration 7.5b)**, then draw the pump from the crankcase, noting how it fits **(see illustration)**. It may be necessary to lever it out to overcome the O-ring on the pump body. Slacken the clamp securing the coolant hose to the body and detach the hose. Remove the O-ring from the rear of the body and discard it as a new one must be used **(see illustration 7.8)**.

Installation

8 Fit the coolant hose onto the body, making sure it is correctly aligned so it will not twist when the body is installed, and tighten the clamp. Apply a smear of engine oil to the new pump body O-ring and fit it into the groove in the body **(see illustration)**. Slide the pump into the crankcase, aligning the slot in the impeller shaft with the tab on the oil pump shaft **(see illustration 7.7)**. If they are slightly misaligned (and the cover is off) turn the impeller as you install the pump until they are felt to locate. If the cover is on, withdraw the pump and turn the inner end of the shaft. Make sure the bolt holes are aligned.

9 Smear the new cover O-ring with grease and fit it into its groove in the cover **(see illustration)**, then fit the cover onto the pump **(see illustration 7.5c)**. Install the bolts and

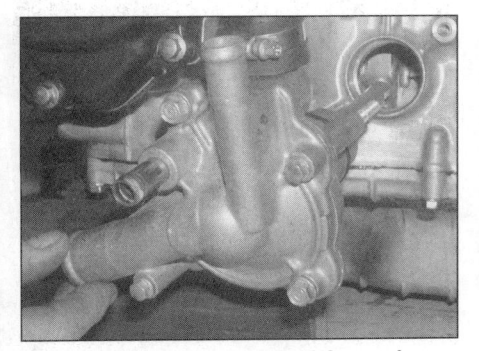

7.7 Draw the pump out of the engine

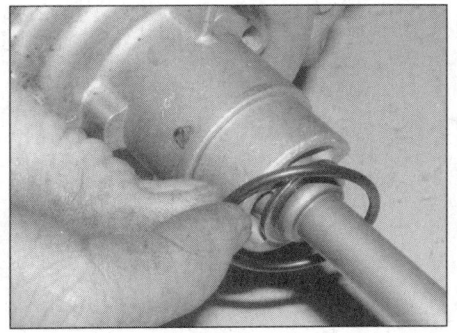

7.8 Fit a new O-ring onto the pump body

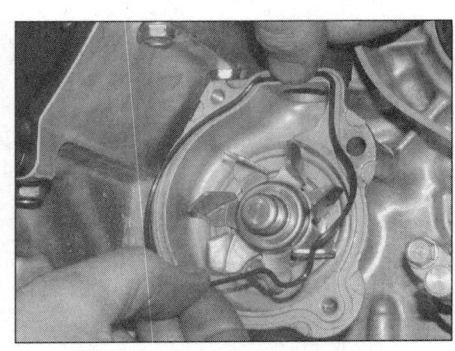

7.9a Fit a new O-ring into the groove

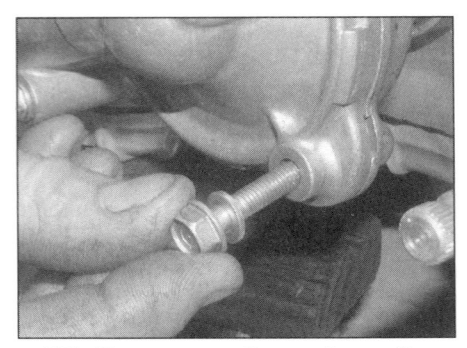

3•8 Cooling system

7.9b Use a new sealing washer on the bottom right bolt

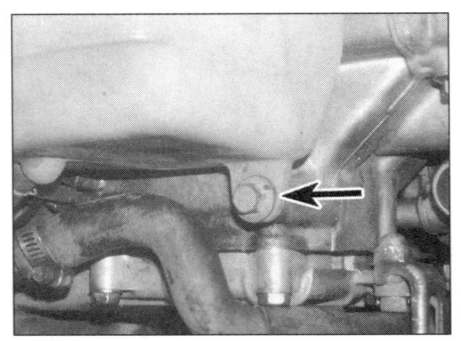

8.2a Unscrew the bottom bolt (arrowed) . . .

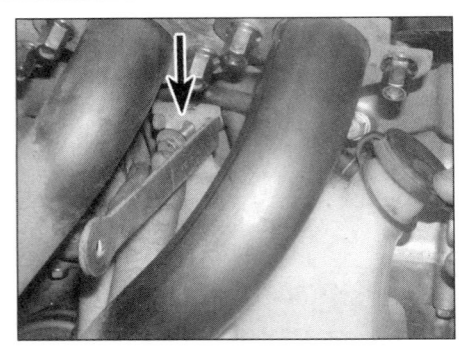

8.2b . . . then the top bolt, noting how it also holds the radiator bracket

tighten them, using a new sealing washer on the bottom right (drain) bolt – the long bolts are for the top right and bottom left holes, and the longest of those for the bottom left **(see illustration)**.
10 Fit the coolant hoses onto the pump cover and secure them with their clamps **(see illustration 7.5a)**.
11 Refill the cooling system (see Chapter 1).

8 Coolant reservoir – removal and installation

Removal

1 The coolant reservoir is located on the left-hand side of the engine at the front. On RR-Y and RR-1 (2000 and 2001) models remove the left-hand fairing side panel and the lower fairing (see Chapter 8). On RR-2 and RR-3 (2002 and 2003) models remove the left-hand fairing side panel (see Chapter 8). Place a suitable container for catching the coolant below the reservoir.
2 Disconnect the reservoir breather/overflow hose from the filler neck. Unscrew the reservoir mounting bolts and manoeuvre the reservoir out **(see illustrations)**. Remove the cap and empty the contents into the container.
3 Disconnect the feed hose from the bottom of the reservoir.

Installation

4 Installation is the reverse of removal. On completion refill the reservoir as described in *Daily (pre-ride) checks*.

9 Coolant hoses and unions – removal and installation

Removal

1 Before removing a hose, drain the coolant (see Chapter 1).
2 Use a screwdriver to slacken the larger-bore hose clamps, then slide them back along the hose and clear of the union spigot. The smaller-bore hoses are secured by spring clamps which can be expanded by squeezing their ears together with pliers.
Caution: The radiator unions are fragile. Do not use excessive force when attempting to remove the hoses.
3 If a hose proves stubborn, release it by rotating it on its union before working it off. If all else fails, cut the hose with a sharp knife. Whilst this means replacing the hose, it is preferable to buying a new radiator.
4 The inlet union to the cylinder block can be removed by unscrewing its bolts **(see illustration)**. If the union is removed, the O-ring must be replaced with a new one. The outlet from the cylinder head goes into the thermostat housing, which is covered in Section 5.

Installation

5 Slide the clamps onto the hose and then work the hose on to its union.

> **HAYNES HiNT** *If the hose is difficult to push on its union, soften it by soaking it in very hot water, or alternatively a little soapy water on the union can be used as a lubricant.*

6 Rotate the hose on its unions to settle it in position before sliding the clamps into place and tightening them securely.
7 If the inlet union to the cylinder block has been removed, fit a new O-ring into the groove, using a dab of grease to hold it in place if necessary. Install the union and tighten the mounting bolts.

9.4 Coolant inlet union bolts (arrowed)

Chapter 4
Fuel and exhaust systems

Contents

Air filter renewal .... see Chapter 1	Fuel supply system – flow rate and pressure check 10
Air filter housing – removal and installation 3	Fuel system – check .... see Chapter 1
Catalytic converter 18	Fuel tank and fuel tap – removal, installation and repair 2
Evaporative emission control (EVAP) system 17	Fuel warning light and sensor – check and renewal 12
Exhaust system – removal and installation 14	General information and precautions 1
Fast idle system wax unit – removal, inspection and installation ... 9	Honda variable intake and exhaust (H-VIX) system 15
Fuel hoses – check and renewal .... see Chapter 1	Idle speed – check and adjustment .... see Chapter 1
Fuel injection system – fault diagnosis and checking 5	Pulse secondary air (PAIR) system 16
Fuel injection system – general information 4	Starter valves – removal, installation and synchronisation 8
Fuel injection and engine management system – check, removal and installation 6	Throttle body assembly – removal and installation 7
	Throttle cables – check and adjustment .... see Chapter 1
Fuel pump – check, removal and installation 11	Throttle cables – removal and installation 13

Degrees of difficulty

Easy, suitable for novice with little experience	**Fairly easy,** suitable for beginner with some experience	**Fairly difficult,** suitable for competent DIY mechanic	**Difficult,** suitable for experienced DIY mechanic	**Very difficult,** suitable for expert DIY or professional

Specifications

Fuel
Grade .... Unleaded. Minimum 91 RON (Research Octane Number) for Europe. Minimum pump octane number 86 for the US
Fuel tank capacity (including reserve) .... 18.0 litres
Reserve volume .... 3.5 litres

Fuel injection system
Throttle bore diameter
  RR-Y and RR-1 (2000 and 2001) models .... 40 mm
  RR-2 and RR-3 (2002 and 2003) models .... 42 mm
Idle speed .... see Chapter 1
Starter valve synchronisation – max. difference between bodies ..... 20 mm Hg
Manifold absolute pressure at idle .... 150 to 250 mm Hg
Fuel pressure at specified idle speed* .... 50 psi (3.5 Bar)
Minimum fuel flow rate .... 188 cc every 10 seconds
Fuel pressure regulator vacuum hose disconnected and plugged

Fuel injection system test data
Note: *All values given are only accurate at 20°C (68°F)*
Cam pulse generator minimum peak voltage output .... 0.7 volts
Ignition pulse generator minimum peak voltage output .... 0.7 volts
Engine coolant temperature (ECT) sensor resistance .... 2.3 to 2.6 K-ohms
Fuel injector resistance
  RR-Y and RR-1 (2000 and 2001) models .... 11.1 to 12.3 ohms
  RR-2 and RR-3 (2002 and 2003) models .... 10.5 to 14.5 K-ohms
Intake air temperature (IAT) sensor resistance .... 1 to 4 K-ohms
Oxygen sensor heater resistance .... 10 to 40 ohms

H-VIX system
Servo motor static resistance .... 5 K-ohms
Servo motor resistance range (see text) .... 0 to 5 K-ohms

Emission control systems
PAIR system control valve resistance .... 20 to 24 ohms
EVAP system control valve resistance .... 30 to 34 ohms

Torque settings
Engine coolant temperature (ECT) sensor .... 23 Nm
Exhaust control valve – pulley cover bolts .... 12 Nm
Exhaust control valve – valve cover bolts .... 12 Nm
Exhaust control valve housing bolts .... 14 Nm
Exhaust control valve pulley nut .... 12 Nm
Exhaust downpipe nuts .... 12 Nm
Fast idle system wax unit mounting screws .... 5 Nm
Fuel pressure regulator bolts .... 10 Nm
Fuel pump assembly mounting plate nuts .... 12 Nm
Fuel rail bolts .... 10 Nm
Fuel supply hose banjo union nut and bolt .... 22 Nm
Oxygen sensor .... 25 Nm

1 General information and precautions

General information

The fuel supply system consists of the fuel tank, fuel pump, filter, fuel hoses, throttle bodies, injectors, and control cables. There is no fuel tap – the fuel pump is switched on and off with the engine via a relay and supplies the fuel. The fuel flows to the fuel rail on the throttle body assembly, and this acts as a reservoir for the fuel injectors, controlled by a pressure regulator which directs excess fuel back to the fuel tank. There is an injector for each cylinder, and these inject fuel into the throttle bodies where it mixes with the air before passing to the intake ducts. The injectors are operated by the Electronic Control Module (ECM) using the information obtained from the various sensors it monitors (refer to Section 4 for more information on the operation of the fuel injection system).

All models have Honda's H-VIX (Honda variable intake and exhaust) system. This controls the flow of air through the air box using a two-position valve and the flow of gases through the exhaust system using a three-position valve. The valves switch at pre-determined engine speeds, and are actuated by cables from a servo motor that is controlled by the ECM.

All models have a low fuel warning light in the instrument cluster which is actuated by a level sensor inside the fuel tank, and comes on when there is approximately 3.5 litres of fuel left.

Many of the fuel system service procedures are considered routine maintenance items and for that reason are covered in Chapter 1.

Precautions

Warning: *Petrol (gasoline) is extremely flammable, so take extra precautions when you work on any part of the fuel system. Always remove the battery (see Chapter 9). Don't smoke or allow open flames or bare light bulbs near the work area, and don't work in a garage where a natural gas-type appliance is present. If you spill any fuel on your skin, rinse it off immediately with soap and water. When you perform any kind of work on the fuel system, wear safety glasses and have a fire extinguisher suitable for a class B type fire (flammable liquids) on hand.*

With the fuel injection system, residual pressure will remain in the fuel feed hose and fuel rail assembly long after the motorcycle was last used. Before disconnecting any fuel line, ensure the ignition is switched OFF then release fuel system pressure (see Section 2). It is vital that no dirt or debris is allowed to enter the fuel tank or the fuel rail assembly whilst the fuel pipes are disconnected. Any foreign matter in the fuel system components could result in injector damage or malfunction. Ensure the ignition is switched OFF before disconnecting or reconnecting any fuel injection system wiring connector. If a connector is disconnected or reconnected with the ignition switched ON, the engine control module (ECM) may be damaged.

Always perform service procedures in a well-ventilated area to prevent a build-up of fumes.

Never work in a building containing a gas appliance with a pilot light, or any other form of naked flame. Ensure that there are no naked light bulbs or any sources of flame or sparks nearby.

Do not smoke (or allow anyone else to smoke) while in the vicinity of petrol (gasoline) or of components containing it. Remember the possible presence of vapour from these sources and move well clear before smoking.

Check all electrical equipment belonging to the house, garage or workshop where work is being undertaken (see the *Safety first!* section of this manual). Remember that certain electrical appliances such as drills, cutters etc, create sparks in the normal course of operation and must not be used near petrol (gasoline) or any component containing it. Again, remember the possible presence of fumes before using electrical equipment.

Always mop up any spilt fuel and safely dispose of the rag used.

Any stored fuel that is drained off during servicing work must be kept in sealed containers that are suitable for holding petrol (gasoline), and clearly marked as such; the containers themselves should be kept in a safe place. Note that this last point applies equally to the fuel tank if it is removed from the machine; also remember to keep its filler cap closed at all times.

Read the *Safety first!* section of this manual carefully before starting work.

2 Fuel tank and fuel tap – removal, installation and repair

Warning: *Refer to the precautions given in Section 1 before starting work.*

Removal

Note: *Due to the arrangement of the fuel delivery and return system, which has no manual tap and no self sealing valves on the hoses, removing the tank involves a certain amount of unavoidable fuel spillage, which is obviously dangerous. Refer to the precautions given in Section 1 before starting work, and have plenty of rag to hand. Before the fuel hoses are disconnected they must be sealed using clamps, and after they have been disconnected the ends must also be blocked (as an added precaution), to prevent the tank from draining itself – a nut and bolt with sealing washers, or a suitably tight section of hose, can be used to block the supply hose banjo eye. If commercially available hose clamps are used, note that there is always a possibility of damaging a hose, especially if it is old, and that new hoses are sometimes too stiff to enable an effective seal to be made. Once the tank has been removed, store it upside down, resting it on some soft rag to prevent damaging the paintwork, and make*

Fuel and exhaust system

2.2a Front mounting bolts (arrowed)

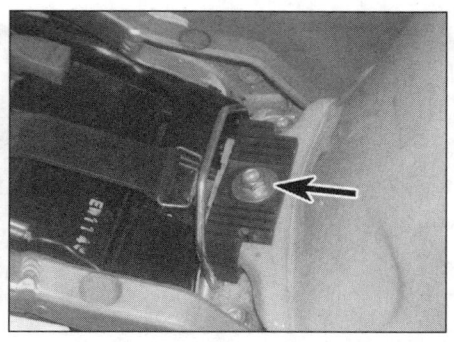

2.2b Rear mounting bolt (arrowed)

2.3a Remove the collars for safekeeping

sure the hose ends are above the level of the tank. Try to time the removal procedure with a near empty tank, which makes it much easier to lift, and means less fuel will be spilt in the event of something not going quite according to plan! To avoid unnecessary spillage, the best thing to do is to completely drain the tank before removing it. Obtain a suitable container for storing the petrol, and release the fuel pressure first, then drain the tank via the fuel return hose, disconnecting it from the fuel rail rather than the tank (see below). It is worth noting that self-sealing fuel hose connectors that fit in the fuel supply and return lines are commercially available through good accessory suppliers – these will make fuel tank removal a much easier and safer task, and negates having to release the pressure first. If you decide to fit them, follow the manufacturer's installation instructions, and make sure they are capable of holding the pressure the system works at – this is listed in the Specifications at the beginning of the Chapter.

1 Make sure the fuel cap is secure. Remove the rider's seat and the air intake duct covers (see Chapter 8). Disconnect the battery negative (–) lead (see Chapter 9).

2 On RR-Y and RR-1 (2000 and 2001) models, to raise the tank, unscrew the front mounting bolts and remove the washers **(see illustration)**. Unscrew the rear mounting bolt and remove the collar, then reinstall the bolt and tighten it slightly **(see illustration)**. Lift the front of the tank slightly and remove the collars from the rubber grommets for safekeeping **(see illustration 2.3a)**. Obtain a suitable prop, such as a block of wood, that will fit securely between the underside of the front of the tank and the top of the frame behind the steering head. Note that a block of wood is better than something like a socket extension bar as it not only provides a wider base for extra stability, but also will not scratch or damage anything. Lift the front of the tank and fit the prop. Do not raise the tank any more than necessary to avoid damaging the rear bracket and rubbers.

3 On RR-2 and RR-3 (2002 and 2003) models, to raise the tank, unscrew the front mounting bolts and remove the washers **(see**

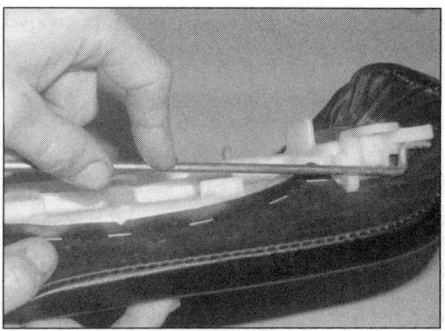

2.3b Release the prop from its clips . . .

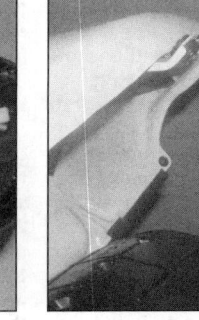

2.3c . . . then raise the tank and locate the prop as shown

illustration 2.2a). Lift the front of the tank slightly and remove the collars from the rubber grommets for safekeeping **(see illustration)**. Remove the prop from the rider's seat **(see illustration)**. Lift the front of the tank and fit one end of the prop into the whole in the steering stem nut, and the other end into one of the mounting bolt holes **(see illustration)**.

4 To remove the tank, the fuel pressure must first be released and the residual fuel in the supply line drained off. Raise the tank as described above. Disconnect the fuel pump and level sensor wiring connector **(see illustration)**. Place a rag and a suitable container for catching the residual fuel under

the fuel supply hose union with the base of the tank, then slowly slacken the banjo bolt until you hear a hissing sound **(see illustration)**. At this point the fuel may spray out, so be ready with the rag to prevent it going everywhere. Once the pressure has been released, tighten the bolt.

5 As mentioned above (see **Note**), it is best to now fully drain the tank. To do this, place some fresh rag under the pressure regulator on the left-hand end of the fuel rail and position a suitable container (make sure it is large enough) very close by so that the return hose end can be placed in it. Release the clamp securing the return hose to the regulator and detach it, and place it

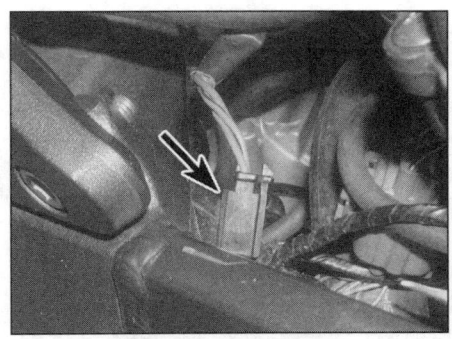

2.4a Disconnect the wiring connector (arrowed)

2.4b Slacken the bolt (arrowed) and catch the fuel with a rag, and tighten the bolt again after the pressure is released

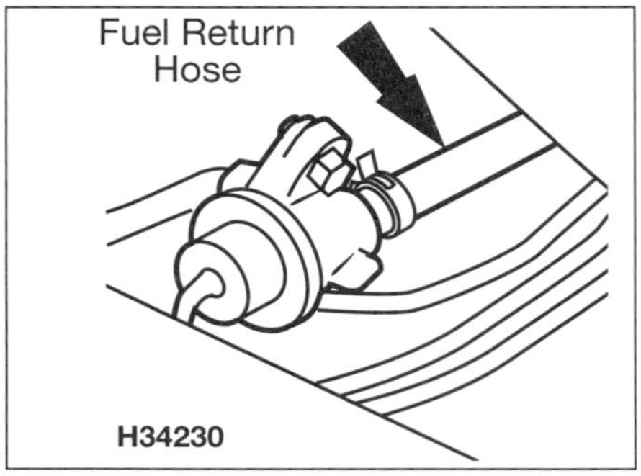

2.5a Release the clamp and detach the return hose from its union

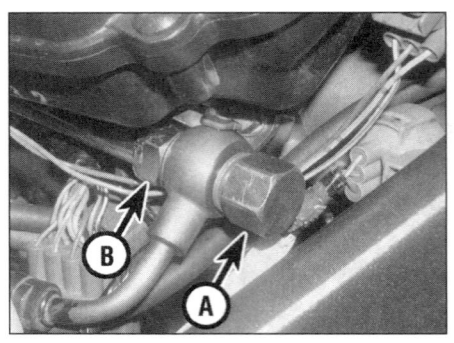

2.5b Unscrew the nut (A) while counter-holding the hex (B) . . .

2.5c . . . as shown

immediately into the container, and allow the tank to drain **(see illustration)**. Now place some fresh rag under the right-hand end of the fuel rail, then unscrew the nut securing the supply hose union to the right-hand end of the fuel rail, counter-holding the base hex on the stud as you do, being prepared for any more residual fuel **(see illustrations)**. Remove the outer sealing washer, then draw the hose off the fuel rail, noting its alignment, and remove the inner sealing washer. Fit the nut back onto the fuel rail stud. Note that access to the base hex is restricted, and depending on your tools it may be necessary to remove the air filter housing – this can be done with the tank in the raised position (see Section 3).

6 If you aren't draining the tank, after releasing the pressure clamp both the supply and return hoses using proper hose clamps **(see illustration)**. Place some fresh rag under the pressure regulator on the left-hand end of the fuel rail, then release the clip securing the return hose to the regulator and detach it and block the end using a suitable bolt or plug (do this as a back-up to the clamp in case it somehow releases itself) **(see illustration 2.5a)**. Now place some fresh rag under the right-hand end of the fuel rail, then unscrew the nut securing the supply hose union to the right-hand end of the fuel rail, counter-holding the base hex on the stud as you do, being prepared for any more residual fuel **(see illustrations 2.5b and 2.5c)**. Remove the outer sealing washer, then draw the hose off the fuel rail, noting its alignment, and block the end as before. Remove the inner sealing washer. Fit the nut back onto the fuel rail stud. Note that access to the base hex is restricted, and depending on your tools it may be necessary to remove the air filter housing – this can be done with the tank in the raised position (see Section 3).

7 Disconnect the breather and overflow hoses from their unions on the tank, again noting which fits where **(see illustration)**. Remove the prop and lower the tank.

8 On RR-Y and RR-1 (2000 and 2001) models unscrew the rear mounting bolt **(see illustration 2.2b)**. On RR-2 and RR-3 (2002 and 2003) models unscrew the tank bracket bolts **(see illustration)**. Carefully lift the tank off the frame and remove it **(see illustration)**. Take care not to lose the rubbers. If there is fuel in the tank, place it upside down on some clean soft rag.

9 Check all the tank rubbers for signs of damage or deterioration and replace them with new ones if necessary.

Installation

10 Installation is the reverse of removal. Before installing the tank, unscrew the banjo bolt securing the supply hose to the tank and detach the hose **(see illustration)**. Fit a new sealing washer on each side of the union then fit the neck of the hose between the lugs and tighten the banjo bolt to the torque setting specified at the beginning of the Chapter **(see**

2.6 Fit a hose clamp as shown on both the supply and return hoses

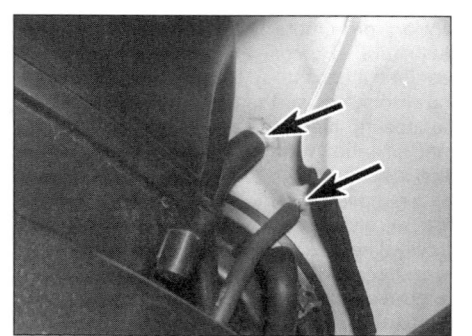

2.7 Pull the breather and overflow hoses (arrowed) off their unions

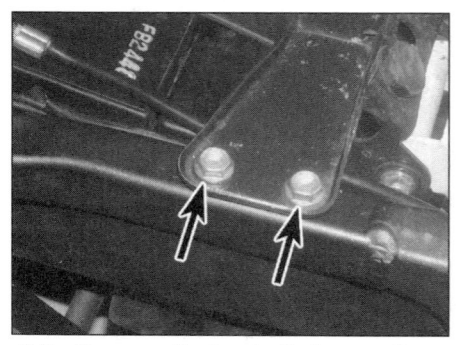

2.8a Unscrew the two bolts (arrowed) on each side

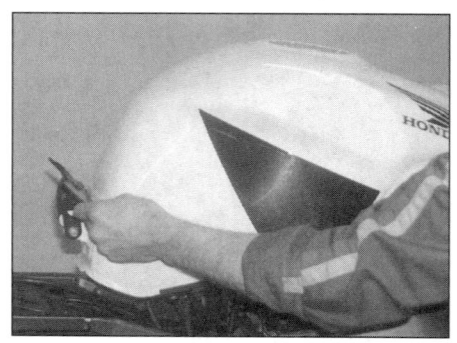

2.8b Carefully lift the tank off the frame and remove it

2.10a Unscrew the bolt and fit new sealing washers

Fuel and exhaust systems 4•5

illustration). If the tank has been resting upside down, or standing on its rear end when full, there is the possibility of fuel having made its way into the breather pipe which will spurt out of the union on the base – be prepared with some rag for this. Once the tank is upright the pipe will fill itself with air. Also use new sealing washers on each side of the supply hose union with the fuel rail (see illustrations). Locate the lug on the union against the fuel rail bracket, and tighten the nut to the specified torque, again counter-holding the base hex on the stud (see illustrations 2.5b and 2.5c). Make sure the return hose clamps are in good condition and use new ones if they have deformed or weakened. Check that the tank mounting rubbers are fitted and in good condition. Make sure all the hoses are securely connected. Start the engine and check that there is no sign of fuel leakage.

Repair

11 All repairs to the fuel tank should be carried out by a professional who has experience in this critical and potentially dangerous work. Even after cleaning and flushing of the fuel system, explosive fumes can remain and ignite during repair of the tank.
12 If the fuel tank is removed from the bike, it should not be placed in an area where sparks or open flames could ignite the fumes coming out of the tank. Be especially careful inside garages where a natural gas-type appliance is located, because the pilot light could cause an explosion.

3 Air filter housing – removal and installation

Removal

1 Remove the air filter (see Chapter 1).
2 Turn the air intake control valve pulley forwards to create slack in the cable and detach the end from the pulley (see Section 15).
3 Disconnect the MAP sensor wiring connector (see illustration).
4 On RR-2 and RR-3 (2002 and 2003) models, disconnect the intake air temperature (IAT) sensor wiring connector (see illustration).
5 Undo the air funnel/filter housing mounting screws and remove the funnels, noting which fits where (see illustration).
6 Displace the housing up off the throttle bodies and disconnect the crankcase breather hose and the PAIR system supply hose and

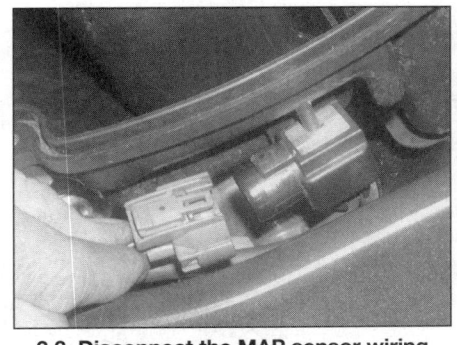

2.10b Locate the neck of the hose between the lugs as shown

2.10c Fit the inner sealing washer . . .

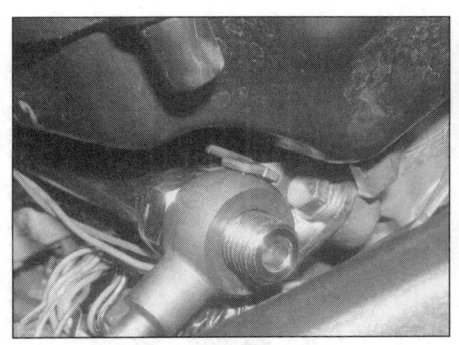

2.10d . . . then the hose . . .

2.10e . . . then the outer sealing washer (arrowed)

3.3 Disconnect the MAP sensor wiring connector

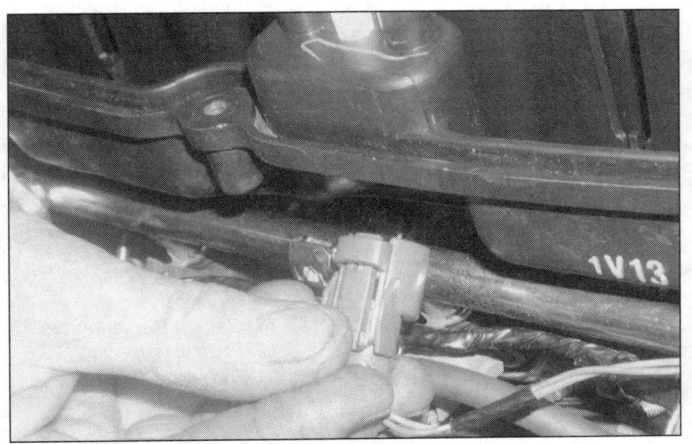

3.4 Disconnect the IAT sensor wiring connector

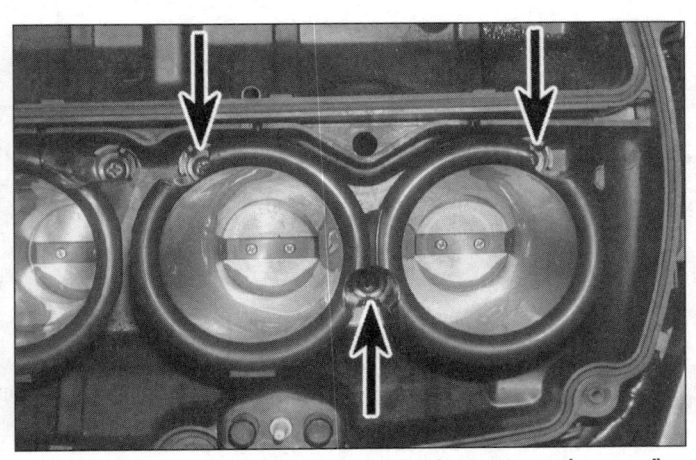

3.5 Each pair of funnels is secured by three screws (arrowed)

4•6 Fuel and exhaust systems

3.6a Detach the hoses (arrowed) – RR-2 and RR-3 (2002 and 2003) model shown

3.6b Unscrew the nut . . .

3.6c . . . and draw the cable out of the housing

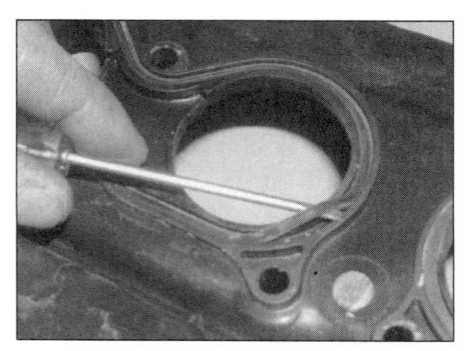

3.7a Fit new O-rings if necessary

3.7b Make sure the funnels are the correct way round

the MAP sensor vacuum hose **(see illustration)**. Note the angle at which the H-VIX cable elbow lies. Unscrew the elbow nut and draw the cable out of the housing **(see illustrations)**. Remove the air filter housing. Cover the throttle bodies with a clean rag.

Installation

7 Installation is the reverse of removal. Check the condition of the throttle body O-rings and use new ones if they are in any way damaged or deteriorated **(see illustration)**. Fit the taller air intake funnels above the middle (Nos. 2 and 3) throttle bodies **(see illustration)**.

8 Check the condition of the various hoses and their clamps and use new ones if they are

1. Ignition switch and HISS receiver
2. PAIR system control valve
3. Fuel pump
4. Intake air temperature (IAT) sensor
5. Throttle position (TP) sensor
6. Manifold absolute pressure (MAP) sensor
7. Engine stop relay
8. Lean angle sensor
9. Fuel cut-off relay
10. Electronic control module (ECM)
11. Speed sensor
12. Ignition pulse generator
13. Engine coolant temperature (ECT) sensor
14. H-VIX system servo motor
15. Fuel pressure regulator
16. Cam pulse generator
17. Fuel injector

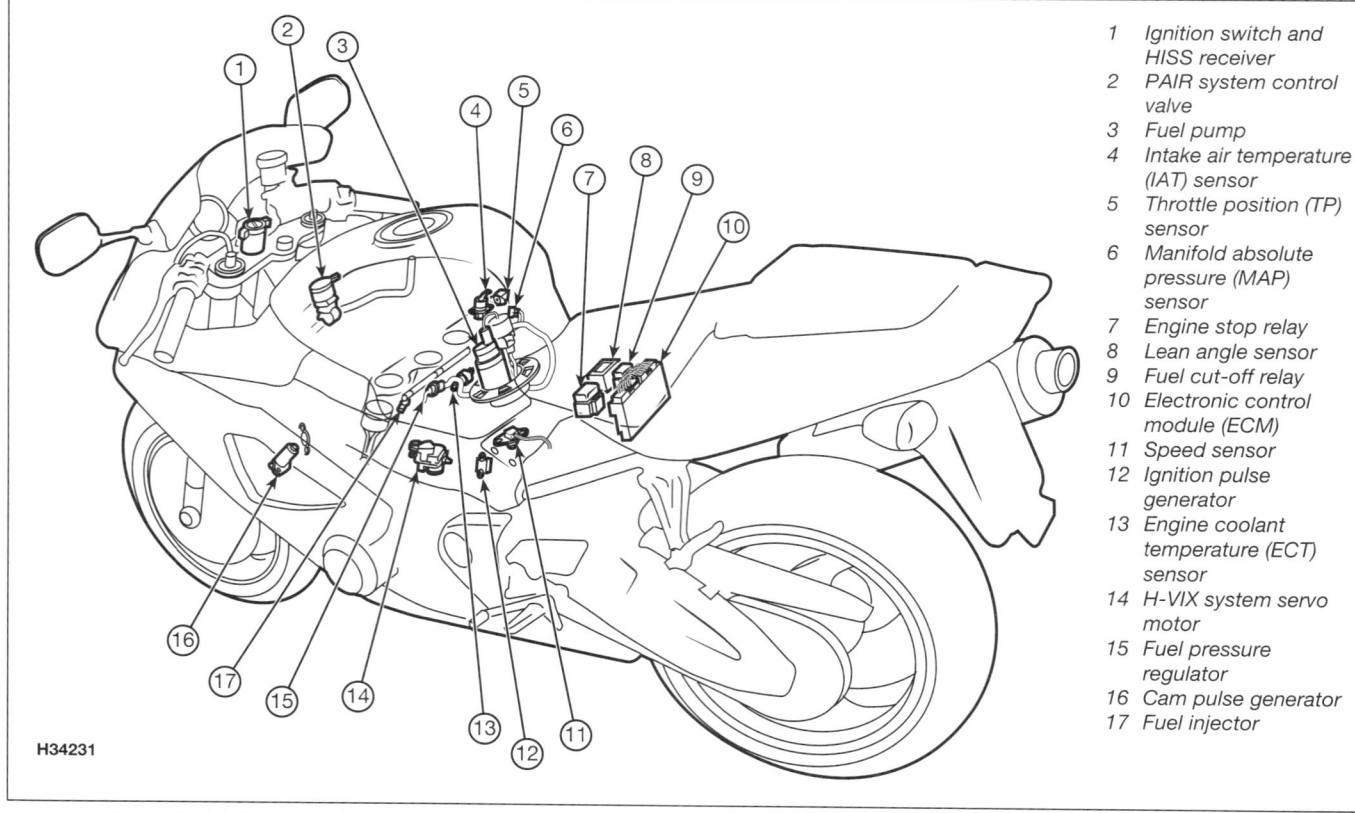

4.1a Fuel injection and engine management system component location – RR-Y and RR-1 (2000 and 2001) models in any way damaged or deteriorated **(see illustration 3.6a)**.

9 Make sure the H-VIX cable routes freely around the engine breather hose – if it presses against the hose it could restrict the movement of the cable.

4 Fuel injection system – general information

1 All models are equipped with Honda's programmed fuel injection (PGM-FI) system. It is controlled by an engine management system which has an electronic control module (ECM) that operates both the injection and ignition systems **(see illustrations)**. The fuel injection and engine management side of the system is covered in this Chapter; refer to Chapter 5 for the ignition system.

2 The fuel pump, contained in the fuel tank, pumps fuel to the fuel rails on the throttle body assembly, via a filter. Fuel supply pressure is controlled by the pressure regulator which keeps the pressure in the fuel rails constant, returning excess fuel to the tank via the return hose. The fuel rails act as a reservoir for the four injectors (one for each cylinder) which are operated by the Engine Control Module (ECM).

3 The engine control module (ECM) monitors signals from the following sensors.

- Throttle position sensor (TPS) – informs the ECM of the throttle position, and the rate of throttle opening or closing.
- Engine coolant temperature (ECT) sensor – informs the ECM of engine temperature. It also actuates the temperature gauge (see Chapter 3).
- Manifold absolute pressure (MAP) sensor – informs the ECM of the engine load by monitoring the pressure in the throttle body inlet tracts.
- Intake air temperature (IAT) sensor – informs the ECM of the temperature of the air entering the throttle body.
- Cam pulse generator – informs the ECM of engine speed and camshaft position.
- Ignition pulse generator – informs the ECM of engine speed and crankshaft position.
- Speed sensor – informs the ECM of the speed of the motorcycle (see Chapter 9).
- Oxygen sensor (only on models with catalytic converter – Germany and California models) – informs the ECM of the oxygen content of the exhaust gases.

4 All the information from the sensors is analysed by the ECM, and from that it determines the appropriate ignition and fuelling requirements of the engine **(see illustration overleaf)**. The ECM controls the fuel injector by varying its pulse width – the length of time the injector is held open – to provide more or less fuel, as appropriate. The mixture strength (fuel/air ratio) is also constantly varied by the ECM, to provide the best setting for cold starting, warm up, idle, cruising, and acceleration. Due to the layout of the engine, the fuelling needs for each cylinder are slightly different and the ECM is programmed to compensate for this; the injection system is fully sequential, with each injector receiving its own operating signal from the ECM.

5 Cold starting and warm up idle speeds are controlled by an 'automatic fast idle system', which basically takes the place of a manual choke lever. A heat sensitive wax-filled unit that has engine coolant circulating around it actuates the starter valve arrangement in the throttle body assembly via a linkage rod. When the coolant is cold the wax unit is contracted and the starter valves are open. As the coolant heats up the wax expands, closing the starter valves. The starter valves allow additional air to bypass the throttle valves when the throttle is closed, and this increases the engine idle speed.

6 If there is an abnormality in any of the readings obtained from any sensor, the ECM enters its back-up mode. In this event, the ECM ignores the abnormal sensor signal, and assumes a pre-programmed value which will allow the engine to continue running (albeit at reduced efficiency). If the ECM enters this back-up mode, or when any faults occur, the malfunction indicator light (MIL) and the fuel injection system (FI) warning light in the

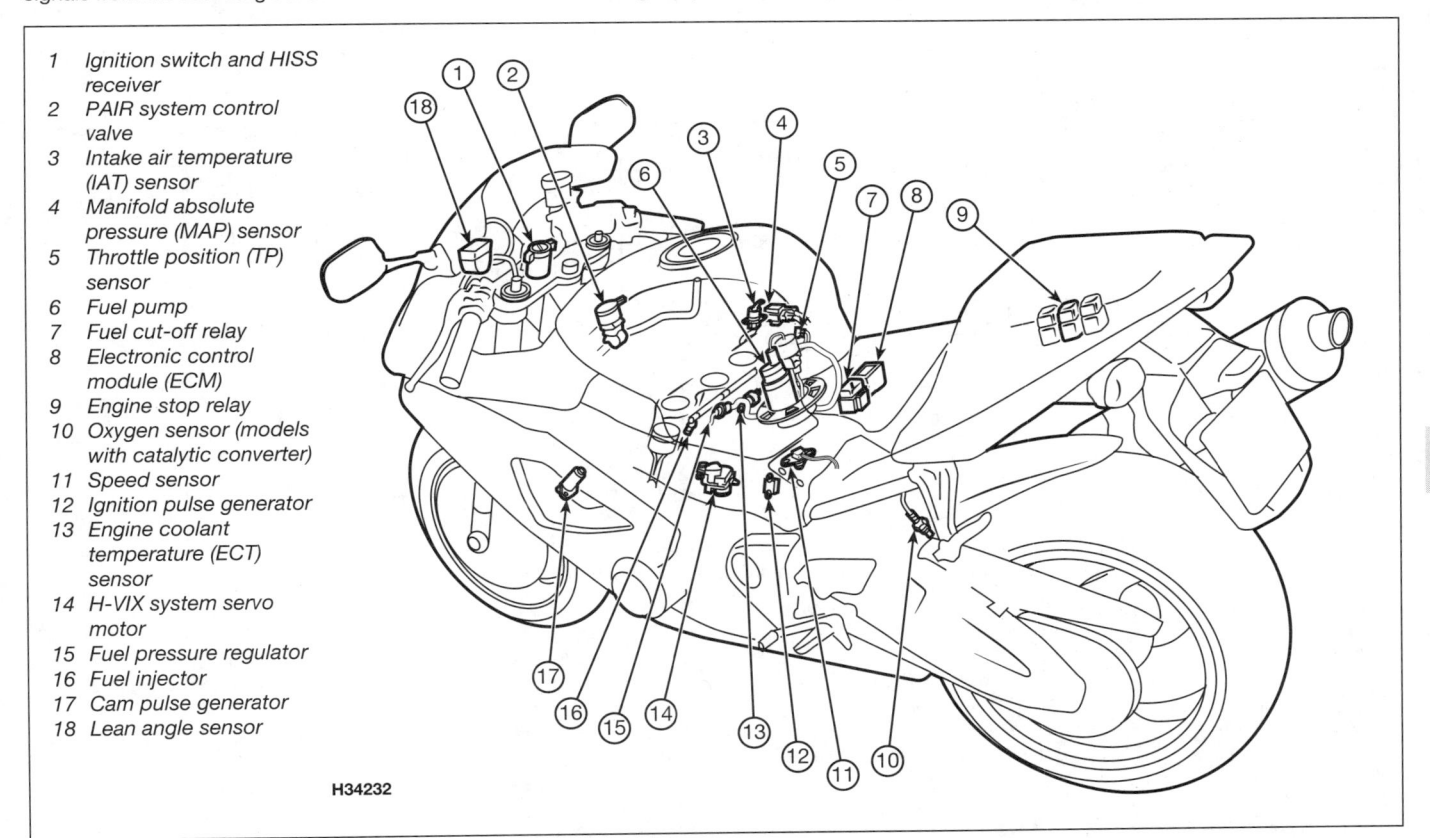

1 Ignition switch and HISS receiver
2 PAIR system control valve
3 Intake air temperature (IAT) sensor
4 Manifold absolute pressure (MAP) sensor
5 Throttle position (TP) sensor
6 Fuel pump
7 Fuel cut-off relay
8 Electronic control module (ECM)
9 Engine stop relay
10 Oxygen sensor (models with catalytic converter)
11 Speed sensor
12 Ignition pulse generator
13 Engine coolant temperature (ECT) sensor
14 H-VIX system servo motor
15 Fuel pressure regulator
16 Fuel injector
17 Cam pulse generator
18 Lean angle sensor

4.1b Fuel injection and engine management system component location – RR-2 and RR-3 (2002 and 2003) models

4•8 Fuel and exhaust systems

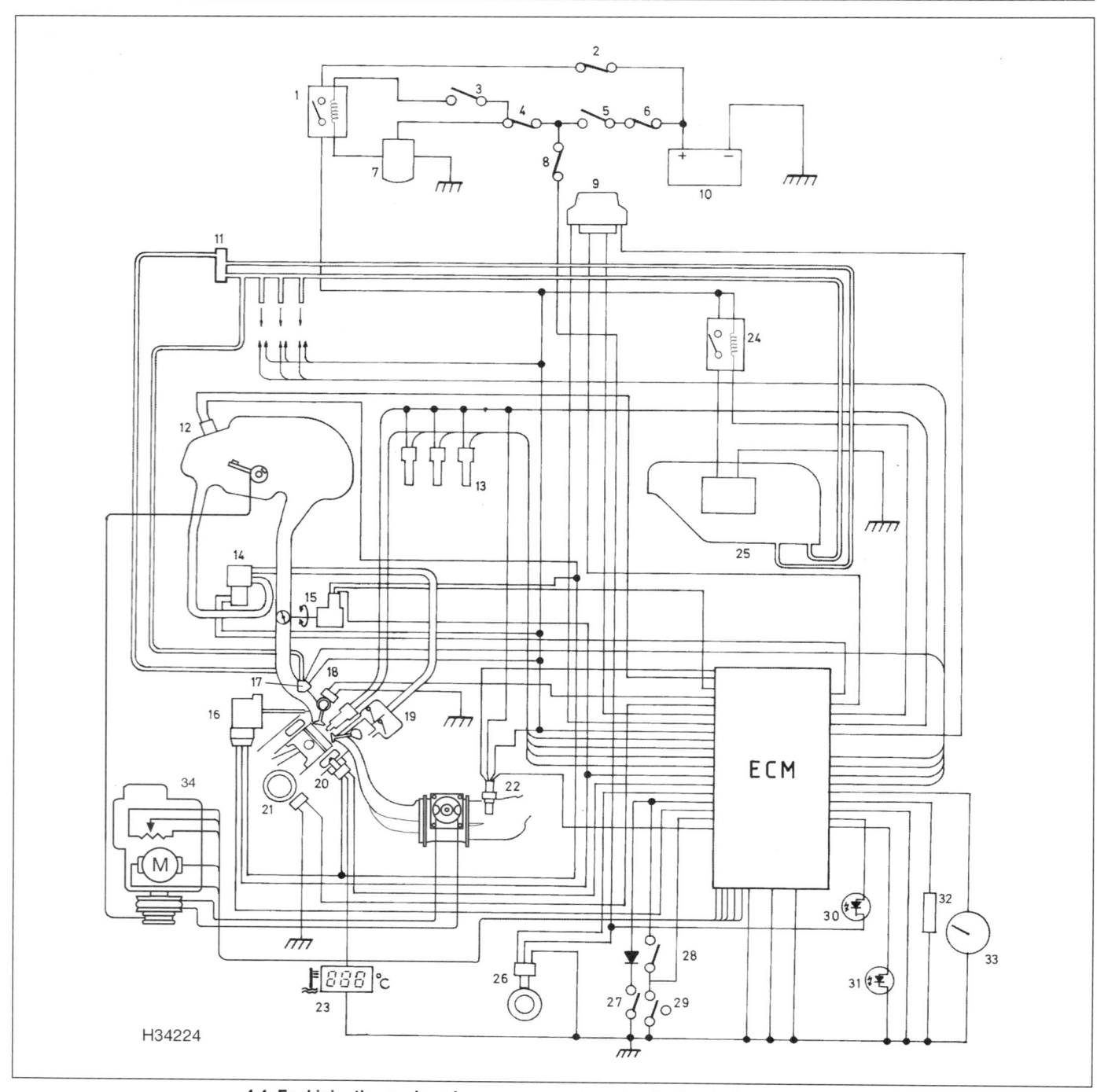

4.4 Fuel injection and engine management system fuel and electrical circuits

1. Engine stop relay
2. Fuse (20A)
3. Engine kill switch
4. Fuse (10A)
5. Ignition switch
6. Main fuse (30A)
7. Lean angle sensor
8. Fuse (10A)
9. HISS system immobiliser receiver (where fitted)
10. Battery
11. Fuel pressure regulator
12. Intake air temperature (IAT) sensor
13. Ignition coil/plug cap
14. PAIR system control valve
15. Throttle position (TP) sensor
16. Manifold absolute pressure (MAP) sensor
17. Fuel injector
18. Cam pulse generator
19. PAIR system reed valve
20. Engine coolant temperature (ECT) sensor
21. Ignition pulse generator
22. Oxygen sensor (models with catalytic converter)
23. Coolant temperature display (LCD)
24. Fuel cut-off relay
25. Fuel pump
26. Speed sensor
27. Neutral switch
28. Clutch switch
29. Sidestand switch
30. Malfunction indicator light (MIL)
31. HISS system immobiliser indicator
32. Service check connector
33. Tachometer
34. H-VIX system servo motor

instrument cluster will come on, and the relevant fault code will be stored in the ECM memory. The fault can be identified using the fault codes which can be accessed using the self-diagnosis function (see Section 5). However if there are certain faults detected in the injectors or the cam or ignition pulse generators, the back-up mode becomes ineffective and the ECM will not allow the engine to run at all. Note that many European models have an immobiliser system (HISS – Honda Ignition Security System) which will not allow the engine to be started unless the correct key is used. A fault in this system should not be confused with a fuel injection system fault. The immobiliser system has its own fault diagnosis function (see Chapter 5).

5 Fuel injection system – fault diagnosis and checking

Fault diagnosis

1 If the red malfunction indicator light (MIL) and the fuel injection system (FI) warning light on the instrument cluster illuminate when the motorcycle is running, a fault has occurred in the fuel injection/ignition system. The engine control module (ECM) will store the relevant fault code in its memory and this code can be read as follows using the self-diagnostic mode of the ECM. While the engine is running above 5000 rpm and the motorcycle is being ridden, the lights will come on and stay on. When the motorcycle is on its sidestand and the engine is running below 5000 rpm, the MIL will flash, the pattern of the flashes indicating the code for the fault the ECM has identified.

2 If the engine can be started, place the motorcycle on its sidestand then start the engine and allow it to idle. Whilst the engine is idling, observe the MIL and FI warning light on the instrument cluster.

3 If the engine cannot be started, or to check for any stored fault codes even though the warning lights have not illuminated, remove the rider's seat (see Chapter 8) to gain access to the fuel injection system service check wiring connector, which is a black 3-pin (2 wire) single-sided connector coming out of the wiring loom on the right-hand side of the battery. Ensure the ignition is switched OFF then bridge the outer terminals of the service check connector with an auxiliary wire **(see illustration)**. With the terminals connected, make sure the kill switch is in the RUN position then turn the ignition ON and observe the red malfunction indicator light (MIL). If there are no stored fault codes, the MIL will come on and stay on. If there are stored fault codes, the MIL will flash.

4 The fuel injection system warning light uses long (1.3 second) and short (0.5 second) flashes to give out the fault code. A long flash

5.3 Short across the connector terminals using an auxiliary wire

is used to indicate the first digit of a double digit fault code (i.e. 10 and above). If a single digit fault code is being displayed (i.e. 0 – 9), there will be a number of short flashes equivalent to the code being displayed. For example, two long (1.3 sec) flashes followed by five short (0.5 sec) flashes indicates the fault code number 25. If there is more than one fault code, there will be a gap before the other codes are revealed (the codes will be revealed in order, starting with the lowest and finishing with the highest). Once all codes have been revealed, the ECM will continuously run through the code(s) stored in its memory, revealing each one in turn with a short gap between them. The fault codes are shown in the table.

Fault code (No. of flashes)	Symptoms	Possible causes
0 – no code (warning light off)	Engine does not start	Blown fuse (main, engine stop, PGM-FI or fuel pump)
		Faulty power supply to electronic control module (ECM)
		Faulty engine stop relay or wiring
		Faulty engine stop switch/open circuit on switch earth (ground) wire
		Faulty ignition switch
		Faulty lean angle sensor or wiring
		Faulty electronic control module (ECM)
0 – no code (warning light off)	Engine runs normally	Blown warning light bulb
		Open or short circuit in warning light wiring
		Faulty electronic control module (ECM)
0 – no code (warning light constantly on)	Engine runs normally	Short circuit in service check connector or wiring
		Faulty electronic control module (ECM)
1	Engine runs normally	Faulty manifold absolute pressure (MAP) sensor or wiring
2	Engine runs normally	Faulty manifold absolute pressure (MAP) sensor or vacuum hose
7	Engine difficult to start at low temperatures	Faulty engine coolant temperature (ECT) sensor or wiring
8	Poor throttle response	Faulty throttle position sensor or wiring
9	Engine runs normally	Faulty intake air temperature (IAT) sensor or wiring
11	Engine operates normally	Faulty speed sensor or wiring
12	Engine does not start	Faulty No. 1 injector or wiring
13	Engine does not start	Faulty No. 2 injector or wiring
14	Engine does not start	Faulty No. 3 injector or wiring
15	Engine does not start	Faulty No. 4 injector or wiring
18	Engine does not start	Faulty cam pulse generator or wiring
19	Engine does not start	Faulty ignition pulse generator or wiring
33	Engine operates normally	Faulty EPROM in electronic control module (ECM)
34	Engine operates normally	Faulty H-VIX system
35	Engine operates normally	Faulty H-VIX system
The following codes are only applicable to models with catalytic converter		
21	Engine operates normally	Faulty oxygen sensor
23	Engine operates normally	Faulty oxygen sensor heating element

Once all the codes have been revealed, switch off the ignition and (where necessary) remove the auxiliary wire from the service check connector. Identify the fault using the table on the previous page, then refer below for checking procedures.

5 Once the fault has been identified and corrected, it will be necessary to reset the system by removing the fault code from the ECM memory. To do this, ensure the ignition is switched OFF then bridge the terminals of the service check connector (see Step 3) **(see illustration 5.3)**. Make sure the kill switch is in the RUN position, then turn the ignition switch ON. Disconnect the auxiliary wire from the service check connector. When the wire is disconnected the malfunction indicator light should come on for about five seconds, during which time the auxiliary wire must be reconnected. The light should start to flash when it is reconnected, indicating that all fault codes have been erased. However if the light flashes twenty times the memory has not been erased and the procedure must be repeated. Turn off the ignition then remove the auxiliary wire. Check the MIL and FI warning light (in some cases it may be necessary to repeat the erasing procedure more than once) then install the seat.

Checking

6 While some of the sensors can be checked using home equipment, there are others which can only be tested using the Honda special electronic diagnostic test pin box which can be plugged into the system. If a fault appears, use the diagnostic function and fault code system described above to work out which component is faulty. First ensure that the relevant system wiring connectors are securely connected and free of corrosion – poor connections are the cause of the majority of problems. Also check the wiring itself for any obvious faults or breaks, and use a continuity tester to check the wiring between the component, its connectors and the ECM, referring to the wiring diagrams at the end of Chapter 9. Next refer to Section 6 to see if there are any other specific checks that can be made on that particular component. If this fails to reveal the cause of the problem, the motorcycle should be taken to a suitably-equipped Honda dealer for testing. They will have access to the test pin box which should locate the fault quickly and simply.

7 Also ensure that the fault is not due to poor maintenance – i.e. check that the air filter element is clean, that the spark plugs are in good condition, that the valve clearances are correctly adjusted, the cylinder compression pressures are correct, and the ignition timing is correct (refer to Chapters 1, 2 and 5). It is also worth removing the sensor(s) in question (see Section 6) and checking that the sensing head is clean and not obstructed by anything. Where there is a vacuum hose to a sensor, make sure it is securely connected at both ends and has no cracks or splits.

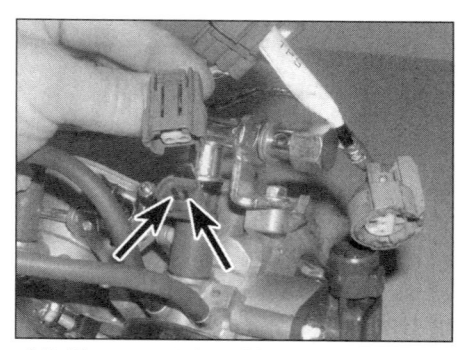

6 Fuel injection and engine management system – check, removal and installation

Caution: Ensure the ignition is switched OFF before disconnecting/reconnecting any fuel injection system wiring connector. If a connector is disconnected/reconnected with the ignition switched ON the engine control module (ECM) could be damaged.

Fuel rail and injectors

⚠ *Warning: Refer to the precautions given in Section 1 before starting work.*

Check

1 Remove the fuel tank and the air filter housing (see Sections 2 and 3).

2 If the engine runs, start it and allow it to idle. Check the operation of each injector using a stethoscope or sounding rod; an injector will emit a 'clicking' noise when functioning. If any injector is silent, either the injector or its wiring harness is faulty. **Note:** *Reconnect the wiring connector to the intake air temperature sensor (on the air filter housing cover) to prevent the ECM detecting a fault while making the check.*

3 If the engine does not run, disconnect the wiring connector from each injector **(see illustration)**. Connect an ohmmeter across the terminals of each injector in turn and measure the resistance. Compare the readings obtained for each injector to that given in the Specifications. Also check that there is no continuity to earth on the black/white wire terminal on the injector. If the resistance of any injector differs greatly from that specified, or there is continuity to earth, a new injector should be installed. Also check for battery voltage at the black/white wire terminal in the wiring connector. If there is no voltage, check the wiring.

Removal

4 Remove the air filter housing (see Section 3). If required, remove the throttle bodies (see Section 7) – this is not essential, but will make the job less fiddly.

5 If the throttle bodies are *in situ*, or if the sub-harness connector was disconnected instead of the individual connectors when removing the throttle bodies, disconnect the wiring connector from each injector **(see illustration 6.3)**. Disconnect the vacuum hose from the fuel pressure regulator **(see illustration 6.14a)**.

6 Unscrew the fuel rail mounting bolts **(see illustration)**. Carefully lift off the fuel rail assembly – the injectors will probably come away with the fuel rail, but could stay in the throttle body **(see illustration)**.

7 Remove the injectors from either the fuel rail or throttle body assembly **(see illustration 6.9)**. Note the correct fitted position of the O-ring and rubber spacer, and the seat if it is on the injector, though it is more likely to remain in the throttle body, in which case dig it out **(see illustration)**; the O-ring and seat must be

6.3 Disconnect the wiring connector and check the resistance between the terminals (arrowed)

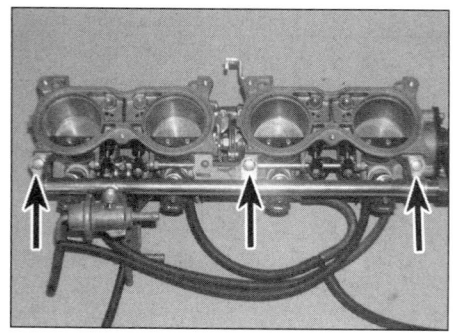

6.6a Unscrew the bolts (arrowed) . . .

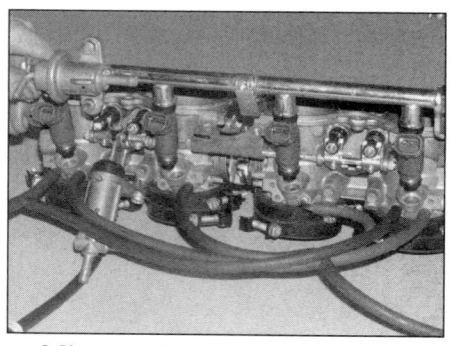

6.6b . . . and remove the fuel rail and injectors

6.7 Remove the O-ring and discard it

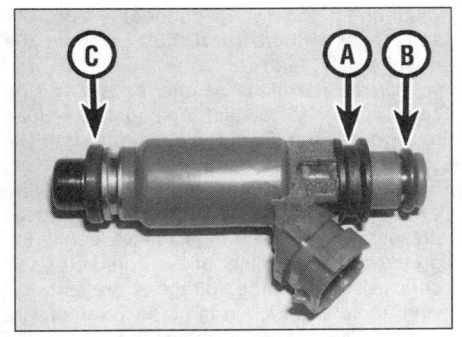

Fuel and exhaust systems 4•11

6.8 Fit a new spacer (A), O-ring (B) and seat (C) onto each injector

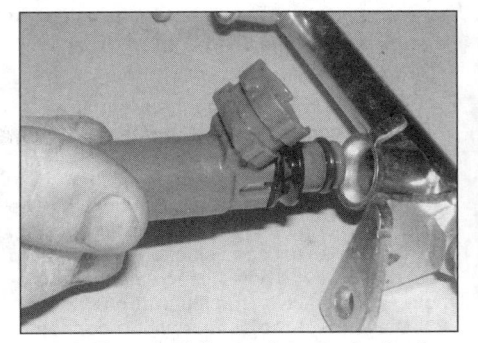

6.9 Ease the injector into the fuel rail

6.10 Install the fuel rail bolts and tighten them to the specified torque

renewed, but the spacer can be reused as long as it is not damaged, deformed or deteriorated.

Installation

8 Lubricate the new seats, spacers and O-rings with a smear of engine oil. Slide the spacer onto the top of each injector, then fit a new O-ring into the groove **(see illustration)**. Fit the seat onto the bottom of the injector, noting that it does not locate in the grooved section, but against its lower rim.

9 Ease the injectors into position in the fuel rail, taking care not to damage the O-rings **(see illustration)**. Make sure the injector wiring connector sockets are all positioned so they will point to the rear when the fuel rail is installed.

10 Install the fuel rail assembly, making sure each injector enters its throttle body and the seats stay in place and locate correctly **(see illustration 6.6b)**. Fit the fuel rail bolts and tighten them to the torque setting specified at the beginning of the Chapter **(see illustration and 6.6a)**.

11 Connect the vacuum tube to the pressure regulator **(see illustrations 6.14a)**. Install the throttle bodies if removed, or reconnect the injector wiring connectors – ensure they are connected correctly **(see illustration 6.3)**.

Fuel pressure regulator

⚠️ **Warning: Refer to the precautions given in Section 1 before starting work.**

Check

12 Check the fuel flow rate and pressure (see Section 10).

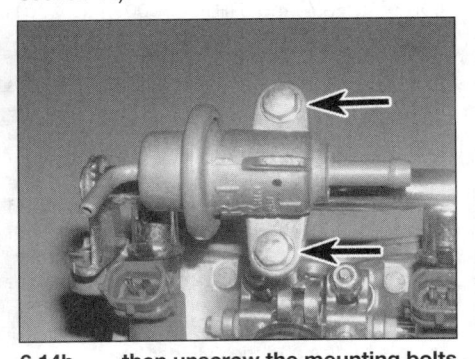

Removal

13 Remove the fuel tank (see Section 2).

14 Detach the vacuum hose from the regulator **(see illustration)**. Unscrew the regulator bolts, holding the fuel rail securely to prevent any distortion, and detach the regulator from the rail **(see illustration)**. Discard the O-ring as a new one must be used.

Installation

15 Fit a new O-ring into the groove in the regulator flange **(see illustration)**. Install the regulator, and tighten the bolts to the torque setting specified at the beginning of the Chapter, again holding the fuel rail to prevent distortion **(see illustration)**. Connect the vacuum hose to its union **(see illustration 6.14a)**.

16 Install the fuel tank (see Section 2).

Throttle position sensor (TPS)

Check

17 The throttle sensor operation can only be checked using the Honda diagnostic test pin box. Its power supply can be checked as follows. Raise the fuel tank (see Section 2). Disconnect the wiring connector from the sensor **(see illustration)**. Connect the positive (+) lead of a voltmeter to the yellow/red terminal of the sensor wiring connector, then connect the negative (–) lead to a good earth. Turn the ignition switch ON and check that a voltage of 4.75 to 5.25 volts is present. If it isn't, there is a fault in the yellow/red wire or the ECM. If voltage was present, now connect

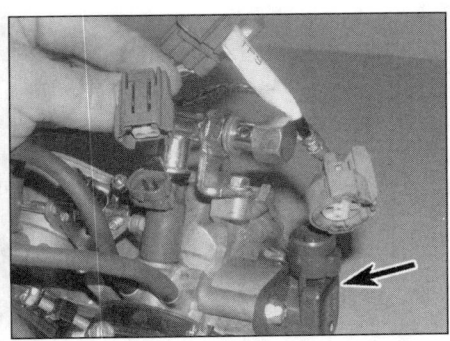

6.14a Detach the vacuum hose ...

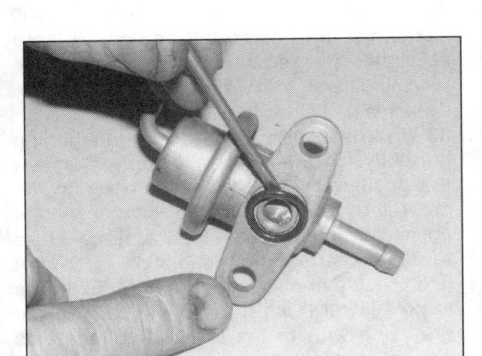

6.14b ... then unscrew the mounting bolts (arrowed)

6.15a Fit a new O-ring into the groove ...

6.15b ... and install the regulator

6.17 Disconnect the wiring connector from the throttle sensor (arrowed)

4•12 Fuel and exhaust systems

the negative lead to the green/orange terminal of the connector and check that the same voltage is present. If it isn't, there is a fault in the green/orange wire or the ECM. If there is voltage, check for continuity to earth in the red/yellow wire. If there is, trace the fault in the wire and repair it. If there isn't, and the sensor is proven good by the Honda tester, then the ECM is faulty.

Removal and installation

18 The throttle sensor is an integral part of the throttle body assembly and is not available separately. If the sensor is faulty, a complete new throttle body assembly will have to be installed, though it is worth checking with your Honda parts specialist whether anything can be done to avoid this.

Engine coolant temperature (ECT) sensor

Note: *The sensor also operates the coolant temperature gauge (see Chapter 3).*

Check

19 Raise or remove the fuel tank (See Section 2). The sensor is mounted in the thermostat housing on the back of the engine.
20 Disconnect the wiring connector from the sensor **(see illustration)**. With the engine cold, connect an ohmmeter across the pink/white and green/orange wire terminals on the sensor and measure its resistance. Compare the reading obtained to that given in the Specifications, noting that the specified value is only valid at 20°C (68°F); the sensor resistance will increase at lower temperatures and decrease at higher temperatures. If the resistance reading differs greatly from that specified, the sensor is probably faulty.
21 If the sensor appears to be functioning correctly, check its power supply. Connect the positive (+) lead of a voltmeter to the pink/white terminal of the sensor wiring connector, then connect the negative (–) lead to a good earth. Turn the ignition switch ON and check that a voltage of 4.75 to 5.25 volts is present. If it isn't, there is a fault in the pink wire or the ECM. If voltage was present, now connect the negative lead to the green/orange terminal of the connector and check that the same voltage is present. If it isn't, there is a fault in the green/orange wire or the ECM. If there is voltage, the ECM is probably faulty.

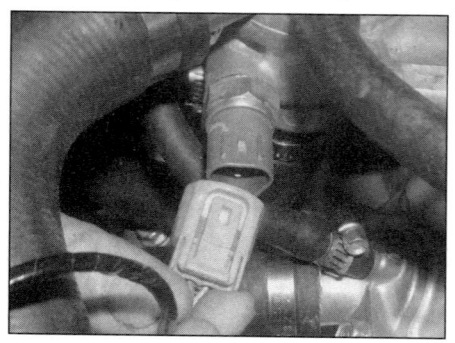

Removal

⚠ Warning: *The engine must be completely cool before carrying out this procedure.*

22 Drain the cooling system (see Chapter 1). Remove the fuel tank (see Section 2). The sensor is mounted in the thermostat housing. To improve access slacken the clamp securing the large bore hose to the thermostat cover and detach it **(see illustration 6.20)**.
23 Disconnect the sensor wiring connector **(see illustration 6.20)**. Unscrew the sensor and remove it from the thermostat housing.

Installation

24 Fit a new sealing washer onto the sensor. Install the sensor and tighten it to the torque setting specified at the beginning of the Chapter. Connect the wiring.
25 Install the fuel tank (see Section 2). Refill the cooling system (see Chapter 1).

Manifold absolute pressure (MAP) sensor

Check

26 The MAP sensor can only be checked using the Honda diagnostic test pin box. However you can raise the fuel tank (see Section 2) and make sure that the vacuum hose to it is securely fixed at both ends, and has no cracks or splits **(see illustration 3.6a)**. The sensor is mounted on the right-hand rear corner (RR-Y and RR-1 (2000 and 2001) models) or right-hand side (RR-2 and RR-3 (2002 and 2003) models) of the air filter housing. If the necessary equipment is available, connect a vacuum gauge into the hose between the throttle bodies and the MAP sensor using an auxiliary three-way joint and some rubber hose, and with the engine idling check that the manifold absolute pressure is as Specified at the beginning of the Chapter. If not, replace the vacuum hose with a new one. If the pressure is out of specification with a new or good hose, check for leaks between the air filter housing, the throttle bodies and the cylinder head.
27 The sensor power supply can be checked as follows. Raise the fuel tank (see Section 2).

Disconnect the wiring connector from the sensor **(see illustration 3.3)**. Connect the positive (+) lead of a voltmeter to the yellow/red terminal of the sensor wiring connector, then connect the negative (–) lead to a good earth. Turn the ignition switch ON and check that a voltage of 4.75 to 5.25 volts is present. If it isn't, there is a fault in the yellow/red wire or the ECM. If voltage was present, now connect the negative lead to the green/orange terminal of the connector and check that the same voltage is present. If it isn't, there is a fault in the green/orange wire or the ECM. If there is voltage, check that the same voltage is present between the light green/yellow terminal of the connector and the green/orange terminal. If it isn't, there is a fault in the light green/yellow wire or the ECM. If there is, and the sensor is proven good by the Honda tester, then the ECM is faulty.

Removal

28 On RR-Y and RR-1 (2000 and 2001) models, raise the fuel tank (see Section 2). Disconnect the wiring connector and detach the vacuum hose. Undo the screw on the underside of the air filter housing and remove the sensor **(see illustration 6.29)**. If access is too restricted for your available tools, remove the air filter housing (see Section 3).
29 On RR-2 and RR-3 (2002 and 2003) models remove the air filter housing (see Section 3). Undo the screw on the underside of the air filter housing and remove the sensor **(see illustration)**.

Installation

30 Installation is the reverse of removal.

Intake air temperature (IAT) sensor

Check

31 Raise the fuel tank (see Section 2). The sensor is mounted in the back of the air filter housing. Disconnect its wiring connector **(see illustration 3.4)**.
32 With the sensor cold, connect an ohmmeter across the sensor terminals and measure its resistance. Compare the reading obtained to that given in the Specifications noting that the specified value is only valid at 20°C (68°F); the sensor resistance will increase at lower temperatures and decrease at higher temperatures. If the resistance reading differs greatly from that specified, the sensor is probably faulty.
33 If the sensor appears to be functioning correctly, check its power supply. Connect the positive (+) lead of a voltmeter to the grey/blue terminal of the sensor wiring connector, then connect the negative (–) lead to a good earth. Turn the ignition switch ON and check that a voltage of 4.75 to 5.25 volts is present. If it isn't, there is a fault in the grey/blue wire or the ECM. If voltage was present, now connect the negative lead to the green/orange terminal of the connector and check that the same voltage is present. If it

6.20 Disconnect the wiring connector. The sensor screws into the thermostat housing

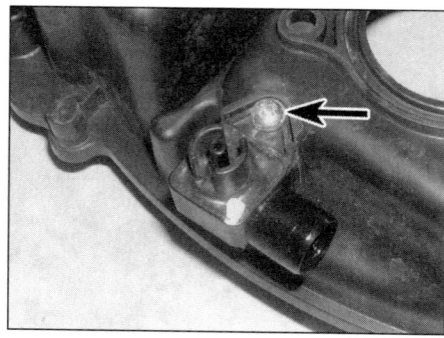

6.29 MAP sensor screw (arrowed)

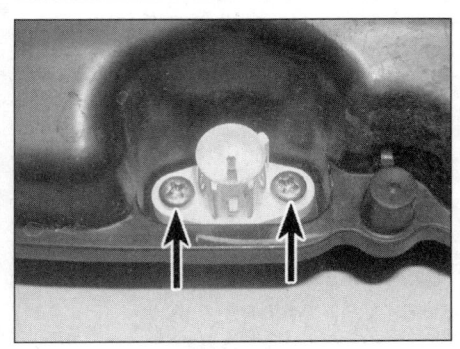

Fuel and exhaust systems 4•13

6.35 IAT sensor screws (arrowed)

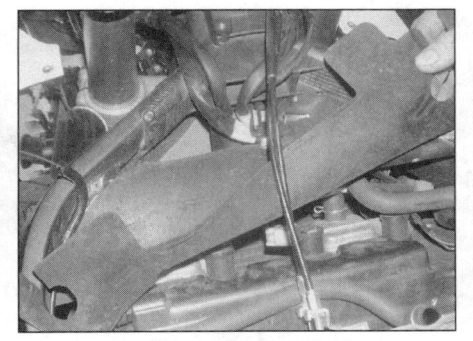

6.37 Raise the heat shield . . .

6.38 . . . to access the wiring connector (arrowed)

isn't, there is a fault in the green/orange wire or the ECM. If there is voltage, the ECM is probably faulty.

Removal

34 On RR-Y and RR-1 (2000 and 2001) models, raise the fuel tank (see Section 2). Disconnect the wiring connector from the sensor, located on the back of the air filter housing cover. Undo the screws and remove the sensor **(see illustration 6.35)**.

35 On RR-2 and RR-3 (2002 and 2003) models remove the air filter housing (see Section 3). Undo the screws securing the sensor on the underside of the air filter housing and remove it **(see illustration)**.

Installation

36 Installation is the reverse of removal.

Cam pulse generator

Check

37 Remove the air filter housing (see Section 3). Lift the rubber heat shield **(see illustration)**.
38 Trace the wiring back from the pulse generator, which is on the left-hand end of the cylinder head, to its 2-pin wiring connector and disconnect it **(see illustration)**. Perform the following check(s).
39 Using an ohmmeter check for continuity first between the grey wire terminal on the sensor side of the connector and earth (ground) and then between the white wire terminal and earth. If there is continuity in either case the cam pulse generator is faulty.
40 Connect the positive (+) lead of a voltmeter and peak voltage adapter arrangement* to the grey terminal on the sensor side of the cam pulse generator connector and the negative (–) lead to the white terminal of the connector. Turn the engine over on the starter motor and note the voltage reading obtained. If this reading is below the specified minimum, the cam pulse generator is faulty.
*Note: *Honda specify their own Imrie diagnostic tester (model 625), or the peak voltage adapter (Pt. No. 07HGJ-0020100) with an aftermarket digital multimeter having an impedance of 10 M-ohm/DCV minimum for this test.*

41 If the cam pulse generator functions correctly then the fault must be in the wiring harness or the ECM, and can be located by a Honda dealer with the test pin box.

Removal

42 Remove the left-hand fairing side panel (see Chapter 8). Remove the air filter housing (see Section 3). Lift the rubber heat shield **(see illustration 6.37)**.
43 Trace the wiring back from the sensor, which is on the left-hand end of the cylinder head, to its 2-pin wiring connector and disconnect it **(see illustration 6.38)**. Feed the wiring through to the pulse generator, noting its routing.
44 Unscrew the bolt securing the cam pulse generator and draw it out of the head **(see illustrations)**. Discard the O-ring.

Installation

45 Fit a new O-ring into the groove in the cam pulse generator body **(see illustration)**, then fit it into the cylinder head and secure it with the bolt **(see illustration 6.44b and 6.44a)**.
46 Feed the wiring back to the connector and reconnect it **(see illustration 6.38)**. Replace the heat shield, then install the air filter housing and the fairing side panel.

Ignition pulse generator

Check

47 Raise the fuel tank (see Section 2). Trace the ignition pulse generator coil wiring from the top of the clutch cover and disconnect it at the red 2-pin wiring connector **(see illustration)**. Perform the following checks.

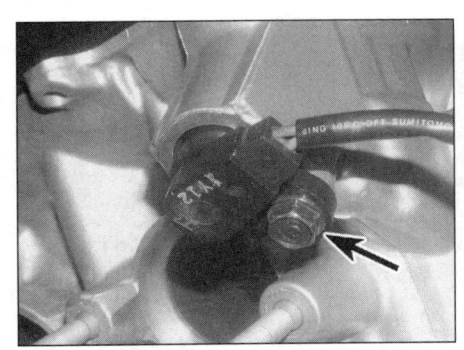

6.44a Unscrew the bolt (arrowed) . . .

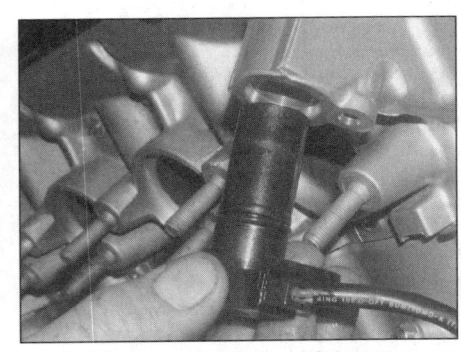

6.44b . . . and withdraw the sensor

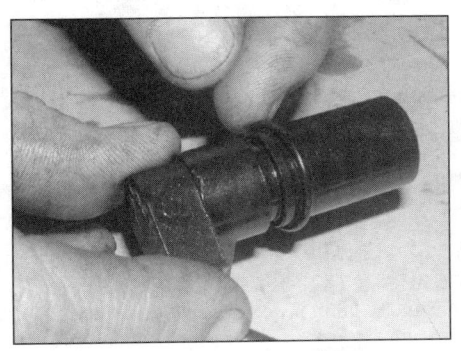

6.45 Fit a new O-ring into the groove

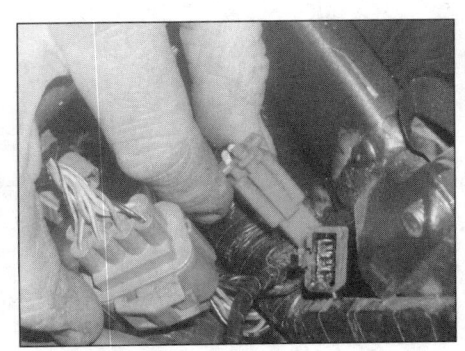

6.47 Disconnect the wiring connector

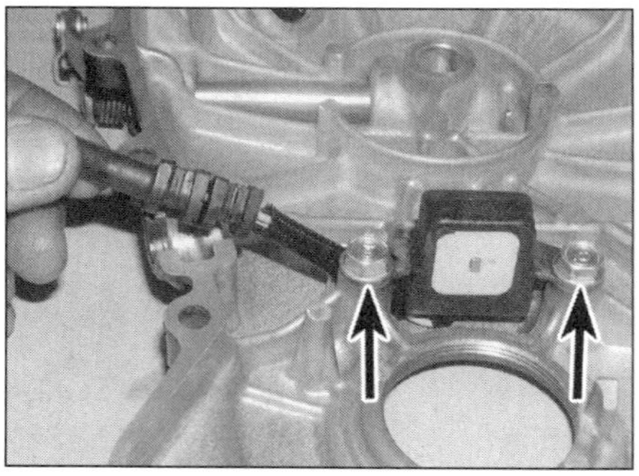

6.52 Unscrew the sensor bolts (arrowed) and free the wiring grommet

48 Using an ohmmeter check for continuity first between the yellow wire terminal on the sensor side of the connector and earth (ground), and then between the white/yellow wire terminal and earth. If there is continuity in either case the ignition pulse generator is faulty.

49 Connect the positive (+) lead of a voltmeter and peak voltage adapter arrangement* to the yellow terminal on the sensor side of the ignition pulse generator connector and the negative (–) lead to the white/yellow terminal of the connector. Turn the engine over on the starter motor and note the voltage reading obtained. If this reading is below the specified minimum, the ignition pulse generator is faulty.

*Note: *Honda specify their own Imrie diagnostic tester (model 625), or the peak voltage adapter (Pt. No. 07HGJ-0020100) with an aftermarket digital multimeter having an impedance of 10 M-ohm/DCV minimum for this test.*

50 If the ignition pulse generator functions correctly then the fault must be in the wiring harness or the ECM, and can be located by a Honda dealer with the test pin box.

Removal

51 Remove the clutch cover (see Chapter 2, Section 16, Steps 1 to 4). The sensor is mounted inside it.

52 Undo the sensor mounting bolts, then free the wiring grommet from the cover and remove the sensor **(see illustration)**.

Installation

53 Remove all traces of sealant from the sensor wiring grommet and clutch cover and apply a smear of fresh sealant to the grommet.

54 Locate the grommet and sensor correctly in the cover and tighten the sensor bolts **(see illustration 6.52)**.

55 Install the clutch cover (see Chapter 2, Section 16, Steps 29 to 31).

Speed sensor

56 See Chapter 9, Section 16.

Lean angle sensor

Check

57 Position the motorcycle on an auxiliary stand so that the motorcycle is level. On RR-Y and RR-1 (2000 and 2001) models, raise the fuel tank (see Section 2) – the lean angle sensor is mounted just below the tank bracket **(see illustration)**. On RR-2 and RR-3 (2002 and 2003) models, remove the fairing (see Chapter 8) – the lean angle sensor is mounted in front of the instrument cluster **(see illustrations)**.

58 With the ignition switch ON, connect the negative (–) lead of a voltmeter to the green terminal of the lean angle sensor connector (with the connector still connected). Connect the voltmeter positive (+) lead first to the white terminal and check that battery voltage (approximately 12 volts) is present, then connect it to the red/white terminal and check that between 0 to 1 volt is present.

59 Switch the ignition OFF. Undo the sensor screws and free it from its mounting (again with the connector still connected). Hold the sensor horizontal and switch the ignition on; the engine stop relay (located next to the sensor on RR-Y and RR-1 (2000 and 2001) models and under the passenger seat on the right-hand side on RR-2 and RR-3 (2002 and 2003) models – **see illustration 6.66a or 6.66b**) should click, indicating the power supply is closed (on). Slowly tilt the sensor to the left whilst listening to the engine stop relay; once the sensor reaches an angle of approximately 60° the relay should be heard to click, indicating the power supply is open (off). Switch the ignition OFF and return the sensor to the horizontal, then switch the ignition back ON again (engine stop relay should click again) and tilt the sensor to the right. The engine stop relay should be heard to click again once the sensor reaches an angle of around 60°.

60 If the voltage readings and/or relay performance are not as given, then it is likely the lean angle sensor is faulty.

Removal

61 On RR-Y and RR-1 (2000 and 2001) models, raise the fuel tank (see Section 2) – the lean angle sensor is mounted just below the tank bracket **(see illustration 6.57a)**. On RR-2 and RR-3 (2002 and 2003) models, remove the fairing (see Chapter 8) – the lean angle sensor is mounted in front of the instrument cluster **(see illustrations 6.57b and 6.57c)**.

62 Disconnect the lean angle sensor wiring connector then undo the screws and remove the sensor from its mounting.

Installation

63 Installation is the reverse of removal. Make sure the sensor is fitted with its UP mark facing upwards.

Engine stop relay

Check

64 Remove the relay (see below).

65 Connect an ohmmeter between the red/white and black/white wire terminals of the relay. Using a 12 volt battery and auxiliary wires, connect the battery positive (+) terminal to the red/orange wire terminal of the relay and the negative (–) terminal to the black wire terminal of the relay and note the meter reading obtained. If the relay is operating correctly there should be continuity (zero resistance) when the battery is connected and no continuity (infinite resistance) when the battery is disconnected. If this is not the case, replace the relay with a new one.

Removal

66 On RR-Y and RR-1 (2000 and 2001) models, raise the fuel tank (see Section 2) – the engine stop relay is mounted just below

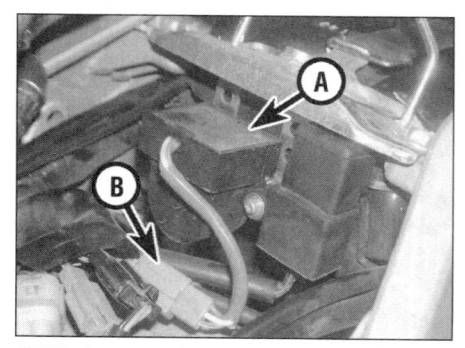

6.57a Lean angle sensor (A) and its wiring connector (B) – RR-Y and RR-1 (2000 and 2001) models

6.57b Lean angle sensor (arrowed) . . .

6.57c . . . and its wiring connector – RR-2 and RR-3 (2002 and 2003) models

Fuel and exhaust systems 4•15

6.66a Engine stop relay (arrowed) – RR-Y and RR-1 (2000 and 2001) models

6.66b Engine stop relay (arrowed) – RR-2 and RR-3 (2002 and 2003) models

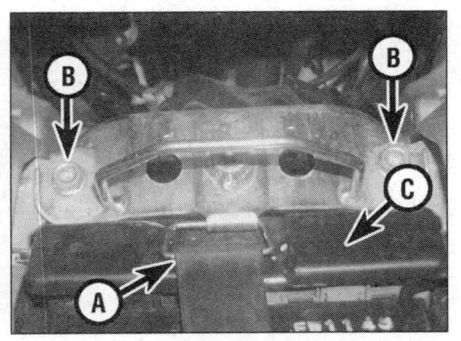

6.72 Release the battery strap (A), then unscrew the bolts (B) and remove the bracket, then remove the ECM cover (C)

the tank bracket **(see illustration)**. On RR-2 and RR-3 (2002 and 2003) models remove the seat cowling (see Chapter 8). The engine stop relay is the middle relay located on the right-hand side of the rear sub-frame **(see illustration)**.

67 Free the relay from its mounting, then disconnect the wiring connector and remove the relay.

Installation

68 Installation is the reverse of removal.

Fuel cut-off relay

Check

69 Remove the relay (see below).
70 Connect an ohmmeter between the black/white and brown wire terminals of the relay. Using a 12 volt battery and auxiliary wires, connect the battery positive (+) terminal to the brown/black wire terminal of the relay and the negative (–) terminal to the other black/white wire terminal of the relay and note the meter reading obtained. If the relay is operating correctly there should be continuity (zero resistance) when the battery is connected and no continuity (infinite resistance) when the battery is disconnected. If this is not the case, replace the relay with a new one.

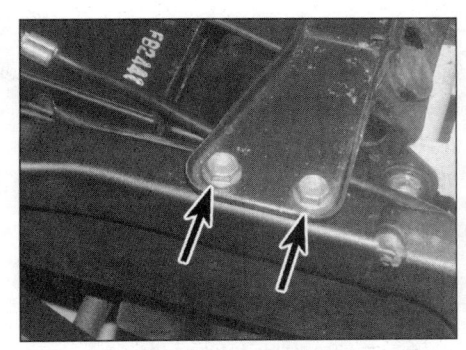

6.73a Unscrew the two bolts (arrowed) on each side and pivot the bracket up

Removal

71 The fuel cut-off relay is located just in front of the ECM. Remove the rider's seat (see Chapter 8).
72 On RR-Y and RR-1 (2000 and 2001) models unscrew the fuel tank mounting bolts and raise the rear of the tank (see Section 2). Release the battery strap then undo the tank bracket bolts and remove the bracket **(see illustration)**. Remove the ECM cover.
73 On RR-2 and RR-3 (2002 and 2003) models unscrew the fuel tank bracket bolts and pivot the bracket up **(see illustration)**.

6.73b Remove the ECM cover

Remove the battery (see Chapter 9), then remove the ECM cover **(see illustration)**.
74 Free the relay from its mounting, then disconnect the wiring connector and remove the relay **(see illustrations)**.

Installation

75 Installation is the reverse of removal.

Engine control module (ECM)

Check

76 The engine control module (ECM) can only be checked using the Honda diagnostic test pin box.

6.74a Free the relay (arrowed) from its mount . . .

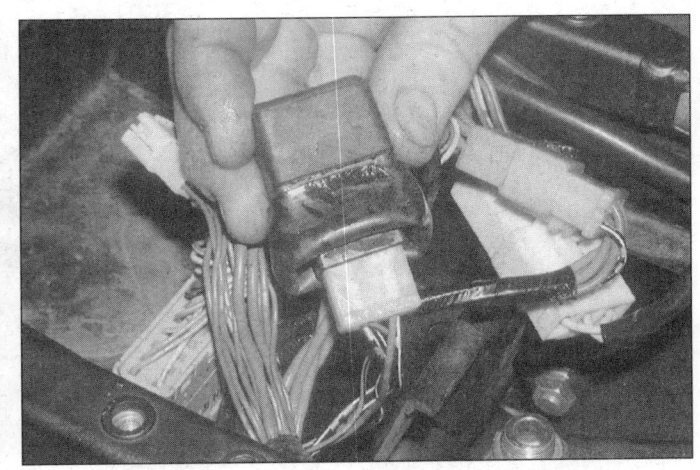

6.74b . . . and disconnect the wiring connector

4•16 Fuel and exhaust systems

Removal

77 Remove the rider's seat (see Chapter 8). Disconnect the battery negative (–) terminal.
78 On RR-Y and RR-1 (2000 and 2001) models unscrew the fuel tank mounting bolts and raise the rear of the tank (see Section 2). Release the battery strap then undo the tank bracket bolts and remove the bracket **(see illustration 6.72)**. Remove the ECM cover.
79 On RR-2 and RR-3 (2002 and 2003) models unscrew the fuel tank bracket bolts and pivot the bracket up **(see illustration 6.73a)**. Remove the battery (see Chapter 9), then remove the ECM cover **(see illustration 6.73b)**.
80 Lift the ECM out of its holder and disconnect the wiring connectors **(see illustration)**.

Installation

81 Installation is the reverse of removal.

Oxygen sensor

Check

82 Apart from the wiring checks that are outlined in Section 5, the operation of the oxygen sensor can only be checked using the Honda diagnostic test pin box.
83 To check the sensor heater, remove the seat (see Chapter 8). Trace the wiring back from the sensor in the exhaust to the wiring connector and disconnect it. Connect an ohmmeter between the white wire terminals on the sensor side of the connector and check that the resistance is between 10 and 40 ohms. Also check that there is no continuity to earth (ground) in either of the white wires. If the resistance is not as specified or if there is continuity to earth, replace the sensor with a new one. Otherwise check for battery voltage between the black/white (+) and black/green (-) terminals on the loom side of the connector with the ignition ON. If there is no voltage, check the wiring, using the wiring diagrams at the end of Chapter 9. Otherwise have the sensor and its circuit tested by a Honda dealer equipped with the diagnostic tester.

Removal

Note: *The oxygen sensor is delicate and will not work if it is dropped or knocked, or if any cleaning materials are used on it. Ensure the exhaust system is cold before proceeding.*
84 On RR-Y and RR-1 (2000 and 2001) models raise the fuel tank (see Section 2). On RR-2 and RR-3 (2002 and 2003) models remove the seat (see Chapter 8).
85 Trace the wiring back from the sensor to the wiring connector and disconnect it. Release the wiring from the clamp securing it to the rear brake reservoir hose.
86 Unscrew the two heel guard/master cylinder mounting bolts and displace the guard. Release the sensor wiring from the clamp on the inside of the guard.
87 Unscrew the oxygen sensor and remove it from the exhaust system.

6.80 Disconnect the wiring connectors and remove the ECM

Installation

88 Installation is the reverse of removal. Tighten the sensor to the torque setting specified at the beginning of the Chapter.

H-VIX servo motor

89 See Section 15.

7 Throttle body assembly – removal and installation

Warning: *Refer to the precautions given in Section 1 before starting work.*

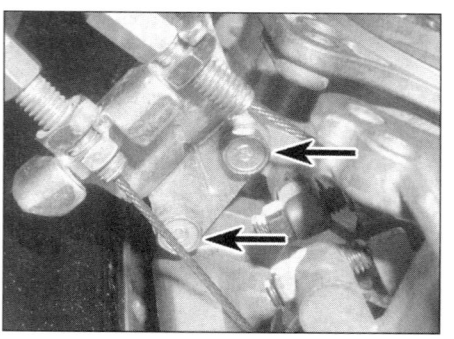

7.2a Unscrew the bolts (arrowed) and displace the bracket . . .

7.3 Free the idle speed adjuster from its holder (arrowed)

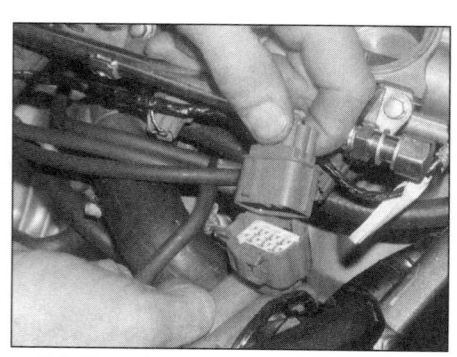

Removal

1 Remove the fuel tank and the air filter housing (Sections 2 and 3). Either drain the coolant (see Chapter 1), or prepare some hose clamps or blanking plugs for the fast idle system wax unit heating system hoses – there are two of them.
2 Unscrew the bolts securing the throttle cable bracket to the throttle body assembly **(see illustration)**. Free both inner cable ends from the throttle pulley **(see illustration)**.
Caution: *Do not snap the throttle cam/valves from fully open to fully closed once the cables have been disconnected because this can lead to engine idle speed problems.*
3 Release the idle speed adjuster from its holder and feed it through to the base of the throttle body assembly **(see illustration)**.
4 Either disconnect the wiring connector from each fuel injector and the throttle position sensor **(see illustrations 6.3 and 6.17)**, or alternatively trace the wiring and disconnect it at the 8-pin sub-loom connector, leaving the injector and throttle position sensor (TPS) wiring connected **(see illustration)**.
5 On California models, disconnect the EVAP system solenoid valve vacuum hose from the five-way hose joint on the throttle body assembly.
6 Slacken the clamps securing the coolant hoses to the fast idle system wax unit **(see**

7.2b . . . then free the cable ends from the pulley

7.4 Throttle body assembly sub-loom wiring connector

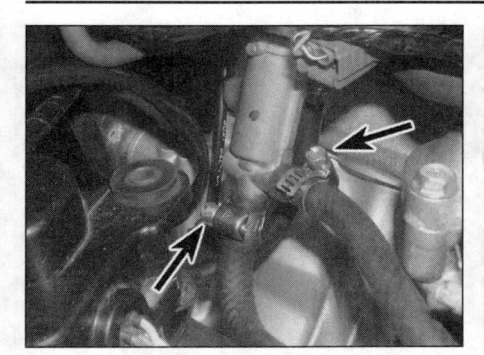

Fuel and exhaust systems 4•17

7.6 Slacken the clamps (arrowed) and detach the hoses

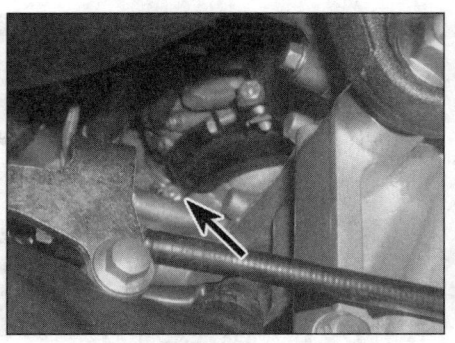

7.7a Slacken the lower clamp screw (arrowed) on each rubber . . .

7.7b . . . and ease the throttle bodies up off the cylinder head

illustration). Detach the hoses, and if the coolant wasn't drained, clamp or plug the ends.

7 Fully slacken the lower clamps on the cylinder head intake manifold rubbers, noting their orientation **(see illustration)**. Ease the throttle body assembly up off the manifold along with the intake rubbers, noting that they are quite a tight fit, and remove the assembly **(see illustration)**.

Caution: Tape over or stuff clean rag into each cylinder head intake after removing the throttle body assembly to prevent anything from falling in.

8 If the intake rubbers shown signs of damage or deterioration new ones must be fitted. Note their orientation and how the clamps locate and are orientated before removing them, and on installation tighten the clamps so that the gap between the ends is 6 to 8 mm **(see illustrations)**. Also check the throttle body vacuum hoses for signs of damage or deterioration and replace any suspect hoses with new ones **(see illustration)**.

Caution: The throttle body assembly must be treated as a sealed unit. With the exception of the fast idle system wax unit screws, NEVER loosen any of the white-painted nuts/bolts/screws on the assembly as these are pre-set at the factory to ensure correct synchronisation of the throttle valves. The only components on the assembly which are serviceable are the starter valves (see Section 8) and the various vacuum hoses.

Caution: NEVER use a solvent-based carburettor cleaner to clean the throttle body assembly. The throttle bores are covered with a molybdenum coating which could be removed by the cleaner.

9 If required, disconnect the injector and throttle position sensor wiring connectors and remove the sub-harness **(see illustrations 6.3 and 6.17)**.

Installation

10 If removed, connect the wiring sub-harness to the injectors and throttle position sensor **(see illustrations 6.3 and 6.17)**. Remove the tape/plugs from the intakes.

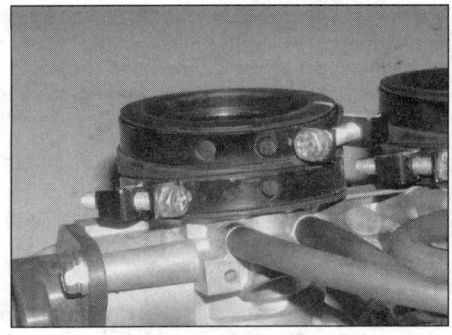

7.8a Note the orientation of the clamps and how they locate . . .

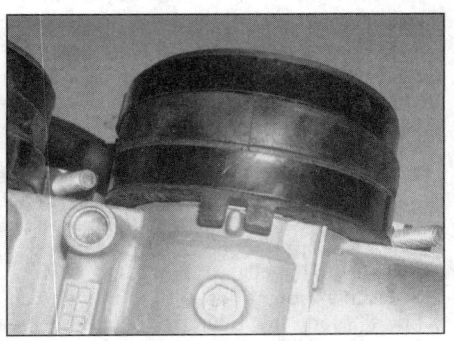

7.8b . . . and how the rubbers locate

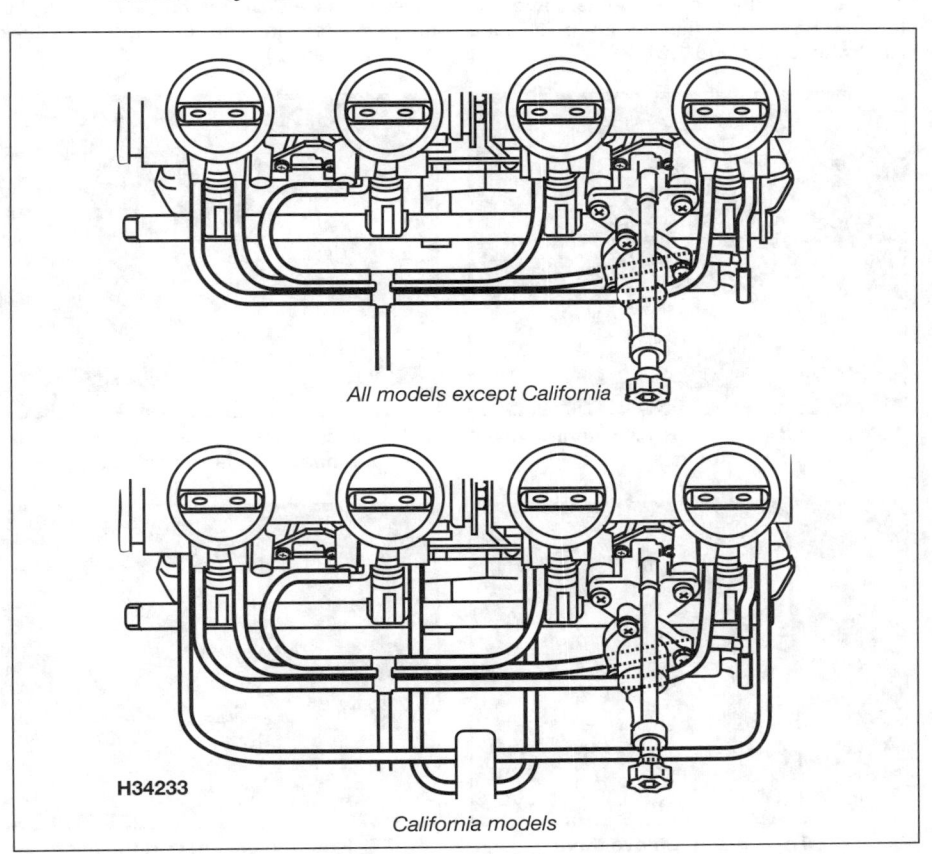

All models except California

California models

7.8c Vacuum hose location and routing

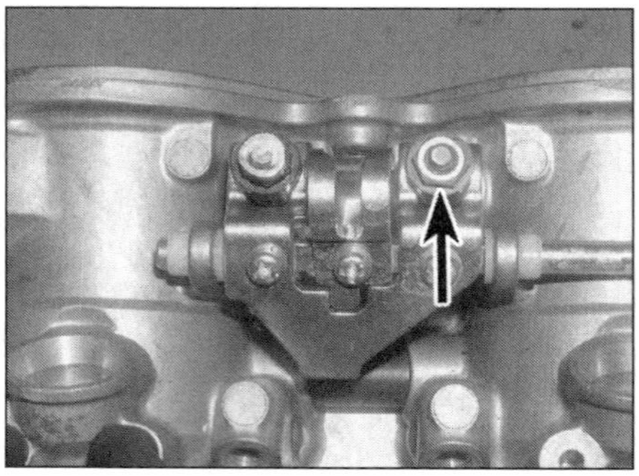

8.3 No. 2 starter valve adjustment nut (arrowed) (wax unit removed) – the No. 1 nut next to it cannot be adjusted

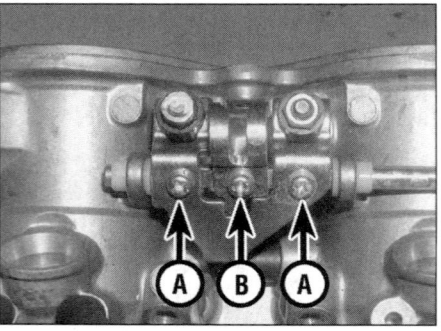

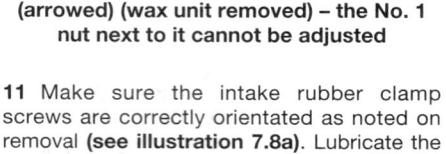

8.5 Starter valve arm screws (A), wax unit link arm screw (B)

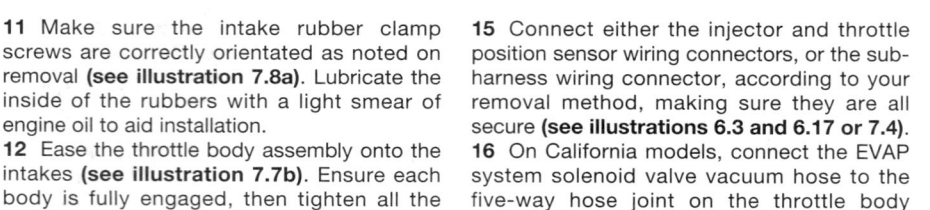

8.6a Undo the two screws...

11 Make sure the intake rubber clamp screws are correctly orientated as noted on removal **(see illustration 7.8a)**. Lubricate the inside of the rubbers with a light smear of engine oil to aid installation.

12 Ease the throttle body assembly onto the intakes **(see illustration 7.7b)**. Ensure each body is fully engaged, then tighten all the clamps so that the gap between the ends is 11 to 13 mm **(see illustration 7.7a)**.

13 Fit the coolant hoses onto the wax unit and tighten the clamps securely **(see illustration 7.6)**.

14 Connect the throttle cable ends to the cam, then fit the cable bracket onto the throttle body and tighten its bolts **(see illustrations 7.2b and 7.2a)**.

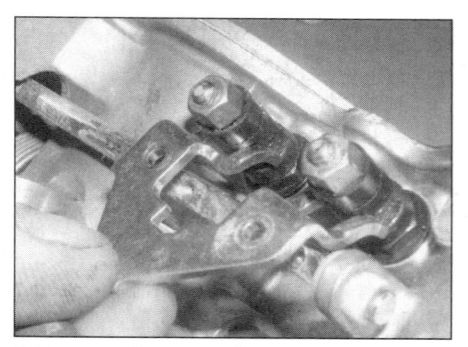

15 Connect either the injector and throttle position sensor wiring connectors, or the sub-harness wiring connector, according to your removal method, making sure they are all secure **(see illustrations 6.3 and 6.17 or 7.4)**.

16 On California models, connect the EVAP system solenoid valve vacuum hose to the five-way hose joint on the throttle body assembly.

17 Fit the idle speed adjuster into its holder **(see illustration 7.3)**.

18 If the cooling system was drained, do not forget to refill it (see Chapter 1). Otherwise, just check the level and top up if necessary (see *Daily (pre-ride) checks*). Install the air filter housing and the fuel tank (Sections 4 and 2).

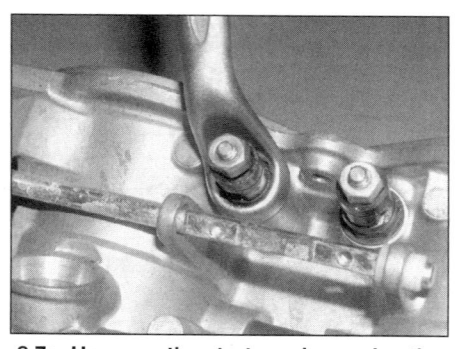

8 Starter valves – removal, installation and synchronisation

Warning: Refer to the precautions given in Section 1 before starting work.

Removal

1 Remove the throttle body assembly (see Section 7).

2 Remove the fuel rail and injectors (see Section 6).

3 Screw the Nos. 2, 3 and 4 cylinder throttle body starter valve synchronisation nuts in until they seat lightly, noting the exact number of turns needed to do so on a piece of paper **(see illustration)**. The valve for the No. 1 cylinder throttle body is the base and is pre-set and fixed.

4 To access the Nos. 1 and 2 starter valves, remove the fast idle system wax unit (see Section 9).

5 Undo the screws securing the arms for the Nos. 1 and 2 starter valves and the screw securing the link arm for the wax unit and remove the arms, noting how they fit **(see illustration)**.

6 Undo the screws securing the arm for the Nos. 3 and 4 starter valves and remove the arm, noting how it fits **(see illustrations)**.

7 Unscrew the starter valve base nuts and remove the four valves from the throttle body, keeping them in their correct fitted order **(see illustrations)**. Use a ring spanner, not a deep socket as this will turn the adjuster nut on the valve as well.

8 To remove the shaft, on RR-Y and RR-1 (2000 and 2001) models remove the E-clip and washer from its left-hand end. On all models, draw the shaft and out from the right-hand end, noting the three collars in the bores that the shaft runs in, and on RR-Y and RR-1 (2000 and 2001) models the spring **(see illustration)**. Remove the collars for safekeeping if they are loose.

9 Check all components for wear and

8.6b ... and remove the arm, noting how it locates

8.7a Unscrew the starter valves using the base nuts (arrowed)...

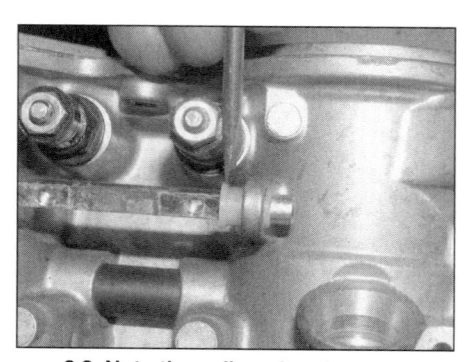

8.7b ... and remove them

8.8 Note the collars that the shaft runs in

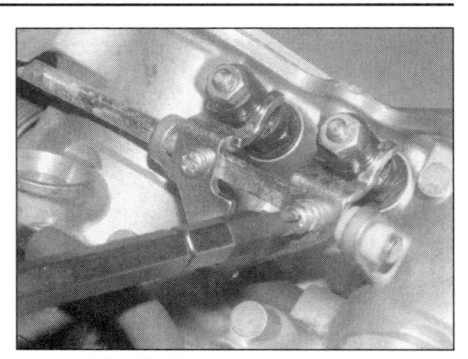

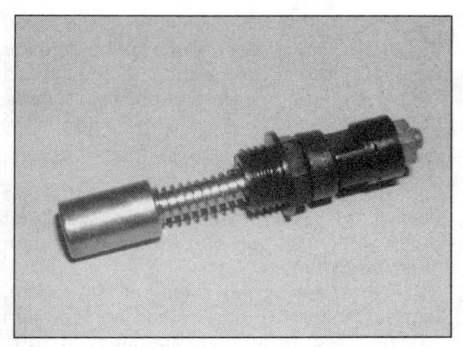

Fuel and exhaust systems 4•19

8.9 Check the action of the starter valve plunger in the body – it should move smoothly and return easily under spring pressure

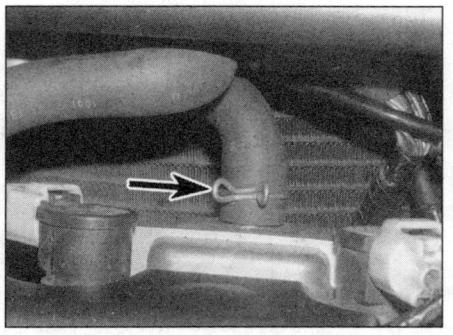

8.21 Release the clamp (arrowed) and detach the hose from the union on each reed valve cover

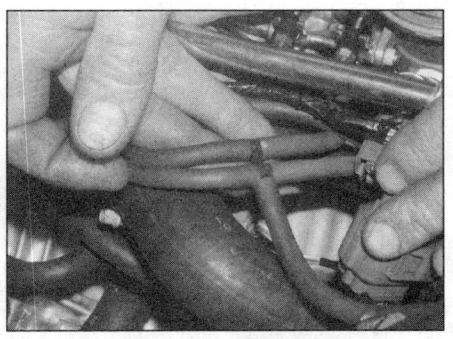

8.22 Detach the four opposed hoses from the joint and connect the gauge hoses to them, making sure the No. 1 gauge goes to the No. 1 throttle body and so on

damage and renew as necessary **(see illustration)**.

Installation

10 Clean the starter valves and throttle body passages using compressed air only. Do not use a carburettor cleaner or any other solvent. *Caution: NEVER use a solvent-based carburettor cleaner to clean the throttle body components. The throttle bores are covered with a molybdenum coating which could be removed by the cleaner.*

11 Install each starter valve in its original location **(see illustration 8.7b)**. Tighten the valve base nuts then check that each valve moves smoothly and easily in its bore by pulling on the valve end, and that it closes fully under spring pressure **(see illustration 8.7a)**.

12 If removed, fit the collars into the shaft bores, noting that the bore adjacent the No. 3 starter valve does not have one, and making sure they are the correct way round **(see illustration 8.8)**. Slide the shaft in from the right-hand end, making sure the collars stay in place. On RR-Y and RR-1 (2000 and 2001) models do not forget to fit the spring, and secure the shaft with the E-clip and washer.

13 Locate the arm for the Nos. 1 and 2 starter valves on the shaft and under the valve ends and secure it with the screws **(see illustration 8.5)**. Make sure the arm ends engage correctly with the flat sides on the valves. Also fit the link arm for the wax unit and secure it with the screw.

14 Locate the arm for the Nos. 3 and 4 starter valves on the shaft and under the valve ends and secure it with the screws **(see illustration 8.6b)**. Again make sure the arm ends engage correctly with the valves **(see illustration 8.6a)**.

15 Check the operation of the starter valve shaft before continuing; it should move smoothly and easily, drawing all the valves out as it turns, and return to the fully closed position under pressure of the return springs.

16 Install the idle system wax unit (see Section 9).

17 Turn the No. 2, 3 and 4 starter valve synchronisation nuts in until they seat lightly, then back each one out by the exact number of turns noted prior to removal **(see illustration 8.3)**.

18 Install the fuel injectors (see Section 6), then install the throttle body assembly (Section 7). On completion check the starter valve synchronisation (see below).

Synchronisation

Note: *This procedure should be carried out if the starter valves have been removed from the throttle body assembly. The procedure does not alter the setting of the throttle valves themselves (these are pre-set at the factory and fixed), but only the starter valves, which control the idle speed when the engine is cold and warming up.*

19 Starter valve synchronisation is simply the process of adjusting the valves so they pass the same amount of fuel/air mixture to each cylinder on cold start and warm-up. This is done by measuring the vacuum produced in each intake duct. Starter valves that are out of synchronisation will result in uneven idling when cold starting and warming the engine. Before synchronising the starter valves, make sure the valve clearances are properly set (see Chapter 1).

20 To properly synchronise the starter valves, you will need a set of vacuum gauges or calibrated tubes to indicate engine vacuum. The equipment used should be suitable for a four cylinder engine and come complete with the necessary adapters and hoses to fit the take-off points. **Note:** *Because of the nature of the synchronisation procedure and the need for special instruments, most owners leave the task to a Honda dealer.*

21 Remove the air filter housing (see Section 3). Lift the rubber heat shield **(see illustration 6.37)**. Detach the air hoses from the PAIR system reed valve cover unions on the valve cover and fit blanking caps in their place **(see illustration)**.

22 On RR-Y and RR-1 (2000 and 2001) models unscrew the vacuum take-off point blanking screws and thread suitable adapters in their place. Connect the gauge hoses to the adapters. On RR-2 and RR-3 (2002 and 2003) models detach the vacuum hoses from the five-way joint **(see illustration)**. Connect the gauge hoses to the vacuum hoses using suitable adapters. Make sure everything is a good fit because any air leaks will result in false readings.

23 Start the engine and adjust the idle speed (see Chapter 1). If using vacuum gauges fitted with damping adjustment, set this so that the needle flutter is just eliminated but so that they can still respond to small changes in pressure.

24 The vacuum readings for the Nos. 2, 3 and 4 cylinders should be the same as the No. 1 cylinder, or at least within the maximum difference specified at the beginning of the Chapter. The No. 1 cylinder starter valve is the base to which all the others are matched, and cannot itself be adjusted. If the vacuum readings vary, adjust each starter valve as required by turning the synchronisation nut on the end of the valve, until the readings are the same **(see illustrations)**. **Note:** *Do not press*

8.24a No. 2 starter valve adjustment nut (arrowed)

8.24b Nos. 3 and 4 starter valves adjustment nuts (arrowed)

hard on the nut whilst adjusting it, otherwise a false reading will be obtained.

25 When the adjustment is complete, recheck the vacuum readings, then adjust the idle speed by turning the throttle stop screw (see Chapter 1) until the idle speed is obtained. Stop the engine.

26 Remove the vacuum gauges and the hose adapters. On RR-Y and RR-1 (2000 and 2001) models thread the blanking screws into place. On RR-2 and RR-3 (2002 and 2003) models fit the vacuum hoses on to the five-way joint **(see illustration 8.22)**. Remove the air filter housing and lift the heat shield, then fit the PAIR system hoses back onto the unions on the reed valve covers and secure them with the clamps **(see illustration 8.21)**. Install the heat shield **(see illustration 6.37)**, the air filter housing, and the fuel tank (see Section 3 and 2).

9 Fast idle system wax unit – removal, inspection and installation

Warning: *Refer to the precautions given in Section 1 before starting work.*

Removal

1 Remove the throttle body assembly (see Section 7).
2 Remove the fuel rail and injectors (see Section 6).

9.3 Undo the screws (arrowed) and pivot the wax unit upwards

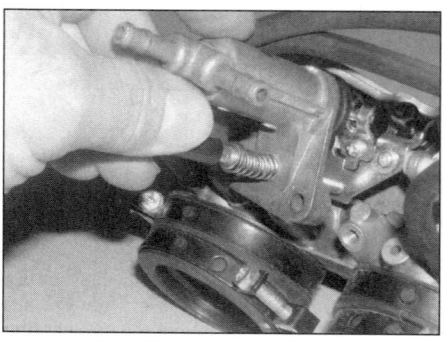

9.11a Pivot the unit down . . .

3 Undo the two screws securing the wax unit and displace it upwards from its mount **(see illustration)**.
4 Turn the starter valve shaft so that the valves are open, then pivot the wax unit up further to release the pushrod pivot piece from the link arm, and remove the wax unit **(see illustration)**.

Inspection

5 Slacken the three wax unit cover screws evenly and in a criss-cross pattern in three stages. Discard the O-ring as a new one must be used. Remove the spring, spring seat and wax element from the body (or in reverse order from the cover, depending where they are). Discard the wax element O-rings.
6 Visually inspect all components for signs of wear and damage.
7 The wax element expands with heat. If you suspect it is not working correctly, place it first in a cold place and check that it is fully retracted. Now gently heat it using a hairdryer or similar and check that it expands. If the element is faulty, on RR-Y and RR-1 (2000 and 2001) models a new one can be installed (it comes as an assembly with a new spring and spring seat). On RR-2 and RR-3 (2002 and 2003) models it is not available separately so a complete new unit must be installed.
8 Check that the pushrod moves smoothly in and out of the body. Check the spring for fatigue and distortion.
9 Fit new O-rings onto the wax element and

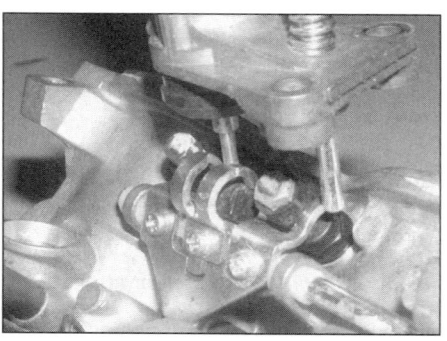

9.4 Open the valves by manually turning the shaft then release the pushrod end from the link arm . . .

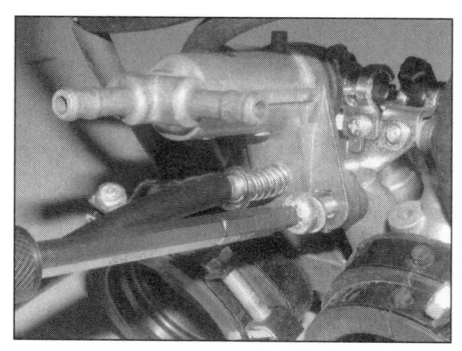

9.11b . . . and secure it with its screws

the cover. Fit the wax element into the body, then fit the spring seat and the spring onto the element. Fit the cover and tighten its screws evenly and in a criss-cross pattern in three stages.

Installation

10 Turn the starter valve shaft so that the valves are open, then fit the wax unit pushrod pivot piece into the link arm **(see illustration 9.4)**.
11 Pivot the wax unit down onto its mount, then install the two screws and tighten them to the torque setting specified at the beginning of the Chapter **(see illustrations)**.
12 Install the injectors and fuel rail (see Section 6).
13 Install the throttle body assembly (see Section 7).

10 Fuel supply system – flow rate and pressure check

Warning: *Refer to the precautions given in Section 1 before starting work.*

1 If there is thought to be a problem with the fuel supply system, the fuel flow rate and fuel pressure can be checked as follows.

Fuel flow rate check

2 Remove the rider's seat (see Chapter 8). Raise and support the fuel tank (see Section 2).
3 Ensure the ignition is switched OFF. Disconnect the wiring connector from the fuel cut-off relay (see Section 6, Steps 71 to 74). Using an auxiliary wire, bridge between the brown and black/white wiring terminals of the connector.
4 Position a wad of rag and a graduated container suitable for holding half a litre of fuel beneath the fuel return hose union on the fuel tank and have a suitable plug ready to block the union. Working quickly to minimise fuel loss, disconnect the return hose from the tank and plug the tank union **(see illustration 11.6)**. Place the end of the return hose into the container and mop up all spilt fuel.
5 Turn on the ignition switch for exactly 10 seconds, catching all the fuel expelled from the return hose in the container, and then turn off the ignition switch. Measure the amount of fuel collected and compare this to the minimum fuel flow amount given in the Specifications.
6 If fuel flow is below the specified minimum there is a problem in the fuel supply system. Likely causes are.
● Blocked/restricted fuel feed or return hose.
● Blocked fuel filter.
● Faulty fuel pressure regulator.
● Faulty fuel pump.
7 On completion, reconnect the return hose to the fuel tank and secure it in position with the clamp **(see illustration 11.6)**.

8 Remove the bridging wire from the fuel cut-off relay and install the relay. Start the engine and check that there is no sign of fuel leakage. If all is well, lower the tank (see Section 2) then install the seat (see Chapter 8).

Fuel pressure check

Note: *A pressure gauge is required for this check. Honda specify the use of their gauge (Pt. No. 07406-0040002 or 3) along with a special banjo bolt and two sealing washers for the gauge to thread onto (part Nos. 90008-PP4-E02, 90428-PD6-003 and 90430-PD6-003). If a different gauge is used an adapter may be needed, either so it can thread onto the special banjo bolt, or that can be used in place of the special bolt. A gauge with a male end of the correct thread size and length could be used in place of the special bolt without an adapter. Two new sealing washers for the supply hose banjo bolt are also required.*

9 Remove the rider's seat and disconnect the battery negative (–) terminal. Raise and support the fuel tank (see Section 2)
10 Disconnect the vacuum hose from the fuel pressure regulator and plug the hose end **(see illustration 6.14a)**.
11 Release the fuel pressure (see Section 2).
12 Remove the banjo bolt and its sealing washer and replace them with the special banjo bolt and the larger sealing washer, then thread the gauge onto the special bolt, using the smaller sealing washer, working quickly to minimise fuel spillage. Securely tighten the gauge to ensure there are no fuel leaks. Mop up any spilt fuel.
13 Connect the battery negative (–) lead then start the engine and allow it to idle at the specified speed. Note the pressure present in the fuel system by reading the gauge, then turn the engine off. Compare the reading obtained to that given in the Specifications.
14 If the fuel pressure is higher than specified, likely causes are.
• Blocked/restricted fuel feed or return hose.
• Faulty fuel pressure regulator.
• Faulty fuel pump (although this is unlikely).
15 If the fuel pressure is lower than specified, likely causes are.
• Leaking fuel hose union/injector.
• Blocked fuel filter.
• Faulty fuel pressure regulator.
• Faulty fuel pump.
16 On completion, disconnect the battery negative (–) lead again. Remove the fuel gauge assembly, then quickly install the original banjo bolt using a new sealing washer on each side of the union, fitting the neck of the hose between the lugs, and tighten the banjo bolt to the torque setting specified at the beginning of the Chapter **(see illustrations 2.10a and 2.10b)**. Reconnect the vacuum hose to the fuel pressure regulator **(see illustration 6.14a)**. Reconnect the battery then start the engine and check that there is no sign of fuel leakage. If all is well, lower the tank (see Section 2)

11 Fuel pump – check, removal and installation

Warning: *Refer to the precautions given in Section 1 before starting work.*

Check

1 The fuel pump is located inside the fuel tank. The fuel pump runs for a few seconds when the ignition is switched ON, to pressurise the fuel system, and then cuts out until the engine is started. Check that it does this. If the pump is thought to be faulty, first check the fuses (see Chapter 9). If they are in good condition proceed as follows.
2 Raise and support the fuel tank (see Section 2).
3 Ensure the ignition is switched OFF then disconnect the fuel pump wiring connector **(see illustration 2.4a)**. Connect the positive (+) lead of a voltmeter to the brown wire terminal on the loom side of the connector and the negative (–) lead to the green wire terminal. Switch the ignition ON whilst noting the reading obtained on the meter.
4 If battery voltage is present for a few seconds, the fuel pump circuit is operating correctly and the fuel pump itself is faulty and must be replaced with a new one.
5 If no reading is obtained, check the fuel pump circuit wiring for continuity and make sure all the connectors are free from corrosion and are securely connected. Repair/replace the wiring as necessary and clean the connectors using electrical contact cleaner. If this fails to reveal the fault, check the following components.
• Engine stop switch (see Chapter 9, Section 21).
• Fuel cut-off relay (see Section 6).
• Engine stop relay (see Section 6).
• Lean angle sensor (see Section 6).
• Engine control module (ECM) (see Section 6).

Removal

6 Remove the fuel tank (see Section 2), and place it upside down. Unscrew the fuel supply hose bolt and remove the hose **(see illustration)**. Release the clamp securing the fuel return hose and detach it from its union.
7 With the fuel tank supported upside-down, unscrew the fuel pump mounting plate nuts **(see illustration 11.6)**. Carefully remove the pump assembly from the tank along with the mounting plate seal **(see illustration)**. Discard the seal – a new one must be used on installation.
8 To remove the pump from its mounting plate, first disconnect the wiring connector **(see illustration)**.
9 Release the fuel hose clamp and detach the hose from the pump **(see illustration)**.
10 Undo the retaining clamp screw, noting the earth (ground) lead, and remove the clamp **(see illustration 11.8)**. Remove the pump,

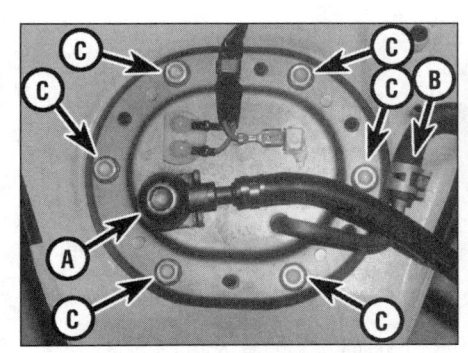

11.6 Fuel supply hose bolt (A), fuel return hose clamp (B), fuel pump plate nuts (C)

11.7 Carefully withdraw the pump assembly, noting its orientation

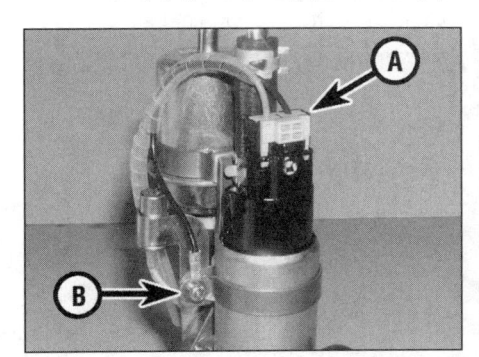

11.8 Disconnect the wiring connector (A). Fuel pump clamp screw (B) – note the earth wire

11.9 Release the clamp (arrowed) and detach the hose

11.10 Check the strainer (arrowed)

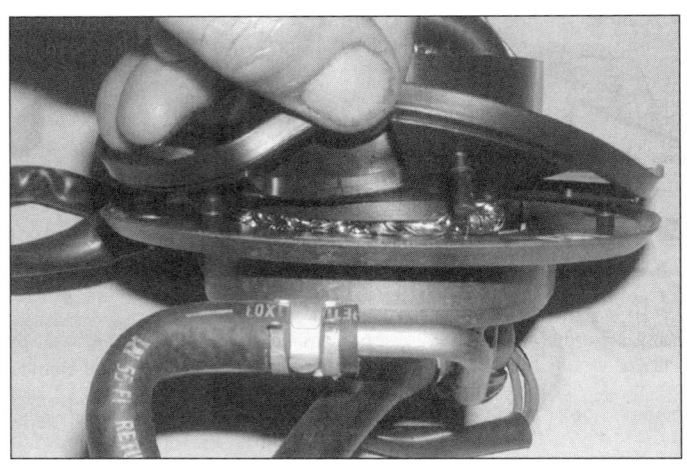

11.14 Fit a new seal, locating the pegs in the holes

noting how it fits. Check the wire strainer in the base of the pump housing for signs of dirt and damage and clean or renew as necessary **(see illustration)**.

Installation

11 If removed, fit the pump into its housing, then secure it with the clamp and tighten the screw, not forgetting to attach the earth wire **(see illustration 11.8)**.
12 Reconnect the fuel hose to the pump and secure it with the clamp **(see illustration 11.9)**.
13 Reconnect the pump wiring connector **(see illustration 11.8)**.
14 Ensure the mounting plate and tank surfaces are clean and dry, then fit the new seal onto the plate making sure its locating pins are all located correctly in the plate holes – the pins and holes are non-symmetrical so it can only fit one way, with the lip on one end designed to tuck in between the gauze strainer and the base **(see illustration)**. Pull each pin from the underside to make sure its lip has pulled through the hole in the base.
15 Manoeuvre the pump assembly into the tank, and seat it in position **(see illustration 11.7)**.
16 Fit the nuts and tighten them finger-tight **(see illustration 11.6)**. Now tighten them evenly and a little at a time in the numerical sequence shown to the torque setting specified at the beginning of the Chapter.
17 Install the fuel tank (see Section 2).

12 Fuel warning light and sensor – check and replacement

Note: *The low fuel warning light operates when there is approximately 3.5 litres of fuel remaining in the tank with the motorcycle upright.*

Check

1 The circuit consists of the sensor, which is an integral part of the fuel pump assembly mounting plate in the fuel tank, and the warning light, which is an LED that is part of the instrument cluster printed circuit board and incorporated in the tachometer. If the system malfunctions first check that the fuses are good (see Chapter 9).
2 If the warning display does not come on when it should, raise the fuel tank (see Section 2). Disconnect the fuel pump assembly wiring connector and bridge between the brown/black and green wire terminals on the loom side of the connector using a jumper wire **(see illustration 2.4a)**. Turn the ignition switch ON – the warning light should come on. If it does, the sensor is faulty and so the pump assembly mounting plate must be replaced with a new one.
3 If the light doesn't come on, remove the fairing (see Chapter 9), and disconnect the instrument cluster wiring connector **(see illustration)**. Check for continuity in the brown/black between the connector and the instrument cluster connector, using the wiring diagrams at the end of Chapter 9. Also check for continuity to earth (ground) in the green wire. If continuity (zero resistance) is not present, locate the break in the wire or faulty connector and repair or replace as required. If continuity exists, then the instrument cluster PCB could be faulty (see Chapter 9).
4 If the warning light is permanently on, raise the fuel tank (see Section 2), then disconnect the brown/black wiring connector from its terminal on the fuel pump mounting plate **(see illustration)**. Turn the ignition switch ON. If the warning light does not come on, replace the pump assembly mounting plate with a new one. If it comes on, check for a short circuit to earth in the brown/black wire between the sensor and the instrument cluster.

Replacement

5 See Chapter 9 for replacement of the instrument cluster.
6 The sensor is an integral part of the fuel pump assembly mounting plate in the fuel tank. Remove the fuel pump from the mounting plate and replace the plate with a new one (see Section 11).

13 Throttle cables – removal and installation

Warning: *Refer to the precautions given in Section 1 before proceeding.*

Removal

1 Remove the air filter housing (see Section 3). Lift the rubber heat shield **(see**

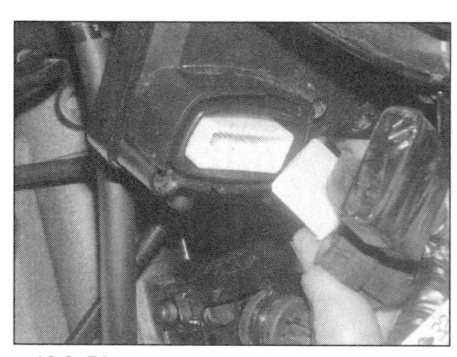

12.3 Disconnect the instrument cluster wiring connector

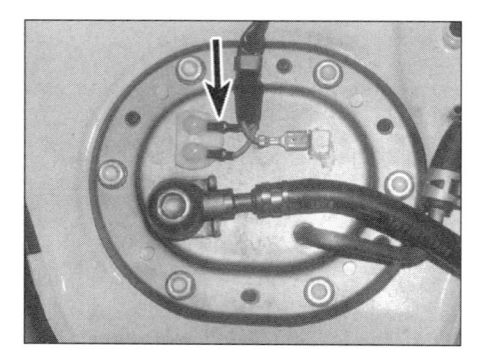

12.4 Disconnect the brown/black wire from its terminal (arrowed)

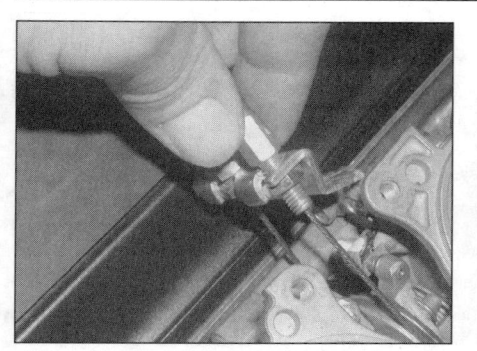

Fuel and exhaust systems 4•23

13.2a Unscrew the hex while keeping the nut captive . . .

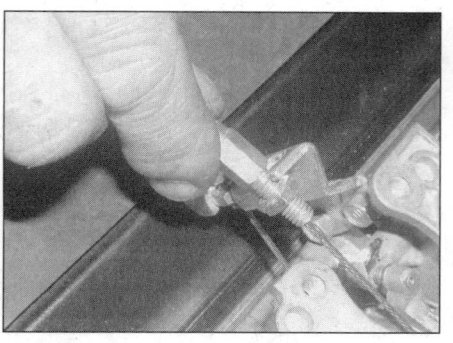

13.2b . . . then slip the cable out of the bracket . . .

13.2c . . . and free the end from the pulley

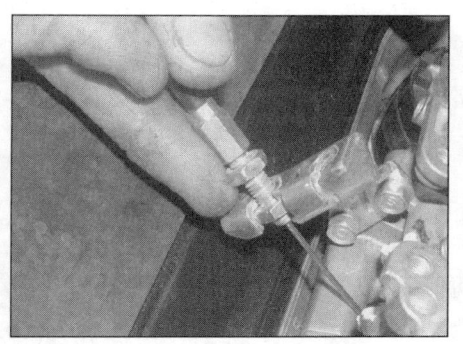

13.2d Thread the locknut up unscrew the adjuster hex if necessary . . .

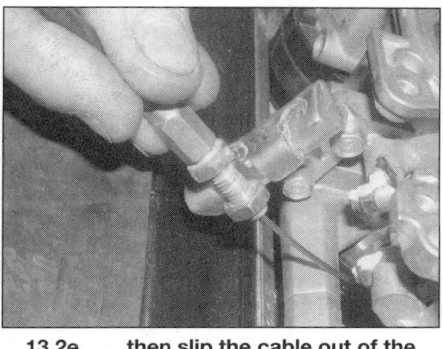

13.2e . . . then slip the cable out of the bracket . . .

13.2f . . . and free the end from the pulley

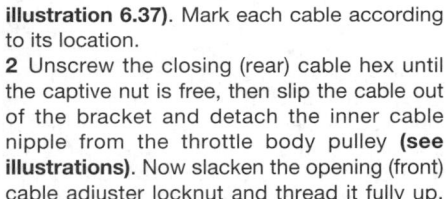

illustration 6.37). Mark each cable according to its location.

2 Unscrew the closing (rear) cable hex until the captive nut is free, then slip the cable out of the bracket and detach the inner cable nipple from the throttle body pulley **(see illustrations)**. Now slacken the opening (front) cable adjuster locknut and thread it fully up, then unscrew the adjuster until the captive nut is free (you may not need to do this if there is already enough clearance to release the captive nut from the bracket – it depends on the setting of the adjuster), then slip the cable out of the bracket and detach the inner cable nipple from the cam **(see illustrations)**. Withdraw the cables from the machine noting their correct routing.

3 Unscrew the cable elbow nuts at the throttle pulley housing, then remove the housing screws and separate the halves **(see illustrations)**. Detach the cable nipples from the pulley, then remove the closing (rear) cable from the housing **(see illustration)**. Thread the opening (front) cable elbow out of the housing and withdraw the cable. Mark each cable to ensure it is connected correctly on installation.

Installation

4 Fit the opening cable elbow into the front socket of the throttle pulley housing and

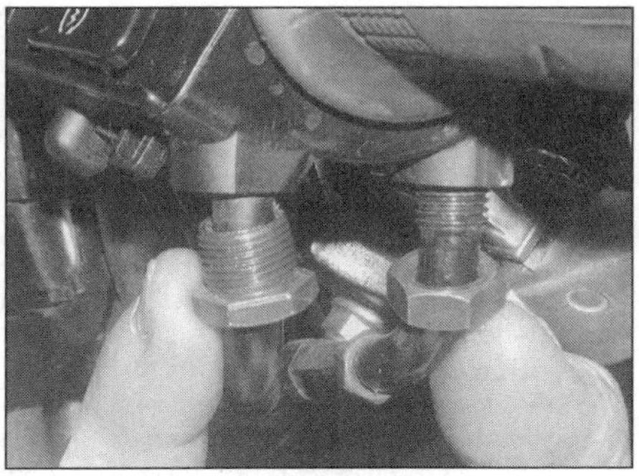

13.3a Unscrew the nuts . . .

13.3b . . . then remove the housing screws . . .

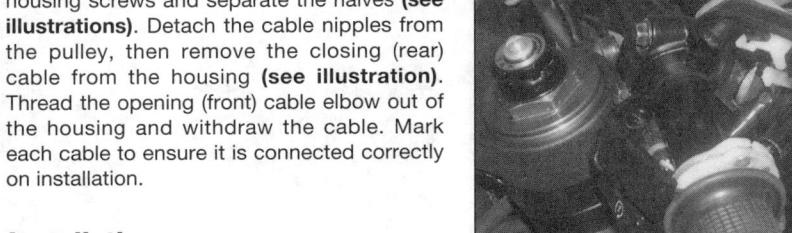

13.3c . . . and separate the halves

13.3d Detach the ends from the pulley and remove the cables as described

thread the elbow into it without becoming tight on the bottom of the threads – the elbow must stay loose so that it aligns itself – then thread the nut onto the elbow, again not so that it is tight **(see illustration 13.3a)**. Fit the closing cable into the rear socket and tighten the nut finger-tight. Lubricate the cable nipples with multi-purpose grease and fit them into the throttle pulley. Assemble the housing onto the handlebar, making sure the pin locates in the hole in the handlebar, then install the screws and tighten them **(see illustrations 13.3c and 13.3b)**. Now tighten both cable elbow nuts.

5 Feed the cables through to the throttle bodies, making sure they are correctly routed. The cables must not interfere with any other component and should not be kinked or bent sharply.

6 Lubricate the opening cable nipple with multi-purpose grease and fit it into the throttle body cam **(see illustration 13.2f)**. Fit the opening cable adjuster into the front of the bracket, locating the bottom nut against the lug so that it is captive, then thread the top nut down the adjuster, but do not yet tighten it **(see illustrations 13.2e and 13.2d)**. Thread the adjuster in or out until the specified amount of cable freeplay is obtained (see Chapter 1). Tighten the locknut against the bracket. Lubricate the closing cable nipple with multi-purpose grease and fit it into the throttle body cam **(see illustration 13.2c)**. Fit the cable into the rear of the bracket, locating the nut against the lug so that it is captive, then thread the hex into the nut until it is tight **(see illustrations 13.2b and 13.2a)**.

7 Operate the throttle to check that it opens and closes freely.

8 Check and adjust the throttle cable freeplay if required (see Chapter 1). Turn the handlebars back-and-forth to make sure the cable doesn't cause the steering to bind.

9 Replace the heat shield **(see illustration 6.37)**, then install the air filter housing (see Section 3).

10 Start the engine and check that the idle speed does not rise as the handlebars are turned. If it does, the throttle cable is routed incorrectly. Correct the problem before riding the motorcycle.

14.1 Unscrew the bolts or nuts (arrowed) according to model

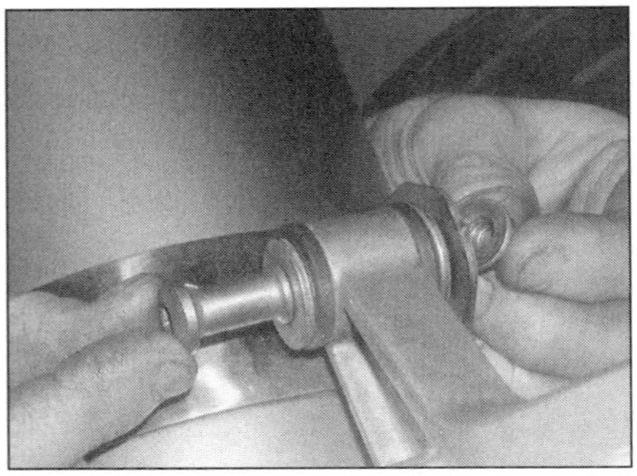

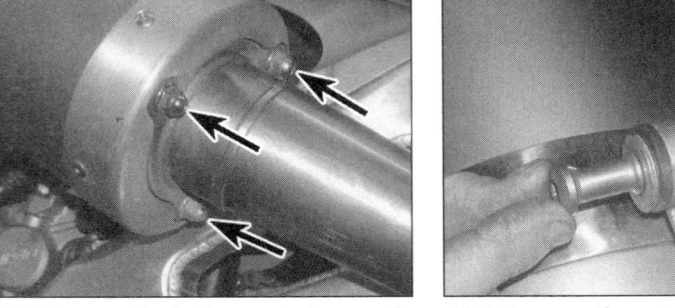

14 Exhaust system – removal and installation

Warning: If the engine has been running the exhaust system will be very hot. Allow the system to cool before carrying out any work.

Removal

Silencer

1 Unscrew the bolts (RR-Y and RR-1 (2000 and 2001) models) or nuts (RR-2 and RR-3 (2002 and 2003) models) securing the silencer to the downpipe assembly **(see illustration)**.

> **HAYNES HiNT** *Exhaust system clamp bolts tend to become corroded and seized. It is advisable to spray them with WD40 or a similar product before attempting to slacken them.*

2 Unscrew the nut and remove the bolt and washer securing the rear of the silencer **(see illustration)**. Release the front of the silencer from the downpipe assembly and remove it.

3 Note the collar which fits into the inside of the rubber on the silencer mounting. Fit a new rubber if the old one is damaged, deformed or deteriorated.

4 Remove the sealing ring between the

14.2 Unscrew the nut and remove the bolt and washer

silencer and downpipe assembly and replace it with a new one.

Complete system

5 Remove the fairing side panels, and on RR-Y and RR-1 (2000 and 2001) models the lower fairing (see Chapter 8).

6 On California and Germany models, refer to Section 6, Steps 84 to 86, and disconnect and release the oxygen sensor wiring.

7 Refer to Section 15 and free the exhaust control valve cable ends from the pulley.

8 Remove the radiator (see Chapter 3). **Note:** *Though it is possible to remove the downpipe assembly with the radiator just displaced from its bottom mount and pivoted forward rather than removed altogether, there is always the possibility of damaging it when unscrewing or tightening the nuts or when manoeuvring the downpipes. As radiators are fairly fragile and easily damaged, and don't take long to remove, it is best to do so.*

9 Either remove the silencer if required (see above, but note that the system can be removed in one piece), or unscrew the nut on the silencer mounting bolt, but do not yet withdraw the bolt **(see illustration 14.2)**.

10 Unscrew the nut on the bolt securing the rear of the downpipe assembly, but do not yet withdraw the bolt **(see illustration)**.

11 Unscrew the nuts securing the header pipes to the cylinder head **(see illustration)**.

12 Support the system, then withdraw the silencer mounting bolt and the downpipe assembly rear mounting bolt with its washer **(see illustration)**. Draw the flanges off the

14.10 Downpipe assembly rear mounting bolt (arrowed)

14.11 Unscrew the header pipe nuts

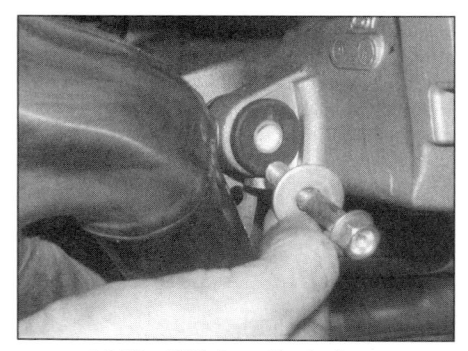

14.12a Withdraw the bolt . . .

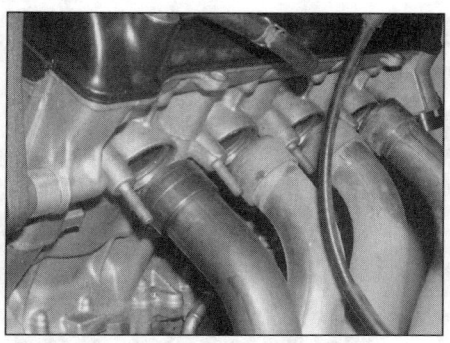

Fuel and exhaust systems 4•25

14.12b . . . then detach the header pipes . . .

14.12c . . . and remove the system

14.12d Remove the old sealing rings and discard them

studs and manoeuvre the downpipe assembly out of the head and remove it **(see illustrations)**. Remove the gasket from each port in the cylinder head and discard them as new ones must be used **(see illustration)**. Note the collar which fits into the inside of the rubber on the downpipe rear mounting. Fit a new rubber if the old one is damaged, deformed or deteriorated.

Installation

13 Installation is the reverse of removal, noting the following:
- Use a new gasket in each cylinder head port **(see illustration)**. Replace any damaged, deformed or deteriorated mounting rubbers with new ones.
- Use a new sealing ring between the downpipe assembly and the silencer.
- Apply a smear of copper grease to all nuts and bolts to prevent them from seizing up.
- Leave all fasteners loose until the entire system has been installed, making alignment of the various sections easier. Tighten the silencer mounting last.
- Tighten the downpipe nuts and the exhaust control valve pulley cover bolts to the torque settings specified at the beginning of the Chapter.
- Run the engine and check the system for leaks.

15 Honda variable intake and exhaust (H-VIX) system

1 The system controls the flow of air through the air box using a two-position valve and the flow of gases through the exhaust system using a three-position valve. The valves switch at pre-determined engine speeds, and are actuated by cables from a servo motor that is controlled by the ECM. Refer to Chapter 1 to check the operation of the system and cable adjustment.

Servo motor

Check

2 To check the servo motor, first remove it (see below).
3 Using a fully charged 12V battery and some jumper leads, connect the positive (+) terminal of the battery to the red wire terminal on the servo connector, and the negative (–) terminal to the blue wire terminal. When the battery is connected, the servo should operate. Disconnect the battery immediately after the test. If the servo does not operate, replace it with a new one.
Caution: Disconnect the battery immediately after the test to prevent possible damage to the servo motor.
4 If the servo now operates, yet did not beforehand, check for voltage at the loom side of the wiring connector using a voltmeter – connect the positive (+) probe of the meter to the red wire terminal on the loom side of the connector, and the negative (–) terminal to the blue wire terminal. With the ignition ON there should be battery voltage. If not, check the connector for loose or corroded terminals, then check the wiring between the connector and the ECM connectors for continuity, referring to the wiring diagrams in Chapter 9. If the wiring is good, the ECM could be faulty.
5 Using an ohmmeter or multimeter set to the K-ohms scale check the static resistance between the yellow/red and green/orange wire terminals on the servo connector. Compare the reading to that specified at the beginning of the Chapter. Now check the variable resistance between the light green/pink wire terminal and the green/orange wire terminal on the servo's connector while turning the servo pulley by hand. Compare the reading range to that specified at the beginning of the Chapter. If the readings from either test are not within the range specified, replace the servo with a new one.

Removal

6 Raise or remove the fuel tank (see Section 2). Trace the wiring from the servo and disconnect it at the wiring connector **(see illustration)**.
7 Remove the air filter element (see Chapter 1). Open the air intake control valve flap by hand to create slack in the cable and detach its end from the pulley **(see illustration)**.

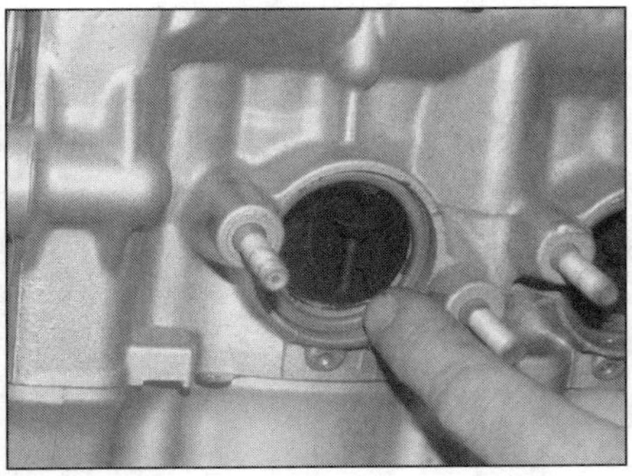

14.13 Use a new sealing ring in each port

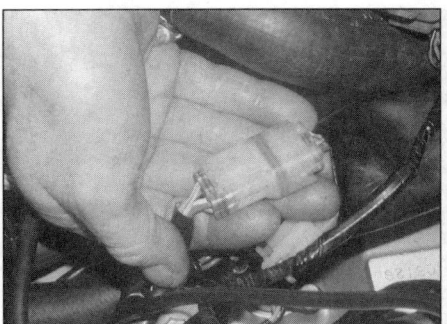

15.6 Disconnect the servo's wiring connector

15.7 Open the flap and detach the cable end from the pulley

15.8a Unscrew the bolts (arrowed) and remove the cover

15.8b Slacken the nuts (arrowed) and free the cables from the bracket and the pulley

8 On RR-Y and RR-1 (2000 and 2001) models remove the lower fairing (see Chapter 8). On all models remove the left-hand fairing side panel (see (Chapter 8). Unscrew the two exhaust control valve pulley cover bolts and remove the cover – you will have to lift the coolant hose to access the top front bolt **(see illustration)**. Slacken the cable locknuts and free the cables from the bracket and pulley, noting which fits where **(see illustration)**.

9 Unscrew the servo mounting bolt, then lift it off its mounting bracket and feed it out between the engine and frame on the left-hand side **(see illustrations)**. Free the cables from the holder on the servo housing, noting which cable fits where, and detach the cable ends from the pulley **(see illustrations)**.

Installation

10 If you are installing the original servo, first connect the wiring connector **(see illustration 15.6)**. Now locate the service check connector – it is an open 3-pin connector located just behind the battery on its right-hand end. Connect between the two outer terminals using a short length of auxiliary wire with bared ends **(see illustration 5.3)**. Turn the ignition ON – the servo should turn, then stop. When it stops, insert a 3 x 28 mm bolt or equivalent pin into the hole in the top of the pulley to lock it in that position. Turn the ignition OFF. Connect the cables to the pulley, making sure they are in their correct position as noted on removal **(see illustrations 15.9e and 15.9d)**. Adjust the cables (see Chapter 1). Remove the bolt or pin, then locate the servo on its bracket and secure it with its bolt **(see illustrations 15.9b and 15.9a)**.

11 If you are installing a new servo, first connect the cables to the pulley, making sure they are in their correct position as noted on removal **(see illustrations 15.9e and 15.9d)**. Adjust the cables (see Chapter 1). Locate the servo on its bracket and secure it with its bolt **(see illustrations 15.9b and 15.9a)**. Connect the wiring connector **(see illustration 15.6)**.

12 Open the air intake control valve flap by hand and connect the cable to the pulley **(see illustration 15.7)**. Install the air filter element (see Chapter 1). Install the fuel tank (see Section 2) and the fairing panels (see Chapter 8).

Air intake control valve

Removal

13 Remove the air filter element (see Chapter 1). Open the air intake control valve flap by hand to create slack in the cable and detach its end from the pulley **(see illustration 15.7)**.

14 Unhook the left-hand return spring end

15.9a Unscrew the bolt (arrowed) . . .

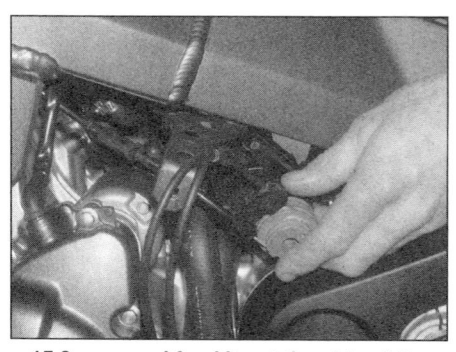

15.9b . . . then lift the servo off its mount . . .

15.9c . . . and feed it out the side of the frame

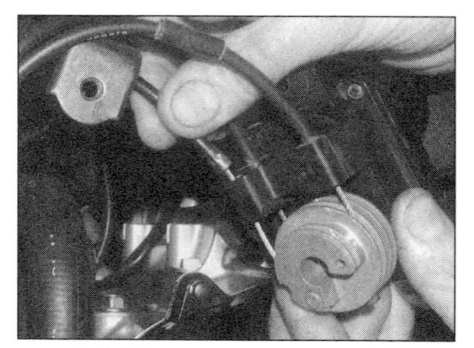

15.9d Free the cables from their holders . . .

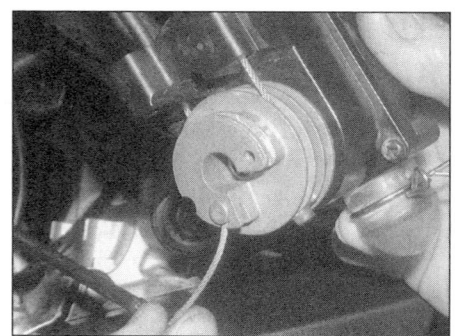

15.9e . . . and detach them from the pulley

Fuel and exhaust systems 4•27

15.14a Unhook the spring on the left-hand end . . .

15.14b . . . and on the right-hand end (arrow)

from the lug on the pulley **(see illustration)**. Unhook the right-hand return spring end from the lug on the shaft **(see illustration)**. When unhooking the springs, note how many turns they unwind – if the same tension is not wound back into the spring on installation the valve will not close properly.

15 Undo the six screws securing the flap assembly and remove the retainer plate and the two outer air guides **(see illustration)**. Lift the flap out of its housing. Remove the lower air guide, noting which way round it fits.

16 Slide the return springs off each end of the shaft and remove the pulley from the left-hand end, noting how they fit.

Installation

17 Installation is the reverse of removal, noting the following:
- Install the lower air guide with the UPPER mark facing up.
- Make sure the cut-out in the pulley locates over the boss on the shaft.
- Make sure the return spring ends locate correctly, and the correct tension is wound into them.
- Install the retainer plate with its UPPER mark facing up.
- On completion check the operation of the valve and the cable freeplay (see Chapter 1).

Exhaust control valve

Check

18 On RR-Y and RR-1 (2000 and 2001) models remove the lower fairing (see Chapter 8). On RR-2 and RR-3 (2002 and 2003) models remove the left-hand fairing side panel (see Chapter 8).

19 Unscrew the two exhaust control valve pulley cover bolts and remove the cover – you will have to lift the coolant hose to access the top front bolt **(see illustration 15.8a)**. Slacken the cable locknuts and free the cable ends from the pulley **(see illustration 15.8b)**.

20 Turn the valve pulley by hand through 180°. If it doesn't turn smoothly, remove the valve and check for a build-up of carbon deposits (see below).

21 Using a small torque wrench on the pulley nut, check the amount of pre-load torque required to turn the valve. It should start to turn at 0.34 Nm. If it doesn't, remove the valve and check for a build-up of carbon deposits (see below).

22 Smear some grease onto the cable ends and attach them to the pulley (see below). Check the operation of the system and the cable freeplay (see Chapter 1).

23 Apply some copper grease to the pulley cover bolt threads. Install the cover and tighten the bolts to the torque setting specified at the beginning of the Chapter **(see illustration 15.8a)**.

Removal

Note: *This sub-section deals with separating the entire valve housing from the exhaust system. The valve itself can be removed from its housing without disturbing the housing and with the exhaust system in place (see Disassembly below). It is only necessary to remove the exhaust system and separate the housing from it if the valve shaft bushes require renewal, or if the housing itself is damaged.*

24 Remove the exhaust system (see Section 14).

25 Unscrew the bolts securing the downpipe assembly to the housing and separate them. It is likely that these bolts will be corroded and tight, so it is wise to spray some penetrating fluid onto them and to give it time to work its way in before undoing the bolts. Check the condition of the gasket and discard it if it is damaged. Honda do not specify to use a new one on installation as a matter of course, though it is advisable to do so whatever the apparent condition of the old one.

26 Unscrew the bolts securing the rear section to the housing and separate them.

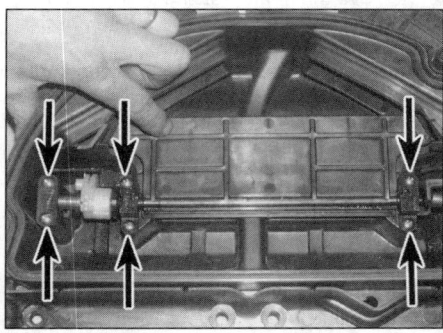

15.15 Undo the screws and remove the plate, the air guides and the flap

27 If required, disassemble the valve (see below), ignoring the first two Steps.

Installation

28 Installation is the reverse of removal, noting the following:
- Use a new gasket on each housing face if necessary.
- Apply a smear of copper grease to the housing bolt threads, and tighten them to the torque setting specified at the beginning of the Chapter.
- Refer to Section 14 for exhaust system installation.
- Check the operation of the system and the cable freeplay (see Chapter 1). Also check the system for leaks.

Disassembly and cleaning

29 On RR-Y and RR-1 (2000 and 2001) models remove the lower fairing (see Chapter 8). On RR-2 and RR-3 (2002 and 2003) models remove the left-hand fairing side panel (see (Chapter 8).

30 Unscrew the two pulley valve cover bolts and remove the cover **(see illustration 15.8a)**. Slacken the cable locknuts and free the cable ends from the pulley, noting which fits where **(see illustration 15.8b)**.

31 Turn the pulley anti-clockwise until the tab

4•28 Fuel and exhaust systems

15.31 Unscrew the nut and remove the pulley

15.32 Remove the retainer, the washers and the spring

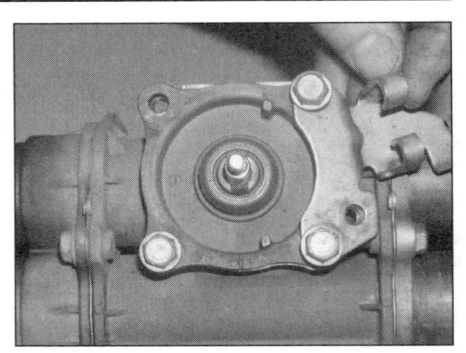

15.33 Unscrew the bolts and remove the bracket, cover and gasket

on its rim locates against the lug on the cover, then unscrew the nut and remove the pulley, noting how it locates **(see illustration)**.

32 Remove the spring retainer, the outer and inner thrust washers, and the spring, noting how they all fit **(see illustration)**.

33 Unscrew the remaining bolts securing the valve cover, noting how two also secure the cable bracket, and remove the cover and its metal gasket **(see illustration)**. Check the condition of the gasket and discard it if it is damaged. Honda do not specify to use a new one on installation as a matter of course, though it is advisable to do so whatever the apparent condition of the old one.

34 Withdraw the valve from the housing **(see illustration)**.

35 Clean the valve and its housing and scrape off any carbon deposits, taking care not to damage the valve shaft bushes.

Caution: Do not apply any cleaning solvent or lubricant to the outer thrust washer and the valve shaft bushes.

Bush renewal

36 Check the condition of the bushes visually (there is one in the housing and one in the cover), then check for excessive clearance between the valve shaft ends and the bushes **(see illustration)**. Replace the bushes with new ones if necessary, referring to *Tools and Workshop Tips* in the Reference Section for removal and installation methods. Measure and/or note the set depth of each bush before removing them and install the new ones so they are the same. The cover bush should project 0.3 to 0.5 mm from the rim of its housing. Note the cap fitted with the housing bush – press the cap and bush out together from the outside of the housing. Install the cap first until it seats, then fit the bush so it sits 0.3 to 0.5 mm below the rim of its housing.

Reassembly

37 Slide the valve into the housing, locating the shaft end in the bush **(see illustration 15.34)**.

38 Fit the cover using a new gasket if necessary, with the index lines on the cover facing forward and down **(see illustrations)**. Install all the cover bolts, including the top front bolt which also secures the pulley cover, and not forgetting to secure the cable bracket with the two rear bolts **(see illustration 15.33)**. Tighten the bolts to the torque setting specified at the beginning of the Chapter.

39 Fit the outer thrust washer onto the spring retainer with its chamfered side facing the retainer. Fit the inner thrust washer onto the outer washer.

40 Fit the spring onto the valve cover, making sure it seats correctly, then fit the retainer assembly onto the spring **(see illustration 15.32)**. Fit the pulley onto the shaft, locating its cut-outs over the tabs, then fit the nut **(see illustrations)**. Turn the pulley clockwise until the tab on its rim locates

15.34 Withdraw the valve from the housing

15.36 Check the bushes (arrowed) as described

15.38a Fit the gasket . . .

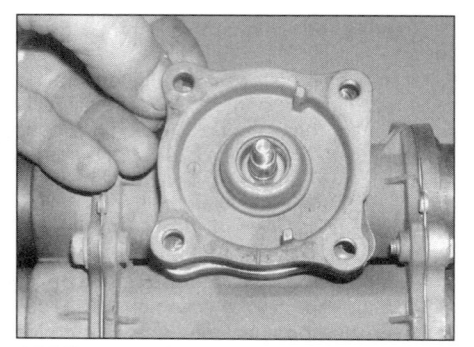

15.38b . . . and the cover, making sure it is the correct way round

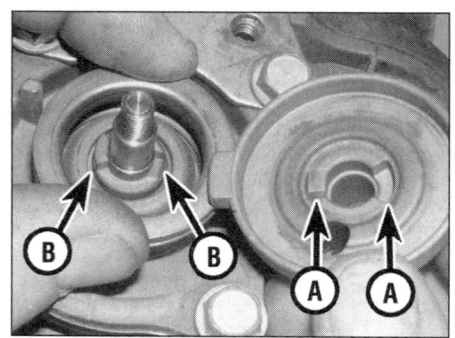

15.40a Fit the pulley, locating the shaped cut-outs (A) over the correspondingly shaped tabs (B) . . .

Fuel and exhaust systems 4•29

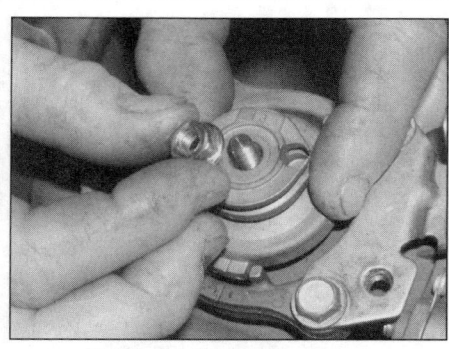

15.40b . . . then fit the nut . . .

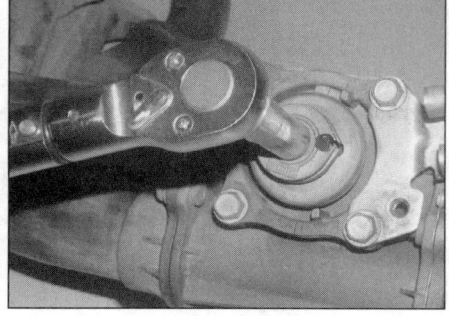

15.40c . . . and tighten it to the specified torque

15.57a With the pulley aligned as shown, fit the bottom cable end into its socket . . .

against the lug on the cover, then tighten the nut to the specified torque setting **(see illustration)**.
41 Refer to Step 21 and check the valve pre-load torque. If it is more than specified, disassemble the valve again.
42 Remove the top front pulley cover bolt. Smear some grease onto the cable ends and attach them to the pulley (see below). Check the operation of the system and the cable freeplay (see Chapter 1).
43 Apply some copper grease to the pulley cover bolt threads. Install the cover and tighten the bolts to the torque setting specified at the beginning of the Chapter **(see illustration 15.8a)**.

Cable renewal

Air intake valve cable

44 Remove the air filter element (see Chapter 1). Open the air intake control valve flap by hand to create slack in the cable and detach its end from the pulley **(see illustration 15.7)**.
45 Trace the wiring from the servo and disconnect it at the wiring connector **(see illustration 15.6)**.
46 Unscrew the servo mounting bolt, then lift it off its mounting bracket **(see illustrations 15.9a and 15.9b)**. Free the cable from the holder on the servo housing and detach the cable end from the pulley **(see illustrations 15.9d and 15.9e)**.
47 Remove the air filter housing (see Section 3), which involves detaching the cable from it.

48 Installation is the reverse of removal.
49 Check the operation of the system and the cable freeplay (see Chapter 1).

Exhaust valve cables

50 On RR-Y and RR-1 (2000 and 2001) models remove the lower fairing (see Chapter 8). On RR-2 and RR-3 (2002 and 2003) models remove the left-hand fairing side panel (see Chapter 8).
51 Unscrew the two exhaust control valve pulley cover bolts and remove the cover **(see illustration 15.8a)**. Slacken the cable locknuts and free the cable ends from the pulley, noting which fits where **(see illustration 15.8b)**.
52 Raise or remove the fuel tank (see Section 2). Trace the wiring from the servo and disconnect it at the wiring connector **(see illustration 15.6)**.
53 Unscrew the servo mounting bolt, then lift it off its mounting bracket **(see illustrations 15.9a and 15.9b)**. Free the cables from the holder on the servo housing, noting which cable fits where, and detach the cable ends from the pulley **(see illustrations 15.9d and 15.9e)**.
54 Withdraw the cables from the machine noting the correct routing of each cable.
55 Lubricate the cable nipples with multi-purpose grease and fit them into the servo pulley. Fit the cables into their holders, making sure they locate correctly.
56 Feed the cables through to the exhaust, making sure they are correctly routed. The

cables must not interfere with any other component and should not be kinked or bent sharply.
57 Lubricate the cable nipples with multi-purpose grease. Fit each cable into its socket in the pulley and in the bracket **(see illustrations)**.
58 Adjust the cable freeplay, then tighten the cable locknuts (see Chapter 1). Check the operation of the system.
59 Apply some copper grease to the pulley cover bolt threads. Install the cover and tighten the bolts to the torque setting specified at the beginning of the Chapter **(see illustration 15.8a)**.

16 Pulse secondary air (PAIR) system

General information

1 To reduce the amount of unburned hydrocarbons released in the exhaust gases, a pulse secondary air (PAIR) system is fitted. The system consists of the control valve (mounted under the front of the air filter housing), the reed valves (fitted in the valve cover) and the hoses linking them. The control valve is actuated electronically by the ECM.
2 Under certain operating conditions, a signal from the ECM opens up the PAIR control valve which then allows filtered air to be drawn through the reed valves and cylinder

15.57b . . . then locate the cable in the bracket

15.57c Realign the pulley as required and fit the upper cable into its socket . . .

15.57d . . . then route it round the top and locate the cable in the bracket

4•30 Fuel and exhaust systems

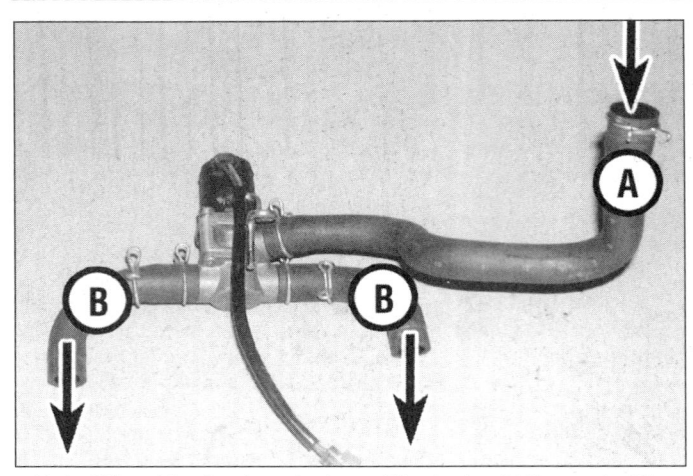

16.4a When blowing into the hose (A), air should flow out of the hoses (B)

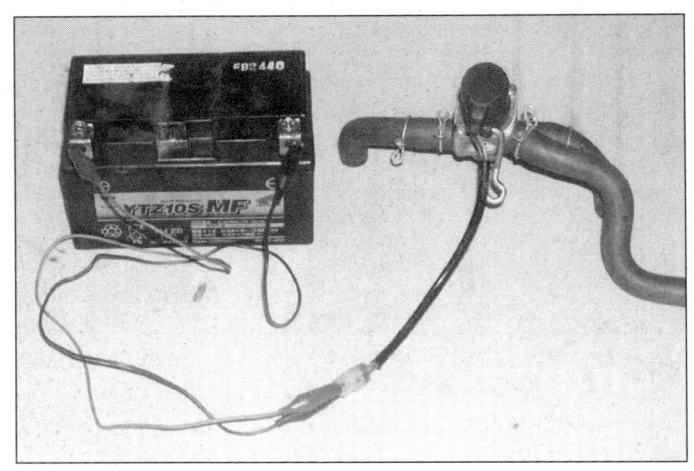

16.4b With a battery connected as shown, no air should flow out of the hoses

head passages and into the exhaust ports. The air mixes with the exhaust gases, causing any unburned particles of the fuel in the mixture to be burnt in the exhaust port/pipes. This process changes a considerable amount of hydrocarbons and carbon monoxide into relatively harmless carbon dioxide and water. The reed valves in the valve cover are fitted to prevent the flow of exhaust gases back up the cylinder head passages and into the air filter housing.

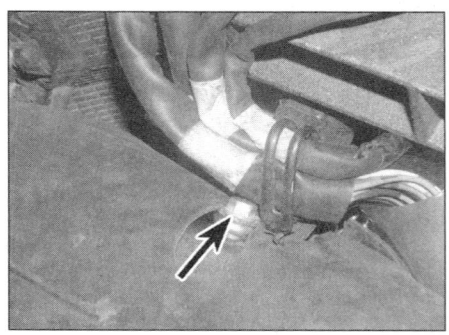

16.8 Unscrew the bolt (arrowed), displace the wiring clamp and remove the heat shield

Testing

Control valve

3 Remove the valve from the motorcycle (see below).
4 Check the operation of the control valve by blowing through the air filter housing hose union; no air should flow through the reed valve hose unions **(see illustration)**. Now connect battery voltage (12 volts) across the valve terminals and repeat the check; air should now flow freely through the valve if it is functioning correctly **(see illustration)**.
5 If an ohmmeter is available, check the resistance of the control valve windings by connecting an ohmmeter between its connector terminals and compare the reading obtained to that given in the Specifications. Replace the valve with a new one if faulty.

Reed valves

6 Remove the air filter housing (see Section 3). Lift the rubber heat shield **(see illustration 6.37)**. Disconnect the hose from each reed valve housing **(see illustration 8.21)**. Attach an auxiliary hose of the correct bore and about a foot long to one of the unions.
7 Check the valve by blowing and sucking on

the auxiliary hose end. Air should flow through the hose only when blown down it and not when sucked back up. If this is not the case the reed valve is faulty. Check the other valve in the same way.

Component renewal

Control valve

8 Remove the air filter housing (see Section 3). Unscrew the bolt securing the wiring guide, the rubber heat shield and the control valve holder **(see illustration)**. Remove the heat shield, noting how it fits **(see illustration 6.37)**.
9 Disconnect the control valve wiring connector **(see illustration)**.
10 Disconnect the hose from each reed valve housing **(see illustration 8.21)**. Remove the control valve with its hoses attached **(see illustration)**. Detach the hoses if required.
11 Installation is the reverse of removal.

Reed valves

12 Remove the air filter housing (see Section 3). Lift the rubber heat shield **(see illustration 6.37)**.
13 To remove either valve, first release the clamp and detach the air hose from its union

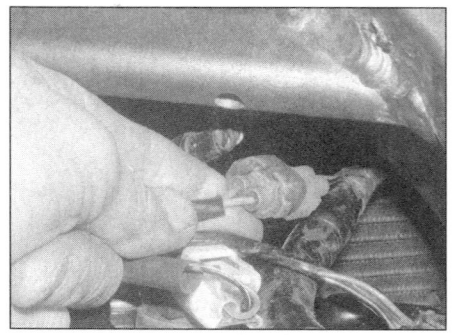

16.9 Disconnect the wiring connector

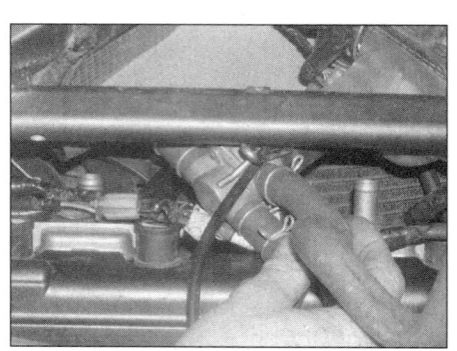

16.10 Manoeuvre the control valve out

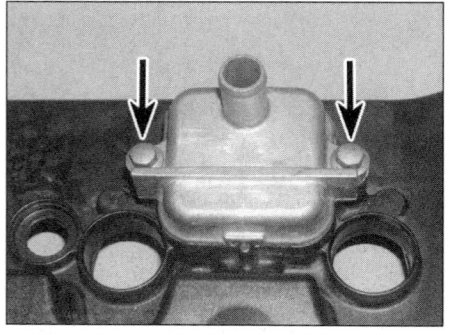

16.13a Unscrew the bolts (arrowed) . . .

Fuel and exhaust systems 4•31

16.13b ... and remove the cover ...

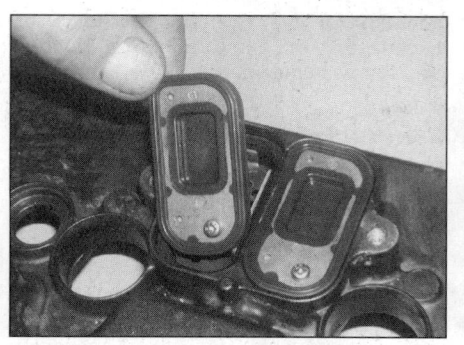

16.13c ... then remove the reed valve ...

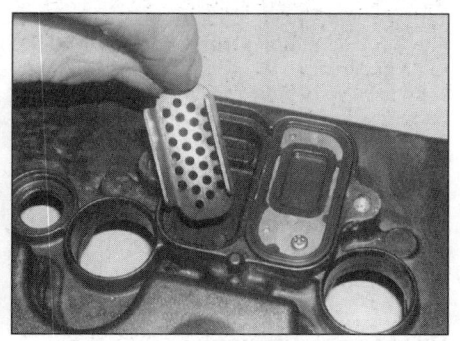

16.13d ... and its base plate

(see illustration 8.21). Unscrew the bolts securing the reed valve cover and remove the cover (see illustrations). Remove the reed valve and the base plate, noting which way around they are fitted (see illustrations).

14 Installation is the reverse of removal. Make sure the reed valve components are clean and correctly fitted.

17 Evaporative emission control (EVAP) system

Note: *This system is fitted to California market models only.*

General information

1 The evaporative emission control system (EVAP) is fitted to minimise the escape of fuel vapour into the atmosphere (see illustration). The fuel tank filler cap is sealed and a charcoal canister collects the fuel vapours generated when the motorcycle is parked and stores them until they can be cleared from the canister, via the control valve, into the throttle body inlet tracts to be burned by the engine during normal combustion. The purge control valve for the fuel tank vapour is opened and closed by the engine control module (ECM).

2 The valve should be tested if there is a problem starting the engine when it is hot.

Testing

Purge control valve

3 Remove the valve from the motorcycle (see below).

4 Check the operation of the control valve by blowing through the inlet (canister hose) union; air should not flow through the valve and out the outlet hose union. Connect battery voltage (12 volts) across the valve terminals and repeat the check; now air should flow through the valve if it is functioning correctly.

5 If an ohmmeter is available, check the resistance of the control valve windings and

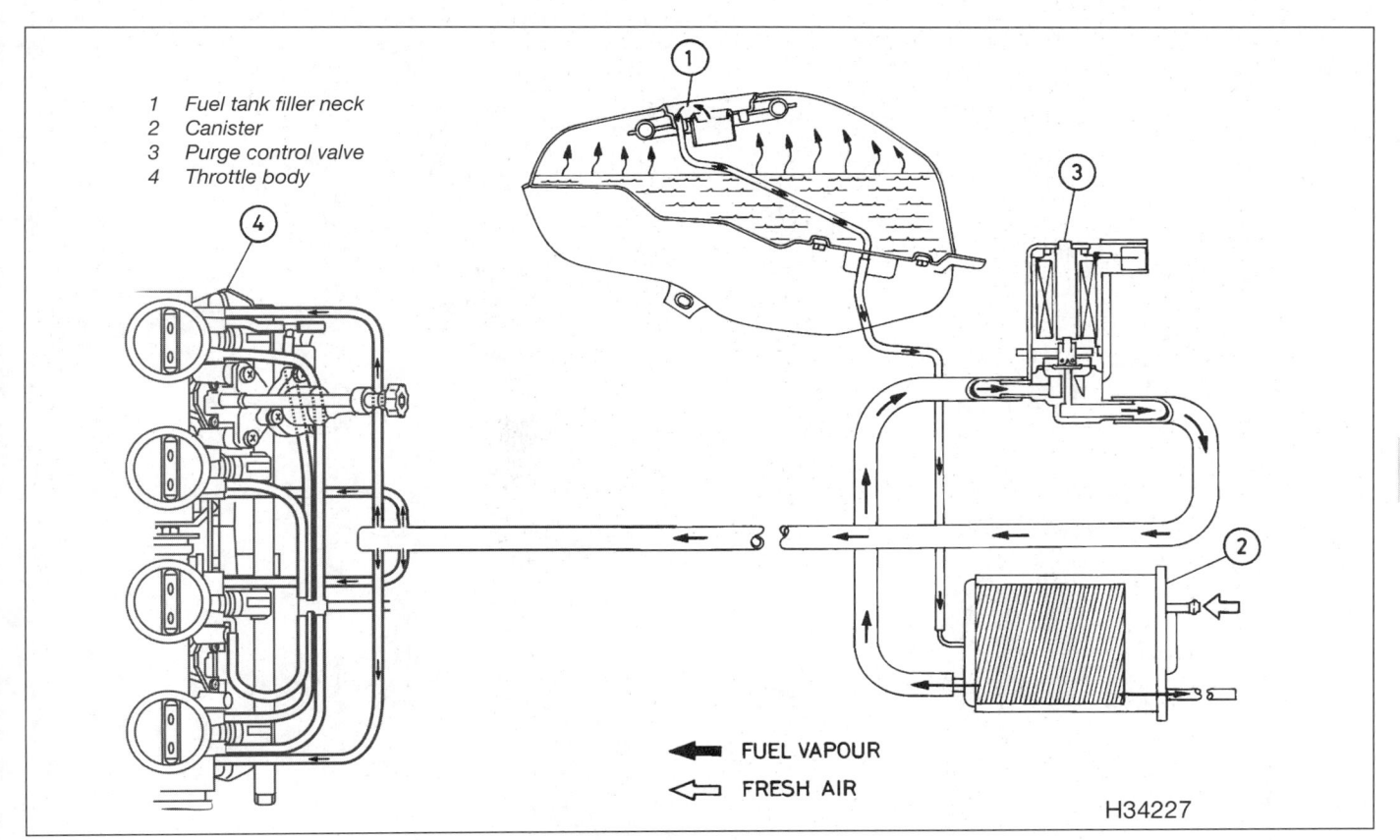

1 Fuel tank filler neck
2 Canister
3 Purge control valve
4 Throttle body

⬅ FUEL VAPOUR
⇐ FRESH AIR

17.1 EVAP system

4•32 Fuel and exhaust systems

compare the reading obtained to that given in the Specifications. Renew the valve if the reading differs.

6 If the valve behaves as described, check for battery voltage using a multimeter across the terminals on the loom side of the valve wiring connector with the ignition ON. If no voltage is present check the wiring.

Charcoal canister

7 No testing of the canister is possible, if it is thought to be faulty a new one must be installed.

Component renewal

Purge control valve

8 Raise the fuel tank (see Section 2). Displace the H-VIX system servo motor (see Section 15).
9 Disconnect the wiring connector and hoses from the valve, noting which fits where, then unscrew the bolts and remove the valve.
10 Installation is the reverse of removal.

Charcoal canister

11 Disconnect the hoses, noting which fits where. Unscrew the canister bracket mounting bolts and remove the canister.
12 Installation is the reverse of removal.

18 Catalytic converter

Note: *A catalytic converter is fitted as standard for the California and German markets only, but is available as an optional extra in some other markets.*

General information

1 A catalytic converter is incorporated in the exhaust system to minimise the level of exhaust pollutants released into the atmosphere.
2 The catalytic converter consists of a canister containing a fine mesh impregnated with a catalyst material, over which the hot exhaust gases pass. The catalyst speeds up the oxidation of harmful carbon monoxide, unburned hydrocarbons and soot, effectively reducing the quantity of harmful products released into the atmosphere via the exhaust gases.
3 The catalytic converter is of the closed-loop type with exhaust gas oxygen content information being fed back to the fuel injection system engine control module (ECM) by the oxygen sensor.
4 The oxygen sensor contains a heating element which is controlled by the ECM. When the engine is cold, the ECM switches on the heating element which warms the exhaust gases as they pass over the sensor. This brings the catalytic converter quickly up to its normal operating temperature and decreases the level of exhaust pollutants emitted whilst the engine warms up. Once the engine is sufficiently warmed up, the ECM switches off the heating element.

5 Refer to Section 14 for exhaust system removal and installation, and Section 6 for oxygen sensor removal and installation information.

Precautions

6 The catalytic converter is a reliable and simple device which needs no maintenance in itself, but there are some facts of which an owner should be aware if the converter is to function properly for its full service life.

- DO NOT use leaded or lead replacement petrol (gasoline) – the additives will coat the precious metals, reducing their converting efficiency and will eventually destroy the catalytic converter.
- Always keep the ignition and fuel systems well-maintained in accordance with the manufacturer's schedule – if the fuel/air mixture is suspected of being incorrect have it checked on an exhaust gas analyser.
- If the engine develops a misfire, do not ride the bike at all (or at least as little as possible) until the fault is cured.
- DO NOT use fuel or engine oil additives – these may contain substances harmful to the catalytic converter.
- DO NOT continue to use the bike if the engine burns oil to the extent of leaving a visible trail of blue smoke.
- Remember that the catalytic converter and oxygen sensor are FRAGILE – do not strike them with tools during servicing work.

Chapter 5
Ignition system

Contents

Clutch switch – check and replacement . . . . .see Chapter 9	Ignition system – check 2
Electronic control module (ECM) – check,	Ignition timing – general information and check 4
removal and installation . . . . .see Chapter 4	Immobiliser system 5
General information 1	Neutral switch – check and replacement . . . . .see Chapter 9
Ignition (main) switch – check, removal and installation .see Chapter 9	Sidestand switch – check and replacement . . . . .see Chapter 9
Ignition HT coils – check, removal and installation 3	Spark plugs – gap check and renewal . . . . .see Chapter 1
Ignition pulse generator – check, removal	Timing rotor . . . . .see Chapter 2
and installation . . . . .see Chapter 4	

Degrees of difficulty

Easy, suitable for novice with little experience	Fairly easy, suitable for beginner with some experience	Fairly difficult, suitable for competent DIY mechanic	Difficult, suitable for experienced DIY mechanic	Very difficult, suitable for expert DIY or professional

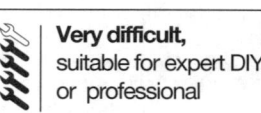

Specifications

General information
Cylinder numbering	1 to 4 from left to right
Firing order	1-2-4-3
Spark plugs	See Chapter 1

Ignition timing
RR-Y and RR-1 (2000 and 2001) models	15° BTDC (F mark) at idle
RR-2 and RR-3 (2002 and 2003) models	13° BTDC (F mark) at idle

Ignition HT coils
Primary winding resistance	1.1 to 1.5 ohms @ 20°C
Secondary winding resistance	11.0 to 11.8 K-ohms @ 20°C
Initial voltage (see text)	Battery voltage (approximately 12 volts)
Minimum peak voltage (see text)	100 volts

Torque setting
Timing inspection cap	18 Nm

5•2 Ignition system

1 General information

All models are fitted with a fully transistorised electronic ignition system, which due to its lack of mechanical parts is totally maintenance free. The system works in conjunction with the fuel injection system (covered in Chapter 4), with both being under the control of the electronic control module (ECM), which effectively 'manages' the engine.

The ignition system comprises the timing rotor, ignition pulse generator, throttle position sensor and ignition HT coils (refer to the wiring diagrams at the end of Chapter 9 for details). The ignition pulse generator and throttle position sensor have dual roles in providing information for both the ignition and the fuelling side of the engine. All the dual role components, including the ECM, are covered in Chapter 4.

The ignition timing rotor, which is on the right-hand end of the crankshaft, has triggers which magnetically actuate the pulse generator coil as the crankshaft rotates. The pulse generator sends a signal to the ECM which then supplies the ignition HT coils with the power necessary to produce a spark at the plugs. A throttle position sensor also supplies the ECM with information that is used in determining the optimum firing point for all conditions. The system incorporates an electronic advance system controlled by the signals from the pulse generator coil and the sensors.

The system uses four HT coils, one for each cylinder. The coils are of the plug top type or 'stick coils' with the coil windings being incorporated in the spark plug cap. This eliminates the need for HT leads and saves space.

The system incorporates a safety interlock circuit which will cut the ignition if the sidestand is extended whilst the engine is running and in gear, or if a gear is selected whilst the engine is running and the sidestand is down. It also prevents the engine from being started if the sidestand is down and the engine is in gear. The engine can be started with the sidestand up when it is in gear as long as the clutch lever is pulled in.

Many models are fitted with an immobiliser system (HISS – Honda Ignition Security System) which will not allow the engine to be started unless the correct key is used. The immobiliser system has its own fault diagnosis function.

Because of their nature, the individual ignition system components can be checked but not repaired. If ignition system troubles occur, and the faulty component can be isolated, the only cure for the problem is to replace the part with a new one. Keep in mind that most electrical parts, once purchased, cannot be returned. To avoid unnecessary expense, make very sure the faulty component has been positively identified before buying a replacement part.

Note that there is no provision for adjusting the ignition timing on these models.

2 Ignition system – check

Warning: *The energy levels in electronic systems can be very high. On no account should the ignition be switched on whilst the plugs or plug caps are being held. Shocks from the HT circuit can be most unpleasant. Secondly, it is vital that the engine is not turned over or run with any of the plug caps removed, and that the plugs are soundly earthed (grounded) when the system is checked for sparking. The ignition system components can be seriously damaged if the HT circuit becomes isolated.*

1 As no means of adjustment is available, any failure of the system can be traced to failure of a system component or a simple wiring fault. Of the two possibilities, the latter is by far the most likely. In the event of failure, check the system in a logical fashion, as described below.

2 Working on one ignition HT coil at a time, disconnect the wiring connector – to access them refer to Section 3, Step 1 **(see illustration)**. Pull the coil off the spark plug **(see illustration)**. Reconnect the wiring connector. Connect the HT coil to a new spark plug and lay the plug against the engine with the threads contacting it. If necessary, hold the spark plug with an insulated tool.

Warning: *Do not remove any of the spark plugs from the engine to perform this check – atomised fuel being pumped out of the open spark plug hole could ignite, causing severe injury! Make sure the plugs are securely held against the engine – if they are not earthed when the engine is turned over, the ECM could be damaged.*

3 Having observed the above precautions, check that the kill switch is in the RUN position and the transmission is in neutral, then turn the ignition switch ON and turn the engine over on the starter motor. If the system is in good condition a regular, fat blue spark should be evident at the plug electrode. If the spark appears thin or yellowish, or is non-existent, further investigation is necessary. Turn the ignition OFF and repeat the check for each HT coil.

4 Ignition faults can be divided into two categories, namely those where the ignition system has failed completely, and those which are due to a partial failure. The likely faults are listed below, starting with the most probable source of failure. Work through the list systematically, referring to the subsequent sections for full details of the necessary checks and tests. **Note:** *Before checking the following items ensure that the battery is fully charged and that all fuses are in good condition.*

- Loose, corroded or damaged wiring connections, broken or shorted wiring between any of the component parts of the ignition system (see Chapter 9).
- Faulty spark plug cap, faulty spark plug, dirty, worn or corroded plug electrodes.
- Faulty ignition (main) switch or engine kill switch (see Chapter 9).
- Faulty neutral, clutch or sidestand switch (see Chapter 9).
- Faulty pulse generator coil or damaged trigger on timing rotor.
- Faulty ignition HT coil(s).
- Faulty throttle position sensor.
- Faulty electronic control module.

5 If the above checks don't reveal the cause of the problem, have the ignition system tested by a Honda dealer.

3 Ignition HT coils – check, removal and installation

Check

1 Remove the rider's seat (see Chapter 8). Disconnect the battery negative (–ve) lead.
2 Remove the air filter housing (see Chap-

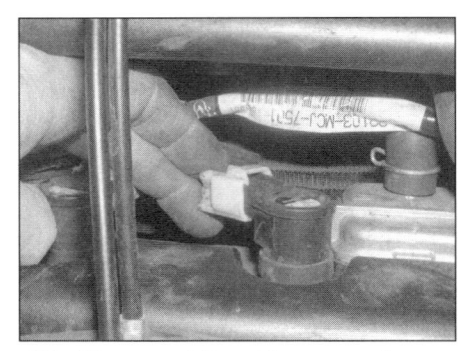

2.2a Disconnect the wiring connector . . .

2.2b . . . and pull the coil/cap off the spark plug

Ignition system 5•3

3.2 Lift the rubber shield to access the coils

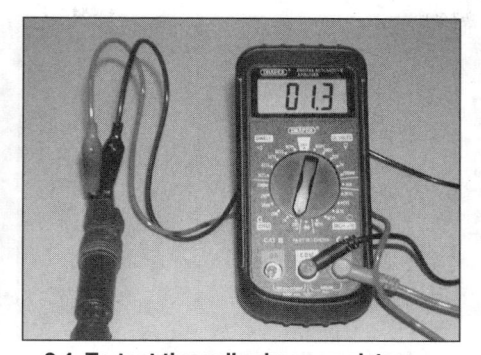

3.4 To test the coil primary resistance, connect the multimeter leads between the connector socket terminals

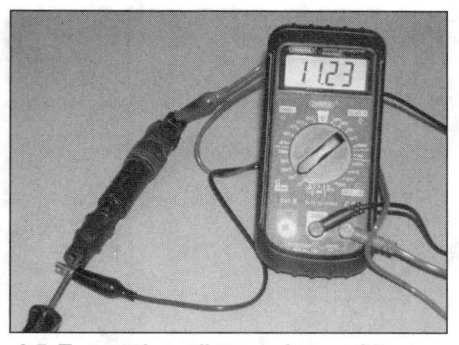

3.5 To test the coil secondary resistance, connect the multimeter leads between one terminal and the spark plug socket

ter 4). Lift the rubber heat shield **(see illustration)**. Check the coils/caps visually for loose or damaged connectors and terminals, cracks and other damage.

3 Disconnect the wiring connector from the HT coil being tested **(see illustration 2.2a)**. Pull the coil off the spark plug **(see illustration 2.2b)**.

4 To check the condition of the primary windings, set a multimeter to the ohms x 1 scale. Connect one meter probe to one terminal in the coil/cap socket and the other probe to the other terminal and measure the resistance **(see illustration)**. If the reading obtained is not within the range given in the Specifications, it is likely that the coil is defective. To confirm this, it must be tested as described below using the specified equipment, or by a Honda dealer.

5 To check the resistance of the secondary windings, set the meter to the K-ohm scale. Connect one meter probe to one of the terminals in the coil/cap socket, and the other to the spark plug contact, using a steel rod or screwdriver as an extension if your probe is not long enough **(see illustration)**. If the reading obtained is not within the range given in the Specifications, it is likely that the coil is defective. To confirm this, it must be tested as described below using the specified equipment, or by a Honda dealer.

6 Honda specify their own Imrie diagnostic tester (model 625), or the peak voltage adapter (Pt. No. 07HGJ-0020100) with an aftermarket digital multimeter having an impedance of 10 M-ohm/DCV minimum, for a complete test. If this equipment is available, reconnect the wiring connector to the coil. Connect the cap to a new spark plug and lay the plug on the engine with the threads contacting it. If necessary, hold the spark plug with an insulated tool. Connect the positive (+) lead of the voltmeter and peak voltage adapter arrangement to the blue/black (No. 1 cylinder coil), yellow/white (No. 2 coil), red/blue (No. 3 coil) or red/yellow (No. 4 coil) wire terminal on the coil connector (inserting the probe from the back), with the wiring connector still securely connected, and

connect the negative (–) lead to a suitable earth (ground) point.

7 Check that the kill switch is in the RUN position and the transmission is in neutral, then turn the ignition switch ON. Note the initial voltage reading on the meter, then turn the engine over on the starter motor and note the ignition coil peak voltage reading on the meter. Once both readings have been noted, turn the ignition switch off and disconnect the meter.

8 If the initial voltage reading is not as specified or the peak voltage readings are lower than the specified minimum then a fault is present somewhere else in the ignition system circuit (see Section 2); note that the peak voltage readings for each coil can be different but each one must exceed the specified minimum.

9 If the initial and peak voltage readings are as specified and the plug does not spark, then the coil is faulty and must be replaced with a new one; the coil is a sealed unit and cannot therefore be repaired.

Removal and installation

10 Remove the rider's seat (see Chapter 8). Disconnect the battery negative (–) lead.

11 Remove the air filter housing (see Chapter 4). Lift the rubber heat shield **(see illustration 3.2)**.

12 Disconnect the wiring connector from the coil **(see illustration 2.2a)**. Pull the coil off the spark plug **(see illustration 2.2b)**.

13 Installation is the reverse of removal.

4 Ignition timing – general information and check

General information

1 Since no provision exists for adjusting the ignition timing and since no component is subject to mechanical wear, there is no need for regular checks; only if investigating a fault such as a loss of power or a misfire should the ignition timing be checked.

2 The ignition timing is checked dynamically (engine running) using a stroboscopic lamp. The inexpensive neon lamps should be adequate in theory, but in practice may produce a pulse of such low intensity that the timing mark remains indistinct. If possible, one of the more precise xenon tube lamps should be used, powered by an external source of the appropriate voltage. Whatever type is used make sure it is capable of picking up the pulse from the low tension side of the coil rather than the high tension, as of course there are no HT leads on this machine. **Note:** *Do not use the machine's own battery as an incorrect reading may result from stray impulses within the machine's electrical system.*

Check

3 Warm the engine up to normal operating temperature then stop it. On RR-Y and RR-1 (2000 and 2001) models remove the lower fairing (see Chapter 8). On RR-2 and RR-3 (2002 and 2003) models remove the right-hand fairing side panel (see Chapter 8).

4 Unscrew the timing inspection cap from the clutch cover **(see illustration)**. Discard the O-ring as a new one must be used.

5 The dynamic timing mark on the rotor which indicates the firing point at idle speed for the No. 1 cylinder is the line next to an

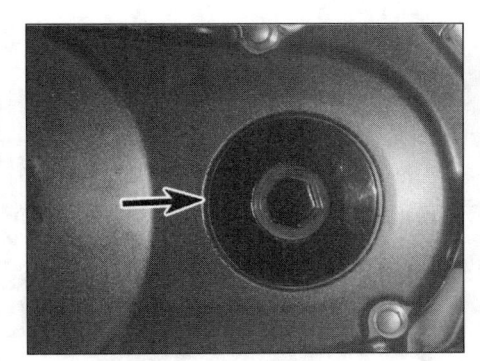

4.4 Unscrew the timing inspection cap (arrowed)

4.5 F mark and static timing mark (arrowed)

4.10 Install the cap using a new O-ring

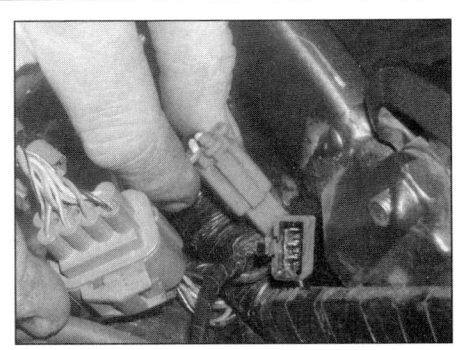

5.5 Disconnect the pulse generator coil wiring connector

F mark (see illustration). The static timing mark with which this should align is the notch in the inspection hole rim.

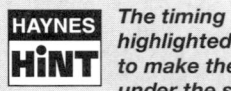

> **HAYNES HiNT**: *The timing marks can be highlighted with white paint to make them more visible under the stroboscope light.*

6 Connect the timing light to the No. 1 cylinder coil/cap blue/black wire.
7 Start the engine and aim the light at the static timing mark.
8 With the machine idling, the line next to the F should align with the static timing mark (see Step 5). Now increase engine speed to approximately 1500 rpm – using the idle speed adjuster will be more accurate than opening the throttle. At this point the dynamic timing mark should move anti-clockwise in relation to the static mark. This confirms the ignition is advancing.
9 As already stated, there is no means of adjustment of the ignition timing on these machines. If the ignition timing is incorrect, or suspected of being incorrect, one of the ignition system components is at fault, and the system must be tested as described in the preceding Sections of this Chapter.
10 When the check is complete, install the timing inspection cap using a new O-ring, and smear it and the cap threads with grease (see illustration). Tighten the cap to the torque setting specified at the beginning of the Chapter.

5 Immobiliser system

General information

1 An immobiliser system (known as HISS – Honda Ignition Security System) is fitted to some models as an anti-theft device. The system will only allow the machine to be started if the correct registered key is used to turn the ignition ON. The system consists of a transponder which is part of the ignition key, a receiver which is fitted around the ignition switch, and the electronic control module (ECM).
2 When the ignition is switched ON, the ECM sends power through the receiver to the transponder. The transponder sends a coded signal back through the receiver to the ECM. If the signal sent by the transponder matches the signal stored in the ECM memory, the immobiliser indicator light in the instrument cluster (marked by a key symbol) comes on for two seconds, then goes out, and the ECM allows the engine to be started. If the key code signal is not recognised, or if there is a fault in the system, the indicator light stays on. If the light stays on, refer to the fault diagnosis and troubleshooting Sections below. Likewise if the light does not come on at all.
3 The ECM can store the codes for up to four registered keys. They keys should be kept separately (i.e. not on the same key-ring) as the proximity of another key to the one being used in the switch can lead to the signal from it being jammed, and the bike will not start. The key has a built in transponder which can be damaged if the key is dropped or knocked, gets too hot, is too close to a magnetic object, or is submerged in water for too long. If all the keys are lost, the ECM must be replaced with a new one, so always make sure you have at least one spare key. If a new key is obtained, it must be registered into the system before the bike can be started.

Key registration procedure

With old ignition switch

Note: *To do this you will need the Honda special tool (Part No. 07XMZ-MBW0100 or 0101 depending on your model) which is a wiring loom adapter that connects to the battery and plugs into the ignition pulse generator wiring connector. If this tool is not available, it does not require rocket science to make up your own, though you will need to purchase the correct connector and some auxiliary wiring and clips that can connect to the terminals of the battery. Otherwise registration must be carried at a Honda dealer with the special tool.*

4 Obtain a new key from a Honda dealer, then have it cut to match the original key.
5 Raise the fuel tank (see Chapter 4). Trace the ignition pulse generator coil wiring from the top of the clutch cover and disconnect it at the red 2-pin wiring connector (see illustration). Connect the special tool wiring connector to the loom side of the connector, then connect the red coloured clip of the tool to the battery positive (+) terminal and the green coloured clip to the battery negative (–) terminal.
6 Turn the ignition switch ON using your original key. The immobiliser indicator light should come on and stay on (if it starts to flash after ten seconds, then there is a fault in the system, which will have gone into fault diagnosis, and the pattern of the flashes it emits should be matched with the fault code (see below)). Now disconnect the red clip from the battery positive terminal and leave it disconnected for at least two seconds, then reconnect it. The indicator should now come on for two seconds, then begin to flash repeatedly four times. This indicates that the system is in registration mode. At this point the registrations of all keys except the one in the switch will have been cancelled, so if you have another spare apart from the new one you want to register, this will also have to be registered.
7 Turn the ignition OFF and remove the original key, placing it well away from the receiver.
8 Insert the new key into the switch and turn it ON. The indicator should now come on for two seconds, then begin to flash repeatedly four times. This indicates that the system has registered the new key. Turn the ignition OFF and remove the key.
9 To register any other spare keys that will have been cancelled, repeat Steps 7 and 8. Up to four keys can be registered.
10 On completion turn the ignition OFF, then remove the special tool and reconnect the ignition pulse generator wiring connector. Now turn the ignition ON using any of the registered keys to return the system to normal mode.
11 Check that all registered keys can start the motorcycle.

With a new ignition switch

Note: *To do this you will need the Honda special tool (Part No. 07XMZ-MBW0100 or 0101 depending on your model) which is a wiring loom adapter that connects to the battery and plugs into the ignition pulse generator*

Ignition system 5•5

wiring connector. If this tool is not available, it does not require rocket science to make up your own, though you will need to purchase the correct connector and some auxiliary wiring and clips that can connect to the terminals of the battery. Otherwise registration must be carried out at a Honda dealer with the special tool.

12 Obtain a new switch and two (or more if you want) new keys (which I presume come with the switch).

13 Remove the faulty switch (see Chapter 9), but retain the receiver to fit with the new switch.

14 Raise the fuel tank (see Chapter 4). Trace the ignition pulse generator coil wiring from the top of the clutch cover and disconnect it at the red 2-pin wiring connector **(see illustration 5.5)**. Connect the special tool wiring connector to the loom side of the connector, then connect the red coloured clip of the tool to the battery positive (+) terminal and the green coloured clip to the battery negative (−) terminal.

15 Place one of the original registered keys for the faulty switch next to the receiver.

16 Connect the new ignition switch to its connector in the wiring loom, but keep it away from the receiver. Turn the new switch ON with one of the new keys. The immobiliser indicator light should come on and stay on, which means the ECM recognises the old key that is next to the receiver (if it starts to flash after ten seconds, then there is a fault in the system, which will have gone into fault diagnosis, and the pattern of the flashes it emits should be matched with the fault code (see below)). Now disconnect the red clip from the battery positive terminal and leave it disconnected for at least two seconds, then reconnect it. The indicator should now come on for two seconds, then begin to flash repeatedly four times. This indicates that the system is in registration mode. At this point the registrations of all keys except the one near the receiver will have been cancelled.

17 Turn the ignition OFF and remove the new key.

18 Install the new ignition switch, then fit the receiver onto it (see Chapter 9).

19 Insert the new key into the switch and turn it ON. The indicator should now come on for four seconds, then begin to flash repeatedly four times. This indicates that the system has registered the new key. If the indicator starts to flash after ten seconds, then there is a fault in the system, which will have gone into fault diagnosis, and the pattern of the flashes it emits should be matched with the fault code (see below). Turn the ignition OFF and disconnect the red clip of the special tool from the battery positive terminal.

20 Turn the ignition ON using the newly registered key. The indicator light should come on for two seconds, then go off.

21 Turn the ignition OFF and reconnect the red clip to the battery positive terminal.

22 Turn the ignition ON using the newly registered key. The indicator light should come on and stay on. Now disconnect the red clip from the battery positive terminal and leave it disconnected for at least two seconds, then reconnect it. The indicator should now come on for two seconds, then begin to flash repeatedly four times. This indicates that the system is in registration mode. At this point the registrations of all old keys (for the faulty switch) are cancelled.

23 Turn the ignition OFF and remove the key, placing it well away from the receiver.

24 Insert the second new unregistered key and turn the ignition ON. The indicator should now come on for four seconds, then begin to flash repeatedly four times. This indicates that the system has registered the second new key. Turn the ignition OFF and remove the key.

25 To register any other new spare keys, repeat Steps 23 and 24. Up to four keys can be registered.

26 On completion turn the ignition OFF, then remove the special tool and reconnect the ignition pulse generator wiring connector. Now turn the ignition ON using any of the registered keys to return the system to normal mode.

27 Check that all newly registered keys can start the motorcycle.

With a new ECM (electronic control module)

28 Obtain a new ECM along with two (or more if you want) new keys. Install the new ECM (see Chapter 4). Have the keys cut to match the original key for your ignition switch.

29 Insert a new key into the switch and turn it ON. The indicator should now come on for two seconds, then begin to flash repeatedly four times. This indicates that the system has registered the new key. If the indicator stays on for ten seconds then starts to flash, then there is a fault in the system, which will have gone into fault diagnosis, and the pattern of the flashes it emits should be matched with the fault code (see below).

30 Turn the ignition OFF and remove the key.

31 Insert the second new key and turn the ignition ON. The indicator should now come on for two seconds, then begin to flash repeatedly four times. This indicates that the system has registered the second new key.

32 Turn the ignition OFF and remove the key.

33 The new ECM will only register two new keys at this stage. If you have a third key that you want to register, refer to Steps 4 to 10 to register it, noting that you will need the special tool mentioned therein.

34 Check that both newly registered keys can start the motorcycle.

Fault diagnosis

Note: *To do this you will need the Honda special tool (Part No. 07XMZ-MBW0100 or 0101 depending on your model) which is a wiring loom adapter that connects to the battery and plugs into the ignition pulse generator wiring connector. If this tool is not available, you can make up your own, though you will need to purchase the correct connector and some auxiliary wiring and clips that can connect to the terminals of the battery.*

35 There are two fault diagnosis modes, one for if there is a fault during normal use, and one for a fault that occurs when registering a new key. Make sure you refer to the correct table below when matching the fault code pattern.

36 If the indicator light has come on and stayed on during normal use, raise the fuel tank (see Chapter 4). Trace the ignition pulse generator coil wiring from the top of the clutch cover and disconnect it at the red 2-pin wiring connector **(see illustration 5.5)**. Connect the special tool wiring connector to the loom side of the connector, then connect the red coloured clip of the tool to the battery positive (+) terminal and the green coloured clip to the battery negative (−) terminal.

37 Turn the ignition switch ON. The indicator light in the tachometer will come on for ten seconds, then start to flash. This means it has entered diagnostic mode, and the pattern of the flashes indicates the fault that has occurred. The pattern repeats continuously. Match the pattern with the fault codes below, making sure you refer to the relevant table. If the indicator stays on after ten seconds and does not flash, then there is no fault logged in the system.

If fault is indicated during normal use		
Flash pattern	Fault	Solution
Two short, one long, one short	Faulty ECM	Install new ECM
Two short, two long	Faulty receiver or wiring	Follow Troubleshooting procedure below
One long, three short	Signal jammed by other key	Place other key well away from receiver
One long, two short, one long	Signal jammed by other key	Place other key well away from receiver
If fault is indicated during key registration		
Flash pattern	Fault	Solution
One short, one long, one short, one long	Key already registered	Use a new or cancelled key
Two short, two long	Faulty receiver or wiring	Follow Troubleshooting procedure below
One short, one long, two short	Key already registered on old ECM	Use a new key

Troubleshooting procedure

Indicator light does not come on when ignition switched ON

38 Check the fuses (see Chapter 9).
39 If the fuses are good, check whether the neutral and oil pressure warning lights have come on.
40 If the lights have not come on, remove the fairing (see Chapter 8). Disconnect the instrument cluster wiring connector **(see illustration)**. Using a voltmeter, connect the positive (+) probe to the black/brown wire terminal on the loom side of the instrument cluster connector and the negative (–) probe to the green wire terminal on the loom side of the connector. With the ignition ON there should be battery voltage. If voltage is present, the instrument cluster is faulty (see Chapter 9). If there is no voltage, check for continuity in the wiring, referring to the wiring diagrams at the end of Chapter 9. The green wire goes to earth (ground).
41 If the lights have come on, on RR-Y and RR-1 (2000 and 2001) models unscrew the fuel tank mounting bolts and raise the rear of the tank (see Chapter 4). Release the battery strap, then undo the tank bracket bolts and remove the bracket **(see illustration)**. On RR-2 and RR-3 (2002 and 2003) models unscrew the fuel tank bracket bolts and pivot the bracket up, then remove the battery (see Chapter 9) **(see illustration)**. On all models remove the ECM cover **(see illustrations)**. Disconnect the ECM 22-pin black wiring connector **(see illustration)**. Using a voltmeter, connect the positive (+) probe to the white/red wire terminal on the loom side of the ECM connector and the negative (–) probe to earth (ground). Turn the ignition ON – there should be battery voltage.
42 If there was no voltage, using a voltmeter, connect the positive (+) probe to the white/red wire terminal on the loom side of the instrument cluster connector and the negative (–) probe to the green wire terminal on the loom side of the instrument cluster connector. Turn the ignition ON – there should be no voltage for two seconds, then there should be battery voltage. If there is no voltage after two seconds, check for continuity in the wiring, referring to the wiring diagrams at the end of Chapter 9. The green wire goes to earth (ground). If no voltage is present, the instrument cluster is faulty (see Chapter 9). If there is voltage, check for continuity in the white/red wire between the indicator unit and the ECM.
43 If there is voltage in Step 41, disconnect the ECM 22-pin black wiring connector. Using a voltmeter, connect the positive (+) probe to the black/white (ECM) wire terminal on the loom side of the ECM connector and the negative (–) probe to earth (ground). Turn the ignition ON – there should be battery voltage. If there is no voltage, check for continuity in the black/white wire, referring to the wiring diagrams at the end of Chapter 9. If voltage is present, check for continuity to earth (ground) in the green wire. If the wiring is good, check the ECM connector for loose, damaged or corroded terminals. If the connector is good, then the ECM could be faulty, and should be checked by a Honda dealer.

Indicator light stays on when ignition switched ON

44 Check that none of the other registered keys are close to the receiver. If they are, remove them and try the ignition again.
45 Turn the ignition ON with a spare key and check the indicator light, which should come on for two seconds, then go out. If it does, the first key is faulty. If it doesn't, perform the fault diagnosis procedure described above. If a fault code is displayed, use the appropriate table to determine the fault and the solution.
46 If no fault code is displayed, or the system does not go into fault diagnosis mode, on RR-Y and RR-1 (2000 and 2001) models unscrew the fuel tank mounting bolts and raise the rear of the tank (see Chapter 4). Release the battery strap, then undo the tank bracket bolts and remove the bracket **(see illustration 5.41a)**. On RR-2 and RR-3 (2002 and 2003) models unscrew the fuel tank bracket bolts and pivot the bracket up, then remove the battery (see Chapter 9) **(see illustration 5.41b)**. On all models remove the ECM cover **(see illustrations 5.41c or 5.41d)**. Disconnect the ECM 22-pin black wiring connector **(see illustration 5.41e)**. Using a voltmeter, connect the positive (+) probe to the white/red wire terminal on the loom side of

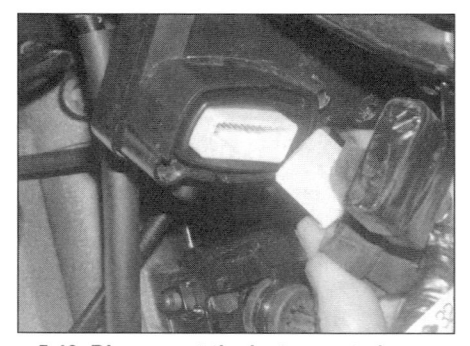

5.40 Disconnect the instrument cluster wiring connector

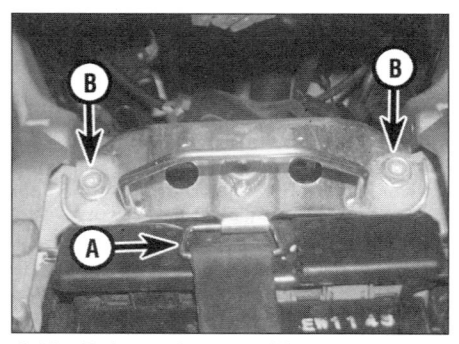

5.41a Release the strap (A), then unscrew the bolts (B) and remove the bracket

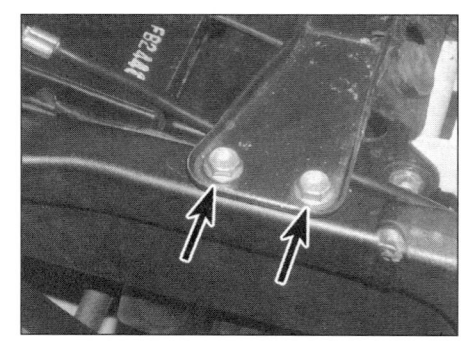

5.41b Unscrew the two bolts (arrowed) on each side and pivot the bracket up

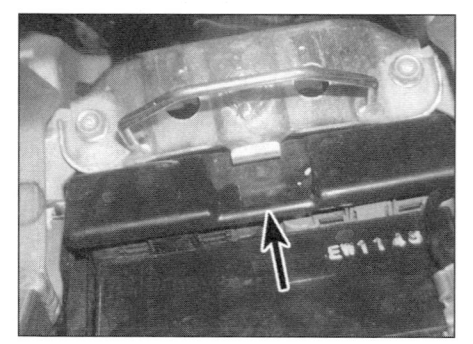

5.41c ECM cover (arrowed) – RR-Y and RR-1 (2000 and 2001) models

5.41d Removing the ECM cover – RR-2 and RR-3 (2002 and 2003) models

5.41e ECM wiring connectors

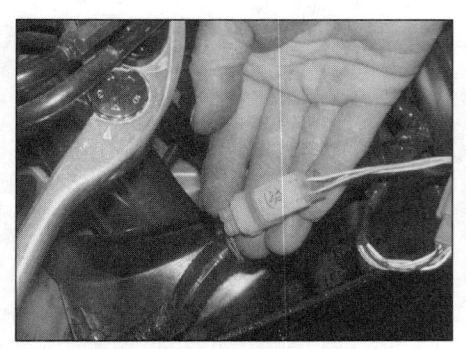

5.52a HISS wiring connector – RR-Y and RR-1 (2000 and 2001) models

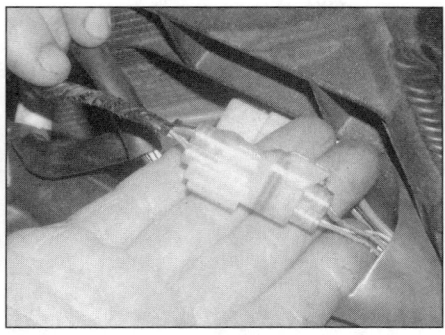

5.52b HISS wiring connector – RR-2 and RR-3 (2002 and 2003) models

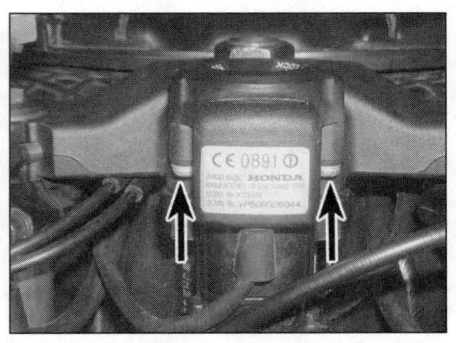

5.52c HISS receiver screws (arrowed)

the connector and the negative (–) probe to earth (ground). Turn the ignition ON – there should be battery voltage. If there is no voltage, check for continuity in the white/red wire between the ECM and the indicator unit.

47 If there is voltage, check for continuity in the yellow and white/yellow wires between the ECM and the ignition pulse generator, referring to the wiring diagrams at the end of Chapter 9. If there is no continuity, trace the fault and repair or replace the wiring as necessary. If there is continuity, the ECM could be faulty and should be taken to a Honda dealer for assessment.

Fault code indicated by flash pattern

48 If the 'two short, two long' flash pattern has been indicated during the fault diagnosis procedure, on RR-Y and RR-1 (2000 and 2001) models remove the right-hand air duct cover (see Chapter 8). On RR-2 and RR-3 (2002 and 2003) models remove the air filter housing (see Chapter 4). Trace the wiring from the receiver on the ignition switch and disconnect it at the 4-pin connector **(see illustrations 5.52a and 5.52b)**. Trace the wiring from the receiver on the ignition switch and disconnect it at the 4-pin connector. Using a voltmeter, connect the positive (+) probe to the yellow/red wire terminal on the loom side of the receiver connector and the negative (–) probe to earth (ground). Turn the ignition ON – there should be approximately 5 volts present. If there is no voltage, check for continuity in the yellow/red wire between the ECM and the receiver, and repair or replace the wiring if there is no continuity.

49 If there is 5 volts present, check for continuity to earth (ground) in the green/orange wire on the loom side of the connector, and repair or replace the wiring if there is no continuity.

50 If the wiring is good, using a voltmeter, connect the positive (+) probe to the pink wire terminal on the loom side of the receiver connector and the negative (–) probe to earth (ground). Turn the ignition ON – there should be approximately 5 volts present. If there is, the receiver is faulty.

51 If there is no voltage, check for continuity in the orange/blue and pink wires between the ECM and the receiver, and repair or replace the wiring if there is no continuity between the connectors, or if there is continuity in either to earth (ground). If the wiring is good, the receiver is faulty.

Replacement

52 To replace the receiver, on RR-Y and RR-1 (2000 and 2001) models remove the right-hand air duct cover (see Chapter 8). On RR-2 and RR-3 (2002 and 2003) models remove the air filter housing (see Chapter 4). Trace the wiring from the receiver on the ignition switch and disconnect it at the 4-pin connector **(see illustrations)**. Undo the screws securing the receiver around the ignition switch and remove the receiver, noting how it fits **(see illustration)**. If you don't have the correct tools to easily access the screws, removing the fairing will help (see Chapter 8), otherwise follow the procedure for removing the top yoke in the ignition switch replacement Section in Chapter 9.

53 To replace the ECM see Chapter 4, Section 6.

Notes

Chapter 6
Frame, suspension and final drive

Contents

Drive chain – removal, cleaning and installation 16
Drive chain and sprockets – check, adjustment
 and lubrication .... see Chapter 1
Footrests, brake pedal and gearchange lever – removal and
 installation 3
Forks – disassembly, inspection and reassembly 8
Forks – oil change 7
Forks – removal and installation 6
Frame – inspection and repair 2
General information 1
Handlebars and levers – removal and installation 5
Handlebar switches – check .... see Chapter 9
Handlebar switches – removal and installation .... see Chapter 9
Rear shock absorber – removal, inspection and installation 11
Rear sprocket coupling/rubber damper – check and replacement . 18
Rear suspension linkage – removal, inspection and installation 12
Sidestand – check .... see Chapter 1
Sidestand – lubrication .... see Chapter 1
Sidestand – removal and installation 4
Sidestand switch – check and replacement .... see Chapter 9
Sprockets – check and replacement 17
Steering head bearings – freeplay check
 and adjustment .... see Chapter 1
Steering head bearings – inspection and replacement 10
Steering head bearings – lubrication .... see Chapter 1
Steering stem – removal and installation 9
Suspension – adjustments 13
Suspension – check .... see Chapter 1
Swingarm – inspection, bearing check and replacement 15
Swingarm – removal and installation 14
Swingarm and suspension linkage bearings –
 lubrication .... see Chapter 1

Degrees of difficulty

| **Easy,** suitable for novice with little experience | **Fairly easy,** suitable for beginner with some experience | **Fairly difficult,** suitable for competent DIY mechanic | **Difficult,** suitable for experienced DIY mechanic | **Very difficult,** suitable for expert DIY or professional |

Specifications

Front forks
Fork oil type .... Pro-Honda SS8 suspension fluid, Honda Ultra Cushion 10W oil or equivalent 10W fork oil

Fork oil capacity
  RR-Y and RR-1 (2000 and 2001) models .... 488 ± 2.5 cc
  RR-2 and RR-3 (2002 and 2003) models .... 513 ± 2.5 cc
Fork oil level*
  RR-Y and RR-1 (2000 and 2001) models .... 90 mm
  RR-2 and RR-3 (2002 and 2003) models .... 73 mm
Fork spring free length (min)
  RR-Y and RR-1 (2000 and 2001) models
    Standard .... 230.5 mm
    Service limit .... 225.9 mm
  RR-2 and RR-3 (2002 and 2003) models
    Standard .... 255.8 mm
    Service limit .... 250.8 mm
Fork tube runout limit .... 0.2 mm

*Oil level is measured from the top of the tube with the fork spring removed and the leg fully compressed.

6•2 Frame, suspension and final drive

Final drive

Drive chain slack and lubricant	See Chapter 1
Drive chain	
RR-Y and RR-1 (2000 and 2001) models	
Type	DID 50VA8C1
Length	108 links
RR-2 and RR-3 (2002 and 2003) models	
Type	DID 50VA8C1 or RK GB50HFOZ5
Length	108 links
Joining link pin projection from side plate (unstaked)	
DID type chain	1.15 to 1.55 mm
RK type chain	1.20 to 1.40 mm
Joining link staked ends diameter	
DID type chain	5.50 to 5.80 mm
RK type chain	5.45 to 5.85 mm
Sprocket sizes	
Front (engine) sprocket	16T
Rear (wheel) sprocket	
Europe models	42T
US and Canada models	43T

Torque settings

Clutch lever bracket clamp bolt(s)	12 Nm
Drive chain slider bolts	9 Nm
Footrest bracket mounting bolts	39 Nm
Footrest holder bolt	44 Nm
Fork damper cartridge bolt	34 Nm
Fork locknut-to-damper rod – RR-2 and RR-3 (2002 and 2003) models	25 Nm
Fork top bolt-to-damper rod – RR-Y and RR-1 (2000 and 2001) models	34 Nm
Fork top bolt-to-fork tube	22 Nm
Fork yoke clamp bolts	
Top yoke bolts	23 Nm
RR-Y and RR-1 (2000 and 2001) models	22 Nm
RR-2 and RR-3 (2002 and 2003) models	23 Nm
Bottom yoke bolts	26 Nm
Front brake master cylinder clamp bolts	12 Nm
Front sprocket bolt	54 Nm
Handlebar clamp bolts	26 Nm
Handlebar end-weight screws	10 Nm
Rear sprocket nuts	64 Nm
Shock absorber	
RR-Y and RR-1 (2000 and 2001) models	
Upper mounting bolt nut	44 Nm
lower mounting bolt nut	44 Nm
RR-2 and RR-3 (2002 and 2003) models	
Upper mounting nut	93 Nm
lower mounting bolt nut	44 Nm
Sidestand bracket bolts	44 Nm
Sidestand pivot bolt	10 Nm
Sidestand pivot bolt nut	29 Nm
Steering head bearing adjuster nut	
RR-Y and RR-1 (2000 and 2001) models	
Initial setting	29 Nm
Final setting	29 Nm
RR-2 (2002) models with caged ball bearings	
Initial setting	30 Nm
Final setting	15 Nm + 90° (See Text)
RR-2 (2002) models with taper roller bearings	
Initial setting	40 Nm
Final setting	18 Nm
RR-3 (2003) models	
Initial setting	30 Nm
Final setting	15 Nm + 90° (See Text)
Steering stem nut	103 Nm
Suspension linkage bolt nuts	
Linkage arm to engine bracket	44 Nm
Linkage arm to linkage plates	44 Nm
Linkage plates to swingarm	44 Nm
Swingarm pivot bolt nut	118 Nm
Swingarm pivot bolt pinch bolts	26 Nm

Frame, suspension and final drive 6•3

1 General information

All models have a box-section twin-spar aluminium frame which uses the engine as a stressed member.

Front suspension is by a pair of upside-down oil-damped telescopic forks. The forks have a cartridge damper and are adjustable for spring pre-load and both rebound and compression damping.

At the rear, a box-section aluminium swingarm acts on a single shock absorber via a three-way linkage. The shock absorber is adjustable for spring pre-load and both rebound and compression damping.

The drive to the rear wheel is by chain and sprockets.

2 Frame – inspection and repair

1 The frame should not require attention unless accident damage has occurred. In most cases, frame replacement is the only satisfactory remedy for such damage. A few frame specialists have the jigs and other equipment necessary for straightening the frame to the required standard of accuracy, but even then there is no simple way of assessing to what extent the frame may have been over stressed.

2 After the machine has accumulated a lot of miles, the frame should be examined closely for signs of cracking or splitting at the welded joints. Loose engine mount bolts can cause ovaling or fracturing of the mounting tabs. Minor damage can often be repaired by welding, depending on the extent and nature of the damage.

3 Remember that a frame which is out of alignment will cause handling problems. If misalignment is suspected as the result of an accident, it will be necessary to strip the machine completely so the frame can be thoroughly checked.

3 Footrests, brake pedal and gearchange lever – removal and installation

Footrests

Removal

1 Remove the split pin and washer from the bottom of the footrest pivot pin, then withdraw the pivot pin and remove the footrest (see illustration). On the rider's footrests, note the fitting of the return spring. On the passenger footrests, note the fitting of the detent plate, ball and spring, and take care not to let the ball and spring ping away when removing the footrest (see illustration).

Installation

2 Installation is the reverse of removal.

Brake pedal

Removal

3 Unscrew the footrest bracket mounting bolts and displace the bracket so that you can access the back of it (see illustration). Make sure you do not strain the brake hose or the brake light switch wiring – to get more slack in the wiring, remove the rider's seat (see Chapter 8), then trace it and disconnect it at the connector.

4 Unhook the brake pedal return spring and the brake light switch spring from the hook on the pedal (see illustration).

5 On RR-Y and RR-1 (2000 and 2001) models, remove the split pin from the clevis pin securing the brake pedal to the master cylinder pushrod, then withdraw the clevis pin. Detach the pushrod from the pedal.

6 On RR-2 and RR-3 (2002 and 2003) models, remove the split pin from the bolt securing the brake pedal to the master cylinder pushrod, then unscrew the nut, remove the washer and withdraw the bolt (see illustration). Detach the pushrod from the pedal.

7 Unscrew the footrest holder bolt and remove the footrest and the brake pedal, noting the thrust washer and wave washer (see illustration).

3.1a Remove the split pin (arrowed) and washer and withdraw the pivot pin from the top, noting the return spring ends – front footrests

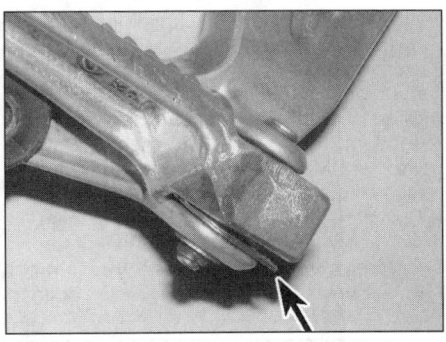

3.1b Note the fitting of the detent plate (arrowed) – rear footrests

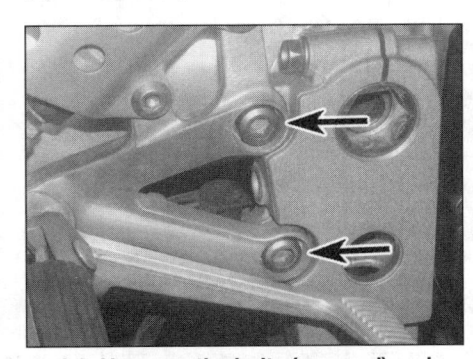

3.3 Unscrew the bolts (arrowed) and displace the bracket

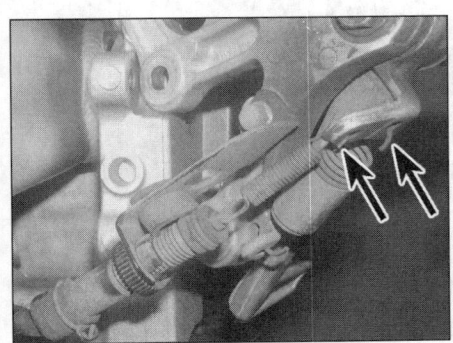

3.4 Unhook the springs (arrowed)

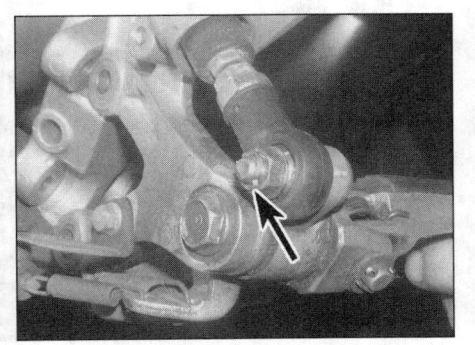

3.6 Remove the split (arrowed). Then unscrew the nut and withdraw the bolt

3.7 Unscrew the bolt (arrowed), withdraw the footrest and remove the pedal

6•4 Frame, suspension and final drive

3.9 Slacken the locknuts (arrowed) and thread the rod out

3.10 Unscrew the bolts and displace the bracket

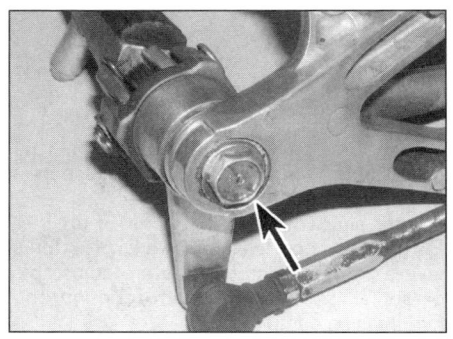

3.11 Unscrew the bolt (arrowed), withdraw the footrest and remove the lever

Installation

8 Installation is the reverse of removal, noting the following:
- Apply molybdenum disulphide oil (a 50/50 mixture of molybdenum disulphide grease and engine oil) to the pedal pivot section on the footrest holder.
- Slide the thrust washer onto the pivot section, then the pedal, then the wave washer. Make sure the footrest holder locates correctly in the bracket.
- Clean the threads of the footrest holder bolt and apply a suitable non-permanent thread locking compound. Tighten the footrest holder bolt and the footrest bracket bolts to the torque settings specified at the beginning of the Chapter.
- Use a new split pin on the clevis pin or bolt securing the brake pedal to the master cylinder pushrod.
- Check the operation of the rear brake light switch (see Chapter 1).

Gearchange lever and linkage

Removal

9 Slacken the gearchange lever linkage rod locknuts, then unscrew the rod and separate it from the lever and the arm (the rod is reverse-threaded on one end and so will simultaneously unscrew from both lever and arm when turned in the one direction) **(see illustration)**. Note how far the rod is threaded into the lever and arm as this determines the height of the lever relative to the footrest.

10 Unscrew the footrest bracket mounting bolts and displace the bracket so that you can access the back of it **(see illustration)**

11 Unscrew the footrest holder bolt and remove the footrest and the gearchange lever, noting the thrust washer and wave washer **(see illustration)**.

Installation

12 Installation is the reverse of removal, noting the following:
- Apply molybdenum disulphide oil (a 50/50 mixture of molybdenum disulphide grease and engine oil) to the lever pivot section on the footrest holder.
- Slide the thrust washer onto the pivot section, then the lever, then the wave washer. Make sure the footrest holder locates correctly in the bracket.
- Clean the threads of the footrest holder bolt and apply a suitable non-permanent thread locking compound. Tighten the footrest holder bolt and the footrest bracket bolts to the torque settings specified at the beginning of the Chapter.
- Adjust the gear lever height as required by screwing the linkage rod in or out of the lever and arm. Tighten the locknuts securely.

4 Sidestand – removal and installation

Removal

1 The sidestand is attached to a bracket that bolts onto the engine bracket. Springs anchored between the stand and its bracket ensure the stand is held in the retracted or extended position. Support the bike on an auxiliary stand.

2 To remove the sidestand without its bracket, first displace the sidestand switch (see Chapter 9). There is no need to disconnect its wiring connector or remove it completely, just let it hang from its wiring. Unhook the stand springs, then unscrew the nut from the pivot bolt **(see illustrations)**. Unscrew the pivot bolt and remove the stand.

3 To remove the complete sidestand

4.2a Unhook the spring ends

4.2b Unscrew the nut (arrowed), then unscrew the bolt from the back

Frame, suspension and final drive 6•5

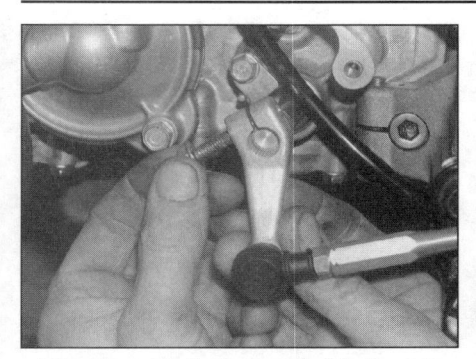

4.3a Unscrew the bolt and slide the arm off the shaft

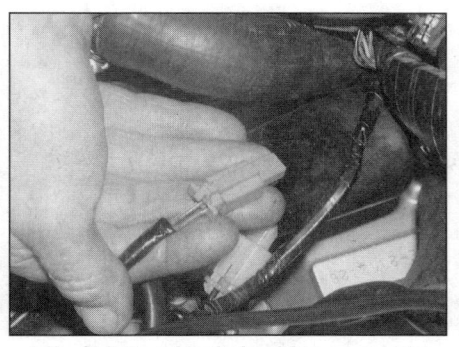

4.3b Sidestand switch wiring connector

4.3c Unscrew the bolts and remove the stand assembly

assembly unscrew the gearchange linkage arm pinch bolt and slide the arm off the shaft, noting how the slot in the arm aligns with the punch mark on the shaft **(see illustration)**. Let the arm dangle so the rod is clear of the sidestand bracket bolts. Either displace the sidestand switch (see Chapter 9), or raise the fuel tank (see Chapter 4), then trace the wiring from the sidestand switch and disconnect it at the green 2-pin wiring connector inside the rubber boot **(see illustration)**. Release the wiring from any clips or guides, noting its routing. Unscrew the sidestand bracket bolts and remove the stand assembly **(see illustration)**.

Installation

4 If the stand was removed without its bracket, apply grease to the pivot bolt shank and tighten to the torque setting specified at the beginning of the Chapter. Fit the nut and tighten it to the specified torque **(see illustration 4.2b)**. Install the sidestand switch (see Chapter 9). Reconnect the springs and check that they hold the stand securely up when not in use – an accident is almost certain to occur if the stand extends while the machine is in motion **(see illustration 4.2a)**.

5 If the complete sidestand assembly was removed, clean the threads of the bolts and apply a suitable non-permanent thread locking compound. Install the assembly and tighten the bolts to the torque setting specified at the beginning of the Chapter **(see illustration 4.3c)**. If displaced, install the sidestand switch (see Chapter 9). Otherwise feed the wiring to the connector and connect it **(see illustration 4.3b)**. Slide the gearchange linkage arm onto the shaft, aligning the slit in the clamp with the punch mark on the end of the shaft, and tighten the bolt **(see illustration 4.3a)**. Check the operation of the switch (see Chapter 1).

5 Handlebars and levers – removal and installation

Right handlebar removal

Note: *To merely displace the handlebar from the top yoke without having to remove all the individual assemblies follow Steps 1, 2 and 4.*

1 Disconnect the wires from the brake light switch **(see illustration)**. Unscrew the two master cylinder assembly clamp bolts and position the assembly clear of the handlebar, making sure no strain is placed on the hydraulic hose **(see illustration)**. Keep the master cylinder reservoir upright to prevent possible fluid leakage.

2 On RR-Y and RR-1 (2000 and 2001) models unscrew the two clutch lever bracket bolts and position the assembly clear of the handlebar – although not part of the right handlebar, this is necessary to access the top yoke left-hand clamp bolt. On RR-2 and RR-3 (2002 and 2003) models slacken the clutch lever bracket bolt and swivel or slide the bracket as required to access the top yoke left-hand clamp bolt **(see illustration)**.

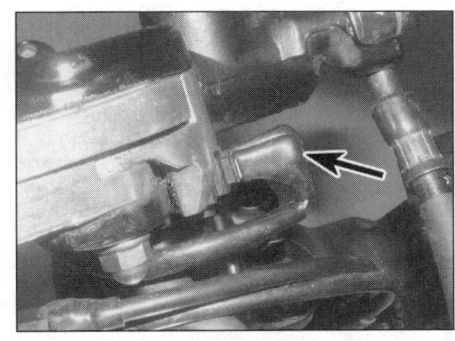

5.1a Disconnect the wiring connectors (arrowed)

5.1b Unscrew the master cylinder clamp bolts (arrowed) and displace the assembly

5.2 Slacken the bolt (arrowed) and move the assembly as required to access the clamp bolt

5.3a Unscrew the top yoke clamp bolts (arrowed) . . .

5.3b . . . then unscrew the steering stem nut . . .

5.3c . . . and ease the yoke up and off the forks

5.4a Undo the screws . . .

5.4b . . . and detach the switch housing

3 Slacken both top yoke clamp bolts (see illustration). Unscrew the steering stem nut, then ease the yoke up and off the forks and lay it aside on some rag (see illustrations). If you want to remove the top yoke from the bike rather than just displace it, trace the wiring from the ignition switch, and where fitted the HISS receiver, and disconnect it/them at the connector(s) – on RR-Y and RR-1 (2000 and 2001) models you will have to remove the right-hand air duct cover to access them (see Chapter 8), and on RR-2 and RR-3 (2002 and 2003) models you will have to remove the air filter housing to access them (see Chapter 4). Release the wiring from any clips or ties and feed it through to the yoke.

4 Unscrew the two handlebar switch housing screws and separate the halves (see illustrations). If required, free the throttle cables from the throttle pulley, creating slack in the cable as necessary using the adjusters (see Chapter 1). To avoid having to do this, note that the throttle pulley can be slid off the end of the handlebar with the cables still attached to it and the housing after the handlebar has been lifted off the fork.

5 Unscrew the handlebar end-weight retaining screw, then remove the weight from the end of the handlebar (see illustration). If the throttle cables have been detached, slide the twistgrip off the handlebar. If a new handlebar is being installed, you need to remove the inner weight from the old bar so it can be used in the new one – to do this, reinstall the end-weight and tighten its screw. Squirt some lubricant (such as WD40) into the inner weight retainer tab hole, then press down on the tab using a screwdriver and twist and pull the end-weight, drawing the inner weight assembly out. Remove the end-weight and discard the retainer as a new one should be used. Check the condition of the rubbers on the inner weight and fit new ones if they are damaged, deformed or deteriorated.

6 Slacken the handlebar clamp bolt, then ease the handlebar up and off the fork (see illustration). If required, slide the throttle twistgrip and switch housing assembly off the handlebar. Note the handlebar stopper ring in the groove in the fork and remove it for safekeeping if required.

Left handlebar removal

Note: *To merely displace the handlebar from the top yoke without having to remove the individual assemblies follow Step 9.*

7 Disconnect the wires from the clutch switch

5.5 Handlebar end-weight screw (arrowed)

5.6 Slacken the clamp bolt (arrowed) and slide the handlebar up and off the fork

Frame, suspension and final drive

5.7a Clutch switch wiring connectors (arrowed) – RR-Y and RR-1 (2000 and 2001) models

5.7b Clutch switch wiring connectors (arrowed) – RR-2 and RR-3 (2002 and 2003) models

(see illustrations). On RR-Y and RR-1 (2000 and 2001) models unscrew the two clutch lever bracket bolts and position the assembly clear of the handlebar. On RR-2 and RR-3 (2002 and 2003) models slacken the clutch lever bracket bolt and swivel or slide the bracket as required to access the top yoke left-hand clamp bolt **(see illustration 5.2)**.

8 To access the top yoke right-hand clamp bolt disconnect the wires from the brake light switch, then unscrew the two master cylinder assembly clamp bolts and position the assembly clear of the handlebar, making sure no strain is placed on the hydraulic hose **(see illustrations 5.1a and 5.1b)**. Keep the master cylinder reservoir upright to prevent possible fluid leakage.

9 Slacken both top yoke clamp bolts **(see illustration 5.3a)**. Unscrew the steering stem nut, then ease the yoke up and off the forks and lay it aside on some rag **(see illustration 5.3b and 5.3c)**. If you want to remove the top yoke from the bike rather than just displace it, trace the wiring from the ignition switch, and where fitted the HISS receiver, and disconnect it/them at the connector(s) – on RR-2 and RR-3 (2002 and 2003) models you will have to remove the air filter housing to access them (see Chapter 4). Release the wiring from any clips or ties and feed it through to the yoke.

10 Either unscrew the two handlebar switch housing screws and separate the halves **(see illustrations)**, or if the handlebars are being displaced with the housing attached, disconnect the wires from the horn **(see illustration)**.

11 If required, unscrew the handlebar end-weight retaining screw, then remove the weight from the end of the handlebar and slide off the grip **(see illustration)**. If the grip has been glued on, you will probably have to slit it with a knife to remove it. If a new handlebar is being installed, you need to remove the inner weight from the old bar so it can be used in the new one – to do this, reinstall the end-weight and tighten its screw. Squirt some lubricant (such as WD40) into the inner weight retainer tab hole, then press down on the tab using a screwdriver and twist and pull the end-weight, drawing the inner weight assembly out. Remove the end-weight and discard the retainer as a new one should be used. Check the condition of the rubbers on the inner weight and fit new ones if they are damaged, deformed or deteriorated.

12 Slacken the handlebar clamp bolt, then ease the handlebar up and off the fork **(see illustrations)**. If required, slide the clutch lever bracket assembly off the handlebar. Note the handlebar stopper ring in the groove in the fork and remove it for safekeeping if required.

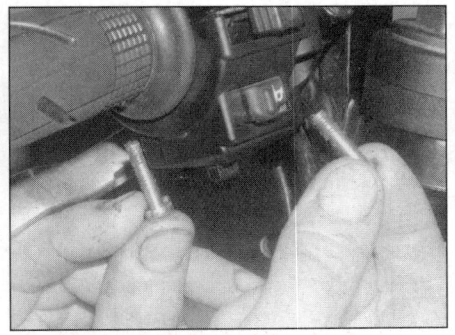

5.10a Either undo the screws . . .

5.10b . . . and detach the switch housing . . .

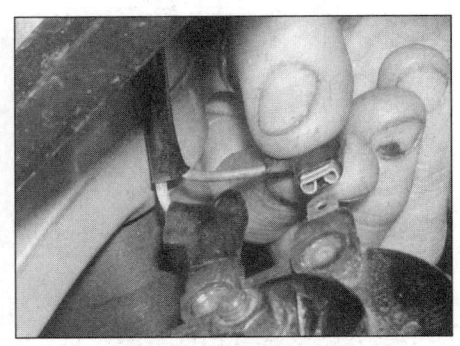

5.10c . . . or disconnect the horn wiring connectors

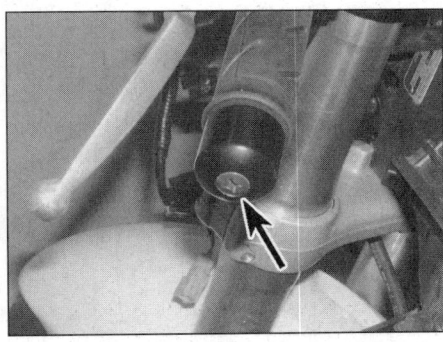

5.11 Handlebar end-weight screw (arrowed)

5.12a Slacken the clamp bolt (arrowed) . . .

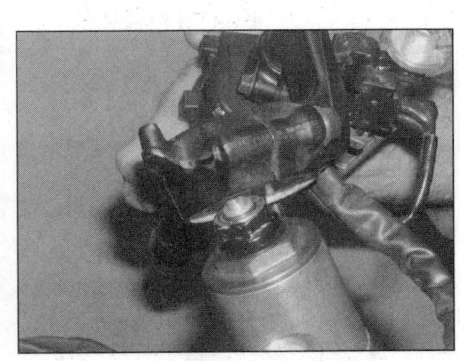

5.12b . . . and slide the handlebar up and off the fork

6•8 Frame, suspension and final drive

5.13a The stopper ring locates inside the lower rim of the handlebar clamp (arrow)

5.13b Locate the lug (A) on each handlebar into its hole (B) in the underside of the yoke

Handlebar installation

13 Installation is the reverse of removal, noting the following.
- If the throttle cables were not detached, slide the throttle twistgrip and cable housing assembly onto the handlebar before fitting the handlebar onto the fork, and smear some grease onto the handlebar.
- If removed, fit the handlebar stopper ring into its groove. When fitting the handlebar onto the fork, slide it down over the ring as far as it will go so the ring seats up inside the bottom **(see illustration 5.13a)**. Do not tighten the handlebar clamp bolts until the top yoke has been installed.

- When fitting the top yoke onto the forks, locate the lug on the top of each handlebar clamp in its hole in the underside of the yoke, so that the handlebars are set in the correct position **(see illustration 5.13b)**. Tighten the steering stem nut first, then the top yoke clamp bolts, then the handlebar clamp bolts, tightening them all to the torque settings specified at the beginning of the Chapter, and lifting the handlebar slightly if necessary so its top surface seats against the bottom of the yoke **(see illustration 5.13c)**.
- When installing the handlebar inner weights, locate the tab on the retainer in the hole in the handlebar.

- When installing the handlebar end-weights, align the boss with the cut-out on the inner weight inside the handlebar. Clean the threads of the end-weight retaining screws and apply a suitable non-permanent thread locking compound. Tighten them to the specified torque setting. If new grips are being fitted, secure them using a suitable adhesive.
- When installing the throttle twistgrip onto the right handlebar, apply some grease to it.
- Make sure the front brake master cylinder clamp is installed with the UP mark facing up, and with the clamp mating surfaces aligned with the punch mark on the top of the handlebar **(see illustration 5.13d)**. Tighten the master cylinder clamp bolts to the specified torque setting, tightening the top bolt first.
- On RR-Y and RR-1 (2000 and 2001) models make sure the front brake master cylinder clamp is installed with the UP mark facing up, and with the clamp mating surfaces aligned with the punch mark on the top of the handlebar. Tighten the clamp bolts to the specified torque setting, tightening the top bolt first.
- On RR-2 and RR-3 (2002 and 2003) models align the punch mark on the clutch lever bracket with the punch mark on the top of the handlebar **(see illustration 5.13e)**. Tighten the clamp bolt to the specified torque setting.
- Make sure the pin in the bottom half of each switch housing locates in its hole in the handlebar **(see illustration 5.13f)**. Tighten the front housing screw first, then the rear.
- Do not forget to reconnect the front brake light switch and clutch switch wiring connectors, and if necessary the horn wiring connectors **(see illustrations 5.1a, 5.7a, 5.7b and 5.10c)**.

Handlebar lever removal

14 To remove the front brake lever, undo the lever pivot screw locknut, then undo the pivot screw and remove the lever **(see illustration)**.
15 To remove the clutch lever, on RR-Y and RR-1 (2000 and 2001) models loosen the

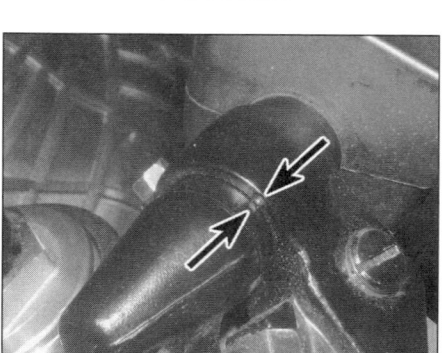

5.13c Tighten the steering stem nut and the other listed bolts to their specified torque setting

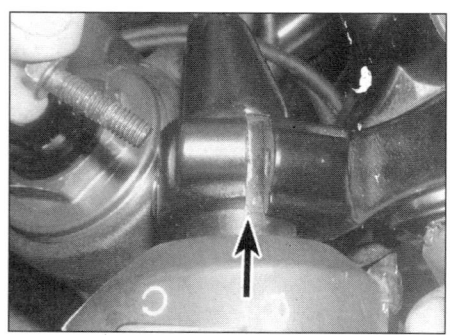

5.13d Align the clamp mating surfaces with the small punch mark (arrowed) in the handlebar

5.13e Align the punch marks (arrowed)

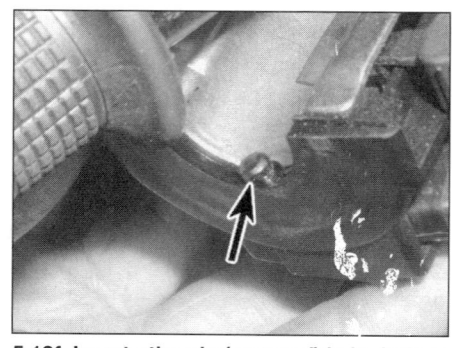

5.13f Locate the pin (arrowed) in its hole in the handlebar

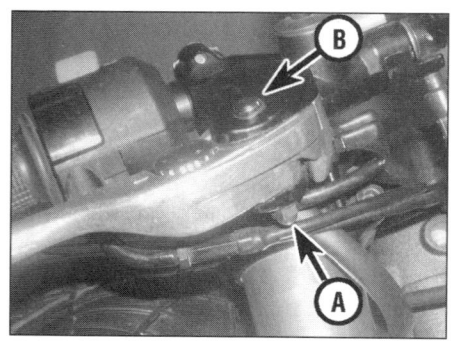

5.14 Undo the nut (A), then undo the pivot screw (B) and remove the lever

Frame, suspension and final drive 6•9

adjuster lockring **(see illustration)**. On all models thread the adjuster into the bracket to provide freeplay in the cable **(see illustration)**.

16 Undo the lever pivot screw locknut, then undo the pivot screw and remove the lever, detaching the cable nipple as you do **(see illustrations)**.

Handlebar lever installation

17 Installation is the reverse of removal. Apply silicone grease to the contact area between the front brake master cylinder pushrod tip and the brake lever. Apply lithium or molybdenum grease to the pivot screw shafts and the contact areas between the lever and its bracket. Apply a spray lubricant such as WD40, or a dry-film Teflon lubricant to the brake lever span adjuster mechanism. Adjust clutch cable freeplay (see Chapter 1).

6 Forks – removal and installation

Caution: Although not strictly necessary, before removing the forks it is recommended that the fairing and fairing panels are removed (see Chapter 8). This will prevent accidental damage to the paintwork.

Removal

1 For best access and to prevent the possibility of damage, remove the fairing (see Chapter 8). Also remove the front mudguard (see Chapter 8).
2 Remove the front wheel (see Chapter 7). Tie the front brake calipers and hoses back so that they are out of the way.
3 Note the routing of any cables, hoses and wiring around the forks.
4 Working on one fork at a time, slacken the fork clamp bolt in the top yoke and the handlebar clamp bolt **(see illustration)**. If the fork is to be disassembled, or if the fork oil is being changed, slacken the fork top bolt now **(see illustration)**. Note the amount of protrusion of the fork above the top yoke.
5 Slacken the fork clamp bolts in the bottom

5.15a On RR-Y and RR-1 (2000 and 2001) models slacken the lockring (A) and thread the adjuster (B) in

5.15b On RR-2 and RR-3 (2002 and 2003) models thread the adjuster in

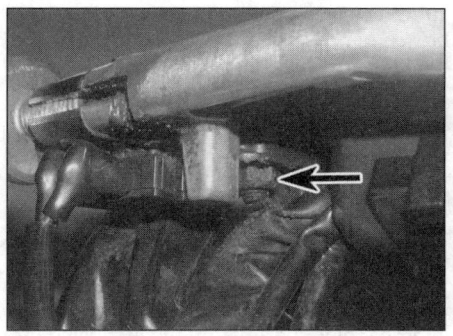

5.16a Unscrew the locknut (arrowed) . . .

5.16b . . . then undo the pivot screw (arrowed) and remove the lever

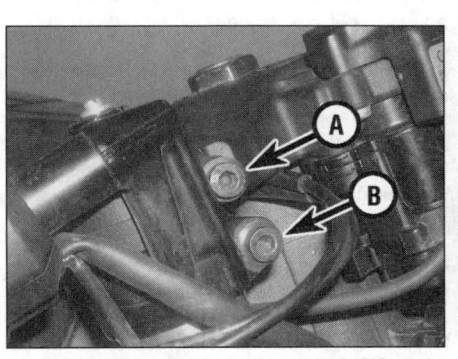

6.4a Slacken the fork clamp bolt (A) and the handlebar clamp bolt (B) . . .

6.4b . . . and if required the fork top bolt

yoke, and remove the fork by twisting it and pulling it downwards, guiding the handlebar off as you do **(see illustrations)**. Note the

handlebar stopper ring in the groove in the fork and remove it for safekeeping if required **(see illustration)**.

6.5a Bottom yoke fork clamp bolts (arrowed)

6.5b Draw the fork down, supporting the handlebar, and out of the yokes

6.5c Remove the stopper ring for safekeeping

6.6 Locate the lug (A) in the hole (B)

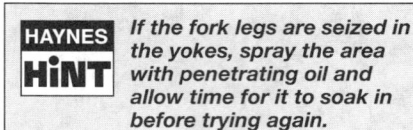

6.8 If necessary tighten the fork top bolt to the specified torque setting

HAYNES HINT *If the fork legs are seized in the yokes, spray the area with penetrating oil and allow time for it to soak in before trying again.*

Installation

6 Remove all traces of corrosion from the fork tube and the yokes. If removed, fit the handlebar stopper ring into its groove (see illustration 6.5c). Slide the fork up through the bottom yoke and the handlebar clamp and into the top yoke, making sure all cables, hoses and wiring are routed on the correct side of the fork (see illustration 6.5b). Locate the lug on the top of the handlebar clamp in its hole in the underside of the yoke (see illustration).
7 Set the amount of protrusion of the fork tube above the top yoke as noted on removal – Honda specify that the joint between the top of the fork tube and the fork top bolt should be set flush with the top of the yoke. On the model we worked on there was actually about 1 mm of fork tube protruding above the top yoke, and this results from pushing the fork as far up as it will go, ensuring that the handlebar stopper ring is snugly up inside the bottom rim of the clamp (see illustration 5.13a), and the top surface of the clamp is butted against the underside of the top yoke. With the fork set as Honda specify, there is either a small gap between handlebar clamp and top yoke, or between the stopper ring and its seat in the rim of the clamp, depending on whether the handlebar butts against the yoke or sits on the ring. Always make sure the forks are set the same on both sides.
8 Tighten the fork clamp bolts in the bottom yoke to the torque setting specified at the beginning of the Chapter (see illustration 6.5a). If the fork has been dismantled or if the fork oil was changed, tighten the fork top bolt to the specified torque setting (see illustration). Now tighten the fork clamp bolt in the top yoke (see illustration 6.4a). Lift the handlebar slightly if necessary so its top surface seats against the bottom of the yoke, then tighten its clamp bolt to the specified torque.
9 Install the front wheel (see Chapter 7) and the front mudguard (see Chapter 8).
10 Check the operation of the front forks and brakes before taking the machine out on the road.

7 Forks – oil change

RR-Y and RR-1 models

1 Remove the forks (see Section 6). Always work on the fork legs separately to avoid interchanging parts and thus causing an accelerated rate of wear.
2 Note the number of lines on the pre-load adjuster that are visible above the top bolt.
3 If the fork top bolt was not slackened with the fork *in situ*, relocate the fork in the yokes and tighten the bottom yoke clamp bolts (see Section 6), or carefully clamp the fork tube in a vice equipped with soft jaws, taking care not to overtighten or score its surface, and slacken (but do not yet unscrew) the top bolt (see illustration 6.4b).
4 Remove the snap-ring from the top of the damper rod. Thread the pre-load adjuster off the damper rod, then hook the adjuster tripod out of the top of the fork using a piece of wire.
5 Unscrew the fork top bolt from the top of the fork tube. The bolt will remain threaded on the damper rod.
6 Slide the fork tube fully down onto the slider (wrap a rag around the top of the tube to minimise oil spillage) while, with the aid of an assistant if necessary, keeping the damper rod fully extended. Push down on the spacer to compress the spring and fit a 17 mm spanner onto the hex on the damper rod, below the top bolt. Counter-hold the hex then unscrew the top bolt and thread it off the damper rod.
7 Remove the washer, the spacer seat if loose (it fits into the top of the spacer and will probable stay with it), and the spacer (see illustration 7.20a). Withdraw the spring from the tube, noting which way up it fits (see illustration 7.20b).
8 Invert the fork leg over a suitable container and pump the fork and damper rod vigorously several times each to expel as much fork oil as possible (see illustration 7.21). Support the fork upside down in the container for a while to allow as much oil as possible to drain, then pump the fork and damper again.
9 Stand the fork upright. Slowly pour in the specified quantity of the specified grade of fork oil (see illustration 7.22a). Fully extend the fork, then cover the top of the tube with your hand and compress the fork as much as possible – this helps to bleed any air. Now pump the fork and damper rod slowly at least ten times each to distribute it evenly. Fully compress the fork tube and damper rod into the slider and measure the oil level, and make any adjustment by adding more or tipping some out until the oil is at the level specified at the beginning of the Chapter (see illustration 7.22b).
10 Pull the damper rod and fork tube out as far as possible then install the spring, with the tapered and closer wound coils at the top (see illustration 7.20b). If removed, fit the spacer seat into the top of the spacer. Keeping the damper rod extended, install the spacer, locating it in the top of the spring (see illustration 7.20a), then fit the washer onto the spacer seat.
11 Fit a new O-ring smeared with fork oil onto the fork top bolt. Push down on the spacer to compress the spring and thread the top bolt onto the damper rod until it seats. Fit a 17 mm spanner onto the hex on the damper rod below the top bolt. Counter-hold the hex then tighten the top bolt, to the specified torque setting if you have a crows-foot wrench. Thread the bolt into the top of the fork tube, keeping the tube fully extended whilst doing so, and making sure it is not cross-threaded. **Note:** *The top bolt can be tightened to the specified torque setting at this stage if the tube is held between the padded jaws of a vice, but do not risk distorting the tube by doing so. A better method is to tighten the top bolt when the fork has been installed in the bike and is held in the bottom yoke, but before the top yoke clamp bolt is tightened* (see illustration 6.8).
12 Fit the pre-load adjuster tripod into the top of the fork, locating its legs into the holes in the top bolt. Smear the pre-load adjuster O-ring with fork oil, then thread the adjuster onto the damper rod and set it with the number of lines noted on removal visible. Fit the snap-ring into its groove in the top of the damper rod.
13 Install the forks (see Section 6).

RR-2 and RR-3 models

14 Remove the forks (see Section 6). Always work on the fork legs separately to avoid interchanging parts and thus causing an accelerated rate of wear (see illustration 8.33).
15 Refer to Section 13 and check the current pre-load setting so the fork can later be reset at the same amount.
16 If the fork top bolt was not slackened with the fork *in situ*, relocate the fork in the yokes and tighten the bottom yoke clamp bolts (see

Frame, suspension and final drive 6•11

7.17a Remove the snap-ring . . .

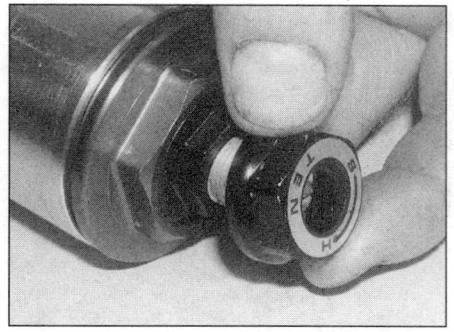

7.17b . . . and the pre-load adjuster hex

7.18 Thread the top bolt out of the tube

Section 6), or carefully clamp the fork tube in a vice equipped with soft jaws, taking care not to overtighten or score its surface, and slacken the top bolt **(see illustration 6.4b)**.

17 Remove the snap-ring from the top of the damper rod **(see illustration)**. Remove the pre-load adjuster hex from the damper rod **(see illustration)**.

18 Unscrew the fork top bolt from the top of the fork tube **(see illustration)**. The bolt will remain on the damper rod, held by the locknut on its top.

19 Counter-hold the top bolt or the damper rod using a spanner on the flats, then unscrew the locknut above the top bolt and thread it off the damper rod **(see illustration)**. Remove the top bolt **(see illustration)**. Note how the cut-outs in the pre-load adjuster plate inside the top bolt locate over the ridges on the damper rod.

20 Remove the spacer, then withdraw the spring from the tube, noting which way up it fits **(see illustrations)**.

21 Invert the fork leg over a suitable container and pump the fork and damper rod vigorously several times each to expel as much fork oil as possible **(see illustration)**. Support the fork upside down in the container for a while to allow as much oil as possible to drain, then pump the fork and damper again.

22 Stand the fork upright. Slowly pour in the specified quantity of the specified grade of fork oil **(see illustration)**. Fully extend the fork, then cover the top of the tube with your hand and compress the fork as much as possible – this helps to bleed any air. Now pump the fork and damper rod slowly at least ten times each to distribute it evenly. Fully compress the fork tube and damper rod into the slider and measure the oil level, and make any adjustment by adding more or tipping some out until the oil is at the level specified at the beginning of the Chapter **(see illustration)**.

23 Pull the damper rod and fork tube out as

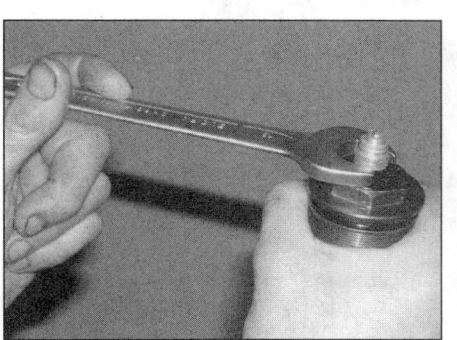

7.19a Unscrew the locknut . . .

7.19b . . . and remove the top bolt

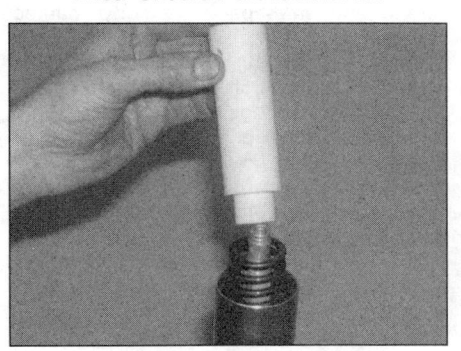

7.20a Remove the spacer . . .

7.20b . . . and the spring

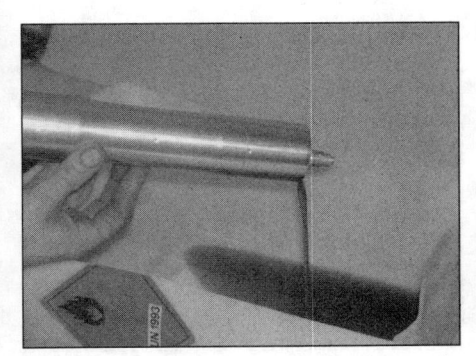

7.21 Invert the fork over a container and pump the tube and the rod to expel the oil

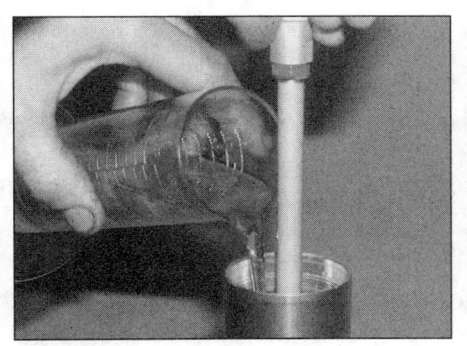

7.22a Pour the oil into the top of the tube and distribute and bleed it as described . . .

7.22b . . . then measure the level

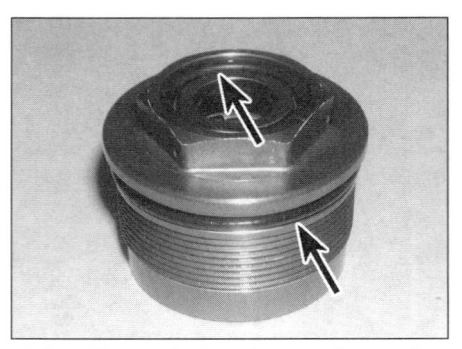

7.24a Fit new O-rings (arrowed) onto the top bolt

7.24b Thread the adjuster plate (arrowed) to the top of the bolt . . .

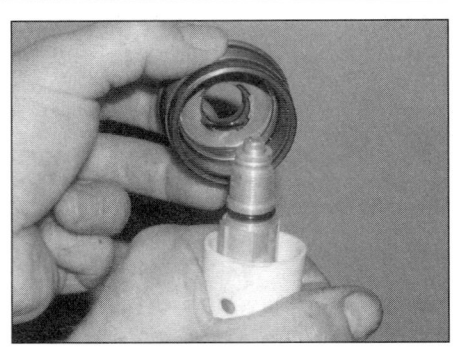

7.24c . . . then locate it onto the damper rod . . .

far as possible then install the spring, with the tapered and closer wound coils at the top (see illustration 7.20b). Keeping the damper rod extended, install the spacer, locating it in the top of the spring (see illustration 7.20a).

24 Fit a new O-ring smeared with fork oil around the fork top bolt, and also into the groove in the top of the bolt (see illustration). Thread the pre-load adjuster plate into the top bolt until it seats lightly (see illustration). Push down on the spacer to compress the spring and fit the top bolt onto the damper rod, aligning the cut-outs in the adjuster plate with the flats on the rod (see illustration). Thread the locknut onto the damper rod (see illustration). Counter-hold the top bolt or the damper rod using a spanner on the flats then tighten the locknut, to the specified torque setting if you have a crows-foot wrench (see illustration 7.19a). Thread the bolt into the top of the fork tube, keeping the tube fully extended whilst doing so, and making sure it is not cross-threaded (see illustration 7.18).

Note: *The top bolt can be tightened to the specified torque setting at this stage if the tube is held between the padded jaws of a vice, but do not risk distorting the tube by doing so. A better method is to tighten the top bolt when the fork has been installed in the bike and is held in the bottom yoke, but before the top yoke clamp bolt is tightened (see illustration 6.8).*

25 Fit a new O-ring into the larger groove in the top of the damper rod (see illustration).

Fit the pre-load adjuster hex onto the top of the damper rod (see illustration 7.17b). Fit the snap-ring into its groove in the top of the damper rod (see illustration 7.17a). Set the pre-load as noted on removal.

26 Install the forks (see Section 6).

8 Forks – disassembly, inspection and reassembly

RR-Y and RR-1 models

Disassembly

1 Remove the forks (see Section 6). Always work on the fork legs separately to avoid interchanging parts and thus causing an accelerated rate of wear. Store all components in separate, clearly marked containers.

2 Before dismantling the fork, slacken the damper cartridge bolt now as there is less chance of the cartridge rotating with it (due to the pressure of the spring). Compress the fork tube in the slider so that the spring exerts maximum pressure on the cartridge head, then slacken the bolt in the base of the fork slider (see illustration). If the bolt does not unscrew, but merely rotates the cartridge inside the fork tube, use an impact driver or air wrench.

3 Note the number of lines on the pre-load adjuster that are visible above the top bolt. If the fork top bolt was not slackened with the fork *in situ*, relocate the fork in the yokes and tighten the bottom yoke clamp bolts (see Section 6), or carefully clamp the fork tube in a vice equipped with soft jaws, taking care not to overtighten or score its surface, and slacken (but do not yet unscrew) the top bolt (see illustration 6.4b).

4 Remove the snap-ring from the top of the damper rod. Thread the pre-load adjuster off the damper rod, then hook the adjuster tripod out of the top of the fork using a piece of wire.

5 Unscrew the fork top bolt from the top of the fork tube. The bolt will remain threaded on the damper rod.

6 Slide the fork tube fully down onto the slider (wrap a rag around the top of the tube to minimise oil spillage) while, with the aid of an assistant if necessary, keeping the damper rod fully extended. Push down on the spacer to compress the spring and fit a 17 mm spanner onto the hex on the damper rod, below the top bolt. Counter-hold the hex then unscrew the top bolt and thread it off the damper rod.

7 Remove the washer, the spacer seat if loose (it fits into the top of the spacer and will probable stay with it), and the spacer (see illustration 7.20a). Withdraw the spring from the tube, noting which way up it fits (see illustration 7.20b).

8 Invert the fork leg over a suitable container and pump the fork and damper rod vigorously several times each to expel as much fork oil

7.24d . . . and secure it with the locknut

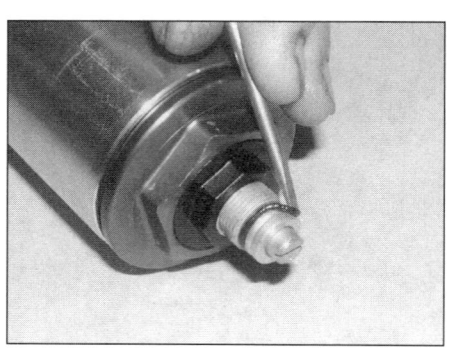

7.25 Fit a new O-ring into the groove

8.2 Slacken the damper rod bolt

Frame, suspension and final drive 6•13

8.9 Unscrew and remove the damper rod bolt

8.11 Prise out the dust seal using a flat-bladed screwdriver

8.12 Prise out the retaining clip using a flat-bladed screwdriver

as possible (see illustration 7.21). Support the fork upside down in the container for a while to allow as much oil as possible to drain, then pump the fork and damper rod again.

9 Remove the previously slackened damper cartridge bolt and its sealing washer from the bottom of the slider (see illustration). Discard the sealing washer as a new one must be used on reassembly.

10 Withdraw the damper assembly from inside the fork tube. The damper cartridge seat may come away with it, otherwise invert the fork and tip it out, noting which way up it fits.

11 Carefully prise out the dust seal from the bottom of the tube to gain access to the oil seal retaining clip (see illustration).

12 Carefully prise out the retaining clip, taking care not to scratch the surface of the tube – slide the tube fully onto the slider to keep any accidental damage below the seal area (see illustration).

13 To separate the slider from the tube it is necessary to displace the bottom bush and oil seal. The top bush should not pass through the bottom bush, and this can be used to good effect. Push the slider gently inwards until it stops. Take care not to do this forcibly. Then pull the slider sharply outwards until the top bush strikes the bottom bush (see illustration). Repeat this operation until the bottom bush and seal are tapped out of the slider (see illustration).

14 With the tube and slider separated, remove the top bush from the slider by carefully levering its ends apart using a screwdriver (see illustration). Slide the bottom bush, the oil seal washer, the oil seal, the retaining clip and the dust seal off the slider, noting which way up they fit (see illustration). Discard the oil seal and the dust seal as new ones must be used.

Inspection

15 Clean all parts in solvent and blow them dry with compressed air, if available. Check the fork slider for score marks, dents, pitting, scratches, flaking of the chrome finish and excessive or abnormal wear. Fit new tubes if any are found. Check the fork seal seat for nicks, gouges and scratches. If damage is evident, leaks will occur. Also check the oil seal washer for damage or distortion and fit a new one if necessary.

16 Check the fork slider for runout using V-blocks and a dial gauge. If the amount of runout exceeds the service limit specified, a new tube should be fitted.

⚠ **Warning: If the slider is bent or exceeds the runout limit, it should not be straightened; replace it with a new one.**

17 Check the fork tube for dents and cracks. Check the fork seal seat and housing for nicks, gouges and scratches. If damage is evident, leaks will occur. Also check the oil seal washer for damage or distortion and fit a new one if necessary.

18 Check the spring for cracks and other damage. Measure the spring free length and compare the measurement to the specifications at the beginning of the Chapter. If it is defective or sagged below the service limit, replace the springs in both forks with new ones. Never renew only one spring.

19 Examine the working surfaces of the two bushes; if worn or scuffed, or if the Teflon outer surface has been worn away to reveal the copper inner surface over more than 75% of the surface area, they must be replaced with new ones.

20 Check the damper assembly for damage

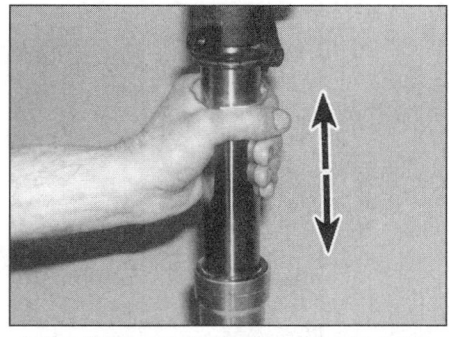

8.13a To separate the fork tube from the slider, pull them apart firmly several times . . .

8.13b . . . the slide hammer effect will displace the oil seal and bush

8.14a Remove the top bush . . .

8.14b . . . then slide the bottom bush, washer, oil seal, retaining clip and dust seal off

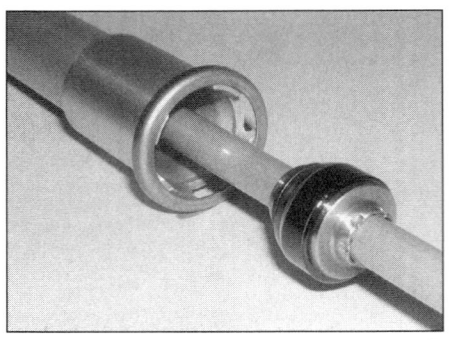

6•14 Frame, suspension and final drive

8.20 Pull the oil lock valve out and check it

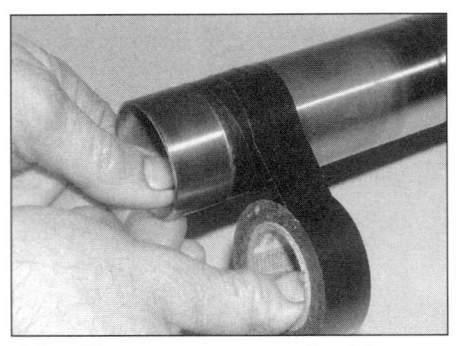

8.21a Wrap some tape over the ridges . . .

8.21b . . . then slide the dust seal . . .

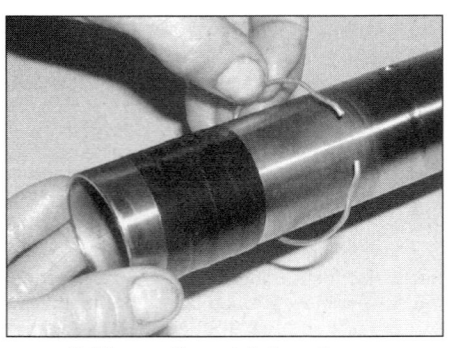

8.21c . . . the retaining clip . . .

8.21d . . . the oil seal . . .

8.21e . . . and its washer onto the shaft

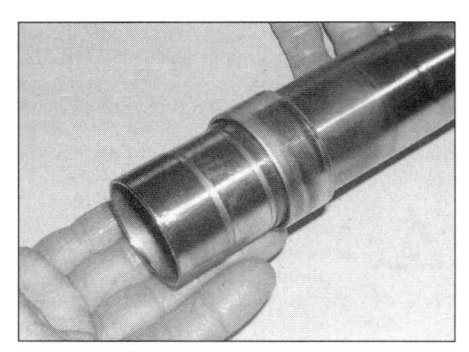

8.21f Slide the top bush on . . .

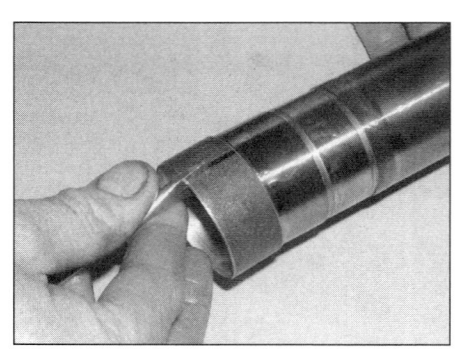

8.21g . . . then fit the bottom bush . . .

8.21h . . . into its groove

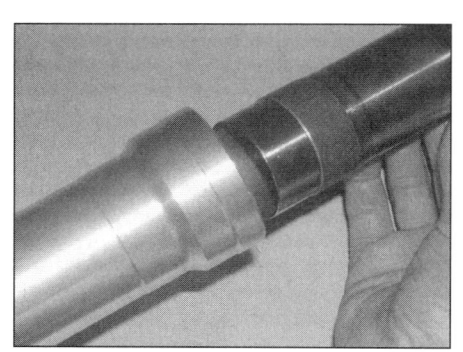

8.22 Fit the slider into the tube

and wear. Pull the oil lock valve out of the top of the cartridge and check it for wear and damage **(see illustration)**. Holding the outside of the cartridge, pump the rod in and out. If any wear or damage is found, or if the rod does not move smoothly in the damper, a new damper assembly must be installed.

Reassembly

21 Wrap some insulating tape over the ridges on the end of the fork slider to protect the lips of the new oil seal as it is installed **(see illustration)**. Apply a smear of the specified clean fork oil to the lips of the oil seal and the inner surface of each bush, then slide the new dust seal, the retaining clip, the oil seal, and the oil seal washer onto the fork slider, making sure that the marked side of the oil seal faces the dust seal **(see illustrations)**. Remove the insulating tape and slide the bottom bush onto the slider, then fit the top bush into its recess **(see illustrations)**.

22 Apply a smear of the specified clean fork oil to the outer surface of each bush, then carefully insert the slider fully into the fork tube **(see illustration)**.

23 Support the fork upside down, then press the bottom bush squarely into its recess in the fork tube as far as possible. Slide the oil seal washer on top of the bush, and keep the oil seal, the retaining clip and the dust seal out of

Frame, suspension and final drive 6•15

8.23 Locate the washer on top of the bush to protect it . . .

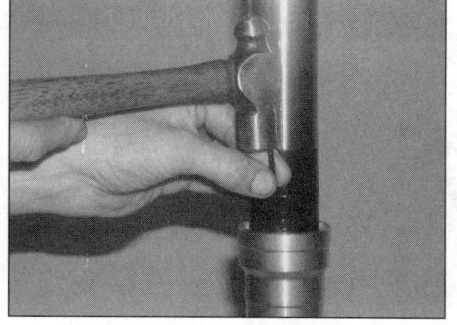

8.24 . . . and drive the bush into place

8.25 Drive the oil seal into place . . .

the way by sliding them up the slider **(see illustration)**. If necessary, tape them to the slider to prevent them from falling down and interfering as the bush is drifted into place.

24 Using either the special service tool (Part No. 07YMD-MCF0100) or a suitable drift, carefully drive the bottom bush fully into its recess – the oil seal washer prevents damaging the edges of the bush **(see illustration)**. If using a drift, wrap tape around it and the fork slider to prevent scratching the chrome. Make sure the bush enters the recess squarely. It is best to make sure that the fork slider is withdrawn as much as possible from the tube so that any accidental scratching is confined to the area that does not affect the oil seal.

25 When the bush is seated fully and squarely in its recess in the tube (remove the washer to check, wipe the recess clean, then reinstall the washer), drive the oil seal into place as described in Step 24 until the retaining clip groove is visible above the seal **(see illustration)**.

26 Once the oil seal is correctly seated, fit the retaining clip, making sure it is correctly located in its groove, then press the dust seal into position **(see illustrations)**.

27 Lay the fork flat and slide the tube fully into the slider. Fit the damper cartridge seat onto the bottom of the cartridge, then slide the damper assembly into the top of the fork tube and into the slider until it seats on the bottom. Fit a new sealing washer onto the cartridge bolt and apply a few drops of a suitable non-permanent thread locking compound **(see illustration)**. Fit the bolt into the bottom of the slider and thread it into the cartridge, tightening it to the torque setting specified at the beginning of the Chapter **(see illustration)**. If the cartridge rotates inside the tube, wait until the fork is fully reassembled before tightening the bolt, and then if necessary turn the fork upside down and compress it as you tighten the bolt. If it still turns, adjust the pre-load to its maximum (see Section 13).

28 Stand the fork upright. Slowly pour in the specified quantity of the specified grade of fork oil **(see illustration 7.22a)**. Fully extend the fork, then cover the top of the tube with your hand and compress the fork as much as possible – this helps to bleed any air. Now pump the fork tube and damper rod slowly at least ten times each to distribute it evenly. Fully compress the tube and damper rod into the slider and measure the oil level, and make any adjustment by adding more or tipping some out until the oil is at the level specified at the beginning of the Chapter **(see illustration 7.22b)**.

29 Pull the damper rod and fork tube out as far as possible then install the spring, with the tapered and closer wound coils at the top **(see illustration 7.20b)**. If removed, fit the spacer seat into the top of the spacer. Keeping the damper rod extended, install the spacer, locating it in the top of the spring, then fit the washer onto the spacer seat **(see illustration 7.20b)**.

30 Fit a new O-ring smeared with fork oil onto the fork top bolt. Push down on the spacer to compress the spring and thread the top bolt onto the damper rod until it seats. Fit a 17 mm spanner onto the hex on the damper rod below the top bolt. Counter-hold the hex then tighten the top bolt, to the specified torque setting if you have a crows-foot wrench. Thread the bolt into the top of the fork tube, keeping the tube fully extended whilst doing so, and making sure it is not cross-threaded. **Note:** *The top bolt can be tightened to the specified torque setting at this stage if the tube is held between the padded jaws of a vice, but do not risk distorting the tube by doing so. A better method is to tighten the top bolt when the fork has been installed in the bike and is held in the bottom yoke, but before the top yoke clamp bolt is tightened* **(see illustration 6.8)**.

8.26a . . . then fit the retaining clip . . .

8.26b . . . and the dust seal

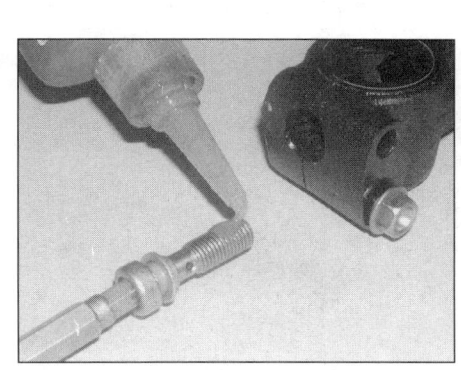

8.27a Fit the bolt using threadlock and a new sealing washer . . .

8.27b . . . and tighten it to the specified torque

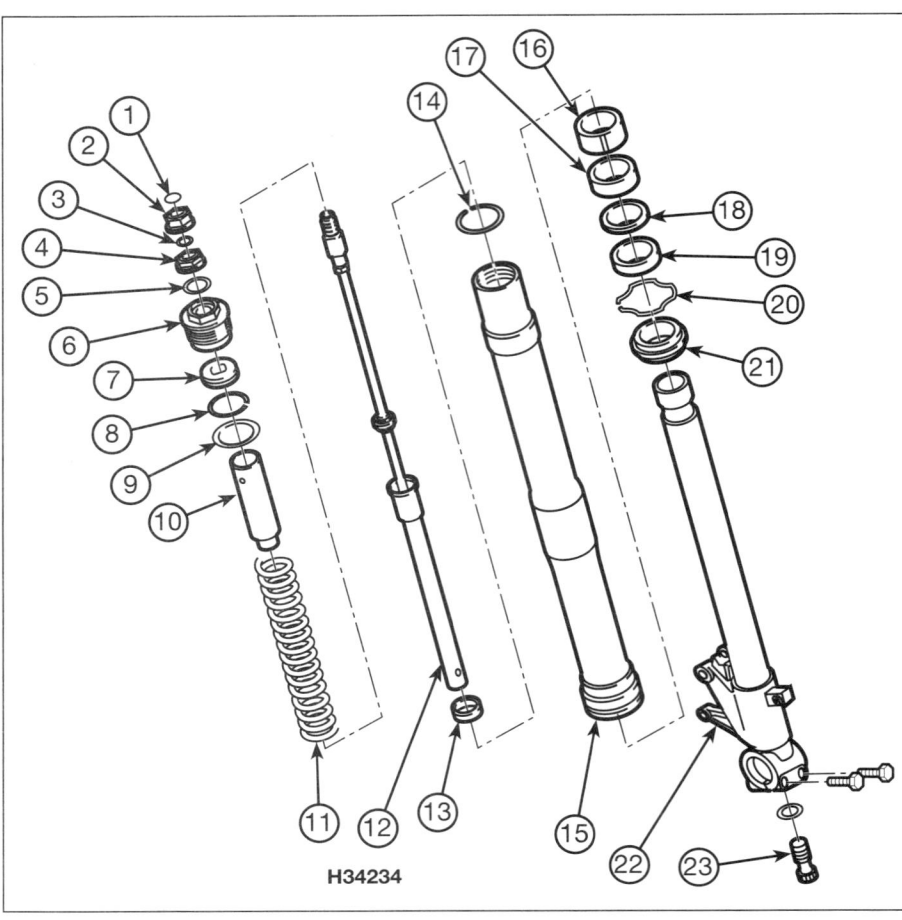

6•16 Frame, suspension and final drive

8.33 Front fork components – RR-2 and RR-3 (2002 and 2003) models

1 Snap-ring
2 Pre-load adjuster hex
3 O-ring
4 Lock nut
5 O-ring
6 Fork top bolt
7 Pre-load adjuster plate
8 Snap-ring
9 Top bolt O-ring
10 Spacer
11 Spring
12 Damper cartridge
13 Damper rod seat
14 Handlebar stopper ring
15 Fork tube
16 Bottom (tube) bush
17 Top (slider) bush
18 Washer
19 Oil seal
20 Retaining clip
21 Dust seal
22 Fork slider
23 Damper cartridge bolt and sealing washer

31 Fit the pre-load adjuster tripod into the top of the fork, locating its legs into the holes in the top bolt. Smear the pre-load adjuster O-ring with fork oil, then thread the adjuster onto the damper rod and set it with the number of lines noted on removal visible. Fit the snap-ring into its groove in the top of the damper rod.

32 Install the forks (see Section 6).

RR-2 and RR-3 models

Disassembly

33 Remove the forks (see Section 6). Always work on the fork legs separately to avoid interchanging parts and thus causing an accelerated rate of wear. Store all components in separate, clearly marked containers **(see illustration)**.

34 Before dismantling the fork, slacken the damper cartridge bolt now as there is less chance of the cartridge rotating with it (due to the pressure of the spring). Compress the fork tube in the slider so that the spring exerts maximum pressure on the cartridge head, then slacken the bolt in the base of the fork slider **(see illustration 8.2)**. If the bolt does not unscrew, but merely rotates the cartridge inside the fork tube, use an impact driver or air wrench.

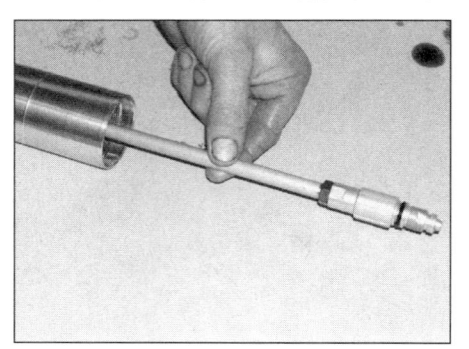

35 Refer to Section 13 and check the current pre-load setting so the fork can later be reset at the same amount.

36 If the fork top bolt was not slackened with the fork *in situ*, relocate the fork in the yokes and tighten the bottom yoke clamp bolts (see Section 6), or carefully clamp the fork tube in a vice equipped with soft jaws, taking care not to overtighten or score its surface, and slacken (but do not yet unscrew) the top bolt **(see illustration 6.4b)**.

37 Remove the snap-ring from the top of the damper rod **(see illustration 7.17a)**. Remove the pre-load adjuster hex from the damper rod **(see illustration 7.17b)**.

38 Unscrew the fork top bolt from the top of the fork tube **(see illustration 7.18)**. The bolt will remain on the damper rod, held by the locknut on its top.

39 Counter-hold the top bolt or the damper rod using a spanner on the flats, then unscrew the locknut above the top bolt and thread it off the damper rod **(see illustration 7.19a)**. Remove the top bolt **(see illustration 7.19b)**. Note how the cut-outs in the pre-load adjuster plate inside the top bolt locate over the ridges on the damper rod.

40 Remove the spacer, then withdraw the spring from the tube, noting which way up it fits **(see illustration 7.20a and 7.20b)**.

41 Invert the fork leg over a suitable container and pump the fork and damper rod vigorously several times each to expel as much fork oil as possible **(see illustration 7.21)**. Support the fork upside down in the container for a while to allow as much oil as possible to drain, then pump the fork and damper rod again.

42 Remove the previously slackened damper cartridge bolt and its sealing washer from the bottom of the slider **(see illustration 8.9)**. Discard the sealing washer as a new one must be used on reassembly.

43 Withdraw the damper assembly from inside the fork tube **(see illustration)**. The damper cartridge seat may come away with it, otherwise invert the fork and tip it out, noting which way up it fits **(see illustration)**.

8.43a Withdraw the damper from the tube . . .

8.43b . . . and tip its seat out

Frame, suspension and final drive 6•17

8.47a Fit the seat onto the bottom of the cartridge . . .

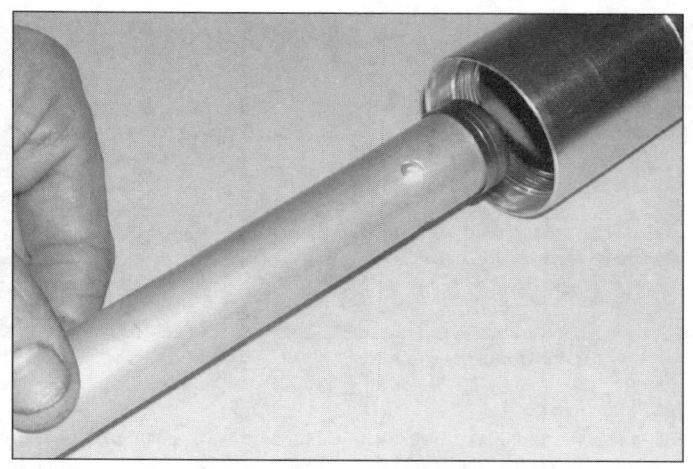

8.47b . . . then slide the damper assembly into the fork slider

44 Refer to Steps 11 to 14 above for the remainder of the disassembly procedure that is common to both models.

Inspection

45 Refer to Steps 15 to 20 above.

Reassembly

46 Refer to Steps 21 to 26 above for the part of the assembly procedure that is common to both models.

47 Lay the fork flat and slide the tube fully into the slider. Fit the damper cartridge seat onto the bottom of the cartridge, then slide the damper assembly into the top of the fork tube and into the slider until it seats on the bottom **(see illustrations)**. Fit a new sealing washer onto the cartridge bolt and apply a few drops of a suitable non-permanent thread locking compound **(see illustration 8.27a)**. Fit the bolt into the bottom of the slider and thread it into the cartridge, tightening it to the torque setting specified at the beginning of the Chapter **(see illustration 8.27b)**. If the cartridge rotates inside the tube, wait until the fork is fully reassembled before tightening the bolt, and then if necessary turn the fork upside down and compress it as you tighten the bolt. If it still turns, adjust the pre-load to its maximum (see Section 13).

48 Stand the fork upright. Slowly pour in the specified quantity of the specified grade of fork oil **(see illustration 7.22a)**. Fully extend the fork, then cover the top of the tube with your hand and compress the fork as much as possible – this helps to bleed any air. Now pump the fork tube and damper rod slowly at least ten times each to distribute it evenly. Fully compress the tube and damper rod into the slider and measure the oil level, and make any adjustment by adding more or tipping some out until the oil is at the level specified at the beginning of the Chapter **(see illustration 7.22b)**.

49 Pull the damper rod and fork tube out as far as possible then install the spring, with the tapered and closer wound coils at the top **(see illustration 7.20b)**. Keeping the damper rod extended, install the spacer, locating it in the top of the spring **(see illustrations 7.20a)**.

50 Fit a new O-ring smeared with fork oil around the fork top bolt, and also into the groove in the top of the bolt **(see illustration 7.24a)**. Thread the pre-load adjuster plate into the top bolt until it seats lightly **(see illustration 7.24b)**. Push down on the spacer to compress the spring and fit the top bolt onto the damper rod, aligning the cut-outs in the adjuster plate with the flats on the rod **(see illustration 7.24c)**. Thread the locknut onto the damper rod **(see illustration 7.24d)**. Counter-hold the top bolt or the damper rod using a spanner on the flats then tighten the locknut, to the specified torque setting if you have a crows-foot wrench **(see illustration 7.19a)**. Thread the bolt into the top of the fork tube, keeping the tube fully extended whilst doing so, and making sure it is not cross-threaded **(see illustration 7.18)**. **Note:** *The top bolt can be tightened to the specified torque setting at this stage if the tube is held between the padded jaws of a vice, but do not risk distorting the tube by doing so. A better method is to tighten the top bolt when the fork has been installed in the bike and is held in the bottom yoke, but before the top yoke clamp bolt is tightened* **(see illustration 6.8)**.

9.2 Unscrew the brake hose guide bolt (arrowed)

51 Fit a new O-ring into the larger groove in the top of the damper rod **(see illustration 7.25)**. Fit the pre-load adjuster hex onto the top of the damper rod **(see illustration 7.17b)**. Fit the snap-ring into its groove in the top of the damper rod **(see illustration 7.17a)**. Set the pre-load as noted on removal.

52 Install the forks (see Section 6).

9 Steering stem – removal and installation

Note: *Two types of head bearing were fitted to RR-2 (2002) models – some have caged ball bearings, as shown in the photographs in this Section, and others have taper roller bearings. Make sure you identify the type of bearing fitted to your model before tightening them if using the specified torque settings. If you are not using a torque wrench, the procedure is the same for both types as it relies on feel, but note that that taper roller bearings should be initially overtightened to pre-load them, then slackened off and reset correctly.*

Removal

1 Remove the fairing (see Chapter 8), and to prevent the possibility of damage should a tool slip raise or remove the fuel tank (see Chapter 4). Also remove the front forks (see Section 6).

2 Remove the horn (see Chapter 9). Unscrew the bolt securing the front brake hose splitter to the bottom yoke and displace it **(see illustration)**. Take care not to strain or knock the brake hoses when removing the steering stem.

3 If the top yoke is being removed from the bike rather than just being displaced, trace the wiring from the ignition switch, and where fitted the HISS receiver, and disconnect it/them at the connector(s) – on RR-Y and

6•18 Frame, suspension and final drive

9.4a Unscrew the nut . . .

9.4b . . . and lift off the top yoke

9.5a Bend down the lockwasher tabs . . .

RR-1 (2000 and 2001) models you will have to remove the right-hand air duct cover to access them (see Chapter 8, Section 4), and on RR-2 and RR-3 (2002 and 2003) models you will have to remove the air filter housing to access them (see Chapter 4). Release the wiring from any clips or ties and feed it through to the yoke.

4 Unscrew the steering stem nut **(see illustration)**. Lift the top yoke up off the steering stem and position it clear, using a rag to protect the tank or other components if it is only being displaced **(see illustration)**.

5 Bend down the tabs on the lockwasher to release it from the adjuster locknut **(see illustration)**. Unscrew and remove the locknut using either your fingers (it shouldn't be tight) or a suitable C-spanner or a drift located in one of the notches **(see illustration)**. Remove the lockwasher from the adjuster nut **(see illustration)**. Inspect the tabs for cracks or signs of fatigue. If there are any, discard the lockwasher and use a new one; otherwise the old one can be reused.

6 Supporting the bottom yoke, unscrew the adjuster nut using either a C-spanner, a peg-spanner or socket, or a drift located in one of the notches **(see illustration)**. Remove the grease seal from the steering stem **(see illustration)**. Check the condition of the grease seal and discard it if it is damaged.

7 Gently lower the bottom yoke and steering stem out of the frame **(see illustration)**.

8 Remove the inner race and bearing from the top of the steering head **(see illustration)**. Remove the bearing from the base of the steering stem **(see illustration 9.9)**. Remove all traces of old grease from the bearings and races and check them for wear or damage as described in Section 9. **Note:** *Do not attempt to remove the races from the steering head or the steering stem unless they are to be replaced with new ones (see Section 10).*

Installation

9 Smear a liberal quantity of multi-purpose grease onto the bearing races, and work some grease well into both the upper and lower bearings. Also smear the grease seal lip, using a new seal if necessary. Fit the lower

9.5b . . . then unscrew the locknut . . .

9.5c . . . and remove the lockwasher

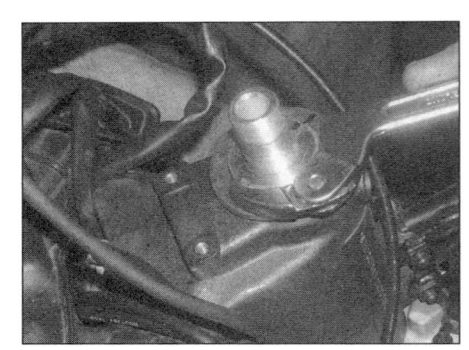

9.6a Unscrew the adjuster nut . . .

9.6b . . . and remove the grease seal

9.7 Draw the bottom yoke/steering stem out of the steering head

9.8 Remove the inner race and bearing

Frame, suspension and final drive 6•19

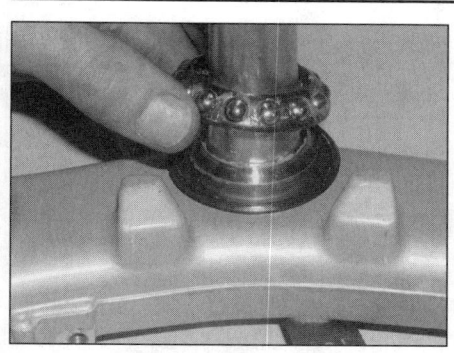

9.9 Fit the lower bearing onto the steering stem

9.10a Fit the upper bearing and its inner race

9.10b Thread the adjuster nut onto the stem

bearing onto the steering stem **(see illustration)**.

10 Carefully lift the steering stem/bottom yoke up through the steering head and support it there **(see illustration 9.7)**. Fit the upper bearing and its inner race into the top of the steering head **(see illustration)**. Install the grease seal **(see illustration 9.6b)**. Apply some clean engine oil to the adjuster nut and thread the nut on the steering stem **(see illustration)**.

11 If the Honda service tool (Part No. 07916-3710101) or a suitable peg spanner (which can be made by cutting castellations into an old socket) is available, tighten the adjuster nut to the initial torque setting specified at the beginning of the Chapter for your model and bearing type, then turn the steering stem through its full lock at least five times, then slacken the adjuster nut and re-tighten it to the final torque setting specified. On RR-2 (2002) models with caged ball bearings and all RR-3 (2003) models now tighten the nut by a further 90° (1/4 turn). Ensure that the steering stem is able to move freely from lock-to-lock following adjustment – if necessary reset the bearing adjustment as described in Chapter 1 after the forks and front wheel and all other components have been installed.

12 If the correct tools are not available, tighten the nut using a C-spanner or drift so that bearing play is eliminated, but the steering stem is able to move freely from lock-to-lock – refer to the procedure in Chapter 1 for details, but note that setting the bearings is a lot easier and more accurate after the forks and wheel are installed as their leverage and inertia need to be taken into account **(see illustration)**. To do it that way, make sure the nut is tight enough to hold the steering stem in the head without any play, then install the forks and wheel, then refer to the procedure in Chapter 1.

Caution: Take great care not to apply excessive pressure because this will cause premature failure of the bearings.

13 When the bearings are correctly adjusted, install the lockwasher, using a new one if the tabs are weakened or cracked, onto the adjuster nut and fit the two bent tabs into the slots in the adjuster nut **(see illustration 9.5c)**. Install the locknut and tighten it finger-tight, then tighten it further (to a maximum of 90°) until the remaining tabs on the lockwasher align with the slots in the locknut **(see illustration 9.5b)**. Hold the adjuster nut to prevent it from moving if necessary. Secure the locknut in position by bending up the other lockwasher tabs into its notches **(see illustration)**.

14 Fit the top yoke onto the steering stem **(see illustration 9.4b)**, then install steering stem nut and tighten it finger-tight **(see illustration 9.4a)**. Temporarily install one of the forks to align the top and bottom yokes, and secure it by tightening the bottom yoke clamp bolts only (see Section 6). Now tighten the steering stem nut to the torque setting specified at the beginning of the Chapter.

15 Install the remaining components in a reverse of the removal procedure, referring to the relevant Sections or Chapters, and to the torque settings specified at the beginning of the Chapter.

16 Carry out a final check of the steering head bearing freeplay as described in Chapter 1, and if necessary re-adjust.

10 Steering head bearings – inspection and replacement

Inspection

1 Remove the steering stem (see Section 9).
2 Remove all traces of old grease from the bearings and races and check them for wear or damage.
3 The outer races should be polished and free from indentations. Inspect the bearing rollers for signs of wear, damage or discoloration, and examine the roller retainer cage for signs of cracks or splits. If there are any signs of wear on any of the above components both upper and lower bearing assemblies must be renewed as a set. Only remove the outer races in the steering head and the lower bearing inner race on the steering stem if they need to be replaced – do not reuse them once they have been removed.

Replacement

4 The outer races are an interference fit in the steering head – tap them from position using a suitable drift **(see illustration)**. Tap firmly

9.12 Tighten the adjuster nut as described

9.13 Bend the lockwasher tabs up into the notches in the lockwasher

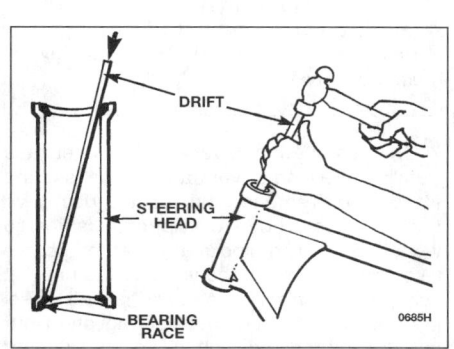

10.4 Drive the bearing races out with a brass drift locating it as shown

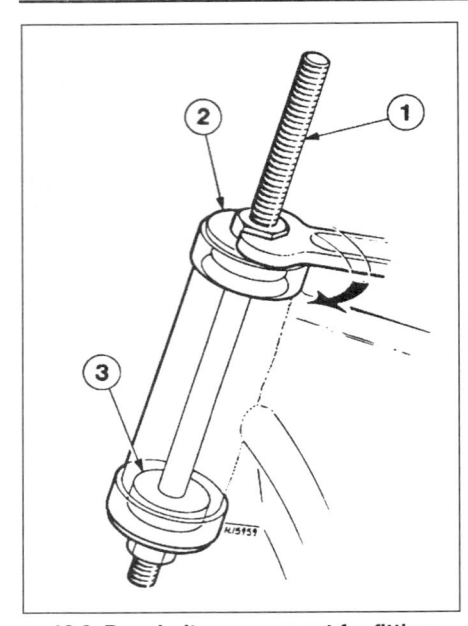

10.6 Drawbolt arrangement for fitting steering stem bearing races

1 Long bolt or threaded bar
2 Thick washer
3 Guide for lower race

and evenly around each race to ensure that it is driven out squarely. Curve the end of the drift slightly to improve access if necessary.

5 Alternatively, remove the races using a slide-hammer type bearing extractor; these can often be hired from tool shops.

6 Press the new outer races into the head using a drawbolt arrangement **(see illustration)**, or drive them in using a large diameter tubular drift. Ensure that the drawbolt washer or drift (as applicable) bears only on the outer edge of the race and does not contact the working surface. Alternatively, have the races installed by a Honda dealer equipped with the bearing race installation tools.

> **HAYNES HINT**
> *Installation of new bearing outer races is made much easier if the races are left overnight in the freezer. This causes them to contract slightly making them a looser fit. Alternatively, use a freeze spray.*

7 Only remove the lower bearing inner race from the steering stem if a new one is being fitted. To remove the race, use two screwdrivers placed on opposite sides to work it free, using blocks of wood to improve leverage and protect the yoke, or tap under it using a cold chisel. If you use the cold chisel method, first thread the steering stem nut onto the top then position the yoke on its front for stability – the nut will protect the threads from the transmitted force of the impact of the chisel. If the race is firmly in place it will be necessary to use a puller **(see illustration)**. Take the steering stem to a Honda dealer if required.

8 Remove the dust seal from the bottom of the stem and replace it with a new one. Smear the new one with grease.

9 Fit the new lower race onto the steering stem. Tap the new race into position using a length of tubing with an internal diameter slightly larger than the steering stem **(see illustration)**.

10 Install the steering stem (see Section 9).

11 Rear shock absorber – removal, inspection and installation

⚠️ **Warning: Do not attempt to disassemble this shock absorber. It is nitrogen-charged under high pressure. Improper disassembly could result in serious injury. No individual components are available for it.**

Removal

1 Support the motorcycle on an auxiliary stand that does not take the weight through

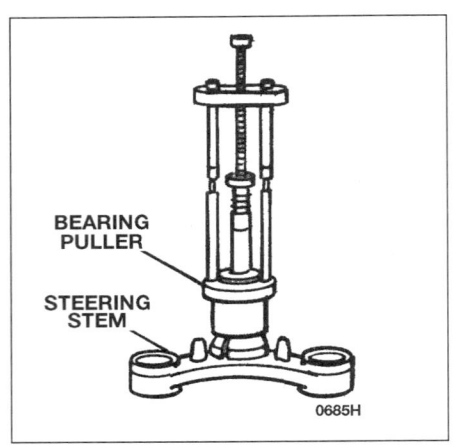

10.7 Remove the lower bearing race using a puller if necessary

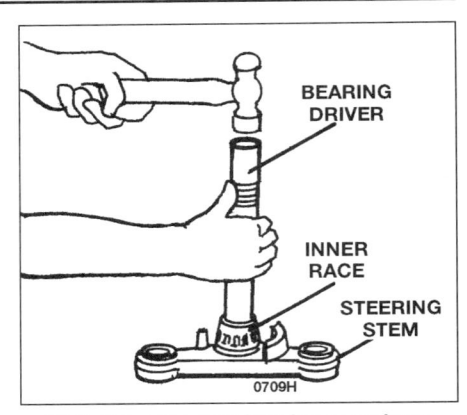

10.9 Drive the new bearing on using a suitable bearing driver or a length of pipe that bears only against the top edge of the inner race and not its working surface

any part of the rear suspension, or by using a hoist – an axle stand under each passenger footrest bracket works well, but whatever is used tie the front brake lever to the handlebar to ensure the wheel is locked so the bike can't roll off. Position a support under the rear wheel or swingarm so that it does not drop when the shock absorber is removed, but also making sure that the weight of the machine is off the rear suspension so that the shock is not compressed.

2 On RR-2 and RR-3 (2002 and 2003) models remove the rider's seat (see Chapter 4).

3 Unscrew the nut and withdraw the bolt securing the linkage plates to the linkage arm **(see illustrations)**. Swing the arm down.

4 Unscrew the nut and withdraw the bolt securing the bottom of the shock absorber to the linkage plates **(see illustrations)**.

5 On RR-Y and RR-1 (2000 and 2001) models unscrew the nut on the shock absorber upper mounting bolt **(see illustration)**. Support the shock absorber, then withdraw the bolt and manoeuvre the shock out of the bottom of the frame **(see illustration 11.6b)**.

6 On RR-2 and RR-3 (2002 and 2003) models support the shock absorber, then unscrew the nut on the upper mounting and manoeuvre the shock out of the bottom of the frame **(see illustrations)**.

11.3a Unscrew the nut . . .

11.3b . . . then withdraw the bolt and swing the arm down

Frame, suspension and final drive 6•21

11.4a Unscrew the nut . . .

11.4b . . . then withdraw the lower mounting bolt

11.5 Shock absorber upper mounting nut (arrowed) – RR-Y and RR-1 (2000 and 2001) models

Inspection

7 Inspect the shock absorber for obvious physical damage and oil leakage, and the coil spring for looseness, cracks or signs of fatigue **(see illustration)**.

8 Withdraw the spacer from the lower mounting **(see illustration)**. Inspect the pivot hardware at the top and bottom of the shock for wear or damage.

9 Lever out the grease seals **(see illustration)**. Check the condition of the bearing and replace it with a new one if necessary. Slip the spacer back into the bearing and check that there is not an excessive amount of freeplay between the two. Renew any components as required.

Refer to *Tools and Workshop Tips* in the Reference Section for more information on bearing checks and replacement methods. When fitting the new bearing make sure it is central in the mount, with a minimum of 8 mm gap between each end and the rim of the mount.

10 Lubricate the needle bearings, spacers and seals with multi-purpose grease. Press the new seals squarely into place **(see illustration)**. Install the spacer **(see illustration 11.8)**.

11 With the exception of the bottom pivot hardware, individual components are not available for the shock absorber. If it is worn or damaged, it must be replaced with a new one. Before disposing of an old shock absorber, you should release the nitrogen gas from the reservoir. To do this, lever the blanking cap off the end of the reservoir using a screwdriver, then point the valve away from you and anyone else (direct it into the ground) and depress the valve. When all the pressure is released, remove the core of the valve using a valve core remover.

⚠️ **Warning:** *Be very careful when releasing the gas pressure – it is possible for fine debris particles to be released with it, and as the pressure is high these could damage you eyes if done carelessly. Always point the valve well away.*

11.6a Unscrew the nut . . .

11.6b . . . and manoeuvre the shock out from the bottom

11.7 Check around the top of the rod for signs of oil (arrow)

11.8 Withdraw the spacer and check the seals and bearing as described

11.9 Lever out the old seals

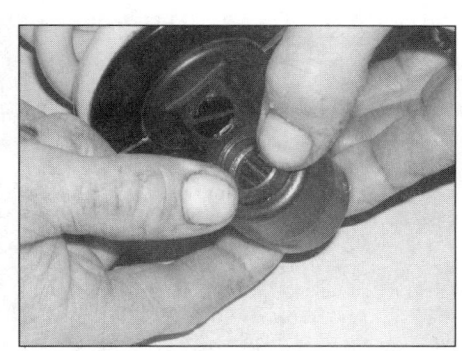

11.10 Press the new seals into place

6•22 Frame, suspension and final drive

Installation

12 Installation is the reverse of removal, noting the following:
- Apply multi-purpose grease to the shock absorber and linkage plate and arm pivot points.
- Do not forget to fit the spacer into the bottom mounting **(see illustration 11.8)**.
- On RR-Y and RR-1 (2000 and 2001) models install the shock absorber with the damping adjuster screws facing the left-hand side.
- On RR-2 and RR-3 (2002 and 2003) models install the shock absorber with the damping adjuster screws facing the right-hand side.
- Install the bolts from the right-hand side and do not tighten the nuts until all bolts (or all components if the suspension linkage has been removed as well) are in position, then tighten them to the torque settings specified at the beginning of the Chapter.

12 Rear suspension linkage – removal, inspection and installation

Removal

1 Support the motorcycle on an auxiliary stand that does not take the weight through any part of the rear suspension, or by using a hoist – an axle stand under each passenger footrest bracket works well, but whatever is used tie the front brake lever to the handlebar to ensure the wheel is locked so the bike can't roll off. Position a support under the rear wheel or swingarm so that it does not drop when the shock absorber is removed, but also making sure that the weight of the machine is off the rear suspension so that the shock is not compressed.

2 Remove the exhaust system (see Chapter 4). Remove the collar from the inside of the downpipe assembly rear mounting, then carefully press the rubber bush out using a blunt tool **(see illustrations)**. If it is difficult to remove, draw the top rim down and spray some lubricant between the bush and the mount, and allow it to work its way down.

3 Unscrew the nut and withdraw the bolt securing the linkage plates to the shock absorber **(see illustrations 11.4a and 11.4b)**. Unscrew the nut and withdraw the bolt securing the linkage plates to the linkage arm **(see illustrations 11.3a and 11.3b)**.

4 Unscrew the nut and withdraw the bolt securing the linkage plates to the swingarm and remove the plates, noting which way round they fit **(see illustrations)**.

5 Unscrew the nut and withdraw bolt securing the linkage arm to the engine bracket and remove the arm, noting which way round it fits **(see illustrations)**.

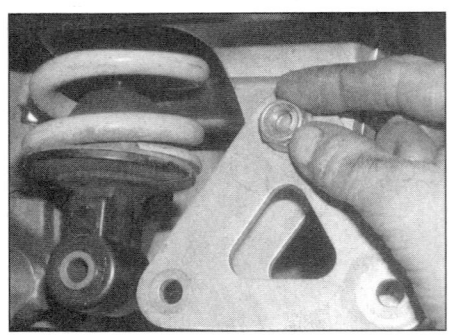

12.2a Withdraw the collar . . .

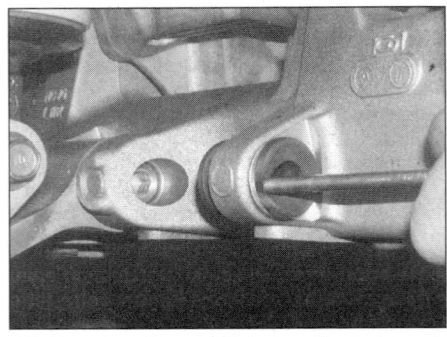

12.2b . . . and carefully press the bush out

12.4a Unscrew the nut . . .

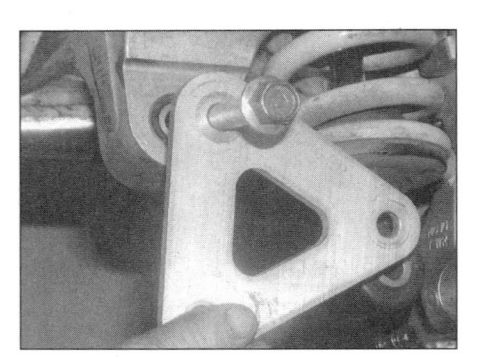

12.4b . . . then withdraw the bolt and remove the plates

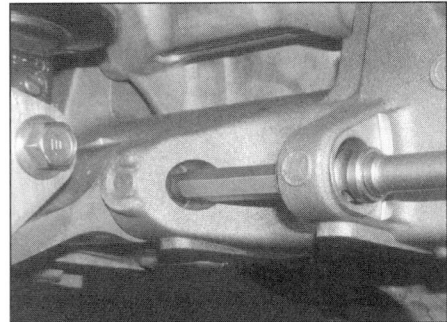

12.5a Accessing the bolt head via the exhaust mounting as shown . . .

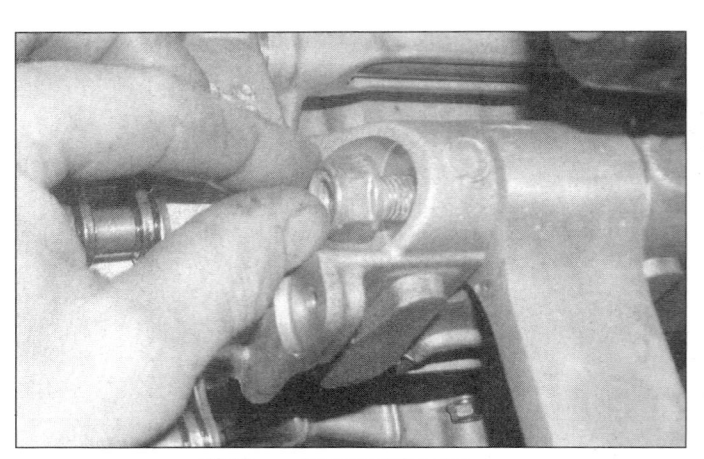

12.5b . . . unscrew the nut . . .

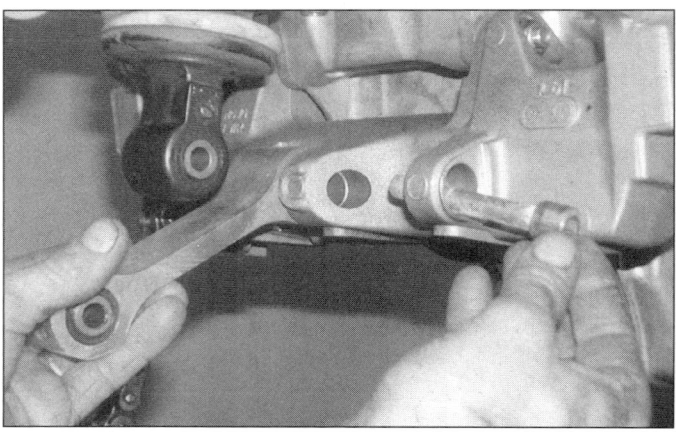

12.5c . . . withdraw the bolt and remove the arm

Frame, suspension and final drive 6•23

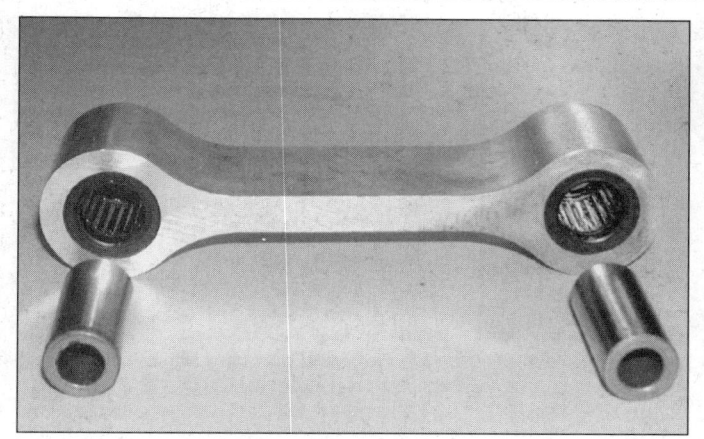

12.6a Withdraw the spacers from the linkage arm . . .

12.6b . . . and swingarm . . .

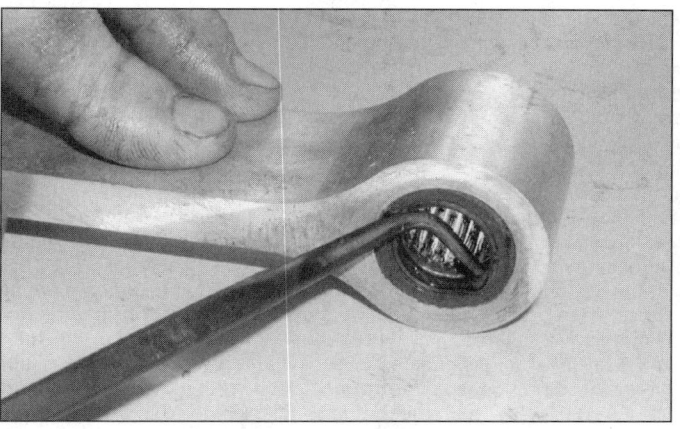

12.6c . . . and lever out the grease seals

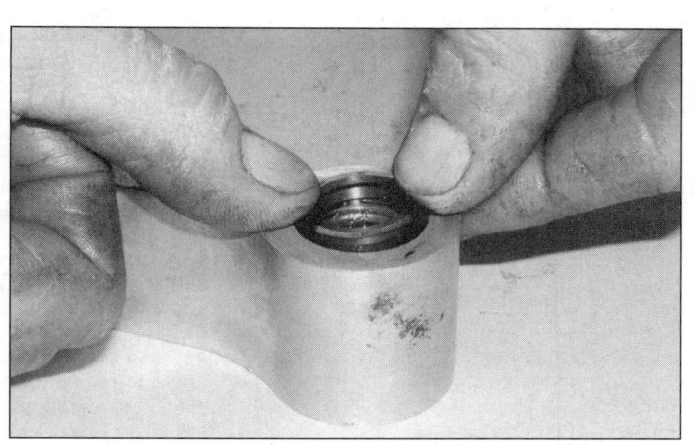

12.11 Press the grease seals in

Inspection

6 Withdraw the spacers from the linkage arm and swingarm, noting any difference in sizes, then lever out the grease seals (see illustrations). Thoroughly clean all components, removing all traces of dirt, corrosion and grease.

7 Inspect all components closely, looking for obvious signs of wear such as heavy scoring, or for damage such as cracks or distortion. Slip each spacer back into its bearing and check that there is not an excessive amount of freeplay between the two components. Renew any components as required.

8 Check the condition of the needle roller bearings in the linkage arm and swingarm. Refer to *Tools and Workshop Tips* (Section 5) in the Reference section for more information on bearings. If the linkage plate bearings in the swingarm need to be renewed, remove the swingarm (see Section 14).

9 Worn bearings can be drifted out of their bores, but note that removal will destroy them; new bearings should be obtained before work commences. The new bearings should be pressed or drawn into their bores rather than driven into position. In the absence of a press, a suitable drawbolt tool can be made up as described in *Tools and Workshop Tips* in the Reference section. When fitting the new bearings make sure they are central in their bores, with a 5.5 to 6.0 mm gap between each end and the rim of the bore.

10 Lubricate the needle bearings, spacers and seals with multi-purpose grease.

11 Press the new seals squarely into place (see illustration). Install the spacers (see illustrations 12.6a and 12.6b).

Installation

12 Installation is the reverse of removal, noting the following:
- Apply multi-purpose grease to the pivot points.
- Install the linkage plates with the arrow at the top, facing the right-hand side and pointing to the front (see illustration 12.4b).
- Insert all bolts from the right-hand side.
- Install the bolts and nuts finger-tight only until all components are in position, then tighten the nuts to the torque settings specified at the beginning of the Chapter.
- Fit a new rubber bush into the exhaust mount if the old one is damaged, deformed or deteriorated. A smear of grease will help ease the bush into the mount.

13 Suspension – adjustments

Front forks

RR-Y and RR-1 models

1 The front forks are adjustable for spring pre-load and both rebound and compression damping. Always make sure both forks are set equally.

2 Spring pre-load is adjusted using a 22 mm spanner on the adjuster flats – one is provided in the toolkit (see illustration 13.6). Turn it clockwise to increase pre-load and anti-clockwise to decrease it. The amount of pre-load is indicated by grooves on the adjuster. The standard position is with the 4th groove aligned with the top of the hex on the fork bolt.

3 Rebound damping is adjusted using a screwdriver in the slot in the top of the damper rod protruding from the pre-load

13.4 Compression damping adjuster (arrowed)

13.6 Spring pre-load adjuster (arrowed)

13.7 Rebound damping adjuster (arrowed)

adjuster **(see illustration 13.7)**. Turn it clockwise to increase damping and anti-clockwise to decrease it. To set the standard position, turn the adjuster fully clockwise until it stops, then turn it anti-clockwise approximately 1 full turn until the punch mark on the adjuster aligns with the reference mark on the directional arrow.

4 Compression damping is adjusted using a screwdriver in the slot in the adjuster which is on the bottom of the fork **(see illustration)**. Turn it clockwise to increase damping and anti-clockwise to decrease it. To set the standard position, turn the adjuster fully clockwise until it stops, then turn it anti-clockwise approximately 1 1/2 turns until the punch mark on the adjuster aligns with the reference mark on the directional arrow.

RR-2 and RR-3 models

5 The front forks are adjustable for spring pre-load and both rebound and compression damping. Always make sure both forks are set equally.
6 Spring pre-load is adjusted using a 22 mm spanner on the adjuster flats – one is provided in the toolkit **(see illustration)**. Turn it clockwise to increase pre-load and anti-clockwise to decrease it. To set the standard position, turn the adjuster fully clockwise until it stops, then turn it anti-clockwise 7 full turns.
7 Rebound damping is adjusted using a screwdriver in the slot in the top of the damper rod protruding from the pre-load adjuster **(see illustration)**. Turn it clockwise to increase damping and anti-clockwise to decrease it. To set the standard position, turn the adjuster fully clockwise until it stops, then turn it anti-clockwise approximately 2 turns until the punch mark on the adjuster aligns with the reference mark on the directional arrow.
8 Compression damping is adjusted using a screwdriver in the slot in the adjuster which is on the bottom of the fork **(see illustration 13.4)**. Turn it clockwise to increase damping and anti-clockwise to decrease it. To set the standard position, turn the adjuster fully clockwise until it stops, then turn it anti-clockwise approximately 2 turns until the punch mark on the adjuster aligns with the reference mark on the directional arrow.

Rear shock absorber

RR-Y and RR-1 models

9 The shock absorber is adjustable for spring pre-load and both rebound and compression damping.
10 Spring pre-load is adjusted using a suitable C-spanner (one is provided in the toolkit) to turn the spring seat on the top of the shock absorber **(see illustration)**. There are nine positions. Position 1 is the lowest setting for light loads, position 4 the standard, and position 9 the highest, for heavy loads. Align the setting required with the adjustment stopper.
11 Rebound damping adjustment is made by turning the adjuster on the bottom of the shock absorber on the left-hand side using a flat-bladed screwdriver **(see illustration)**. To increase the damping, turn the adjuster clockwise. To decrease the damping, turn the adjuster anti-clockwise. To set the standard position, turn the adjuster clockwise until it stops, then turn it anti clockwise 2 turns until the punch mark on the adjuster aligns with the index mark on the shock absorber.
12 Compression damping adjustment is made by turning the adjuster on the top of the shock absorber on the left-hand side using a flat-bladed screwdriver **(see illustration)**. To increase the damping, turn the adjuster clockwise. To decrease the damping, turn the adjuster anti-clockwise. To set the standard position, turn the adjuster clockwise until it stops, then turn it anti-clockwise 1 turn until the punch mark on the adjuster aligns with the index mark on the shock absorber.

RR-2 and RR-3 models

13 The shock absorber is adjustable for spring pre-load and both rebound and compression damping.
14 Spring pre-load is adjusted using a suitable C-spanner (one is provided in the toolkit) to turn the spring seat on the top of the

13.10 Spring pre-load adjuster (arrowed)

13.11 Rebound damping adjuster (arrowed)

13.12 Compression damping adjuster (arrowed)

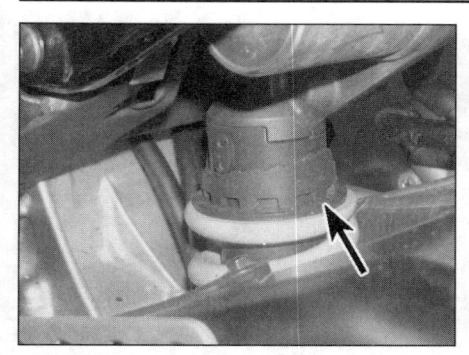

Frame, suspension and final drive 6•25

13.14 Spring pre-load adjuster (arrowed)

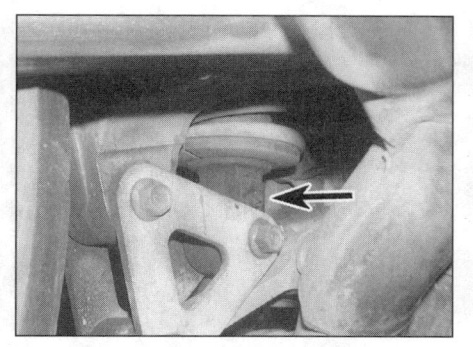

13.15 Rebound damping adjuster (arrowed)

13.16 Compression damping adjuster (arrowed)

shock absorber **(see illustration)**. There are nine positions. Position 1 is the lowest setting for light loads, position 4 the standard, and position 9 the highest, for heavy loads. Align the setting required with the adjustment stopper.

15 Rebound damping adjustment is made by turning the adjuster on the bottom of the shock absorber on the right-hand side using a flat-bladed screwdriver **(see illustration)**. To increase the damping, turn the adjuster clockwise. To decrease the damping, turn the adjuster anti-clockwise. To set the standard position, turn the adjuster clockwise until it stops, then turn it anti clockwise 2 turns until the punch mark on the adjuster aligns with the index mark on the shock absorber.

16 Compression damping adjustment is made by turning the adjuster on the top of the shock absorber on the right-hand side using a flat-bladed screwdriver **(see illustration)**. To increase the damping, turn the adjuster clockwise. To decrease the damping, turn the adjuster anti-clockwise. To set the standard position, turn the adjuster clockwise until it stops, then turn it anti clockwise 2 turns until the punch mark on the adjuster aligns with the index mark on the shock absorber.

14 Swingarm – removal and installation

Note: *To unscrew the swingarm pivot bolt nut you need a 24 mm hex key, and to do it up you will need a second one to counter-hold the bolt head as the pinch bolts cannot be tightened before the pivot bolt. 24 mm hex keys are not readily available, but it is worth trying an automotive or commercial vehicle tool supplier – sump keys are sometimes that big. Alternatively Honda can supply the tools (part no. 07930-KA50100), or alternatively a tool can be fabricated using a 24 mm head bolt with two nuts threaded onto it and locked very tightly together. To counter-hold the pivot bolt head on installation a 24 mm nut inserted half-way into the head and held on the protruding section using a spanner or pipe wrench should suffice – it worked well for us.*

Alternatively weld a bar onto the nut to act as a handle.

Removal

1 For best access, remove the exhaust system (see Chapter 4). Remove the rear wheel (see Chapter 7). Remove the front sprocket (see Section 17).

2 On RR-Y and RR-1 (2000 and 2001) models undo the screw securing each rear brake hose guide to the swingarm. Displace the rear brake caliper and bracket assembly from the swingarm (there is no need to disconnect the brake hose), noting how it locates, and tie it to the frame, making sure no strain is placed on the hose.

3 On RR-2 and RR-3 (2002 and 2003) models unscrew the bolts securing the rear hugger to

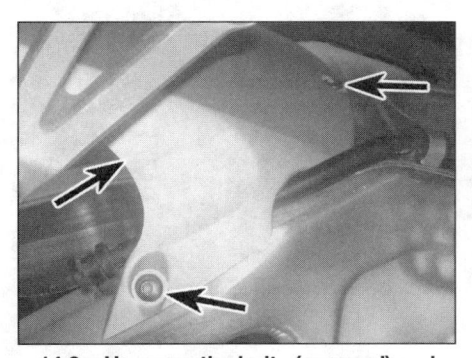

14.3a Unscrew the bolts (arrowed) and remove the hugger

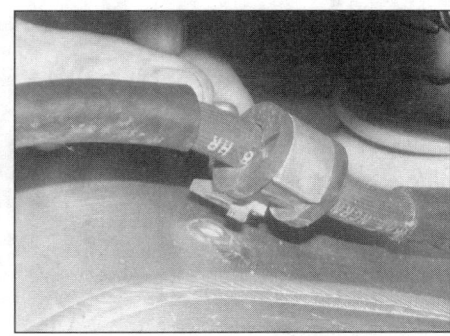

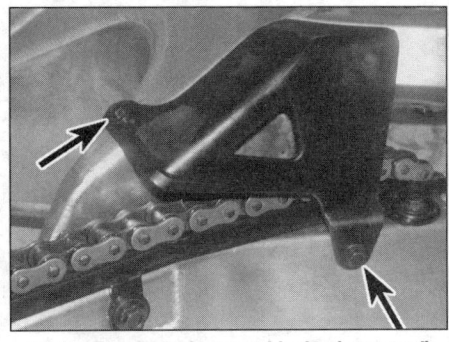

14.4a Front chain guard bolts (arrowed)

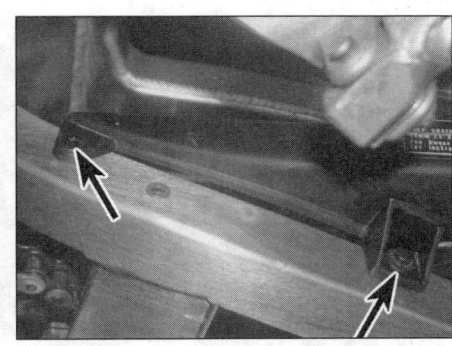

the swingarm and remove it, noting how one bolt secures the brake hose rear guide **(see illustration)**. Undo the screw securing the brake hose front guide to the swingarm **(see illustration)**. Displace the rear brake caliper and bracket assembly from the swingarm (there is no need to disconnect the brake hose), noting how it locates, and tie it to the frame, making sure no strain is placed on the hose.

4 If required unscrew the bolts securing the front chain guard, and the trim clip and bolt securing the rear chain guard, and remove them – this can be done after removing the swingarm **(see illustrations)**.

5 Support the swingarm, then unscrew the nut and withdraw the bolt securing the suspension linkage plates to the swingarm

14.3b Undo the screw securing the other brake hose guide, noting how it locates

14.4b Rear chain guard trim clip and bolt (arrowed)

6

6•26 Frame, suspension and final drive

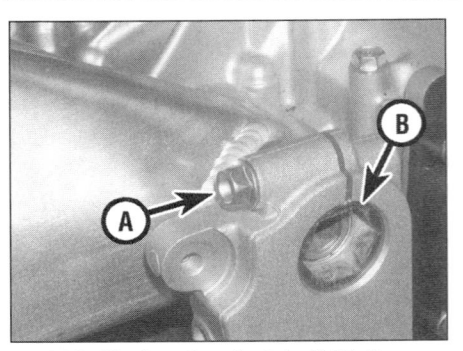

14.6 Slacken the pinch bolt (A), then unscrew the pivot bolt nut (B)

14.7a Slacken the pinch bolt (arrowed) . . .

14.7b . . . then withdraw the pivot bolt . . .

(see illustrations 12.4a and 12.4b). Unscrew the nut and withdraw the bolt securing the bottom of the shock absorber to the linkage plates, then swing the linkage assembly down (see illustrations 11.4a and 11.4b).

6 Slacken the pinch bolt on the right-hand end of the swingarm pivot (see illustration). Unscrew the nut on the right-hand end of the pivot bolt using one of the tools described in the **Note** at the beginning of this section.

7 Slacken the pinch bolt on the left-hand end of the swingarm pivot (see illustration). Withdraw the pivot bolt then manoeuvre the swingarm clear of the shock absorber and out of the frame (see illustrations).

8 Remove the chain slider from the swingarm if necessary, noting the collars that fit with the rear bolts (see illustration). If it is badly worn or damaged, it should be replaced with a new one – there are some wear limit arrows on the front. Inspect all pivot components for wear or damage as described in Section 15.

Installation

9 Remove the collar from the inner end of each swingarm pivot, noting which fits where (see illustrations). Also withdraw the spacer from the suspension linkage plate pivot (see illustration 12.6b). Clean off all old grease, then lubricate the grease seals, bearings, collars, spacer and the pivot bolt with multi-purpose grease. Insert the collars and spacer.

10 If removed, install the chain slider, making sure it locates correctly at the front and over the side mounting bolt lug (see illustration 14.8). Apply a suitable non-permanent thread locking compound to the slider mounting bolts, then fit them with their collars on the rear bolts and tighten them to the torque setting specified at the beginning of the Chapter.

11 Offer up the swingarm and have an assistant hold it in place (see illustration 14.7c). Make sure the drive chain is looped over the front of the swingarm. Slide the pivot bolt through from the left-hand side (see illustration).

14.7c . . . and remove the swingarm

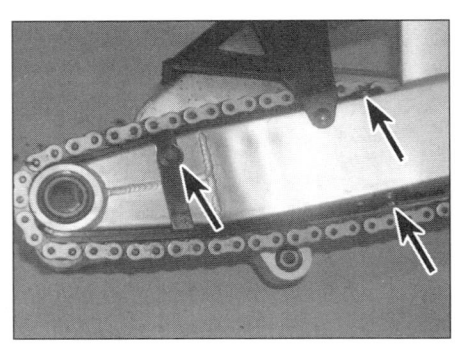

14.8 Chain slider is secured by three bolts (arrowed)

14.9a Remove the right-hand pivot collar . . .

14.9b . . . and the left-hand pivot collar

14.11 Slide the pivot bolt through

Frame, suspension and final drive 6•27

12 Thread the nut onto the pivot bolt and tighten it to the torque setting specified at the beginning of the Chapter, counter-holding the pivot bolt head as described previously (see illustrations).
13 Now tighten both the pinch bolts to the specified torque setting (see illustration 14.6 and 14.7a).
14 Align the rear suspension linkage plates with the shock absorber, then install the bolt and tighten the nut to the specified torque setting (see illustrations 11.4b and 11.4a).
15 Align the swingarm with the rear suspension linkage plates, then install the bolt and tighten the nut to the specified torque setting (see illustrations 12.4b and 12.4a).
16 Install the chain guards if removed (see illustrations 14.4a and 14.4b).
17 Fit the rear brake caliper and bracket assembly onto the swingarm, locating the slot in the bracket onto the lug on the swingarm. Fit the brake hose guides, along with the rear hugger on RR-2 and RR-3 (2002 and 2003) models, and tighten their screws and/or bolts.
18 Install the front sprocket (see Section 17) and the rear wheel (see Chapter 7). If removed, install the exhaust system (see Chapter 4).
19 Check and adjust the drive chain slack (see Chapter 1). Check the operation of the rear suspension and brake before taking the machine on the road.

15 Swingarm – inspection, bearing check and replacement

Inspection

1 Remove the swingarm (see Section 14).
2 Thoroughly clean the swingarm, removing all traces of dirt, corrosion and grease.
3 Inspect the swingarm closely, looking for obvious signs of wear such as heavy scoring, and cracks or distortion due to accident damage. Any damaged or worn component must be replaced.
4 Check the swingarm pivot bolt for straightness by rolling it on a flat surface such as a piece of plate glass (first wipe off all old grease and remove any corrosion using fine emery cloth). Renew the pivot bolt if it is bent.

Bearing check and replacement

5 Remove the collar from the inner end of each swingarm pivot, noting which fits where (see illustrations 14.9a and 14.9b). Lever out the grease seal from each side of the pivot (see illustration). New ones should be used.
6 Refer to Tools and Workshop Tips in the Reference section and check the bearings – on RR-Y and RR-1 (2000 and 2001) models there are two caged ball bearings separated by a spacer in the right-hand pivot, and a needle bearing in the left-hand pivot. On RR-2 and RR-3 (2002 and 2003) models, there is a caged ball bearing and a needle bearing in the right-hand pivot, and a needle bearing in the left-hand pivot. Clean them and inspect them for wear or damage. If the bearings do not run smoothly and freely or if there is excessive freeplay, they must be replaced with new ones – refer to the Reference Section for removal and installation methods. The caged ball bearing(s) in the right-hand pivot are held by a circlip (see illustration). The needle bearing(s) must be renewed if removed – they cannot be reused.
7 Lubricate the bearings and grease seal lips with multi-purpose grease. Press the new seals into place (see illustration).

14.12a Tighten the pivot bolt nut to the specified torque . . .

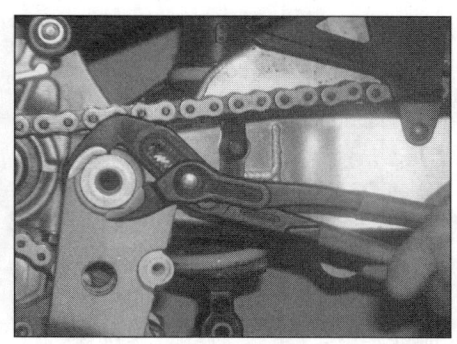

14.12b . . . counter-holding the bolt head as described

16 Drive chain – removal, cleaning and installation

Note: The original equipment drive chain fitted to these models has a staked-type master (joining) link which can be disassembled using either Honda service tool, Pt. No. 07HMH-MR1010C for US models or 07HMH-MR10103 for all other models, or one of several commercially-available drive chain cutting/staking tools. Such chains can be recognised by the master joining link side plate's identification marks (and usually its different colour), as well as by the staked ends of the link's two pins which look as if they have been deeply centre-punched, instead of peened over as with all the other pins.

Removal

1 Support the motorcycle on an auxiliary stand so that the rear wheel is off the ground. Locate the joining link in a suitable position to work on by rotating the back wheel. Slacken the drive chain as described in Chapter 1.
2 If required, unscrew the bolts securing the front chain guard, and the trim clip and bolt securing the rear chain guard, and remove them (see illustrations 14.4a and 14.4b).
3 Remove the front sprocket cover (see Section 17).
4 Split the chain at the joining link using the chain cutter, following carefully the

15.5 Lever the grease seals out of the swingarm

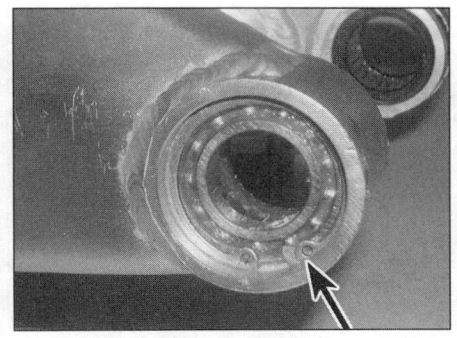

15.6 A circlip (arrowed) secures the caged ball bearing(s)

15.7 Press the new seals into place

manufacturer's operating instructions (see also Section 8 in *Tools and Workshop Tips* in the Reference Section) **(see illustration)**. Remove the chain from the bike, noting its routing around the swingarm.

Cleaning

5 Refer to Chapter 1, Section 1, for details of routine cleaning with the chain installed on the sprockets.

6 If the chain is extremely dirty remove it from the motorcycle and soak it in paraffin (kerosene) for approximately five or six minutes, then clean it using a soft brush. *Caution: Don't use gasoline (petrol), solvent or other cleaning fluids which might damage its internal sealing properties. Don't use high-pressure water. Remove the chain, wipe it off, then blow dry it with compressed air immediately. The entire process shouldn't take longer than ten minutes – if it does, the O-rings in the chain rollers could be damaged.*

Installation

⚠️ *Warning: NEVER install a drive chain which uses a clip-type master (split) link. Use ONLY the correct service tools to secure the staked-type of master link – if you do not have access to such tools, have the chain replaced by a dealer service department or bike repair shop to be sure of having it securely installed.*

7 Fit the drive chain through the swingarm and around the sprockets, leaving the two ends in a convenient position to work on.

8 Refer to Section 8 in *Tools and Workshop Tips* in the Reference Section. Install the new joining link from the inside with the four O-rings correctly located between the link plate and side plate. Install the new side plate with its identification marks facing out. Measure the amount that the joining link pins project from the side plate and check they are within the measurements specified at the beginning of the Chapter. Stake the new link using the drive chain cutting/staking tool, following carefully the instructions of both the chain manufacturer and the tool manufacturer. DO NOT reuse old joining link components.

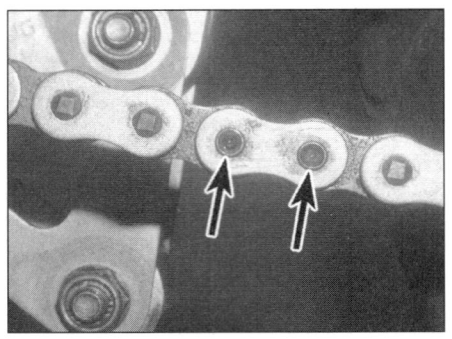

16.4 The master joining link is identified by the different pin ends (arrowed)

9 After staking, check the joining link and staking for any signs of cracking. If there is any evidence of cracking, the joining link, O-rings and side plate must be replaced. Measure the diameter of the staked ends in two directions and check that it is evenly staked and within the measurements specified at the beginning of the Chapter. Check that the link pivots freely.

10 Install the sprocket cover (see Section 17).

11 Install the chain guards if removed **(see illustrations 14.4a and 14.4b)**.

12 On completion, adjust and lubricate the chain following the procedures described in Chapter 1.

17 Sprockets – check and replacement

Check

1 Unscrew the bolts securing the front sprocket cover and remove the cover and the guide plate **(see illustration)**.

2 Check the wear pattern on both sprockets (see Chapter 1, Section 1). If the sprocket teeth are worn excessively, replace the chain and both sprockets as a set. Whenever the sprockets are inspected, the drive chain should be inspected also (see Chapter 1). Always renew the chain and sprockets as a set – worn sprockets can ruin a new drive chain and *vice versa*.

17.1 Unscrew the bolts (arrowed) and remove the cover

3 Adjust and lubricate the chain following the procedures described in Chapter 1.

Replacement

Front sprocket

4 Unscrew the bolts securing the front sprocket cover and remove the cover and the guide plate **(see illustration 17.1)**.

5 Have an assistant apply the rear brake, then unscrew the sprocket bolt and remove the washer **(see illustration)**.

6 Fully slacken the drive chain as described in Chapter 1. If the rear sprocket is being renewed as well, remove the rear wheel now to give full slack (see Chapter 7). Otherwise disengage the chain from the rear sprocket if required to provide more slack.

7 Slide the chain and sprocket off the shaft then slip the sprocket out of the chain **(see illustration)**.

8 Engage the new sprocket with the chain, making sure the marked side is facing out, and slide it on the shaft **(see illustration 17.7)**.

9 If the rear wheel was removed, change the sprocket now and install the wheel (see Chapter 7). If the chain was merely disengaged, fit it back onto the rear sprocket. In all cases take up the slack in the chain.

10 Install the sprocket bolt with its washer and tighten it to the torque setting specified at the beginning of the Chapter, using the rear brake to prevent the sprocket turning **(see illustrations)**.

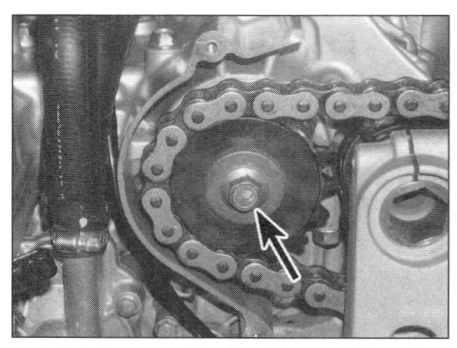

17.5 Unscrew the bolt (arrowed) and remove the washer

17.7 Draw the sprocket off the shaft and disengage the chain

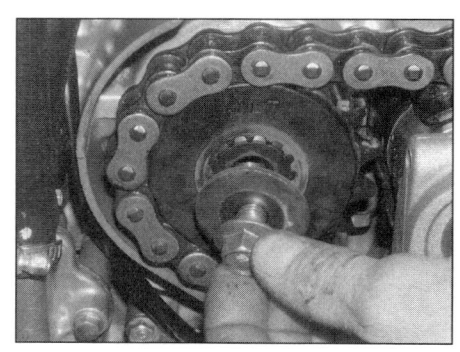

17.10a Install the bolt with its washer ...

Frame, suspension and final drive 6•29

17.10b ... and tighten it to the specified torque

17.11 Fit the guide plate and cover

11 Fit the guide plate and sprocket cover and tighten its bolts **(see illustration)**. Adjust and lubricate the chain following the procedures described in Chapter 1.

Rear sprocket

12 Remove the rear wheel (see Chapter 7).
13 Unscrew the nuts securing the sprocket to the hub assembly and remove the washers. Remove the sprocket, noting which way round it fits.
14 Fit the sprocket onto the hub with the stamped mark facing out. Install the nuts with their washers, and tighten them evenly and in a criss-cross sequence to the torque setting specified at the beginning of the Chapter.
15 Install the rear wheel (see Chapter 7).

18 Rear sprocket coupling/rubber dampers – check and replacement

1 Remove the rear wheel (see Chapter 7). Check for play between the sprocket coupling and the wheel hub by turning the sprocket. Any play indicates worn rubber damper segments.

Caution: Do not lay the wheel down on the disc as it could become warped. Lay the wheel on wooden blocks so that the disc is off the ground.

2 Lift the sprocket coupling away from the wheel leaving the rubber dampers in position **(see illustration)**. Note the spacer inside the coupling – it should be a tight fit but remove it if it is likely to drop out. Check the coupling for cracks or any obvious signs of damage. Also check the sprocket studs for wear or damage.
3 Lift the rubber damper segments from the wheel and check them for cracks, hardening and general deterioration **(see illustration)**. Renew them as a set if necessary.
4 Check the condition of the hub O-ring – if it is damaged, deformed or deteriorated replace it with a new one and smear it with oil **(see illustration)**. Otherwise clean it and smear it with oil.
5 Checking and replacement procedures for the sprocket coupling bearing are in Chapter 7.
6 Installation is the reverse of removal. Make sure the spacer is still correctly installed in the coupling, or install it if it was removed.
7 Install the rear wheel (see Chapter 7).

18.2 Lift the sprocket coupling out of the wheel

18.3 Check the rubber dampers . . .

18.4 . . . and the O-ring

Notes

Chapter 7
Brakes, wheels and tyres

Contents

Brake discs – inspection, removal and installation 3	Rear brake caliper – removal, overhaul and installation 5
Brake fluid level check . . . . .see *Daily (pre-ride) checks*	Rear brake master cylinder – removal, overhaul and installation . . . 7
Brake light switches – check and replacement . . . . .see Chapter 9	Rear wheel – removal and installation 13
Brake pads – renewal 2	Tyres – general information and fitting 15
Brake pad wear check . . . . .see Chapter 1	Tyres – pressure, tread depth and
Brake hoses – inspection and replacement 8	condition . . . . .see *Daily (pre-ride) checks*
Brake system – bleeding and fluid change 9	Wheels – general check . . . . .see Chapter 1
Brake system check . . . . .see Chapter 1	Wheel bearings – check . . . . .see Chapter 1
Front brake calipers – removal, overhaul and installation 4	Wheel bearings – renewal 14
Front brake master cylinder – removal, overhaul and installation . . . 6	Wheels – alignment check 11
Front wheel – removal and installation 12	Wheels – inspection and repair 10
General information 1	

Degrees of difficulty

Easy, suitable for novice with little experience	**Fairly easy,** suitable for beginner with some experience	**Fairly difficult,** suitable for competent DIY mechanic	**Difficult,** suitable for experienced DIY mechanic	**Very difficult,** suitable for expert DIY or professional

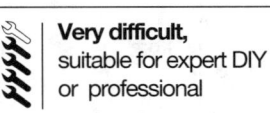

Specifications

Front brakes

Brake fluid type	DOT 4
Caliper bore ID	
RR-Y and RR-1 (2000 and 2001) models	
Upper bore	
Standard	33.960 to 34.010 mm
Service limit	34.02 mm
Lower bore	
Standard	30.250 to 30.280 mm
Service limit	30.29 mm
RR-2 and RR-3 (2002 and 2003) models	
Upper bore	
Standard	32.025 to 32.035 mm
Service limit	32.05 mm
Lower bore	
Standard	30.250 to 30.280 mm
Service limit	30.29 mm
Caliper piston OD	
RR-Y and RR-1 (2000 and 2001) models	
Upper piston	
Standard	33.802 to 33.835 mm
Service limit	33.794 mm
Lower piston	
Standard	30.082 to 30.115 mm
Service limit	30.074 mm
RR-2 and RR-3 (2002 and 2003) models	
Upper piston	
Standard	31.965 to 31.998 mm
Service limit	31.953 mm
Lower piston	
Standard	30.082 to 30.115 mm
Service limit	30.074 mm

Front brakes (continued)
Master cylinder bore ID
  RR-Y and RR-1 (2000 and 2001) models
    Standard .... 19.050 to 19.093 mm
    Service limit .... 19.105 mm
  RR-2 and RR-3 (2002 and 2003) models
    Standard .... 17.460 to 17.503 mm
    Service limit .... 17.515 mm
Master cylinder piston OD
  RR-Y and RR-1 (2000 and 2001) models
    Standard .... 19.018 to 19.034 mm
    Service limit .... 19.006 mm
  RR-2 and RR-3 (2002 and 2003) models
    Standard .... 17.321 to 17.367 mm
    Service limit .... 17.309 mm
Disc thickness
  Standard .... 4.5 mm
  Service limit .... 3.5 mm
Disc maximum runout .... 0.3 mm

Rear brake
Brake fluid type .... DOT 4
Caliper bore ID
  Standard .... 38.180 to 38.230 mm
  Service limit .... 38.240 mm
Caliper piston OD
  Standard .... 38.098 to 38.148 mm
  Service limit .... 38.09 mm
Master cylinder bore ID
  Standard .... 15.870 to 15.913 mm
  Service limit .... 15.925 mm
Master cylinder piston OD
  Standard .... 15.827 to 15.854 mm
  Service limit .... 15.815 mm
Disc thickness
  Standard .... 5.0 mm
  Service limit .... 4.0 mm
Disc maximum runout .... 0.3 mm

Wheels
Maximum wheel runout (front and rear)
  Axial (side-to-side) .... 2.0 mm
  Radial (out-of-round) .... 2.0 mm
Maximum axle runout (front and rear) .... 0.20 mm

Tyres
Tyre pressures .... see *Daily (pre-ride)* checks
Tyre sizes*
  Front 120/70-ZR17 58W
  Rear 190/50-ZR17 73W
*Refer to the owners handbook or the tyre information label on the swingarm for approved tyre brands.

Torque settings
Brake caliper bleed valves .... 6 Nm
Brake disc bolts
  Front .... 20 Nm
  Rear .... 42 Nm
Brake hose banjo bolts .... 34 Nm
Front axle bolt .... 78 Nm
Front axle clamp bolts .... 22 Nm
Front brake caliper body joining bolts .... 23 Nm
Front brake caliper mounting bolts .... 30 Nm
Front brake master cylinder clamp bolts .... 12 Nm
Front brake pad retaining pins .... 18 Nm
Rear axle nut .... 113 Nm
Rear brake caliper front slider pin .... 27 Nm
Rear brake caliper rear mounting bolt/slider pin .... 22 Nm
Rear brake pad retaining pin .... 18 Nm

Brakes, wheels and tyres 7•3

1 General information

All models covered in this manual are fitted with cast alloy wheels designed for tubeless tyres only. Both front and rear brakes are hydraulically operated disc brakes.

On all models the front brakes have twin opposed piston calipers and the rear brake has a single piston sliding caliper.

Caution: *If an hydraulic brake line is loosened, the entire system must be disassembled, drained, cleaned and then properly filled and bled upon reassembly. Do not use solvents on internal brake components. Solvents will cause the seals to swell and distort. Use only clean brake fluid or denatured alcohol for cleaning. Use care when working with brake fluid as it can injure your eyes and it will damage painted surfaces and plastic parts.*

2 Brake pads – renewal

> **Warning:** *The dust created by the brake system may contain asbestos, which is harmful to your health. Never blow it out with compressed air and don't inhale any of it. An approved filtering mask should be worn when working on the brakes.*

Front brake pads

Note 1: *Honda recommend using new caliper bolts when the old ones are removed. This is because the bolts are pre-treated with a locking compound. It is possible, however, to clean up the old bolts and reinstall them using a suitable non-permanent thread locking compound that is commercially available.*

Note 2: *If the pads are being removed for cleaning and inspection purposes only and are not going to be renewed, the calipers do not need to be displaced – ignore Steps 1 and 3 and remove the pads with the calipers in situ. The calipers only need to be displaced for the purpose of pushing the pistons back in to create the clearance for new pads to be installed (which are obviously much thicker than worn ones).*

2.1a Slacken the brake pad retaining pins (arrowed) . . .

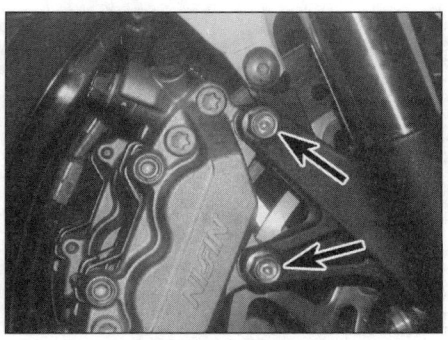

2.1b . . . then unscrew the mounting bolts (arrowed) and slide the caliper off the disc

1 Slacken the pad retaining pins **(see illustration)**. Unscrew the caliper mounting bolts and slide the caliper off the disc **(see illustration)**.

2 Unscrew and remove the pad retaining pins (pressing down on the pad spring makes the pins easier to withdraw), then lift off the pad spring, noting how it fits, and remove the pads **(see illustrations)**.

3 If new pads are being installed, the pistons will have to pushed back into their bores to allow for the increased friction material thickness of new pads. Before doing this, clean around the exposed section of each piston to remove any dirt or debris stuck to them that could cause the seals to be damaged. Loosely fit the old pads back into the caliper, then push the pistons as far back into the caliper as possible using a piece of wood, a metal bar or a screwdriver inserted between the old pads, using a rag to protect the caliper body. You can do this with the pads removed, but then only use wood or you fingers to avoid marking the pistons **(see illustration)**. If any of the pistons are stuck or are difficult to push back, disassemble the caliper and overhaul it (see Section 4). It may be necessary to remove the master cylinder reservoir cover, diaphragm plate and diaphragm, and to siphon out some fluid (see *Daily (pre-ride) checks*). If the pistons are difficult to push back, attach a length of clear hose to the bleed valve and place the open end in a suitable container, then open the valve and try again. Take great care not to draw any air into the system. If in doubt, bleed the brakes afterwards (see Section 9).

4 Inspect the surface of each pad for contamination and check whether the friction material has worn beyond its service limit (see Chapter 1, Section 3). If either pad is worn to or beyond the service limit, is fouled with oil or grease, or is heavily scored or damaged by dirt and debris, both sets of pads must be renewed as a set. Unevenly worn pads are indicative of a seized or sticky piston or pistons – if this is the case disassemble the caliper and overhaul it (see Section 4).

5 If the pads are in good condition clean them carefully, using a fine wire brush which is completely free of oil and grease to remove all traces of road dirt and corrosion. Using a pointed instrument, clean out the grooves in the friction material and dig out any embedded particles of foreign matter. Any areas of glazing may be removed using emery cloth. Spray with a dedicated brake cleaner to remove any dust. It is also worth spraying the inside of the caliper to remove any dust there, and also to spray the discs.

6 Check the condition of the brake disc (see Section 3).

7 Remove all traces of corrosion from the pad pins and check they are not bent or damaged. Smear the pins, the backs of the pads and the leading and trailing edges of the backing

2.2a Unscrew the pins and remove the spring . . .

2.2b . . . and the pads

2.3 Press the pistons in as described to make clearance for new pads

7•4 Brakes, wheels and tyres

2.8 Locate the pad spring onto the caliper with the arrow pointing up then install one pin

2.9 Slide the caliper onto the disc and install the bolts

2.12a Unscrew the pin plug . . .

material with copper-based grease, making sure that none gets on the friction material.

8 Insert the pads into the caliper so that the friction material of each pad faces the disc **(see illustration 2.2b)**. Fit the pad spring onto the pads, making sure the arrow points to the top of the caliper. Insert one pad retaining pin through the hole in the outer pad, then press down on the pad spring and push the pin through the spring and the hole in the inner pad **(see illustration)**. Tighten the pin finger-tight. Install the second pin in the same way **(see illustration 2.2a)**. If the caliper was not displaced, tighten the pad retaining pins to the specified torque **(see illustration 2.1a)**.

9 If the caliper was displaced, slide it onto the disc, making sure the pads sit squarely on either side **(see illustration)**. Install the caliper mounting bolts, either using new ones or cleaning and applying a suitable non-permanent thread locking compound to the threads of the old ones (see **Note 1**), and tighten them to the torque setting specified at the beginning of the Chapter. Now tighten the pad retaining pins to the specified torque **(see illustration 2.1a)**.

10 Top up the master cylinder reservoir if necessary (see *Daily (pre-ride) checks*).

11 Operate the brake lever several times to bring the pads into contact with the disc. Repeat the procedure on the other front brake caliper. Check the operation of the brake before riding the motorcycle. **Note:** *New pads will not give full braking efficiency until they have bedded-in. Be prepared for this, and avoid hard braking as far as possible for the first hundred miles (150 km) or so after pad renewal.*

Rear brake pads

12 Unscrew the pad retaining pin plug, then slacken the pad retaining pin **(see illustrations)**. Unscrew the caliper rear mounting bolt/slider pin **(see illustration)**.

13 Unscrew the pad retaining pin **(see illustration)**. Pivot the back of the caliper up off the disc and remove the pads, noting how they fit **(see illustrations)**. Note the pad spring in the top of caliper and the pad guide on the caliper bracket and remove them if required for cleaning or replacement, noting how they fit.

14 Inspect the pads as described in Steps 4 to 7 above. Also check the slider pins and boots (see Section 5, Step 12) – slide the caliper off its bracket to do this **(see illustration 5.5)**.

15 If new pads are being installed, the piston will have to pushed back into its bore to allow for the increased friction material thickness of new pads. Slide the caliper off its bracket **(see illustration 5.5)**. Clean around the exposed section of the piston to remove any dirt or debris stuck to it that could cause the seals to be damaged. Now press the piston back into its bore – you can use your hands to do this, though a piece of wood against the piston in conjunction with some grips and a bit of card to protect the caliper works very well, as does a proper piston-pushing tool. It may be

2.12b . . . then slacken the pin (arrowed)

2.12c Unscrew the rear bolt/slider pin

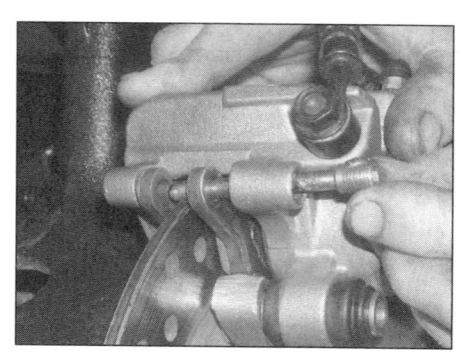

2.13a Remove the retaining pin . . .

2.13b . . . then pivot the caliper up . . .

2.13c . . . and remove the pads

Brakes, wheels and tyres 7•5

2.16a Make sure the backing material and shim are correctly fitted

3.3a The minimum thickness is marked on each disc

3.3b Checking disc thickness

necessary to remove the master cylinder reservoir cover, diaphragm plate and diaphragm, and to siphon out some fluid (see *Daily (pre-ride) checks*). If the pistons are difficult to push back, attach a length of clear hose to the bleed valve and place the open end in a suitable container, then open the valve and try again. Take great care not to draw any air into the system. If in doubt, bleed the brake afterwards (see Section 9). Slide the caliper back onto the bracket.

16 Make sure the backing material and shim are correctly attached to the pad where fitted **(see illustration)**. Seat the pads in the caliper bracket so that the friction material of each pad faces the disc, making sure the leading edges locate correctly against the guide **(see illustration 2.13c)**. Pivot the caliper down over the pads and onto the disc, making sure the pads stay in position **(see illustration 2.13b)**. Push down on the top of the caliper so the pads compress the spring and insert the pad pin when the holes are aligned **(see illustration 2.13a)**. Lightly tighten the pin. Install the rear mounting bolt/slider pin and tighten it to the torque setting specified at the beginning of the Chapter **(see illustration 2.12c)**. Now tighten the pad pin to the specified torque setting. Install the pad pin plug **(see illustration 2.12a)**.

17 Top up the master cylinder reservoir if necessary (see *Daily (pre-ride) checks*).

18 Operate the brake lever several times to bring the pads into contact with the disc. Check the operation of the brake before riding the motorcycle. **Note:** *New pads will not give full braking efficiency until they have bedded-in. Be prepared for this, and avoid hard braking as far as possible for the first hundred miles (150 km) or so after pad renewal.*

3 Brake discs – inspection, removal and installation

Note: *Honda recommend using new disc mounting bolts when the old ones are removed. This is because the bolts are pre-treated with a locking compound. It is possible, however, to clean up the old bolts*

and reinstall them using a suitable non-permanent thread locking compound that is commercially available.

Inspection

1 Visually inspect the surface of the disc for score marks and other damage. Light scratches are normal after use and won't affect brake operation, but deep grooves and heavy score marks will reduce braking efficiency and accelerate pad wear. If a disc is badly grooved it must be machined or replaced with a new one.

2 To check disc runout, position the bike on an auxiliary stand so that the wheel being checked is off the ground. Mount a dial gauge to a fork leg or on the swingarm, according to wheel, with the plunger on the gauge touching the surface of the disc about 10 mm (1/2 in) from the outer edge. Rotate the wheel and watch the indicator needle, comparing the reading with the limit listed in the Specifications at the beginning of the Chapter. If the runout is greater than the service limit, check the wheel bearings for play (see Chapter 1). If the bearings are worn, replace them with new ones (see Section 14) and repeat this check. It is also worth removing the disc (see below) and checking for built-up corrosion between the disc and the hub (see Step 6) as this will cause runout. If the runout is still excessive, the disc will have to be replaced with a new one, although machining by an engineer may be possible.

3 The disc must not be machined or allowed to wear down to a thickness less than the service limit as listed in this Chapter's Specifications and as marked on the disc itself **(see illustration)**. Check the thickness of the disc using a micrometer **(see illustration)**. If the thickness of the disc is less than the service limit, it must be replaced with a new one.

Removal

4 Remove the wheel (see Section 12 or 13). **Caution:** *Do not lay the wheel down and allow it to rest on the disc – the disc could become warped. Set the wheel on wood blocks so the disc doesn't support the weight of the wheel.*

5 Mark the alignment of the disc to the wheel, so it can be installed in the same position. Unscrew the disc retaining bolts, loosening them a little at a time in a criss-cross pattern to avoid distorting the disc, then remove the disc from the wheel **(see illustrations)**.

Installation

6 Before installing the disc, make sure there is no dirt or corrosion where the disc seats on the hub, particularly right in the angle of the seat, as this will not allow the disc to sit flat when it is bolted down and it will appear to be warped when checked or when using the brake.

7 Fit the disc onto the wheel, making sure the directional arrow is on the outside and pointing in the direction of normal (i.e. forward) rotation. Align the previously applied marks (if you're reinstalling the original disc).

8 Install the disc mounting bolts, either using

3.5a Front disc mounting bolts

3.5b Rear disc mounting bolts

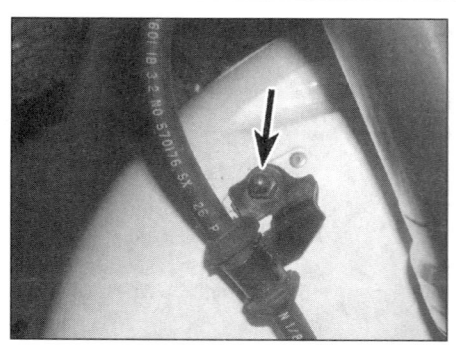

7•6 Brakes, wheels and tyres

4.3a Unscrew the nut (arrowed) and displace the guide

4.3b Brake hose banjo bolt (arrowed)

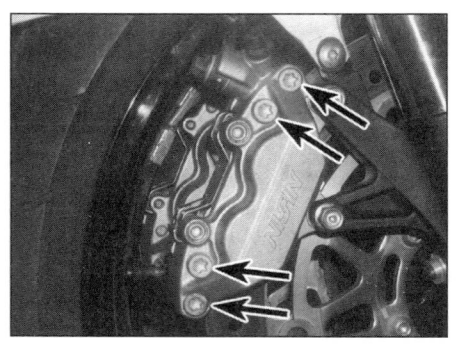

4.4 Brake caliper body joining bolts (arrowed)

new ones or cleaning and applying a suitable non-permanent thread locking compound to the threads of the old ones (see **Note** at the beginning of this section), and tighten them evenly in a criss-cross pattern to the torque setting specified at the beginning of the Chapter **(see illustration 3.5a or 3.5b)**. Clean off all grease from the brake disc(s) using acetone or brake system cleaner. If a new brake disc has been installed, remove any protective coating from its working surfaces.

9 Install the wheel (see Section 12 or 13). When installing a new disc always fit new brake pads (see Section 2).

10 Operate the brake lever and pedal several times to bring the pads into contact with the disc. Check the operation of the brakes carefully before riding the bike.

4 Front brake calipers – removal, overhaul and installation

⚠️ **Warning: The dust created by the brake pads may contain asbestos, which is harmful to your health. Never blow it out with compressed air and don't inhale any of it. An approved filtering mask should be worn when working on the brakes. Do not, under any circumstances, use petroleum-based solvents to clean brake parts. Use clean brake fluid, brake cleaner or denatured alcohol only. Use care when working with brake fluid as it can injure your eyes and it will damage painted surfaces and plastic parts – cover these with rag. Disassembly, overhaul and reassembly of the brake calipers must be done in a spotlessly clean work area to avoid contamination and possible failure of the hydraulic system components.**

Note 1: *Honda recommend using new mounting bolts and caliper body joining bolts when the old ones are removed. This is because the bolts are pre-treated with a locking compound. It is possible, however, to clean up the old bolts and reinstall them using a suitable non-permanent thread locking compound that is commercially available.*

Note 2: *If the entire front brake system is being overhauled (i.e. master cylinder as well as calipers), or if you intend to change the brake fluid as part of the caliper overhaul (which is advisable), drain the brake fluid completely from the system (see Section 9), as opposed to retaining the old fluid within it by blocking the hose as described (Step 3).*

1 If the caliper is leaking fluid, if the brake pads are wearing unevenly, or the pistons do not move smoothly or are tight in their bores, then caliper overhaul is required.

2 Before disassembling the caliper, read through the entire procedure and make sure that you have the correct seal kit. Also, you will need some new DOT 4 brake fluid, and some clean rags.

Removal

3 If the calipers are just being displaced and not completely removed or overhauled, do not disconnect the brake hoses, but unscrew the nut securing each hose guide to the front mudguard **(see illustration)**. If the calipers are being completely removed or overhauled, remove the brake hose banjo bolt and detach the hose, noting the alignment with the caliper **(see illustration)**. Either plug the hose using another suitable short piece of hose fitted through the eye of the banjo union (it needs to be a fairly tight fit to seal it properly), clamp it using a hose clamp, block it using a suitable bolt with sealing washers and a capped (domed) nut, or wrap a plastic bag tightly around, the object being to minimise fluid loss and prevent dirt entering the system. Whatever you do, also cover the end of the hose in rag, just in case. Discard the sealing washers as new ones must be used on installation. **Note:** *If you are planning to overhaul the caliper and don't have a source of compressed air to blow out the pistons, just loosen the banjo bolt at this stage and retighten it lightly. The bike's hydraulic system can then be used to force the pistons out of the body once the pads have been removed. Disconnect the hose once the pistons have been sufficiently displaced.*

4 If the calipers are being overhauled, slacken the caliper body joining bolts slightly, then nip them up again so they are secure but not tight **(see illustration)**. Remove the brake pads (see Section 2), following the procedure that incorporates removing the caliper.

5 If the calipers are just being displaced, unscrew the mounting bolts and slide the caliper off the disc **(see illustration 2.1b)**.

Overhaul

6 Clean the exterior of the caliper with denatured alcohol or brake system cleaner **(see illustration)**.

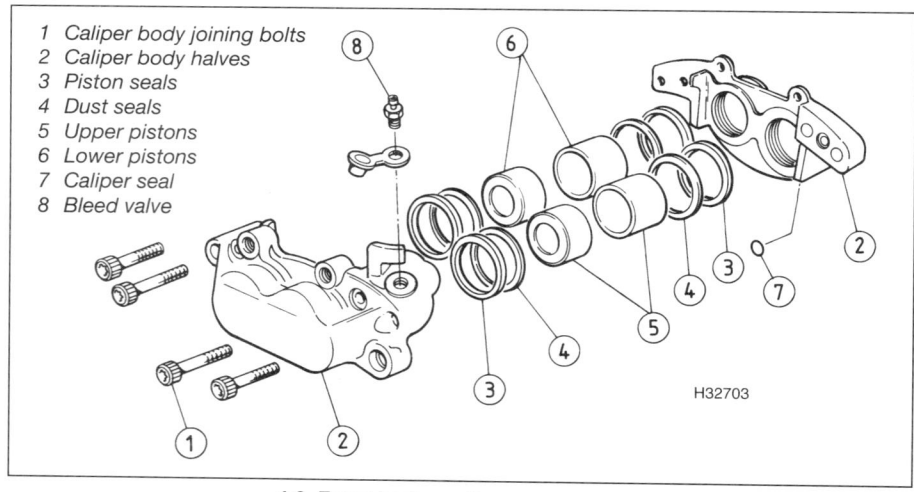

1 Caliper body joining bolts
2 Caliper body halves
3 Piston seals
4 Dust seals
5 Upper pistons
6 Lower pistons
7 Caliper seal
8 Bleed valve

4.6 Front brake caliper components

Brakes, wheels and tyres

7 Displace the pistons as far as possible from the caliper body, either by pumping them out by operating the front brake lever, or by forcing them out using compressed air, until they just meet in the middle. If the compressed air method is used, place a piece of cardboard or thin plywood between the pistons to act as a cushion, then use compressed air directed into the fluid inlet to force the pistons out of the body. Use only low pressure to ease the pistons out and make sure all pistons are displaced at the same time. If the air pressure is too high and the pistons are forced out, the caliper and/or pistons may be damaged.

⚠️ **Warning: Never place your fingers in front of the pistons in an attempt to catch or protect them when applying compressed air, as serious injury could result.**

8 Unscrew the caliper body joining bolts and separate the caliper halves. Mark each piston and caliper body with a felt marker to ensure that the pistons can be matched to their original bores on reassembly, then remove the pistons. Remove the caliper seal from either half of the caliper body and discard it as a new one must be used.

Caution: Do not try to remove the pistons by levering them out, or by using pliers or any other grips.

9 Using a wooden or plastic tool, remove the dust seals from the caliper bores. Discard them as new ones must be used on installation. If a metal tool is being used, take great care not to damage the caliper bores.

10 Remove and discard the piston seals in the same way.

11 Clean the pistons and bores, paying attention to the seal grooves, with clean brake fluid of the specified type. If compressed air is available, use it to dry the parts thoroughly (make sure it's filtered and unlubricated).

⚠️ **Warning: Do not, under any circumstances, use a petroleum-based solvent to clean brake parts.**

12 Inspect the caliper bores and pistons for signs of corrosion, nicks and burrs and loss of plating. If surface defects are present, the caliper and/or pistons must be renewed. If the necessary measuring equipment is available, compare the dimensions of the pistons and bores to those specified at the beginning of the Chapter, renewing any component that is worn beyond its service limit. If the caliper is in bad shape the master cylinder should also be checked.

13 Lubricate the new piston seals with clean brake fluid and fit them into their grooves in the caliper bores. Note that different sizes of bore and piston are used (see Specifications), and care must therefore be taken to ensure that the correct size seals are fitted to the correct bores. The same applies when fitting the new dust seals and pistons.

14 Lubricate the new dust seals with silicone grease and fit them into their grooves in the caliper bores.

15 Lubricate the pistons with clean brake fluid and fit them closed-end first into the caliper bores. Using your thumbs, push the pistons all the way in, making sure they enter the bore squarely.

16 Lubricate the new caliper seal with clean brake fluid and install it in the fluid passage orifice in the caliper body, then join the body halves together. Install the caliper body joining bolts, either using new ones or cleaning and applying a suitable non-permanent thread locking compound to the threads of the old ones (see **Note** at the beginning of this section), and tighten them to the torque setting specified at the beginning of the Chapter (note that they can be fully tightened after the calipers have been installed if required, to prevent having to hold the calipers by hand – just tighten them as much as possible before installing them).

Installation

17 If the calipers have been overhauled, install the brake pads (see Section 2). If not, slide the caliper onto the disc, making sure the pads sit squarely on either side **(see illustration 2.9)**. Install the caliper mounting bolts, either using new ones or cleaning and applying a suitable non-permanent thread locking compound to the threads of the old ones (see **Note** at the beginning of this section), and tighten them to the torque setting specified at the beginning of the Chapter.

18 If the calipers have been overhauled, and if not already done, tighten the body joining bolts to the specified torque **(see illustration 4.4)**.

19 If detached, connect the brake hose to the caliper, using new sealing washers on each side of the fitting. Align the hose as noted on removal **(see illustration 4.3b)**. Tighten the banjo bolt to the torque setting specified at the beginning of the Chapter.

20 Locate the brake hose guide on the front mudguard and secure it with the nut **(see illustration 4.3a)**.

21 Fill up the master cylinder reservoir with DOT 4 brake fluid (see *Daily (pre-ride) checks*) and fill or bleed the hydraulic system as required (see Section 9).

22 Check for leaks and thoroughly test the operation of the brakes before riding the motorcycle on the road.

5 Rear brake caliper – removal, overhaul and installation

⚠️ **Warning: The dust created by the brake pads may contain asbestos, which is harmful to your health. Never blow it out with compressed air and don't inhale any of it. An approved filtering mask should be worn when working on the brakes. Do not, under any circumstances, use petroleum-based solvents to clean brake parts. Use clean brake fluid, brake cleaner or denatured alcohol only. Use care when working with brake fluid as it can injure your eyes and it will damage painted surfaces and plastic parts – cover these with rag. Disassembly, overhaul and reassembly of the brake caliper must be done in a spotlessly clean work area to avoid contamination and possible failure of the hydraulic system components.**

Note: *If the entire rear brake system is being overhauled (i.e. master cylinder as well as caliper), or if you intend to change the brake fluid as part of the caliper overhaul (which is advisable), drain the brake fluid completely from the system (see Section 9), as opposed to retaining the old fluid within it by blocking the hose as described (Step 3).*

1 If the caliper is leaking fluid, if the piston does not move smoothly or is tight in its bore, then caliper overhaul is required.

2 Before disassembling the caliper, read through the entire procedure and make sure that you have the correct seal kit. Also, you will need some new DOT 4 brake fluid, and some clean rags.

Removal

3 If the caliper is just being displaced and not completely removed or overhauled, do not disconnect the brake hose. If the caliper is being completely removed or overhauled, remove the brake hose banjo bolt and detach the hose, noting the alignment with the caliper **(see illustration)**. Either plug the hose using another suitable short piece of hose fitted through the eye of the banjo union (it needs to be a fairly tight fit to seal it properly), clamp it using a hose clamp, block it using a suitable bolt with sealing washers and a capped (domed) nut, or wrap a plastic bag tightly around, the object being to minimise fluid loss and prevent dirt entering the system. Whatever you do, also cover the end of the hose in rag, just in case. Discard the sealing washers as new ones must be used on installation. **Note:** *If you are planning to overhaul the caliper and don't have a source of compressed air to blow out the piston, just loosen the banjo bolt at this stage and retighten it lightly. The bike's hydraulic system*

5.3 Brake hose banjo bolt (arrowed)

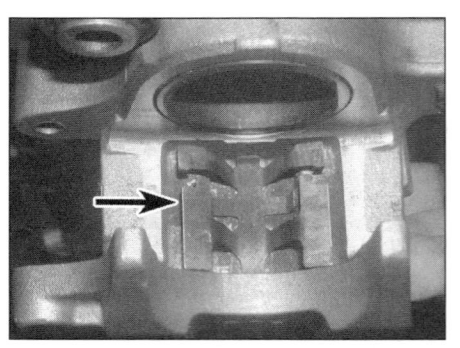

7•8 Brakes, wheels and tyres

5.4a Remove the pad spring (arrowed) from the caliper . . .

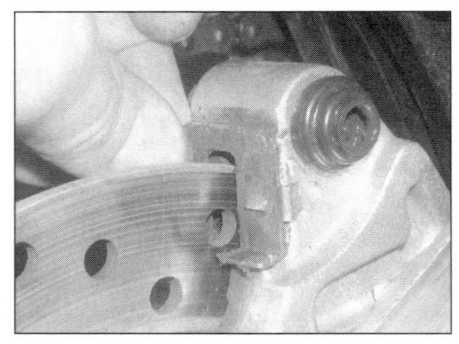

5.4b . . . and the guide from the bracket, noting how they fit

5.5 Slide the front slider pin out of its boot in the caliper bracket

can then be used to force the piston out of the body once the pads have been removed. Disconnect the hose once the piston have been sufficiently displaced.

4 If the caliper is being overhauled, remove the brake pads (see Section 2), then slide the caliper off the bracket and remove it **(see illustration 5.5)**. Remove the pad spring from the caliper, and if required the guide from the bracket, noting how they fit **(see illustrations)**.

5 If the caliper is just being displaced, unscrew the caliper rear mounting bolt/slider pin **(see illustration 2.12c)**, then pivot the rear of the caliper up and slide it off its bracket **(see illustration)**. If required, remove the pad spring from the caliper and the guide from the bracket, noting how they fit **(see illustrations 5.4a and 5.4b)**.

Overhaul

6 Clean the exterior of the caliper with denatured alcohol or brake system cleaner **(see illustration)**.

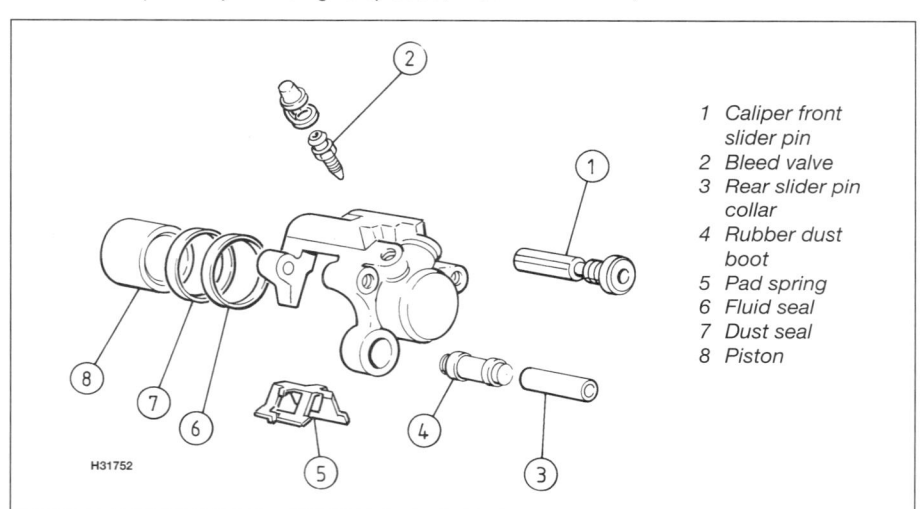

7 Remove the piston from the caliper body, either by pumping it out by operating the brake pedal, or by using compressed air. If the compressed air method is used, place a wad of rag over the piston to act as a cushion, then use compressed air directed into the fluid inlet to force the pistons out of the body. Use only low pressure to ease the piston out and make sure. If the air pressure is too high and the piston is forced out, the caliper and/or piston may be damaged.

⚠ **Warning: Never place your fingers in front of the piston in an attempt to catch or protect it when applying compressed air, as serious injury could result.**

Caution: Do not try to remove the pistons by levering them out, or by using pliers or any other grips.

8 Using a wooden or plastic tool, remove the dust seal from the caliper bore. Discard it as a new one must be used on installation. If a metal tool is being used, take great care not to damage the caliper bore.

9 Remove and discard the piston seal in the same way.

10 Clean the piston and bore, paying attention to the seal grooves, with clean brake fluid of the specified type. If compressed air is available, use it to dry the parts thoroughly (make sure it's filtered and unlubricated).

⚠ **Warning: Do not, under any circumstances, use a petroleum-based solvent to clean brake parts.**

11 Inspect the caliper bore and piston for signs of corrosion, nicks and burrs and loss of plating. If surface defects are present, the caliper and/or piston must be renewed. If the necessary measuring equipment is available, compare the dimensions of the piston and bore to those given in the Specifications Section of this Chapter, renewing any component that is worn beyond the service limit. If the caliper is in bad shape the master cylinder should also be checked.

12 Remove the collar from the rear slide pin rubber boot and remove the boots from the caliper and the bracket **(see illustrations)**. Clean off all traces of corrosion and hardened grease from the collar, boots and pins. Renew

1 Caliper front slider pin
2 Bleed valve
3 Rear slider pin collar
4 Rubber dust boot
5 Pad spring
6 Fluid seal
7 Dust seal
8 Piston

5.6 Rear brake caliper components

5.12a Remove the collar then remove the boot (arrowed) from the caliper . . .

5.12b . . . and from the bracket (arrowed)

Brakes, wheels and tyres 7•9

5.19a Pivot the caliper down . . .

5.19b . . . making sure the pads locate correctly against the guide

the rubber boots if they are damaged, deformed or deteriorated. Apply a smear of silicone based grease to the collar, boots and slider pins. Fit the boots into their bores, then fit the collar into the rear boot.

13 Lubricate the new piston seal with clean brake fluid and install it in its groove in the caliper bore.

14 Lubricate the new dust seal with silicone grease and install it in its groove in the caliper bore.

15 Lubricate the piston with clean brake fluid and install it closed-end first into the caliper bore. Using your thumbs, push the piston all the way in, making sure it enters the bore squarely.

Installation

16 If the caliper has not been overhauled, remove the collar from the rear slider pin rubber boot and remove the boots from the caliper and the bracket (see illustrations 5.12a and 5.12b). Clean off all traces of corrosion and hardened grease from the collar, boots and pins. Renew the rubber boots if they are damaged, deformed or deteriorated. Apply a smear of silicone based grease to the collar, boots and slider pins. Fit the boots into their bores, then fit the collar into the rear boot.

17 Make sure that the pad spring and pad guide are correctly fitted (see illustrations 5.4a and 5.4b).

18 If the caliper was overhauled, slide the caliper onto the bracket (see illustration 5.5) and leave it with the rear pivoted up, then install the brake pads (see Section 2).

19 If the caliper was just displaced, slide it onto the bracket (see illustration 5.5), then pivot it down, making sure the pads locate on each side of the disc, and the front edges locate correctly against the guide (see illustrations). Install the rear mounting bolt/slider pin and tighten it to the torque setting specified at the beginning of the Chapter (see illustration 2.12c).

20 If detached, connect the brake hose to the caliper, using new sealing washers on each side of the fitting. Align the hose as noted on removal (see illustration 5.3). Tighten the banjo bolt to the torque setting specified at the beginning of the Chapter.

21 Fill up the master cylinder reservoir with DOT 4 brake fluid (see *Daily (pre-ride) checks*) and fill or bleed the hydraulic system as required (see Section 9).

22 Check for leaks and thoroughly test the operation of the brake before riding the motorcycle.

6 Front brake master cylinder – removal, overhaul and installation

Warning: *Do not, under any circumstances, use petroleum-based solvents to clean brake parts. Use clean brake fluid, brake cleaner or denatured alcohol only. Use care when working with brake fluid as it can injure your eyes and it will damage painted surfaces and plastic parts – cover surrounding components with rag. Disassembly, overhaul and reassembly of the brake master cylinder must be done in a spotlessly clean work area to avoid contamination and possible failure of the hydraulic system components.*

Note: *If the entire front brake system is being overhauled (i.e. calipers as well as master cylinder), or if you intend to change the brake fluid as part of the master cylinder overhaul (which is advisable), drain the brake fluid completely from the system (see Section 9), as opposed to retaining the old fluid within it by blocking the hose as described (Step 6).*

1 If the master cylinder is leaking fluid, or if the lever does not produce a firm feel when the brake is applied, and bleeding the brakes does not help (see Section 9), and the calipers and hydraulic hoses and unions are all in good condition, then master cylinder overhaul is required.

2 Before disassembling the master cylinder, read through the entire procedure and make sure that you have the correct rebuild kit. Also, you will need some new DOT 4 brake fluid, some clean rags and internal circlip pliers. **Note:** *To prevent damage to the paint from spilled brake fluid, always cover the fuel tank when working on the master cylinder.*

Removal

Note: *If the master cylinder is being displaced from the handlebar and not being removed completely or overhauled, follow Steps 4 and 7 only.*

3 If the system hasn't been drained (see **Note** at the beginning of this section), loosen, but do not remove, the screws holding the reservoir cap in place (see illustration). Access to the cover screws is partially restricted by the windshield, so if a short or angled screwdriver is not available, remove the windshield (see Chapter 8). Unscrew the nut and withdraw the bolt securing the reservoir to its bracket and detach it, noting how the peg locates in the

6.3a Slacken the reservoir cover screws (arrowed) . . .

7•10 Brakes, wheels and tyres

6.3b ... then unscrew the nut (arrowed), withdraw the bolt, displace the reservoir and empty it

6.3c Release the clamp (arrowed) and detach the hose

6.4 Disconnect the brake light switch wiring connectors (arrowed)

6.6 Brake hose banjo bolt (arrowed)

6.7 Front master cylinder clamp bolts (arrowed)

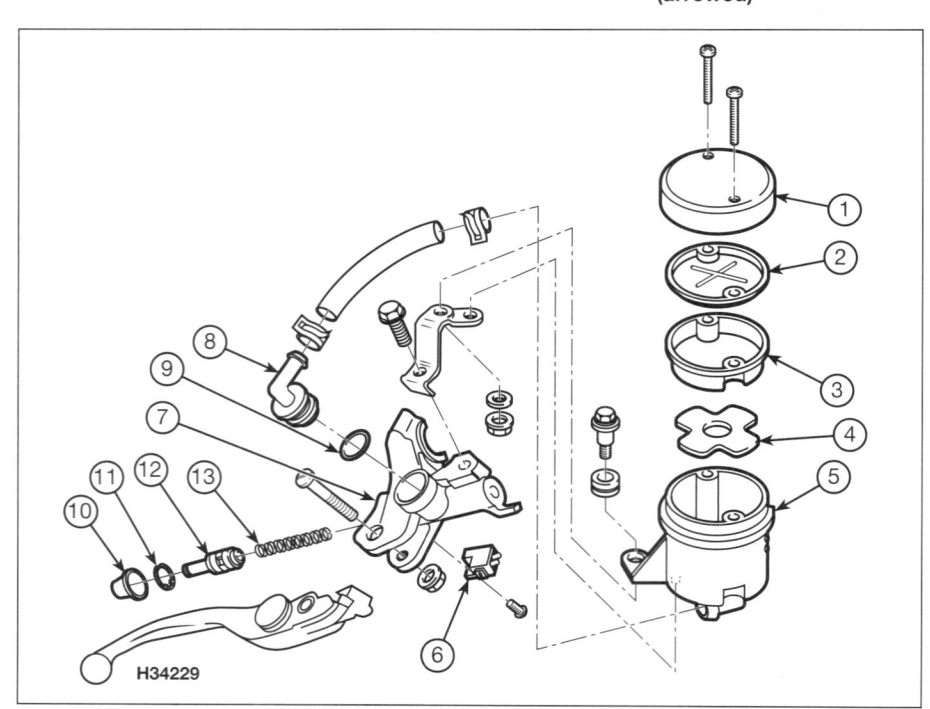

hole **(see illustration)**. Remove the cap, diaphragm plate and diaphragm then empty the reservoir into a suitable container, preventing the surface float from dropping out as you do. Release the clamp and detach the hose from its union on the master cylinder, being prepared with a rag to catch any residue fluid, and remove the reservoir **(see illustration)**. Wipe any remaining fluid out with a clean rag. If required unscrew the bolt securing the reservoir bracket to the master cylinder and remove it. Inspect the reservoir hose for cracks or splits and replace it with a new one if necessary.

4 Disconnect the electrical connectors from the brake light switch **(see illustration)**.

5 Remove the front brake lever (see Chapter 6, Section 5).

6 Unscrew the brake hose banjo bolt and separate the hose from the master cylinder, noting its alignment **(see illustration)**. Discard the two sealing washers as they must be replaced with new ones. Either plug the hose using another suitable short piece of hose fitted through the eye of the banjo union (it needs to be a fairly tight fit to seal it properly), clamp it using a hose clamp, block it using a suitable bolt with sealing washers and a capped (domed) nut, or wrap a plastic bag tightly around, the object being to minimise fluid loss and prevent dirt entering the system. Whatever you do, also cover the end of the hose in rag, just in case.

7 Unscrew the master cylinder clamp bolts, then lift the master cylinder and reservoir away from the handlebar, noting how the top mating surfaces of the clamp align with the punch mark on the top of the handlebar **(see illustration)**.

Caution: Do not tip the master cylinder upside down or brake fluid will run out.

8 If required undo the brake light switch screw and remove the switch, noting how it fits.

Overhaul

9 Carefully remove the dust boot from the end of the master cylinder and from around the piston, noting how it locates **(see illustration)**.

10 Push the piston in and, using circlip pliers,

6.9 Front master cylinder components

1 Reservoir cap
2 Diaphragm plate
3 Diaphragm
4 Surface float
5 Reservoir
6 Brake light switch
7 Master cylinder
8 Reservoir hose union
9 O-ring
10 Rubber boot
11 Circlip
12 Piston assembly (including cup and seal)
13 Spring

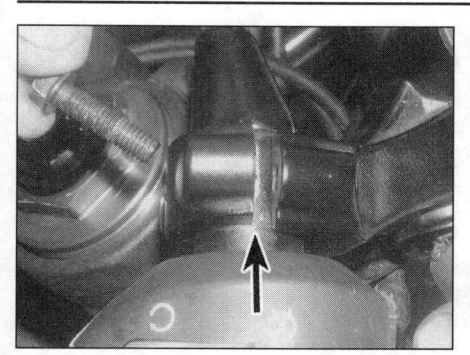

6.21 Align the clamp mating surfaces with the punch mark (arrowed)

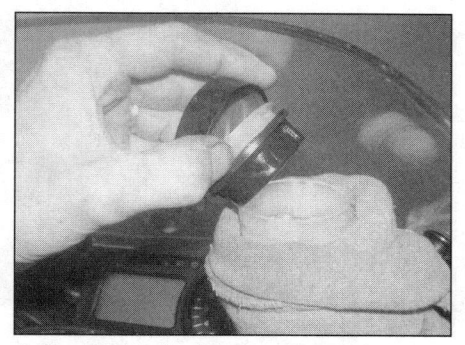

6.27a Fit the diaphragm, plate and cap . . .

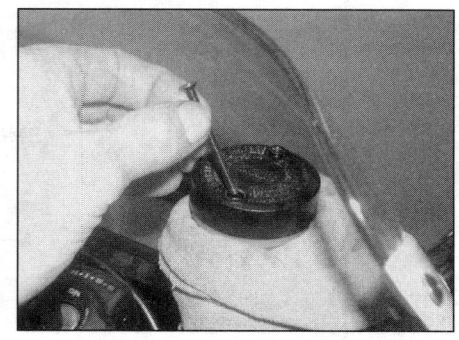

6.27b . . . then install the screws

remove the circlip from its groove in the master cylinder and slide out the piston assembly and the spring, noting how they fit. If they are difficult to remove, apply low pressure compressed air to the fluid outlet. Remove the cup and seal from the piston. Lay the parts out in order as you remove them to prevent confusion during reassembly.

11 If required displace the dust cover from the reservoir hose union then release the circlip and detach the union from the master cylinder. Remove the O-ring and discard it – note that a new O-ring is not included in the rebuild kit so must be obtained separately.

12 Clean all parts with clean brake fluid. If compressed air is available, use it to dry the parts thoroughly (make sure it's filtered and unlubricated).

Warning: Do not, under any circumstances, use a petroleum-based solvent to clean brake parts.

13 Check the master cylinder bore for corrosion, scratches, nicks and score marks. If the necessary measuring equipment is available, compare the dimensions of the piston (just behind the seal) and bore to those given in the Specifications Section of this Chapter. If damage or wear to the bore is evident, the master cylinder must be replaced with a new one. If the master cylinder is in poor condition, then the calipers should be checked as well. Check that the fluid inlet and outlet ports in the master cylinder are clear.

14 If removed smear the new reservoir hose union O-ring with new brake fluid and fit it into the groove on the union. Locate the union and secure it with the circlip, making sure it is seated in its groove. Fit the dust cover.

15 The dust boot, circlip, piston, seal, cup and spring are included in the rebuild kit. Use all of the new parts, regardless of the apparent condition of the old ones. If the cup and seal are not already on the piston, fit them according to the layout of the old ones.

16 Lubricate the cup, seal and piston with new brake fluid.

17 Slide the spring into the master cylinder, then fit the piston with the narrow end facing out. Make sure the lips on the cup and seal do not turn inside out when they enter the bore.

Depress the piston and install the new circlip, making sure that it locates in the groove in the master cylinder.

18 Install the rubber dust boot, making sure it is seated properly in the groove in the master cylinder and around the groove in the tip of the piston.

19 Inspect the reservoir rubber diaphragm and replace it with a new one it if it is damaged, deformed or deteriorated.

Installation

20 Locate the brake light switch on the underside of the master cylinder and secure it with the screw.

21 Attach the master cylinder to the handlebar and fit the clamp with its UP mark facing up **(see illustration 6.7)**. Align the top mating surfaces of the clamp with the small punch mark on the top of the handlebar, then tighten the top bolt first, then the bottom bolt to the torque setting specified at the beginning of the Chapter **(see illustration)**.

22 Connect the brake hose to the master cylinder, using new sealing washers on each side of the union, and aligning the hose as noted on removal **(see illustration 6.6)**. Tighten the banjo bolt to the torque setting specified at the beginning of the Chapter.

23 Install the brake lever (see Chapter 6, Section 5).

24 Connect the brake light switch wiring **(see illustration 6.4)**.

25 If removed, fit the reservoir bracket onto the master cylinder and tighten its bolt. Mount the reservoir on the bracket, locating the peg in the hole, and secure it with the bolt **(see illustration 6.3b)**. Connect the hose to the union and secure it with the clamp **(see illustration 6.3c)**. Check that the hose is secure and clamped at the reservoir end as well. If the clamps have weakened, use new ones.

26 Fill up the master cylinder reservoir with DOT 4 brake fluid (see *Daily (pre-ride) checks*) and fill or bleed the hydraulic system as required (see Section 9).

27 Fit the rubber diaphragm, making sure it is correctly seated, the diaphragm plate and the cap onto the reservoir, and secure the cap with its screws **(see illustrations)**.

28 Check the operation of the front brake and brake light before riding the motorcycle.

7 Rear brake master cylinder – removal, overhaul and installation

Warning: Do not, under any circumstances, use petroleum-based solvents to clean brake parts. Use clean brake fluid, brake cleaner or denatured alcohol only. Use care when working with brake fluid as it can injure your eyes and it will damage painted surfaces and plastic parts – cover surrounding areas with rag. Disassembly, overhaul and reassembly of the brake master cylinder must be done in a spotlessly clean work area to avoid contamination and possible failure of the hydraulic system components.

Note: *If the entire rear brake system is being overhauled (i.e. caliper as well as master cylinder), or if you intend to change the brake fluid as part of the master cylinder overhaul (which is advisable), drain the brake fluid completely from the system (see Section 9), as opposed to retaining the old fluid within it by blocking the hose as described (Step 4).*

1 If the master cylinder is leaking fluid, or if the lever does not produce a firm feel when the brake is applied, and bleeding the brakes does not help (see Section 9), and the caliper and hydraulic hoses and unions are all in good condition, then master cylinder overhaul is required.

2 Before disassembling the master cylinder, read through the entire procedure and make sure that you have the correct rebuild kit. Also, you will need some new DOT 4 brake fluid, some clean rags and internal circlip pliers. **Note:** *To prevent damage to the paint from spilled brake fluid, always cover the surrounding components when working on the master cylinder.*

Removal

3 If the system hasn't been drained (see **Note** above), unscrew the bolt securing the master

7•12 Brakes, wheels and tyres

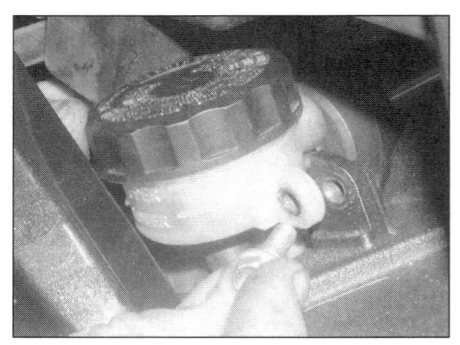

7.3a Unscrew the bolt (arrowed) and displace the reservoir . . .

7.3b . . . then remove the cap, plate and diaphragm and empty it

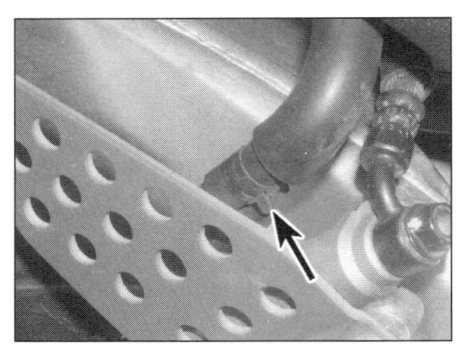

7.3c Release the clamp (arrowed) and detach the hose

7.4 Brake hose banjo bolt (arrowed)

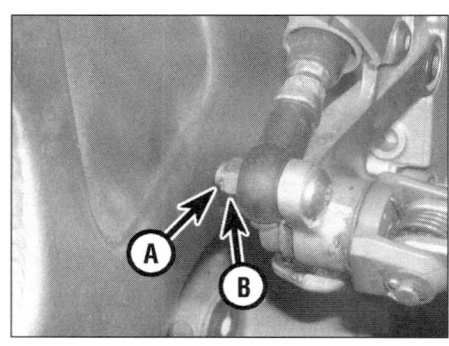

7.6a Remove the split pin (A), then counter-hold the nut (B) . . .

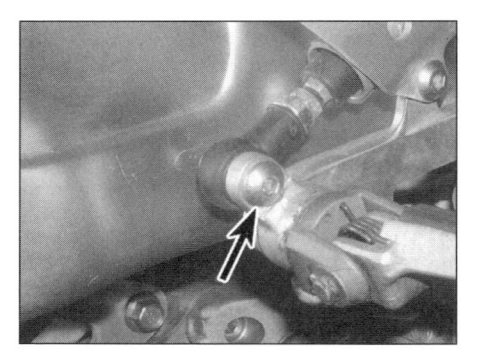

7.6b . . . and unscrew the bolt (arrowed)

cylinder reservoir and manoeuvre it round the back of the rear sub-frame **(see illustration)**. Unscrew the cap and remove the diaphragm plate and diaphragm, then empty the reservoir into a suitable container **(see illustration)**. Release the clamp and detach the hose from its union on the master cylinder, being prepared with a rag to catch any residue fluid, and remove the reservoir **(see illustration)**. Wipe any remaining fluid out with a clean rag. Inspect the reservoir hose for cracks or splits and replace it with a new one if necessary.

4 Unscrew the brake hose banjo bolt and separate the hose from the master cylinder, noting its alignment **(see illustration)**. Discard the sealing washers as they must be replaced with new ones. Either plug the hose using another suitable short piece of hose fitted through the eye of the banjo union (it needs to be a fairly tight fit to seal it properly), clamp it using a hose clamp, block it using a suitable bolt with sealing washers and a capped (domed) nut, or wrap a plastic bag tightly around, the object being to minimise fluid loss and prevent dirt entering the system. Whatever you do, also cover the end of the hose in rag, just in case.

5 On RR-Y and RR-1 (2000 and 2001) models, remove the split pin from the inner end of the clevis pin, then withdraw the clevis pin.

6 On RR-2 and RR-3 (2002 and 2003) models, remove the split pin from the inner end of the pushrod joint piece bolt, then counter-hold the nut and unscrew the bolt, noting the washers **(see illustrations)**.

7 Unscrew the two bolts securing the master cylinder to the footrest bracket, noting how they also secure the heel guard, then remove the guard and the master cylinder. On models equipped with a catalytic converter, release the oxygen sensor wiring from the guard.

Overhaul

8 If necessary, not how far the clevis or joint piece is threaded up the pushrod, then slacken its locknut and thread it off **(see illustration)**.

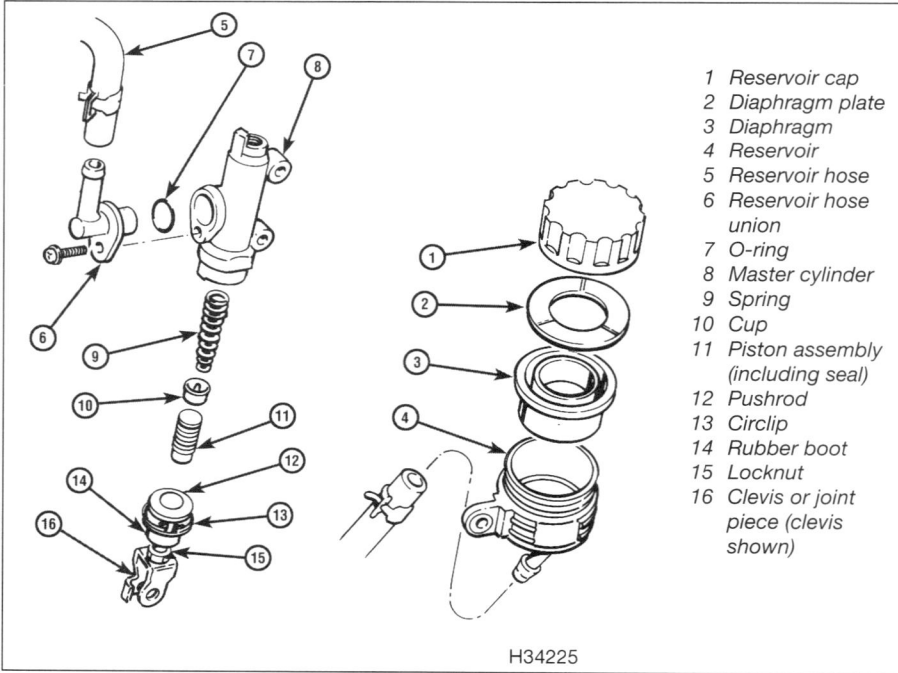

7.8 Rear master cylinder components

1 Reservoir cap
2 Diaphragm plate
3 Diaphragm
4 Reservoir
5 Reservoir hose
6 Reservoir hose union
7 O-ring
8 Master cylinder
9 Spring
10 Cup
11 Piston assembly (including seal)
12 Pushrod
13 Circlip
14 Rubber boot
15 Locknut
16 Clevis or joint piece (clevis shown)

7.11 Reservoir hose union screw (arrowed)

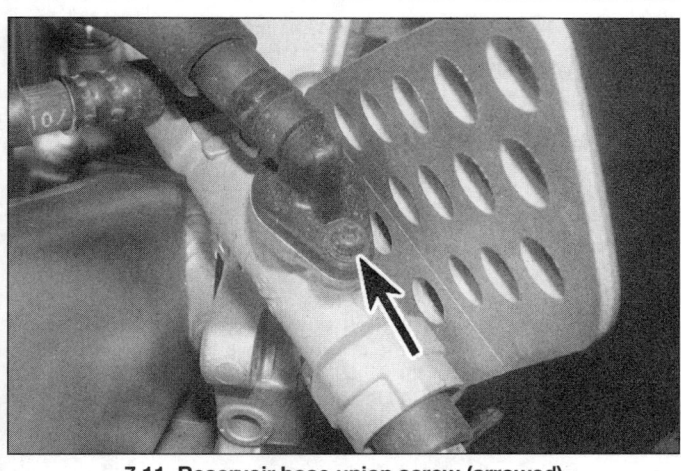

8.2 Flex the brake hoses and check for cracks, bulges and leaking fluid

9 Dislodge the rubber dust boot from the base of the master cylinder and from around the pushrod, noting how it locates.

10 Push the pushrod in and, using circlip pliers, remove the circlip from its groove in the master cylinder and slide out the piston assembly and the spring, noting how they fit. If they are difficult to remove, apply low pressure compressed air to the fluid outlet. Lay the parts out in order as you remove them to prevent confusion during reassembly.

11 If required, undo the screw securing the fluid reservoir hose union and detach it from the master cylinder **(see illustration)**. Discard the O-ring as a new one must be used.

12 Clean all of the parts with clean brake fluid.

> ⚠ **Warning: Do not, under any circumstances, use a petroleum-based solvent to clean brake parts.**

13 Check the master cylinder bore for corrosion, scratches, nicks and score marks. If the necessary measuring equipment is available, compare the dimensions of the piston (just behind the seal) and bore to those given in the Specifications Section of this Chapter. If damage or wear to the bore is evident, the master cylinder must be replaced with a new one. If the master cylinder is in poor condition, then the caliper should be checked as well.

14 The dust boot, circlip, piston, seal, cup and spring are included in the rebuild kit. Use all of the new parts, regardless of the apparent condition of the old ones. If the seal is not already on the piston, fit it according to the layout of the old one.

15 Fit the cup over the end of the spring so that its inner raised section fits into the outer coil on the spring. Lubricate the cup, seal and piston with clean brake fluid.

16 Slide the spring into the master cylinder with the cup facing out, then fit the piston into the master cylinder, making sure it is the correct way round with the seal on the outer end. Make sure the lips on the cup and seal do not turn inside out when they enter the bore.

17 Apply some silicone grease to the end of the pushrod and fit it into the master cylinder. Depress the pushrod, then install the new circlip, making sure it is properly seated in the groove.

18 Install the rubber dust boot, making sure it is seated properly in the groove in the master cylinder and around the pushrod.

19 If removed, fit a new O-ring onto the fluid reservoir hose union, then fit the union into the master cylinder and secure it with its screw.

20 If removed, thread the locknut and clevis or joint piece onto the master cylinder pushrod, setting them as noted on removal. On RR-Y and RR-1 (2000 and 2001) models Honda specify the distance between the eye in the clevis and the lower mounting bolt hole should be 74 to 76 mm. On RR-2 and RR-3 (2002 and 2003) models there is no specification. Tighten the locknut securely against the clevis or joint piece.

Installation

21 Mount the heel guard and master cylinder onto the footrest bracket and tighten the bolts. On models equipped with a catalytic converter, locate the oxygen sensor wiring in its guide on the guard.

22 On RR-Y and RR-1 (2000 and 2001) models, align the pushrod clevis with the brake pedal, then insert the clevis pin and secure it with a new split pin. Bend the ends of the pin up to lock it.

23 On RR-2 and RR-3 (2002 and 2003) models, align the pushrod joint piece with the brake pedal, then insert the bolt with its washer and secure it with the washer, nut and a new split pin **(see illustrations 7.6b and 7.6a)**. Bend the ends of the pin up to lock it.

24 Connect the brake hose to the master cylinder, using new sealing washers on each side of the union. Align the hose as noted on removal and tighten the banjo bolt to the specified torque setting **(see illustration 7.4)**.

25 Connect the reservoir hose to its union on the master cylinder and secure it with the clamp **(see illustration 7.3c)**. Check that the hose is secure and clamped at the reservoir end as well. If the clamps have weakened, use new ones. Mount the reservoir on its lug, but without fitting the diaphragm, plate and cap, and lightly tighten the bolt – the reservoir will have to be dismounted after the filling/bleeding process, but it is unwise to leave it free as it is too easy for it to spill its contents.

26 Fill up the master cylinder reservoir with DOT 4 brake fluid (see *Daily (pre-ride) checks*) and fill or bleed the hydraulic system as required (see Section 9).

27 Check the operation of the brake and brake light carefully before riding the motorcycle.

8 Brake hoses – inspection and replacement

Inspection

1 Brake hose condition should be checked regularly and new hoses should be installed at the specified interval (see Chapter 1).

2 Twist and flex the hoses while looking for cracks, bulges and seeping fluid **(see illustration)**. Check extra carefully around the areas where the hoses connect with the banjo fittings, as these are common areas for failure.

3 Inspect the metal joints and the banjo union fittings connected to the brake hoses. If the fittings are rusted, scratched or cracked, replace the hoses with new ones.

Replacement

> ⚠ **Warning: Use care when working with brake fluid as it can injure your eyes and it will damage painted surfaces and plastic parts.**

4 Before removing a hose, drain all old fluid from the system being worked on (see Section 9).

7•14 Brakes, wheels and tyres

5 The brake hoses have banjo union fittings on each end **(see illustrations 7.4, 6.6, 5.3, and 4.3b)**. Cover the surrounding area with plenty of rag then unscrew the banjo bolt at each end of the hose, noting its alignment. Free the hose from any clips or guides and remove it – when removing the rear hose on RR-2 and RR-3 (2002 and 2003) models, you will have to remove the rear hugger first. Discard the sealing washers on the hose banjo unions.

6 Position the new hose, making sure it isn't twisted or otherwise strained, and abut the tab on the hose union with the lug on the component casting, where present. Otherwise align the hose as noted on removal **(see illustrations 7.4, 6.6, 5.3, and 4.3b)**. Install the hose banjo bolts using new sealing washers on both sides of the unions. Tighten the banjo bolts to the torque setting specified at the beginning of this Chapter. Make sure the hoses are correctly aligned and routed clear of all moving components.

7 Refill the system reservoir with new DOT 4 brake fluid (see *Daily (pre-ride) checks*) and bleed the air from the system (see Section 9). Check the operation of the brakes carefully before riding the motorcycle.

9 Brake system – bleeding and fluid change

Note: *Honda recommend using a commercially available vacuum-type brake bleeding tool* **(see illustration)**. *If bleeding the system using the conventional method does not work sufficiently well, it is advisable to obtain a bleeder and repeat the procedure detailed below, following the manufacturer's instructions for using the tool. If the tool is not available, take the machine to a Honda dealer.*

⚠️ **Warning: Use care when working with brake fluid as it can injure your eyes and it will damage painted surfaces and plastic parts.**

Bleeding

1 Bleeding the brakes is simply the process

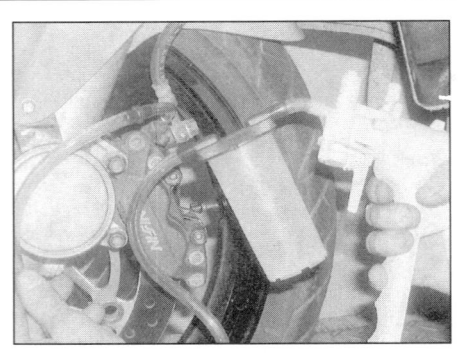

9.0 Bleeding the brakes using a commercial vacuum-operated bleeding tool

of removing all the air bubbles from the brake fluid reservoirs, the hoses and the brake calipers. Bleeding is necessary whenever a brake system hydraulic connection is loosened, when a component or hose is replaced, or when a master cylinder or caliper is overhauled. Leaks in the system may also allow air to enter, but leaking brake fluid will reveal their presence and warn you of the need for repair.

2 To bleed the brakes, you will need some new DOT 4 brake fluid, a length of clear vinyl or plastic tubing, a small container partially filled with clean brake fluid, some rags and a ring spanner to fit the brake caliper bleed valves.

3 Cover the fuel tank, fairing panels, front mudguard and other painted components to prevent damage in the event that brake fluid is spilled.

4 Remove the reservoir cap, diaphragm plate and diaphragm (see *Daily (pre-ride) checks*) and slowly pump the brake lever or pedal a few times, until no air bubbles can be seen floating up from the holes in the bottom of the reservoir. Doing this bleeds the air from the master cylinder end of the line. Loosely refit the reservoir cover. When working on the rear brake, displace the master cylinder to remove the cap, then remount it without its cap during the procedure to prevent the possibility of spilling its contents **(see illustrations 7.3a and 7.3b)**.

5 Pull the dust cap off the bleed valve **(see illustrations)**. Attach one end of the clear vinyl or plastic tubing to the bleed valve and submerge the other end in the brake fluid in the container **(see illustration)**.

6 Remove the reservoir cap and check the fluid level. Do not allow the fluid level to drop below the lower mark during the bleeding process.

7 Carefully pump the brake lever or pedal three or four times and hold it in (front) or down (rear) while opening the caliper bleed valve. When the valve is opened, brake fluid will flow out of the caliper into the clear tubing and the lever will move toward the handlebar or the pedal will move down.

8 Retighten the bleed valve, then release the brake lever or pedal gradually. Repeat the process until no air bubbles are visible in the brake fluid leaving the caliper and the lever or pedal is firm when applied. On completion, disconnect the bleeding equipment, then tighten the bleed valve to the torque setting specified at the beginning of the chapter and install the dust cap.

9 Install the diaphragm, plate and cap, and when working on the rear brake remount the reservoir **(see illustrations 7.3b and 7.3a)**. Wipe up any spilled brake fluid and check the entire system for leaks.

> **HAYNES HiNT**
> *If it's not possible to produce a firm feel to the lever or pedal the fluid my be aerated. Let the brake fluid in the system stabilise for a few hours and then repeat the procedure when the tiny bubbles in the system have settled out. Also check to make sure that there are no 'high-spots' in the brake hose in which an air bubble can become trapped – this will occur most often in an incorrectly mounted hose union, but can also arise through bleeding the brakes while some of the brake system components are at such an angle to encourage this. Reversing the angle or displacing and moving the offending component around will normally dislodge any trapped air.*

9.5a Front brake caliper bleed valve (arrowed)

9.5b Rear brake caliper bleed valve (arrowed)

9.5c To bleed the brakes, you need a spanner, a short section of clear tubing, and a clear container half-filled with brake fluid

Brakes, wheels and tyres 7•15

Changing the fluid

10 Changing the brake fluid is a similar process to bleeding the brakes and requires the same materials, plus a suitable tool for siphoning the fluid out of the hydraulic reservoir (such as a syringe, though if one isn't available it is no problem to displace the reservoir and tip the fluid out as described in Section 6 or 7). Ensure that your container is large enough to take all the old fluid when it is flushed out of the system.

11 Follow Steps 3, 4 and 5, but after removing the reservoir cap, diaphragm plate and diaphragm siphon or tip the old fluid out of the reservoir (if you want to tip the contents out of the front reservoir displace it from its bracket **(see illustration 6.3b)**). Fill the reservoir with new brake fluid, then follow Step 7.

12 Retighten the bleed valve, then release the brake lever or pedal gradually. Keep the reservoir topped-up with new fluid to above the LOWER level at all times or air may enter the system and greatly increase the length of the task. Repeat the process until new fluid can be seen emerging from the bleed valve.

> **HAYNES HINT**: Old brake fluid is invariably much darker in colour than new fluid, making it easy to see when all old fluid has been expelled from the system.

13 Disconnect the hose, then tighten the bleed valve to the specified torque setting and install the dust cap.

14 Top-up the reservoir, install the diaphragm, plate and cap, and when working on the rear brake remount the reservoir **(see illustrations 7.3b and 7.3a)**. Wipe up any spilled brake fluid and check the entire system for leaks.

15 Check the operation of the brakes before riding the motorcycle.

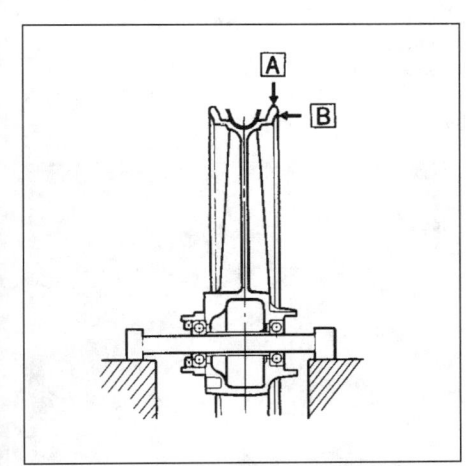

Draining the system

16 Draining the brake fluid is again a similar process to bleeding the brakes. The quickest and easiest way is to use a commercially available vacuum-type brake bleeding tool (see **Note** at the beginning of this section) – follow the manufacturer's instructions. Otherwise follow the procedure described above for changing the fluid, but quite simply do not put any new fluid into the reservoir – the system simply fills itself with air instead.

10 Wheels – inspection and repair

1 Position the motorcycle on an auxiliary stand so that the wheel being checked is raised off the ground. Clean the wheels thoroughly to remove mud and dirt that may interfere with the inspection procedure or mask defects. Make a general check of the wheels (see Chapter 1) and tyres (see *Daily (pre-ride) checks*).

2 To check axial (side-to-side) runout, attach a dial gauge to the fork slider or the swingarm and position its stem against the side of the rim **(see illustration)**. Spin the wheel slowly and check the amount of runout at the rim. To accurately check radial (out of round) runout with the dial gauge, remove the wheel from the machine, and the tyre from the wheel. With the axle clamped in a vice and the dial gauge positioned on the top of the rim, rotate the wheel and check the runout.

3 An easier, though slightly less accurate, method is to attach a stiff wire pointer to the fork slider or the swingarm and position the end a fraction of an inch from the wheel (where the wheel and tyre join). If the wheel is true, the distance from the pointer to the rim will be constant as the wheel is rotated. **Note:** *If wheel runout is excessive, check the wheel bearings and axle very carefully before replacing.*

4 Visually inspect the wheels for cracks, flat spots on the rim, and other damage. Look very closely for dents in the area where the tyre bead contacts the rim. Dents in this area may prevent complete sealing of the tyre against the rim, which leads to deflation of the tyre over a period of time. If damage is evident, or if runout in either direction is excessive, the wheel will have to be replaced with a new one. Never attempt to repair a damaged cast alloy wheel.

11 Wheels – alignment check

1 Misalignment of the wheels, which may be due to a cocked rear wheel or a bent frame or fork yokes, can cause strange and possibly serious handling problems. If the frame or yokes are at fault, repair by a frame specialist or replacement with new parts are the only alternatives.

2 To check the alignment you will need an assistant, a length of string or a perfectly straight piece of wood and a ruler. A plumb bob or other suitable weight will also be required.

3 Place the bike on an auxiliary stand. Measure the width of both tyres at their widest points. Subtract the smaller measurement from the larger measurement, then divide the difference by two. The result is the amount of offset that should exist between the front and rear tyres on both sides.

4 If a string is used, have your assistant hold one end of it about halfway between the floor and the rear axle, touching the rear sidewall of the rear tyre.

5 Run the other end of the string forward and pull it tight so that it is roughly parallel to the floor **(see illustration)**. Slowly bring the string into contact with the front sidewall of the rear tyre, then turn the front wheel until it is parallel with the string. Measure the distance from the front tyre sidewall to the string.

6 Repeat the procedure on the other side of the motorcycle. The distance from the front tyre sidewall to the string should be equal on both sides.

7 As previously mentioned, a perfectly straight length of wood or metal bar may be

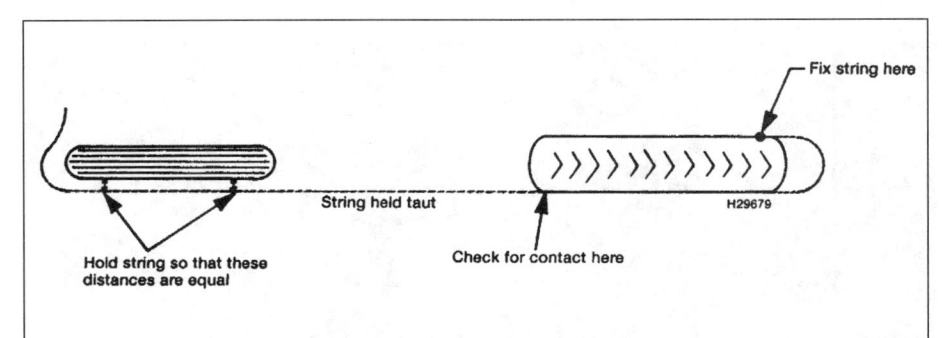

10.2 Check the wheel for radial (out-of-round) runout (A) and axial (side-to-side) runout (B)

11.5 Wheel alignment check using string

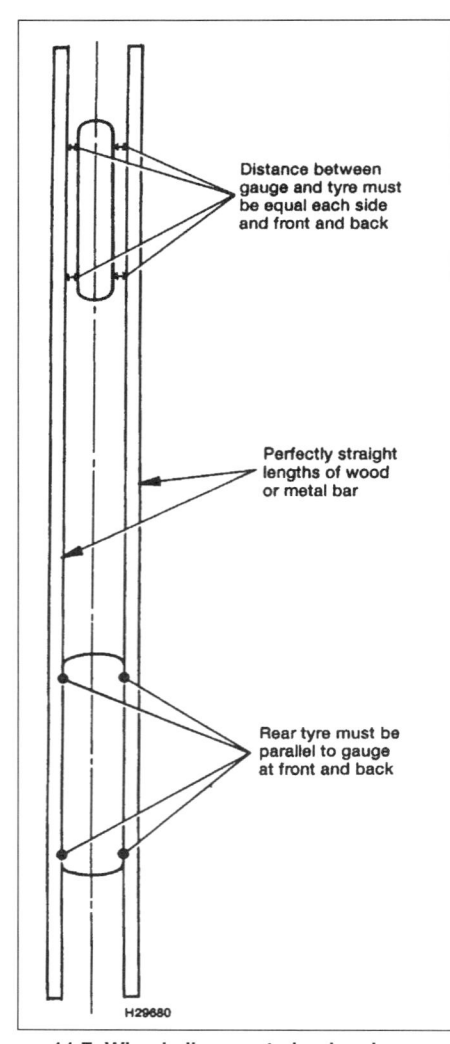

7•16 Brakes, wheels and tyres

11.7 Wheel alignment check using a straight edge

12.4 Slacken the axle clamp bolts (A), then unscrew the axle bolt (B)

12.5a Slacken the axle clamp bolts (arrowed) . . .

substituted for the string **(see illustration)**. The procedure is the same.

8 If the distance between the string and tyre is greater on one side, or if the rear wheel appears to be cocked, refer to Chapter 1 and check that the chain adjuster markings are in the same position on each side of the swingarm.

9 If the front-to-back alignment is correct, the wheels still may be out of alignment vertically.

10 Using a plumb bob, or other suitable weight, and a length of string, check the rear wheel to make sure it is vertical. To do this, hold the string against the tyre upper sidewall and allow the weight to settle just off the floor. When the string touches both the upper and lower tyre sidewalls and is perfectly straight, the wheel is vertical. If it is not, place thin spacers under one leg of the stand until it is.

11 Once the rear wheel is vertical, check the front wheel in the same manner. If both wheels are not perfectly vertical, the frame and/or major suspension components are bent.

12 Front wheel – removal and installation

Removal

1 Position the motorcycle on an auxiliary stand so that the front wheel is off the ground. Always make sure the motorcycle is properly supported. If a support is being placed under the engine, remove the lower fairing (RR-Y and RR-1 (2000 and 2001) models) or fairing side panels (RR-2 and RR-3 (2002 and 2003) models), and the exhaust system to prevent the possibility of damaging them (see Chapters 8 and 4).

2 Displace the front brake calipers (see Section 4). Support the calipers with a cable tie or a bungee cord so that no strain is placed on the hydraulic hoses. There is no need to disconnect the hoses from the calipers. **Note**: *Do not operate the front brake lever with the calipers removed.*

3 If required, remove the front mudguard (see Chapter 8). Whether you need to or not depends on how high the front wheel is off the ground.

4 Slacken the axle clamp bolts on the bottom of the right-hand fork, then unscrew the axle bolt from the right-hand end of the axle **(see illustration)**.

5 Slacken the axle clamp bolts on the bottom of the left-hand fork **(see illustration)**. Take the weight of the wheel, then withdraw the axle from the left-hand side, using a screwdriver inserted through the holes in the end of the axle as a handle, and carefully lower the wheel **(see illustration)**.

6 Remove the shouldered wheel spacer from the right-hand side of the wheel and the plain spacer from the left-hand side **(see illustrations)**.

Caution: Don't lay the wheel down and allow it to rest on a disc – the disc could become warped. Set the wheel on wood blocks so the disc doesn't support the weight of the wheel.

7 Check the axle for straightness by rolling it on a flat surface such as a piece of plate glass (first wipe off all old grease and remove any

12.5b . . . then withdraw the axle and remove the wheel

12.6a Remove the shouldered right-hand spacer . . .

12.6b . . . and the plain left-hand spacer

Brakes, wheels and tyres 7•17

12.12a Install the axle bolt and tighten it to the specified torque . . .

12.12b . . . counter-holding the axle head using a nut as shown

13.2 Unscrew the axle nut and remove the washer and adjustment marker (arrowed)

corrosion using fine emery cloth). If the equipment is available, place the axle in V-blocks and measure the runout using a dial gauge. If the axle is bent or the runout exceeds the limit specified, replace it with a new one.
8 Check the condition of the grease seals and wheel bearings (see Section 14).

Installation

9 Apply a smear of grease to the inside of the wheel spacers, and also to the outside where they fit into the wheel. Fit the shouldered spacer into the right-hand side of the wheel and the plain spacer into the left-hand side **(see illustration 12.6a and 12.6b)**. Each side of the wheel can be identified using the directional arrow cast into one of the spokes near the rim. The arrow denotes the normal direction of wheel rotation.
10 Manoeuvre the wheel into position between the fork sliders, making sure the directional arrows on the tyre and brake discs are pointing in the normal direction of rotation (check both to make sure they concur, especially if you have just had a new front tyre fitted – the tyre could have been fitted the wrong way round). Apply a thin coat of grease to the axle.
11 Lift the wheel into place, making sure the spacers remain in position. Slide the axle in from the left-hand side **(see illustration 12.5b)**.
12 Install the axle bolt and tighten it to the torque setting specified at the beginning of the Chapter **(see illustration)**. Use a 24 mm nut inserted into the left-hand end of the axle and a spanner on the nut to counter-hold it if necessary **(see illustration)**.
13 Tighten the axle clamp bolts on the bottom of the right-hand fork to the specified torque setting **(see illustration 12.4)**.
14 Lower the front wheel to the ground, then install the brake calipers (see Section 4, and the **Note** therein regarding the caliper mounting bolts). If removed, install the front mudguard (see Chapter 8).
15 Apply the front brake a few times to bring the pads back into contact with the discs. Move the motorcycle off its stand, apply the front brake and pump the front forks a few times to settle all components in position. Now tighten the axle clamp bolts on the bottom of the left-hand fork to the specified torque **(see illustration 12.5a)**.
16 Check for correct operation of the brakes before riding the motorcycle. Check that there is at least 0.7 mm clearance between each front brake disc and the caliper bracket – make the check using a feeler gauge.

13 Rear wheel – removal and installation

Removal

1 Position the motorcycle on an auxiliary stand so that the rear wheel is off the ground. Always make sure the motorcycle is properly supported.
2 Unscrew the axle nut and remove the washer and right-hand chain adjustment marker **(see illustration)**.
3 Take the weight of the wheel, then withdraw the axle from the left-hand side, bringing the left-hand chain adjustment marker with it, and lower the wheel to the ground **(see illustration)**. Disengage the chain from the sprocket and remove the wheel from the swingarm **(see illustration)**. If the axle is difficult to withdraw, either drift it through, making sure you don't damage the threads, or preferably create some slack in the chain (see Chapter 1).
4 Note how the rear brake caliper bracket locates against the swingarm, and support it so that it will not fall off. If required, displace it from the swingarm, noting how it fits, and tie it to the passenger footrest bracket, making sure no strain is placed on the hose **(see illustration)**.
Caution: Do not lay the wheel down and allow it to rest on the disc or the sprocket – they could become warped. Set the wheel on wood blocks so the disc or the sprocket doesn't support the weight of the wheel. Do not operate the brake pedal with the wheel removed.
5 Remove the shouldered spacer from the left-hand side of the wheel and the plain

13.3a Withdraw the axle . . .

13.3b . . . then lower the wheel to the ground and disengage the chain

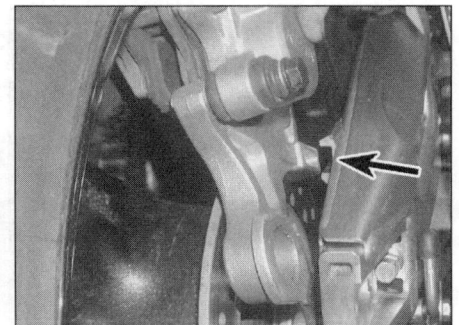

13.4 Note how the caliper bracket locates onto the swingarm (arrow)

7•18 Brakes, wheels and tyres

13.5a Remove the shouldered spacer from the left-hand side . . .

13.5b . . . and the plain spacer from the right-hand side

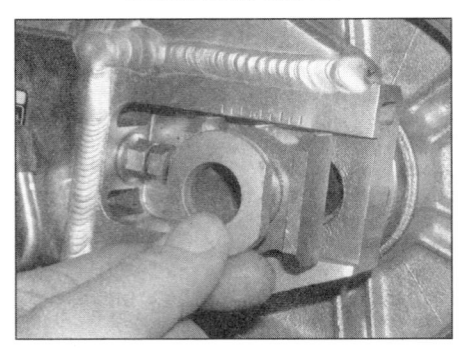

13.11a Locate the axle head as shown

13.11b Tighten the nut to the specified torque

spacer from the right-hand side **(see illustrations)**.

6 Check the axle for straightness by rolling it on a flat surface such as a piece of plate glass (if the axle is corroded, first remove the corrosion with fine emery cloth). If the equipment is available, place the axle in V-blocks and check the runout using a dial indicator. If the axle is bent or the runout exceeds the limit specified at the beginning of the Chapter, replace it with a new one.

7 Check the condition of the grease seals and wheel bearings (see Section 14).

Installation

8 Apply a smear of grease to the inside of the wheel spacers, and also to the outside where they fit into the wheel. Fit the shouldered spacer into the left-hand side of the wheel and the plain spacer into the right-hand side **(see illustration 13.5a and 13.5b)**. If displaced, locate the brake caliper bracket onto the swingarm **(see illustration 13.4)**.

9 Manoeuvre the wheel into position between the ends of the swingarm, making sure the directional arrows on the tyre and wheel are pointing in the normal direction of rotation (check both to make sure they concur, especially if you have just had a new rear tyre fitted – the tyre could have been fitted the wrong way round). Apply a thin coat of grease to the axle. Make sure the brake caliper bracket is still correctly positioned against the swingarm.

10 If removed, slide the left-hand chain adjustment marker onto the axle. Engage the drive chain with the sprocket and lift the wheel into position **(see illustration 13.3b)**. Make sure the spacers and caliper bracket remain correctly in place, and that the brake disc fits squarely into the caliper, with the pads positioned correctly either side of the disc. If there is not enough slack in the chain to align the wheel with the swingarm and chain adjusters, thread the adjuster bolts into the swingarm.

11 Install the axle from the left **(see illustration 13.3a)**, making sure the spacers and caliper bracket remain correctly installed, and locate the flat edges of the axle head between the flat ends of the left-hand adjustment marker **(see illustration)**. Check that everything is correctly aligned. Fit the right-hand adjustment marker onto the end of the axle, then fit the washer and axle nut but leave it loose **(see illustration 13.2)**. Check and adjust the drive chain slack (see Chapter 1). On completion tighten the axle nut to the torque setting specified at the beginning of the Chapter **(see illustration)**.

12 Operate the brake pedal several times to bring the pads into contact with the disc. Check the operation of the rear brake carefully before riding the bike.

14 Wheel bearings – renewal

Front wheel bearings

Note: *Always renew the wheel bearings in pairs. Never renew the bearings individually. Avoid using a high pressure cleaner on the wheel bearing area.*

1 Remove the wheel, and if not already done remove the spacers from it (see Section 12).

2 Rotate the inner race of each bearing to check their operation. If either inner race doesn't turn smoothly, has rough spots or is noisy, replace both bearings with new ones.

3 Set the wheel on blocks so as not to allow the weight to rest on a brake disc. Prise out the seal on each side of the wheel using a flat-bladed screwdriver, taking care not to damage the rim of the hub **(see illustration)**. Discard the seals as new ones must be used.

4 Using a metal rod (preferably a brass drift punch) inserted through the centre of the upper bearing, tap evenly around the inner race of the lower bearing to drive it from the hub **(see illustrations)**. The bearing spacer will also come out.

14.3 Lever out the grease seal on each side

14.4a Knock out the bearings using a drift . . .

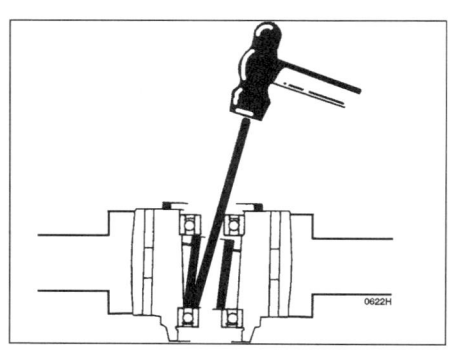

14.4b . . . locating it as shown

Brakes, wheels and tyres 7•19

14.6 A socket can be used to drive in the bearing

14.8a Fit the grease seal and press or tap it into place...

14.8b ...using a piece of wood ensures the seal sits flush with the rim

5 Lay the wheel on its other side so that the remaining bearing faces down. Drive the bearing out of the wheel using the same technique as above.

HAYNES HiNT *Refer to Tools and Workshop Tips for more information about bearings.*

6 Thoroughly clean the hub area of the wheel. Install one new bearing into the recess in the hub, with the marked or sealed side facing outwards. Using the old bearing, a bearing driver or a socket large enough to contact the outer race of the bearing, drive it in until it's completely seated **(see illustration)**.

7 Turn the wheel over and install the bearing spacer. Drive the other new bearing into place as described above.

8 Apply a smear of grease to the lips of the new seals, then press them into the wheel **(see illustration)**. Gently drive them into place if necessary using a seal or bearing driver, a suitable socket or a flat piece of wood **(see illustration)**. As the seals sit flush with the top surface of their housing, using a piece of wood as shown will automatically set them flush without the risk of setting them too deep and having to lever them out again.

9 Clean off all grease from the brake discs using acetone or brake system cleaner then install the wheel (see Section 12).

Rear wheel bearings

Note: *Always renew the wheel bearings in pairs. Never renew the bearings individually. Avoid using a high pressure cleaner on the wheel bearing area.*

10 Remove the rear wheel, and if not already done remove the spacers from it (see Section 13). Lift the sprocket coupling out of the wheel, taking care not to lose the spacer if it is loose **(see illustration)**.

11 Rotate the inner race of each bearing to check their operation. If either inner race doesn't turn smoothly, has rough spots or is noisy, replace both bearings with new ones.

12 Set the wheel on blocks so as not to allow the weight of the wheel to rest on the brake disc.

13 Lever out the grease seal on the right-hand side of the wheel using a flat-bladed screwdriver, taking care not to damage the rim of the hub **(see illustration)**. Discard the seal as a new one should be used.

14 Using a metal rod (preferably a brass drift punch) inserted through the centre of the right-hand bearing, tap evenly around the inner race of the left-hand bearing to drive it

14.10 Lift the sprocket coupling out of the wheel

from the hub **(see illustrations 14.4a and 14.4b)**. The bearing spacer will also come out.

15 Lay the wheel on its other side so that the remaining bearing faces down. Drive the bearing out of the wheel using the same technique as above **(see illustration)**.

16 Thoroughly clean the hub area of the wheel. First install the new right-hand bearing into its recess in the hub, with the marked or sealed side facing outwards. Using the old bearing, a bearing driver or a socket large enough to contact the outer race of the

14.13 Lever out the grease seal

14.15 Drive the bearings out as described

7•20 Brakes, wheels and tyres

14.17 A socket can be used to drive in the bearings

14.18 Check the O-ring and fit a new one if necessary

14.19 Fit the grease seal and press or tap it into place – using a piece of wood ensures the seal sits flush with the rim

14.23 Lever out the grease seal

14.24a Drive the spacer out from the outside, making sure the socket locates on the spacer rim and not on the bearing inner race . . .

14.24b . . . and remove it . . .

14.25 . . . then drive out the bearing

14.27 Support the bearing on a socket when driving in the spacer

bearing, drive it in squarely until it's completely seated.

17 Turn the wheel over and install the bearing spacer. Drive the new left-hand side bearing into place as described above **(see illustration)**.

18 Check the condition of the hub O-ring and replace it with a new one it if it is damaged, deformed or deteriorated **(see illustration)**.

19 Apply a smear of grease to the lips of the new grease seal, and press it into the right-hand side of the wheel, using a seal or bearing driver, a suitable socket or a flat piece of wood to drive it into place if necessary **(see illustration)**. As the seal sits flush with the top surface of its housing, using a piece of wood as shown will automatically set it flush without the risk of setting it too deep and having to lever it out again.

20 Clean off all grease from the brake disc using acetone or brake system cleaner. Fit the sprocket coupling assembly onto the wheel, ensuring that its spacer is in place **(see illustration 14.10)**. Install the wheel (see Section 13).

Sprocket coupling bearing

21 Remove the rear wheel, and if not already done remove the spacers from it (see Section 13). Lift the sprocket coupling out of the wheel, noting how it fits **(see illustration 14.10)**.

22 Rotate the inner race of the bearing to check its operation. If the inner race doesn't turn smoothly, has rough spots or is noisy, a new bearing must be installed.

23 Using a flat-bladed screwdriver, lever out the grease seal from the outside of the coupling **(see illustration)**.

24 Remove the spacer from the inside of the coupling bearing, noting which way round it fits – it should be a tight fit and so will probably have to be driven out using a suitable socket or piece of tubing **(see illustrations)**. Support the coupling on blocks of wood to do this, and make sure the socket or tubing locates only on the spacer rim and not the bearing inner race.

25 Support the coupling on blocks of wood and drive the bearing out from the inside using a bearing driver or socket **(see illustration)**.

26 Thoroughly clean the bearing recess then install the new bearing into the coupling, with the marked or sealed side facing out. Using the old bearing, a bearing driver or a socket large enough to contact the outer race of the bearing, drive it in until it is completely seated.

27 Fit the spacer into the inside of the coupling, making sure it is the correct way round and fits squarely into the bearing **(see illustration 14.24b)**. Drive it into place if it is tight, supporting the inner race of the bearing on a suitable socket as you do to prevent it from being driven out at the same time **(see illustration)**.

28 Check the condition of the hub O-ring and replace it with a new one if it is damaged,

deformed or deteriorated **(see illustration 14.18)**.

29 Apply a smear of grease to the lips of the new seal, and press it into the coupling, using a seal or bearing driver, a suitable socket or a flat piece of wood to drive it into place if necessary **(see illustration)**. As the seal sits flush with the top surface of their housing, using a piece of wood as shown will automatically set it flush without the risk of setting it too deep and having to lever it out again **(see illustration)**.

30 Check the sprocket coupling/rubber dampers (see Chapter 6).

31 Clean off all grease from the brake disc using acetone or brake system cleaner. Fit the sprocket coupling into the wheel **(see illustration 14.10)**, then install the wheel (see Section 13).

15 Tyres – general information and fitting

General information

1 The wheels fitted to all models are designed to take tubeless tyres only. Tyre sizes are given in the Specifications at the beginning of this chapter.

14.29a Fit the grease seal and press or tap it into place . . .

14.29b . . . using a piece of wood ensures the seal sits flush with the rim

2 Refer to the *Daily (pre-ride) checks* listed at the beginning of this manual for tyre maintenance.

Fitting new tyres

3 When selecting new tyres, refer to the tyre information label on the swingarm and the tyre options listed in the owners handbook. Ensure that front and rear tyre types are compatible, the correct size and correct speed rating; if necessary seek advice from a Honda dealer or tyre fitting specialist **(see illustration)**.

4 It is recommended that tyres are fitted by a motorcycle tyre specialist rather than attempted in the home workshop. This is particularly relevant in the case of tubeless tyres because the force required to break the seal between the wheel rim and tyre bead is substantial, and is usually beyond the capabilities of an individual working with normal tyre levers. Additionally, the specialist will be able to balance the wheels after tyre fitting.

5 Refer to the advice in the owner's manual supplied with the motorcycle concerning the repair of punctured tyres and for details of safe reduced speeds when riding on repaired tyres. Also seek advice from a Honda dealer or motorcycle tyre fitting specialist on the tyre's suitability for repair.

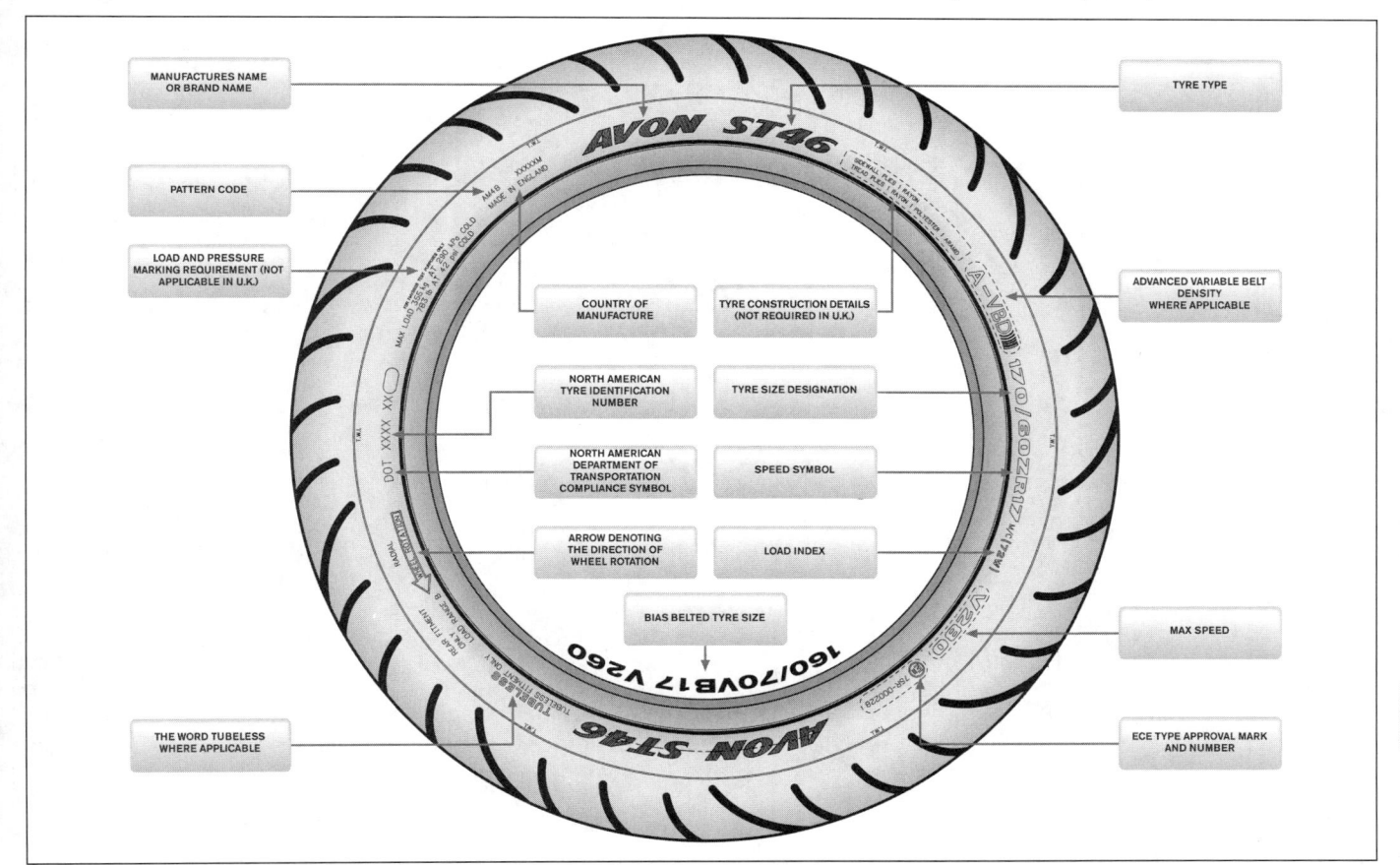

15.3 Common tyre sidewall markings

Chapter 8
Bodywork

Contents

Fairing and body panels – removal and installation 4
Front mudguard – removal and installation 5
General information 1
Seat – removal and installation 2
Rear view mirrors – removal and installation 3
Windshield – removal and installation 6

Degrees of difficulty

Easy, suitable for novice with little experience	Fairly easy, suitable for beginner with some experience	Fairly difficult, suitable for competent DIY mechanic	Difficult, suitable for experienced DIY mechanic	Very difficult, suitable for expert DIY or professional

1 General information

This Chapter covers the procedures necessary to remove and install the body parts. Since many service and repair operations on these motorcycles require the removal of body parts, the procedures are grouped here and referred to from other Chapters.

In the case of damage to the body parts, it is usually necessary to remove the broken component and replace it with a new (or used) one. The material that the body panels are composed of doesn't lend itself to conventional repair techniques. There are however some shops that specialise in 'plastic welding', so it may be worthwhile seeking the advice of one of these specialists before consigning an expensive component to the bin.

When attempting to remove any body panel, first study it closely, noting any fasteners and associated fittings, to be sure of returning everything to its correct place on installation. In some cases the aid of an assistant will be required when removing panels, to help avoid the risk of damage to paintwork. Once the evident fasteners have been removed, try to withdraw the panel as described but DO NOT FORCE IT – if it will not release, check that all fasteners have been removed and try again. Where a panel engages another by means of tabs, be careful not to break the tab or its mating slot or to damage the paintwork. Remember that a few moments of patience at this stage will save you a lot of money in replacing broken fairing panels!

When installing a body panel, first study it closely, noting any fasteners and associated fittings removed with it, to be sure of returning everything to its correct place. Check that all fasteners are in good condition, including all trim nuts or clips and damping/rubber mounts; any of these must be replaced if faulty before the panel is reassembled. Check also that all mounting brackets are straight and repair or replace them if necessary before attempting to install the panel. Where assistance was required to remove a panel, make sure your assistant is on hand to install it. Tighten the fasteners securely, but be careful not to overtighten any of them or the panel may break (not always immediately) due to the uneven stress.

Where trim clips are used, to release them unscrew the centre of the clip, then pull the body of the clip out of the panel **(see illustration)**. When installing them, unscrew the centre of the clip and insert it in the panel then push the centre fully into the body **(see illustration)**. As they are made of plastic, the threads easily become worn in which case the centres may not unscrew. If this happens, lever the centre out of the body using a small screwdriver and replace the trim clip with a new one.

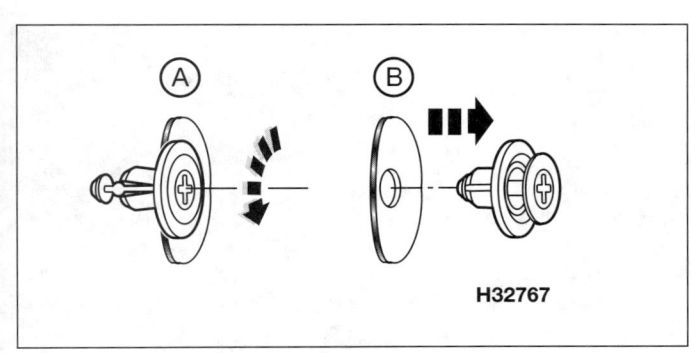

1.5a Unscrew the centre of the clip (A) then pull out the complete clip (B)

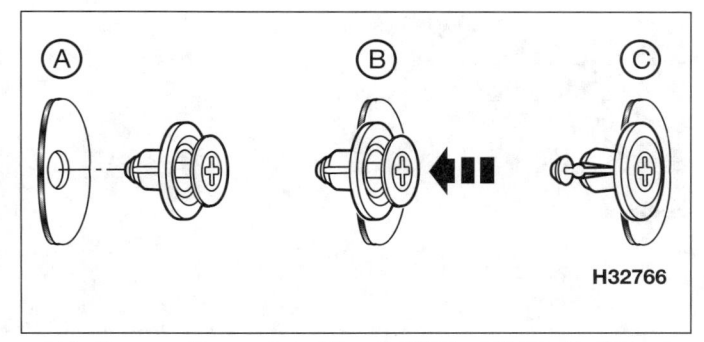

1.5b Unscrew the centre of the clip (A) then fit the clip (B) and push the centre in to lock it (C)

8•2 Bodywork

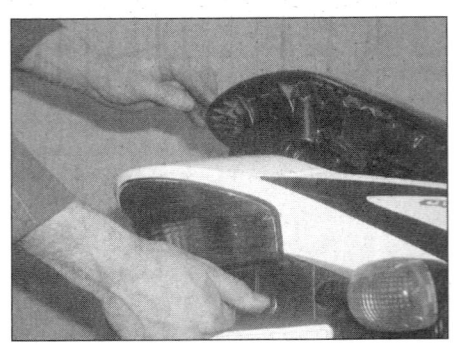

2.1 Unlock the seat and allow it to open – on RR-2 and RR-3 models the lock is at the front of the seat

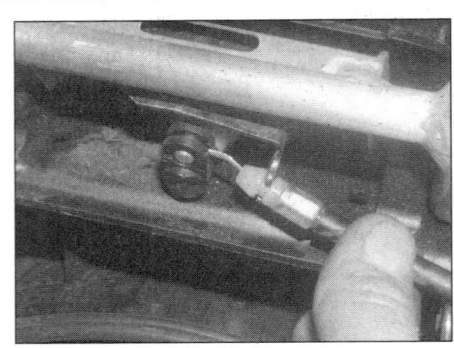

2.2a Free the cable from its holder . . .

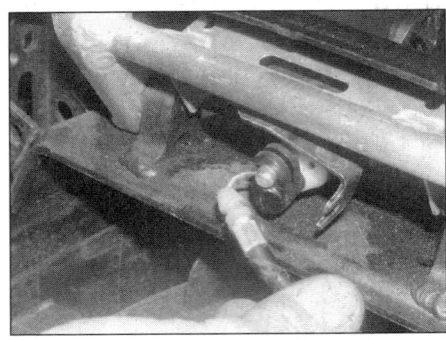

2.2b . . . and detach the end from the latch

2.2c Unscrew the nuts (arrowed) and remove the seat

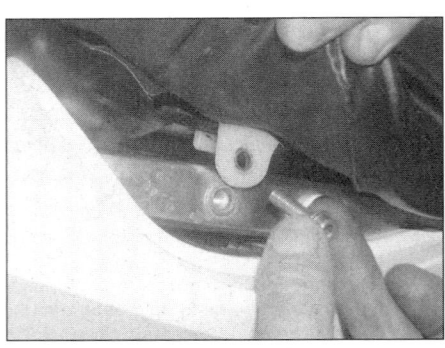

2.3a On RR-Y and RR-1 (2000 and 2001) models the bolts are on the side . . .

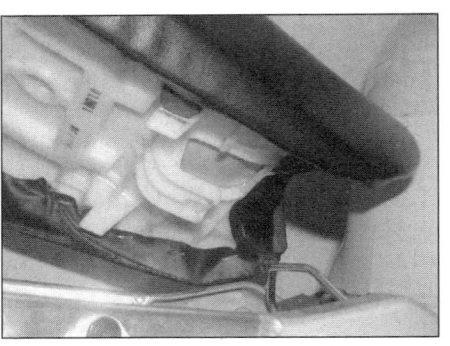

2.3b . . . and the tab locates under the bar on the top of the tank bracket

2 Seat – removal and installation

Removal

Passenger seat

1 To raise the seat, insert the ignition key into the seat lock and turn it clockwise – the seat is spring loaded and will automatically raise **(see illustration)**.

2 To remove the seat, first raise it. On RR-2 and RR-3 (2002 and 2003) models, free the lock cable from its guide and detach the end from the latch **(see illustrations)**. Unscrew the two nuts on the underside of the seat, noting how they also secure the passenger grab-strap, and remove the seat **(see illustration)**.

Rider's seat

3 Lift the seat padding on each rear corner to access the bolts and unscrew them. Draw the seat back and up to remove it, noting how it locates **(see illustrations)**.

Installation

4 Installation is the reverse of removal. Make sure the tab on the front of the rider's seat locates correctly under the bar on the top of the fuel tank bracket on RR-Y and RR-1 (2000 and 2001) models, and under the bracket itself on RR-2 and RR-3 (2002 and 2003) models. Push down on the rear of the passenger seat to engage the latch.

3 Rear view mirrors – removal and installation

1 Unscrew the two bolts and remove the mirror **(see illustration)**. Check the condition of the rubber washers that fit between the fairing and the fairing stay and replace them

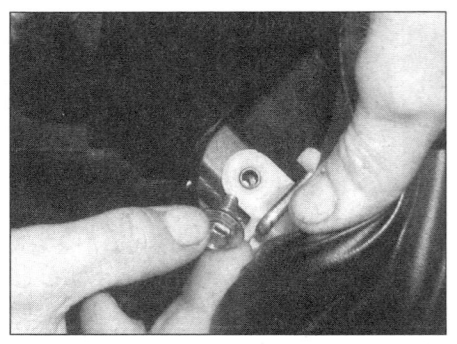

2.3c On RR-2 and RR-3 (2002 and 2003) models the bolts are on the side . . .

2.3d . . . and the tab locates under the tank bracket

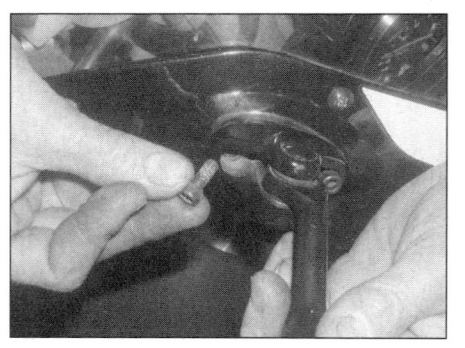

3.1a Unscrew the bolts and remove the mirror

Bodywork 8•3

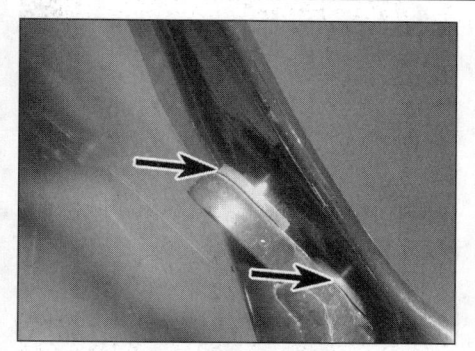

3.1b Check the rubber washers (arrowed) as described

4.2a Remove the fasteners (arrowed) from the right-hand side ...

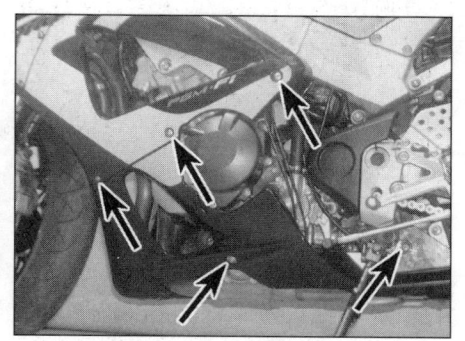

4.2b ... and the left-hand side ...

with new ones if they are damaged, deformed or deteriorated **(see illustration)**.
2 Installation is the reverse of removal.

4 Fairing and body panels – removal and installation

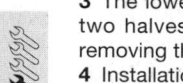

RR-Y and RR-1 models

Lower fairing

1 Remove the top trim clip from each side of the lower fairing front section.
2 Undo the five screws and bolts on each side, then disengage the lower fairing from the fairing side panels and the fairing inner panel at the front and manoeuvre it out from under the bike **(see illustrations)**.
3 The lower fairing can be separated into its two halves and front section if required by removing the various trim clips and screws.
4 Installation is the reverse of removal. Make sure the slots in the top of the front section engage correctly with the tab on each fairing inner panel.

Fairing side panels

5 Remove the bottom trim clip from the fairing inner panel and the top trim clip from each side of the lower fairing front section.
6 Undo the three screws securing the fairing side panel to the lower fairing **(see illustration)**.
7 Undo the three screws securing the fairing side panel to the fairing and the screw securing it to the air duct cover **(see illustration)**.
8 Carefully pull the top of the panel away from the air duct cover to release the peg from the grommet and remove the panel, noting how it engages with the air duct cover, fairing and lower fairing **(see illustrations)**.
9 Installation is the reverse of removal.

Air duct covers

10 Remove the two trim clips securing the inner face of the cover to the fairing inner panel, and the screw securing the cover to the fairing side panel **(see illustration)**.

4.2c ... and manoeuvre the lower fairing out from under the bike

4.6 Undo the three screws (arrowed) ...

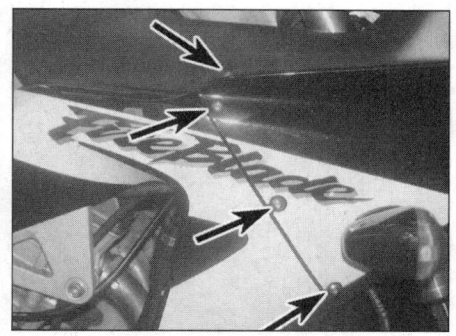

4.7 ... and the four screws (arrowed) ...

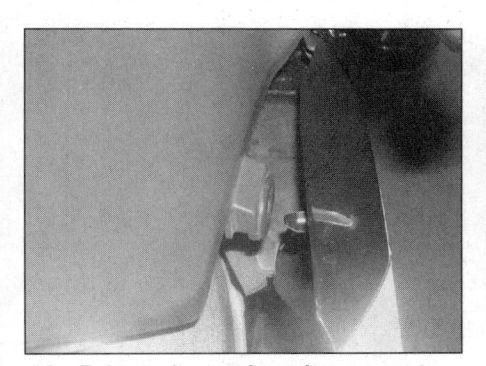

4.8a Release the peg from the grommet ...

4.8b ... and remove the panel

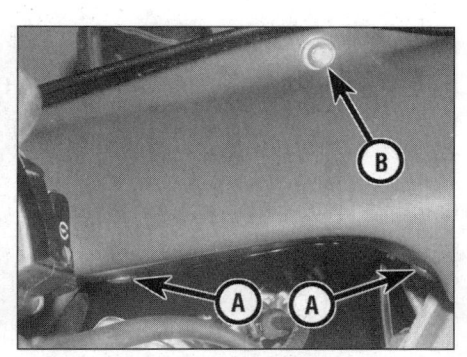

4.10 Release the trim clips (A) and undo the screw (B) ...

8

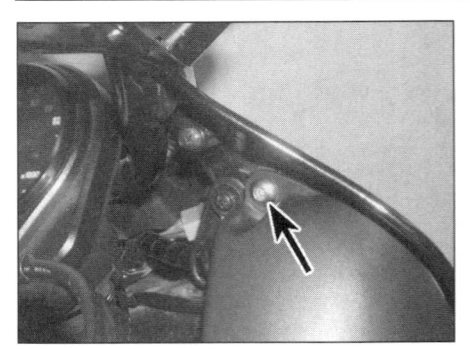

8•4 Bodywork

4.11 . . . and the screw (arrowed)

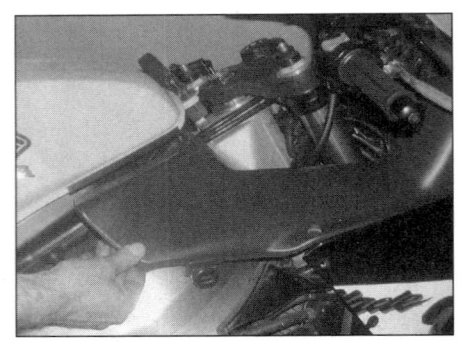

4.12 Remove the cover, noting how it locates

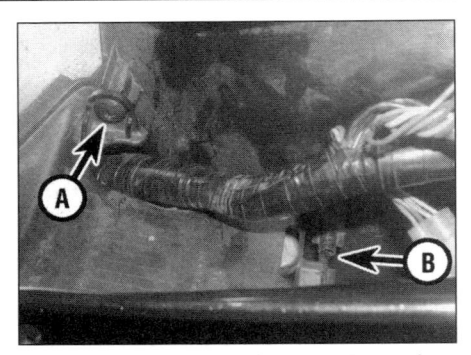

4.15 Undo the screw (A) and release the peg from its socket (B). Also note the routing of the wiring loom

11 Undo the screw securing the air duct cover to the fairing **(see illustration)**.

12 Carefully pull the top of the fairing side panel away from the air duct cover to release the peg from the grommet and remove the air duct cover, noting how it engages with the fairing, fairing side panel and fuel tank **(see illustration 4.8a)**. If it is difficult to disengage the cover from the fuel tank, unscrew the tank front mounting bolts and lift it slightly.

13 Installation is the reverse of removal.

Heat guards

14 Remove the air duct cover (see above).

15 Undo the screw securing the guard to the fairing inner panel and remove the guard, noting how the peg at the front locates in the hole, and on the right-hand side how the wiring loom is routed in the right-hand guard **(see illustration)**.

16 Installation is the reverse of removal.

Fairing

17 Remove the air duct covers and the right-hand heat guard (see above).

18 Remove the trim clip from the fairing inner panel **(see illustration)**. Undo the three screws securing the fairing to the fairing side panel **(see illustration 4.7)**.

19 Remove the windshield (see Section 6).

20 Disconnect the headlight, sidelight and turn signal wiring connectors **(see illustrations)**. Depending on how the sidelight wiring has been routed previously the connector may be difficult to access, in which case disconnect it after the fairing has been displaced and drawn forwards slightly. Free the wiring loom from its clip on the right-hand end of the headlight assembly **(see illustration)**.

21 Remove the rear view mirrors (see Section 3).

22 Undo the screw in the centre of the fairing at the front, noting the washer **(see illustration)**.

23 Carefully draw the fairing forwards off its stay and remove it, noting the routing of the wiring loom on the right-hand side and how the pegs locate in the grommets in the stay

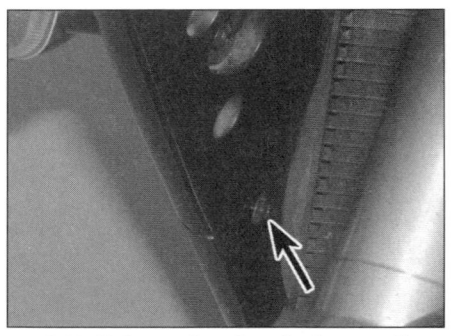

4.18 Release the trim clip (arrowed)

4.20a Disconnect the headlight wiring connectors (arrowed) . . .

4.20b . . . the sidelight wiring connector (when accessible) . . .

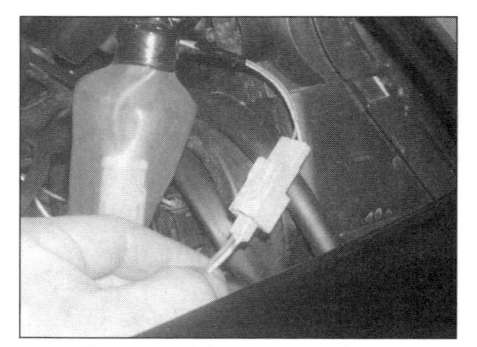

4.20c . . . and the turn signal wiring connector on each side

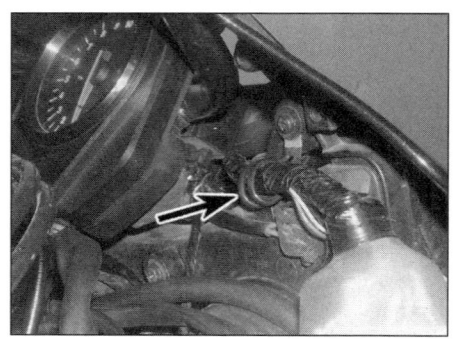

4.20d Release the wiring from the clip (arrowed)

4.22 Undo the screw . . .

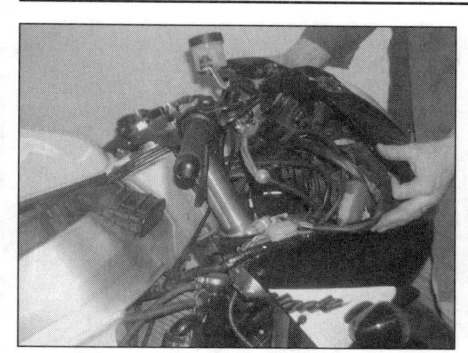

Bodywork 8•5

4.23 ... and draw the fairing off the bike

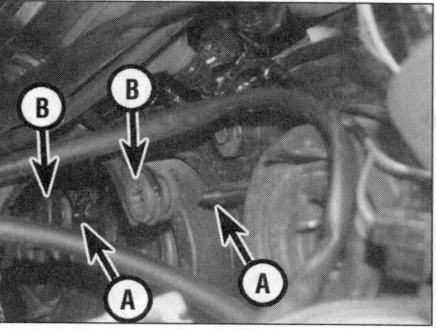

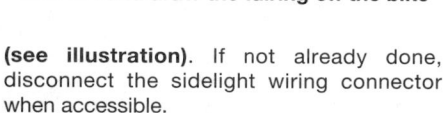

4.24 Make sure the pegs (A) locate in the grommets (B)

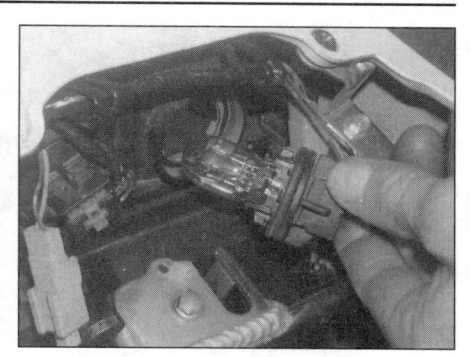

4.26 Release the bulbholders ...

(see illustration). If not already done, disconnect the sidelight wiring connector when accessible.

24 Installation is the reverse of removal. Make sure the fairing pegs locate correctly in the grommets on the fairing stay (see illustration).

Seat cowling

25 Remove both seats (see Section 2).
26 Release the bulbholders from the tail light by twisting them anti-clockwise and drawing them out (see illustration).
27 Remove the two trim clips on each side (see illustration). Undo the two screws at the back (see illustration). Unscrew the bungee-hook bolt on each side, noting the collar and spacer (see illustration).
28 Carefully draw the cowling off the bike, pulling each side out and up in turn to create clearance, and holding the sprung seat base down to clear the middle section – if available have an assistant do this (see illustration).
29 Installation is the reverse of removal.

RR-2 and RR-3 models

Fairing side panels

30 Remove the trim clip from the fairing inner panel (see illustration).
31 Remove the four trim clips securing the fairing side panels together on the underside – you may need to remove accumulated road dirt first.
32 Remove the air duct cover.
33 Undo the two screws securing the fairing side panel to the fairing and the screw securing the top of the panel to the frame (see illustration).

34 Undo the three screws securing the back and bottom of the fairing side panel to the frame (see illustration).

4.27a ... and the trim clips (arrowed) on each side ...

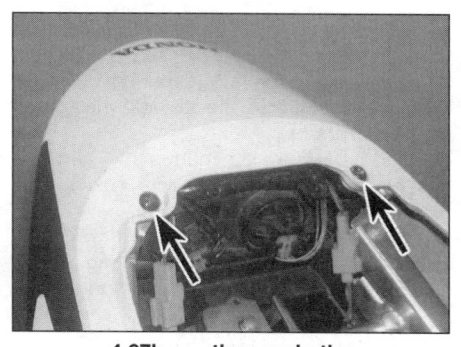

4.27b ... then undo the screws (arrowed) ...

4.27c ... and the bolt on each side ...

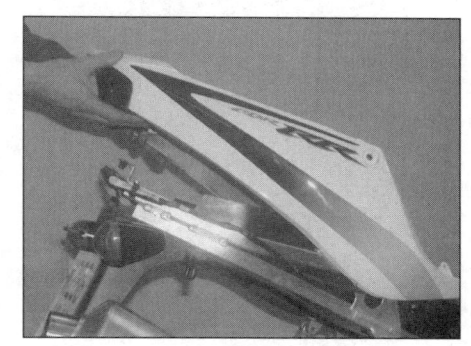

4.28 ... and carefully remove the seat cowling

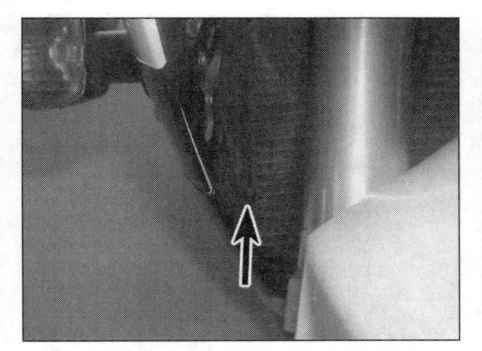

4.30 Release the trim clip (arrowed)

4.33 Undo the screws (arrowed)

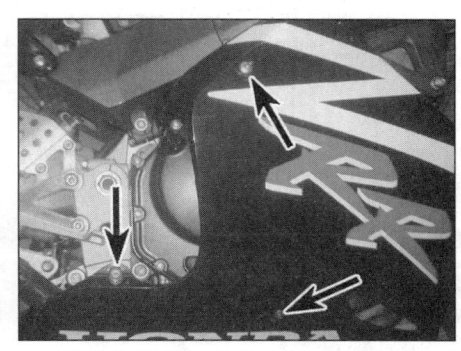

4.34 Undo the screws (arrowed) ...

8•6 Bodywork

4.35a ... and remove the panel ...

4.35b ... noting how it engages

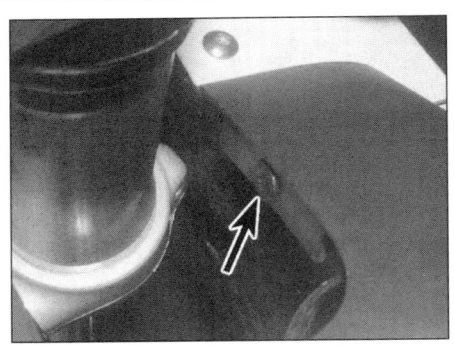

4.37 Release the trim clip (arrowed)

35 Carefully remove the panel, noting how it engages with the fairing **(see illustrations)**.
36 Installation is the reverse of removal.

Air duct covers

37 Remove the trim clip securing the inner face of the cover to the fairing inner panel **(see illustration)**.
38 Undo the screw securing the air duct cover to the fairing and the screw securing it to the fuel tank **(see illustration)**.
39 Carefully remove the air duct cover, noting how it engages with the fairing, fairing side panel and fuel tank **(see illustrations)**. If it is difficult to disengage the cover from the fuel tank, unscrew the tank front mounting bolts and lift it slightly.
40 Installation is the reverse of removal.

Fairing

41 Remove the air duct covers (see above).
42 Remove the trim clip from the fairing inner panel **(see illustration 4.30)**.
43 Undo the middle screw on each side of the windshield, noting the washers **(see illustration)**.
44 Disconnect the headlight, sidelight and turn signal wiring connectors. Depending on how the sidelight wiring has been routed previously the connector may be difficult to access, in which case disconnect it after the fairing has been displaced and drawn forwards slightly.
45 Remove the rear view mirrors (see Section 3).
46 Undo the screw securing each side of the fairing at the front to the fairing stay **(see illustration)**.
47 Carefully draw the fairing forwards off its stay and remove it, noting the routing of the wiring loom on the right-hand side and how the pegs locate in the grommets in the stay **(see illustration)**. If not already done, disconnect the sidelight wiring connector when accessible.

4.38 Undo the screws (arrowed) ...

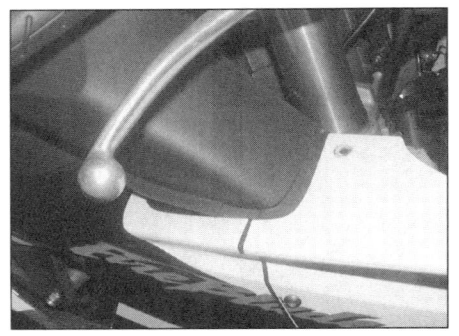

4.39a ... and remove the cover ...

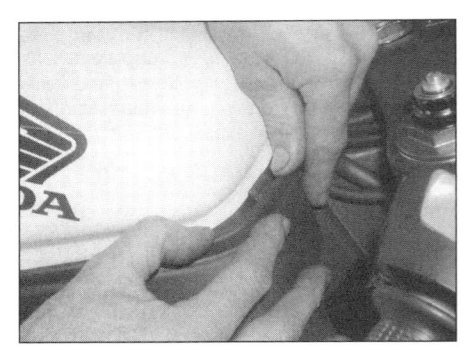

4.39b ... noting how it engages with the tank

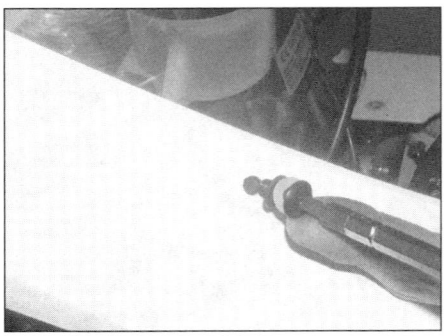

4.43 Undo the screw on each side

4.46 Undo the screw (arrowed) on each side ...

4.47 ... and draw the fairing off the bike

Bodywork 8•7

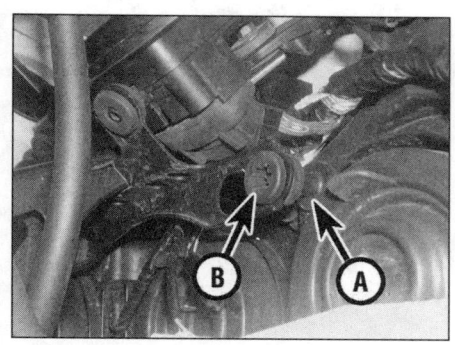

4.48 Make sure the peg (A) on each side locates in the grommet (B)

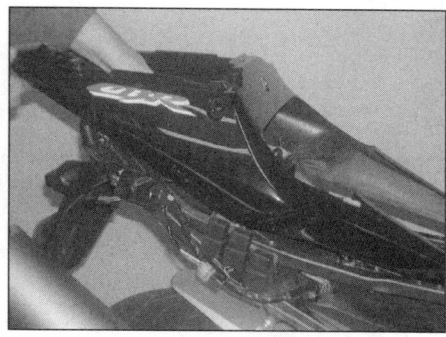

4.52 Carefully remove the seat cowling

48 Installation is the reverse of removal. Make sure the fairing pegs locate correctly in the grommets on the fairing stay, the wiring loom routes correctly over the fairing inner panel, and the inner panel locates correctly around the radiator filler cap **(see illustration)**.

Seat cowling

49 Remove both seats (see Section 2).
50 Disconnect the tail light wiring connector.
51 Remove the two trim clips on each side **(see illustration 4.27a)**. Undo the two screws at the back **(see illustration 4.27b)**. Unscrew the bungee-hook bolt on each side, noting the collar and spacer **(see illustration 4.27c)**.
52 Carefully draw the cowling off the bike, pulling each side out and up in turn to create clearance, and holding the sprung seat base down to clear the middle section – if available have an assistant do this **(see illustration)**.
53 Installation is the reverse of removal.

5 Front mudguard – removal and installation

1 Unscrew the nut securing each hose guide to the front mudguard and displace the hoses **(see illustration)**.
2 Unscrew the two bolts securing each side of the mudguard to each fork slider **(see illustration)**. Draw the mudguard forwards, noting how it fits, and remove it **(see illustration)**.
3 Installation is the reverse of removal.

6 Windshield – removal and installation

1 Undo the screws securing the windshield, noting the various washers. Carefully remove the windshield, noting how it engages with the fairing.
2 Installation is the reverse of removal.

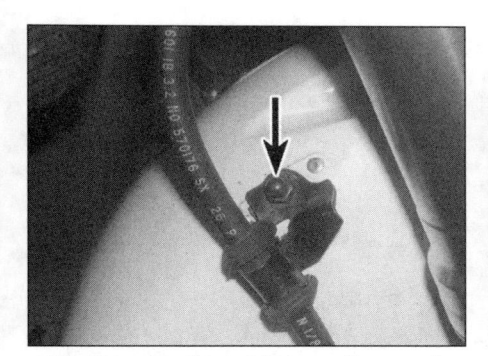

5.1 Unscrew the nut (arrowed) on each side

5.2a Unscrew the bolts (arrowed) . . .

5.2b . . . and remove the mudguard

8

Notes

Chapter 9
Electrical system

Contents

Alternator – check, removal and installation	32
Battery – charging	4
Battery – removal, installation, inspection and maintenance	3
Brake light switches – check and replacement	14
Brake/tail light bulb and license plate bulb – replacement	9
Charging system – leakage and output test	31
Charging system testing – general information and precautions	30
Clutch switch – check and replacement	24
Diode – check and replacement	25
Electrical system – fault finding	2
Fuses – check and replacement	5
General information	1
Handlebar switches – check	20
Handlebar switches – removal and installation	21
Headlight aim – check and adjustment	see Chapter 1
Headlight bulb and sidelight bulb – replacement	7
Headlight assembly – removal and installation	8
Horn – check and replacement	26
Ignition (main) switch – check, removal and installation	19
Ignition system components	see Chapter 5
Instrument and warning light bulbs – replacement	17
Instrument cluster – removal and installation	15
Instruments and speed sensor – check and replacement	16
Lighting system – check	6
Neutral switch – check, removal and installation	22
Oil pressure switch – check, removal and installation	18
Regulator/rectifier – check and replacement	33
Sidestand switch – check and replacement	23
Speed sensor – check and replacement	16
Starter motor – check, disassembly, inspection and reassembly	29
Starter motor – removal and installation	28
Starter relay – check and replacement	27
Tail light assembly – removal and installation	10
Turn signal bulbs – replacement	12
Turn signal assemblies – removal and installation	13
Turn signal circuit – check	11

Degrees of difficulty

Easy, suitable for novice with little experience	Fairly easy, suitable for beginner with some experience	Fairly difficult, suitable for competent DIY mechanic	Difficult, suitable for experienced DIY mechanic	Very difficult, suitable for expert DIY or professional

Specifications

Battery
Capacity .... 12 V, 8.6 Ah
Voltage
  Fully-charged .... 13.0 to 13.2 V
  Uncharged .... below 12.3 V
Charging rate
  Normal .... 0.9 A for 5 to 10 hrs
  Quick .... 4.0 A for 0.5 hr
Current leakage .... 0.2 mA (max)

Alternator
Stator coil resistance .... 0.1 to 1.0 ohms
Output .... 421 W @ 5000 rpm

Regulator/rectifier
Regulated voltage output .... 14.0 to 15.5 V @ 5000 rpm

Starter motor
Brush length
  RR-Y and RR-1 (2000 and 2001) models
    Standard .... 12.0 to 13.0 mm
    Service limit (min) .... 4.5 mm
  RR-2 and RR-3 (2002 and 2003) models
    Standard .... 10.0 to 10.5 mm
    Service limit (min) .... 3.5 mm

Instruments
Tachometer peak voltage (see Text) .... 10.5 V min.

Fuses
RR-Y and RR-1 (2000 and 2001) models
    Main .... 30 A
    PGM-FI (fuel injection system) .... 20 A
    Others .... 10 A x 5, 20 A x 1
RR-2 and RR-3 (2002 and 2003) models
    Main .... 30 A
    PGM-FI (fuel injection system) .... 20 A
    Others .... 10 A x 3, 20 A x 2

Bulbs
Headlights
    HI beam .... 55 W x 2 halogen
    LO beam .... 55 W halogen
Sidelights (where fitted) .... 5.0 W x 2
Brake/tail lights
    RR-Y and RR-1 (2000 and 2001) models .... 21/5 W x 2
    RR-2 and RR-3 (2002 and 2003) models .... LED
License plate light (where fitted) .... 5.0 W
Turn signal lights
    European spec .... 21 W x 4
    US and Canada spec
        Front .... 23/8 W x 2
        Rear .... 21 W x 2
Instrument lights .... LED
Turn signal indicator light .... LED x 2
HI beam indicator light .... LED
Neutral indicator light .... LED
Oil pressure indicator light .... LED
PGM-FI malfunction indicator light .... LED
Immobiliser indicator light .... LED

Torque settings
Alternator rotor bolt .... 103 Nm
Alternator stator bolts .... 12 Nm
Footrest bracket mounting bolts .... 39 Nm
Ignition switch bolts .... 26 Nm
Neutral switch .... 12 Nm
Oil pressure switch .... 12 Nm
Sidestand switch bolt .... 10 Nm
Steering stem nut .... 103 Nm
Top yoke fork clamp bolts
    RR-Y and RR-1 (2000 and 2001) models .... 22 Nm
    RR-2 and RR-3 (2002 and 2003) models .... 23 Nm

1 General information

All models have a 12 volt electrical system charged by a three-phase alternator with a separate regulator/rectifier.

The regulator maintains the charging system output within the specified range to prevent overcharging, and the rectifier converts the ac (alternating current) output of the alternator to dc (direct current) to power the lights and other components and to charge the battery. The alternator rotor is mounted on the left-hand end of the crankshaft.

The starter motor is mounted on the top of the crankcase behind the cylinders on the left-hand side. The starting system includes the motor, the battery, the relay and the various wires and switches. Some of the switches are part of a starter interlock system which prevents the engine from being started if the sidestand is down and the engine is in gear. The engine can be started with the sidestand up when it is in gear as long as the clutch lever is pulled in. The system will also cut the engine should the sidestand extend while the engine is running and in gear – see Chapter 1 for further information and checks on the system.

Note: *Keep in mind that electrical parts, once purchased, often cannot be returned. To avoid unnecessary expense, make very sure the faulty component has been positively identified before buying a replacement part.*

2 Electrical system – fault finding

Warning: To prevent the risk of short circuits, the ignition (main) switch must always be OFF and the battery negative (–ve) terminal should be disconnected before any of the bike's other electrical components are disturbed. Don't forget to reconnect the terminal securely once work is finished or if battery power is needed for circuit testing.

1 A typical electrical circuit consists of an electrical component, the switches, relays, etc, related to that component and the wiring

Electrical system 9•3

and connectors that link the component to the battery and the frame. To aid in locating a problem in any electrical circuit, refer to the wiring diagrams at the end of this Chapter.

2 Before tackling any troublesome electrical circuit, first study the wiring diagram thoroughly to get a complete picture of what makes up that individual circuit. Trouble spots, for instance, can often be narrowed down by noting if other components related to that circuit are operating properly or not. If several components or circuits fail at one time, chances are the fault lies either in the fuse or in the earth (ground) connection, as several circuits are often routed through the same fuse and earth (ground) connections. Earthing points are recognisable by a number of wires all routing into one connector that is either screwed directly to the frame or is secured to the engine by one of its bolts.

3 Electrical problems often stem from simple causes, such as loose or corroded connections or a blown fuse. Prior to any electrical fault finding, always visually check the condition of the fuse, wires and connections in the problem circuit. Intermittent failures can be especially frustrating, since you can't always duplicate the failure when it's convenient to test. In such situations, a good practice is to clean all connections in the affected circuit, whether or not they appear to be good. All of the connections and wires should also be wiggled to check for looseness which can cause intermittent failure.

4 If testing instruments are going to be utilised, use the wiring diagram to plan where you will make the necessary connections in order to accurately pinpoint the trouble spot.

5 The basic tools needed for electrical fault finding include a battery and bulb test circuit or a continuity tester, a test light, and a jumper wire. A multimeter capable of reading volts, ohms and amps is a very useful alternative and performs the functions of all of the above, and is necessary for performing more extensive tests and checks where specific voltage, current or resistance values are needed.

> **HAYNES HiNT** *Refer to Fault Finding Equipment in the Reference section for details of how to use electrical test equipment.*

3 Battery – removal, installation, inspection and maintenance

Caution: Be extremely careful when handling or working around the battery. The electrolyte is very caustic and an explosive gas (hydrogen) is given off when the battery is charging.

Removal and installation

1 Make sure the ignition is switched OFF. Remove the rider's seat (see Chapter 8).
2 Release the battery strap **(see illustration)**.
3 Unscrew the negative (–ve) terminal bolt first and disconnect the lead from the battery **(see illustration)**. Lift up the red insulating cover to access the positive (+ve) terminal, then unscrew the bolt and disconnect the lead. Lift the battery from the bike **(see illustration)**.
4 On installation, clean the battery terminals and lead ends with a wire brush or knife and emery paper. Reconnect the leads, connecting the positive (+ve) terminal first.

> **HAYNES HiNT** *Battery corrosion can be kept to a minimum by applying a layer of petroleum jelly to the terminals after the cables have been connected. There are also dedicated sprays commercially available.*

5 Fit the battery strap. Install the seat (see Chapter 8).

Inspection and maintenance

6 The battery fitted to all models covered in this manual is of the maintenance free (sealed) type, therefore requiring no scheduled maintenance. However, the following checks should still be regularly performed.

7 Check the battery terminals and leads for tightness and corrosion. If corrosion is evident, unscrew the terminal screws and disconnect the leads from the battery, disconnecting the negative (–ve) terminal first, and clean the terminals and lead ends with a wire brush or knife and emery paper. Reconnect the leads, connecting the negative (–ve) terminal last, and apply a thin coat of petroleum jelly to the connections to slow further corrosion.

8 Keep the battery case clean to prevent current leakage, which can discharge the battery over a period of time (especially when it sits unused). Wash the outside of the case with a solution of baking soda and water. Rinse the battery thoroughly, then dry it.

9 Look for cracks in the case and replace the battery with a new one if any are found. If acid has been spilled on the frame or battery box, neutralise it with a baking soda and water solution, dry it thoroughly, then touch up any damaged paint.

10 If the motorcycle sits unused for long periods of time, disconnect the cables from the battery terminals, negative (–ve) terminal first. Refer to Section 4 and charge the battery once every month to six weeks.

11 Check the condition of the battery by measuring the voltage present at the battery terminals. Connect the voltmeter positive (+ve) probe to the battery positive (+ve) terminal, and the negative (–ve) probe to the battery negative (–ve) terminal. When fully-charged there should be 13.0 to 13.2 volts present. If the voltage falls below 12.3 volts remove the battery (see above), and recharge it as described below in Section 4.

4 Battery – charging

Caution: Be extremely careful when handling or working around the battery. The electrolyte is very caustic and an explosive gas (hydrogen) is given off when the battery is charging.

3.2 Release the battery strap

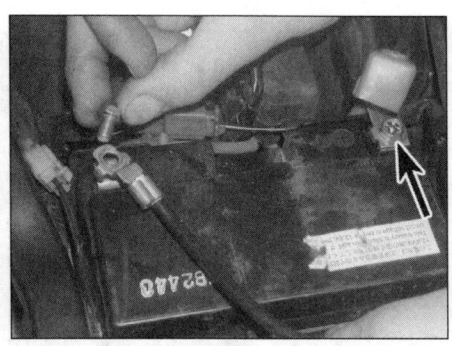

3.3a Disconnect the negative lead first, then disconnect the positive lead (arrowed) . . .

3.3b . . . and remove the battery

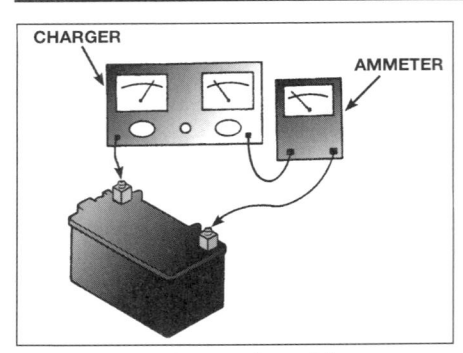

9•4 Electrical system

4.2 If the charger doesn't have an ammeter built in, connect one in series as shown. DO NOT connect the ammeter between the battery terminals or it will be ruined

1 Remove the battery (see Section 3). Connect the charger to the battery, making sure that the positive (+ve) lead on the charger is connected to the positive (+ve) terminal on the battery, and the negative (–ve) lead is connected to the negative (–ve) terminal.
2 Honda recommend that the battery is charged at the normal rate specified at the beginning of the Chapter. Exceeding this figure can cause the battery to overheat, buckling the plates and rendering it useless. Few owners will have access to an expensive current controlled charger, so if a normal domestic charger is used check that after a possible initial peak, the charge rate falls to a safe level **(see illustration)**. If the battery becomes hot during charging **stop**. Further charging will cause damage. **Note:** *In emergencies the battery can be charged at the quick rate specified. However, this is not recommended and the normal charging rate is by far the safer method of charging the battery.*
3 If the recharged battery discharges rapidly if left disconnected it is likely that an internal short caused by physical damage or sulphation has occurred. A new battery will be required. A sound item will tend to lose its charge at about 1% per day.
4 Install the battery (see Section 3).
5 If the motorcycle sits unused for long periods of time, charge the battery once every month to six weeks and leave it disconnected.

5 Fuses – check and replacement

1 The electrical system is protected by fuses of different ratings. All except the main fuse, and on RR-Y and RR-1 (2000 and 2001) models the PGM-FI (fuel injection system) fuse, are housed in the fusebox, which is located under the seat behind the battery **(see illustration)**. The main fuse is integral with the starter relay, which is also behind the battery. On RR-Y and RR-1 (2000 and 2001) models the PGM-FI fuel injection system fuse is located in its own housing next to the fusebox.
2 To access all fuses, remove the rider's seat (see Chapter 8). To access the fusebox fuses unclip the fusebox lid **(see illustration)**. To access the main fuse, disconnect the starter relay wiring connector **(see illustration)**. On RR-Y and RR-1 (2000 and 2001) models to access the PGM-FI fuse unclip the fuseholder cap **(see illustration)**.
3 The fuses can be removed and checked visually. If you can't pull the fuse out with your fingertips, use a pair of suitable pliers **(see illustration)**. A blown fuse is easily identified by a break in the element **(see illustration)**. Each fuse is clearly marked with its rating and must only be replaced by a fuse of the correct rating. A spare fuse of each rating except the main fuse is housed in the fusebox, and a spare main fuse is housed in the starter relay holder **(see illustration 5.2b)**. If a spare fuse is used, always replace it so that a spare of each rating is carried on the bike at all times.

⚠ **Warning:** *Never put in a fuse of a higher rating or bridge the terminals with any other substitute, however temporary it may be. Serious damage may be done to the circuit, or a fire may start.*

4 If the new fuse blows immediately check the wiring circuit very carefully for evidence of a short-circuit. Look for bare wires and chafed, melted or burned insulation.

5.1 The fusebox (arrowed) is behind the battery

5.2a Unclip the lid to access the fuses

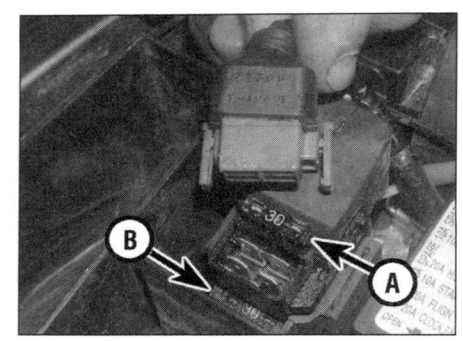

5.2b Remove the starter relay wiring connector to access the main fuse (A). Spare main fuse (B)

5.2c PGM-FI fuse holder (arrowed) – RR-Y and RR-1 (2000 and 2001) models

5.3a Remove the fuse to check it

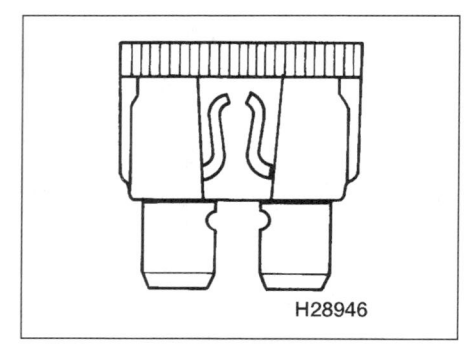

5.3b A blown fuse can be identified by a break in its element

Electrical system 9•5

5 Occasionally a fuse will blow or cause an open-circuit for no obvious reason. Corrosion of the fuse ends and fusebox terminals may occur and cause poor fuse contact. If this happens, remove the corrosion with a wire brush or emery paper, then spray the fuse end and terminals with electrical contact cleaner.

6 Lighting system – check

1 The battery provides power for operation of the headlight, tail light, brake light and instrument cluster lights. If none of the lights operate, always check battery voltage before proceeding. Low battery voltage indicates either a faulty battery or a defective charging system. Refer to Section 3 for battery checks and Sections 30 and 31 for charging system tests. Also, check the condition of the fuses. Note that if there is more than one problem at the same time, it is likely to be a fault relating to a multi-function component, such as one of the fuses governing more than one circuit, or the ignition switch.

Headlight

2 All models have three single filament bulbs – the outer bulbs work on HI beam, and the centre bulb works on LO beam. If either or both headlight beams fail to work, first check the fuse (see Section 5), and then the bulb(s) (see Section 7). If they are good, use jumper wires to connect the bulb in question directly to the battery terminals. If the light comes on, the problem lies in the wiring or connectors, the HI beam relay or the switches in the circuit. Refer to Section 20 for the switch testing procedures, and also to the wiring diagrams at the end of this Chapter.

3 If the HI beams do not work and the relay is suspected of being faulty, the easiest way to tell is to substitute it with another one, if available. On RR-Y and RR-1 (2000 and 2001) models remove the fairing to access the relay – it is mounted in front of the instrument cluster on the left-hand side **(see illustration)**. On RR-2 and RR-3 (2002 and 2003) models remove the seat cowling to access the relay – it is the rear relay on the right-hand side **(see illustration)**. If the beam then works, the old relay is faulty. If a substitute is not available, remove the suspect one and test it as follows: set a multimeter to the ohms x 1 scale and connect it across the relay's blue/black and black/red wire terminals. There should be no continuity (infinite resistance). Using a fully-charged 12 volt battery and two insulated jumper wires, connect the positive (+ve) terminal of the battery to the blue wire terminal on the relay, and the negative (–ve) terminal to the green wire terminal. At this point the relay should be heard to click and the meter read 0 ohms (continuity). If this is the case the relay is good. If the relay does not click when battery voltage is applied and indicates no continuity (infinite resistance) across its terminals, it is faulty and must be replaced with a new one.

4 If the relay is good, check for battery voltage at the black/red wire terminal on the relay wiring connector with the ignition ON. If there is no voltage, check the wiring between the relay wiring connector and the ignition switch, via the fusebox, then check the switch itself (see Section 19). If voltage is present, check that there is continuity to the headlight wiring connector in the blue/black wire, and continuity to earth (ground) in the green (RR-Y and RR-1 (2000 and 2001) models) or black (RR-2 and RR-3 (2002 and 2003) models) wire from the headlight connector. Also check for battery voltage at the blue wire terminal on the relay wiring connector with the ignition ON, the light switch ON and the dimmer switch set to HI. If voltage is present, check for continuity to earth (ground) in the green wire from the relay wiring connector. Repair or renew the wiring or connectors as necessary.

5 If the LO beam does not work, check for battery voltage at the white (RR-Y and RR-1 (2000 and 2001) models) or grey (RR-2 and RR-3 (2002 and 2003) models) wire terminal on the headlight wiring connector with the ignition ON, the light switch ON and the dimmer switch set to LO. If voltage is present, check for continuity to earth (ground) in the green (RR-Y and RR-1 (2000 and 2001) models) or black (RR-2 and RR-3 (2002 and 2003) models) wire from the relay wiring connector. Repair or renew the wiring or connectors as necessary.

Tail light

RR-Y and RR-1 models

6 If both tail light bulbs fail to work, first check the fuse (see Section 5). If the fuse is good, or if only one bulb fails, remove the bulb(s) (see Section 9), and test it/them by connecting the bulb in question directly to the battery terminals using jumper wires. Alternatively use a continuity tester to check for continuity in the filament. In either case make sure you use the correct terminals on the bulb – you are checking between the brown/white and green wire terminals (match them from the connector). If the bulb comes on or the filament shows continuity, the problem lies in the wiring or connectors, or the switches in the circuit. Otherwise the bulb is faulty and must be replaced with a new one.

7 If the bulb is good, check for battery voltage at the brown/white wire terminal on each tail light wiring connector with the ignition and light switches ON. If voltage is present, check for continuity to earth (ground) in the green wire from the wiring connector. If no voltage is indicated, check the wiring and connectors in the tail light circuit, via the fusebox and the handlebar switch. Refer to Sections 19 and 20 for the switch testing procedures, and also to the wiring diagrams at the end of this Chapter.

RR-2 and RR-3 models

8 If the tail light fails to work, first check the fuse (see Section 5). If the fuse is good, disconnect the tail light wiring connector and check for battery voltage at the brown/white wire terminal on the connector with the ignition and light switches ON. If voltage is present, check for continuity to earth (ground) in the green wire from the wiring connector. If no voltage is indicated, check the wiring and connectors between the tail light and the ignition switch, via the fusebox and the handlebar switch, then check the ignition switch itself (see Section 19). Refer to the wiring diagrams at the end of this Chapter.

9 If the power, wiring and connectors are good, or if only one or some of the tail light LEDs have failed leaving others working, then the tail light unit is faulty and must be replaced with a new one – individual LEDs are not available.

Brake light

RR-Y and RR-1 models

10 If both brake light bulbs fail to work, first check the fuse (see Section 5). If the fuse is good, or if only one bulb fails, remove the bulb(s) (see Section 9), and test it/them by connecting the bulb in question directly to the battery terminals using jumper wires. Alternatively use a continuity tester to check for continuity in the filament. In either case make sure you use the correct terminals on the bulb – you are checking between the green/yellow and green wire terminals (match them from the connector). If the light comes

6.3a Headlight relay (arrowed) – RR-Y and RR-1 (2000 and 2001) models

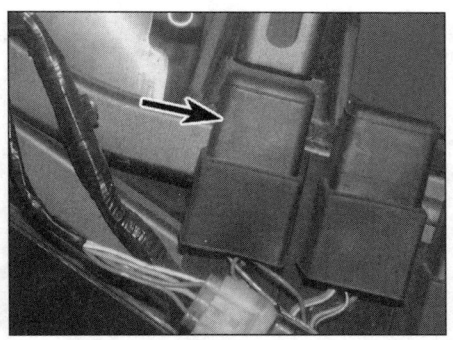

6.3b Headlight relay (arrowed) – RR-2 and RR-3 (2002 and 2003) models

on or the filament shows continuity, the problem lies in the wiring or connectors, or the switches. Otherwise the bulb is faulty and must be replaced with a new one.

11 If the bulb is good, check for battery voltage at the green/yellow wire terminal on each tail light wiring connector, first with the front brake lever on, then with the rear brake pedal on. If voltage is present with one brake on but not the other, then the switch or its wiring is faulty. If voltage is present in both cases, check for continuity to earth (ground) in the green wire from the wiring connectors. If no voltage is indicated, check the wiring and connectors between the brake light and the brake switches, the fusebox, and the ignition switch, then check the switches themselves. Refer to Section 14 for the switch testing procedures, and also to the wiring diagrams at the end of this Chapter.

RR-2 and RR-3 models

12 If the brake light fails to work, first check the fuse (see Section 5). If the fuse is good, disconnect the tail light wiring connector and check for battery voltage at the green/yellow wire terminal on the connector, first with the front brake lever on, then with the rear brake pedal on. If voltage is present with one brake on but not the other, then the switch or its wiring is faulty. If voltage is present in both cases, check for continuity to earth (ground) in the green wire from the wiring connector. If no voltage is indicated, check the wiring and connectors between the brake light and the brake switches, the fusebox, and the ignition switch, then check the switches themselves. Refer to Section 14 for the switch testing procedures, and also to the wiring diagrams at the end of this Chapter.

13 If the power, wiring and connectors are good, or if only one or some of the brake light LEDs have failed leaving others working, then the tail light unit is faulty and must be replaced with a new one – individual LEDs are not available.

License plate light – RR-2 and RR-3 models

14 If the license plate light bulb fails to work, first check the fuse (see Section 5). If the fuse is good, remove the bulb (see Section 9), and test it by connecting it directly to the battery terminals using jumper wires. Alternatively use a continuity tester to check for continuity in the filament. If the bulb comes on or the filament shows continuity, the problem lies in the wiring or connectors. Otherwise the bulb is faulty and must be replaced with a new one.

15 If the bulb is good, check for battery voltage at the brown/white wire terminal on the wiring connector with the ignition and light switches ON. If voltage is present, check for continuity to earth (ground) in the green wire from the wiring connector. If no voltage is indicated, check the wiring and connectors in the tail light circuit. If other lights in the system have not come on as well, check the wiring via the fusebox and the handlebar switch, and check the switches themselves. Refer to Sections 19 and 20 for the switch testing procedures, and also to the wiring diagrams at the end of this Chapter.

Instrument and warning lights

10 See Section 17.

Turn signals

11 See Section 11.

7 Headlight bulb and sidelight bulb – replacement

Note: *The headlight bulbs are of the quartz-halogen type. Do not touch the bulb glass as skin acids will shorten the bulb's service life. If the bulb is accidentally touched, it should be wiped carefully when cold with a rag soaked in methylated spirit and dried before fitting.*

⚠ **Warning: Allow the bulb time to cool before removing it if the headlight has just been on.**

Headlight

1 Reaching behind the back of the headlight, disconnect the wiring connector from the bulb in question, then remove the rubber dust cover, noting how it fits **(see illustrations)**.

2 Release the bulb retaining clip, noting how it fits, then remove the bulb **(see illustrations)**.

3 Carefully pull the bulb off the socket adapter, then fit the new bulb onto the adapter, bearing in mind the information in the **Note** above **(see illustration)**.

4 Fit the new bulb into the headlight, making sure it locates correctly, and secure it in position with the retaining clip.

> **HAYNES HiNT** *Always use a paper towel or dry cloth when handling new bulbs to prevent injury if the bulb should break and to increase bulb life.*

5 Install the dust cover, making sure it is correctly seated and with the TOP mark at the top, and connect the wiring connector.

6 Check the operation of the headlight.

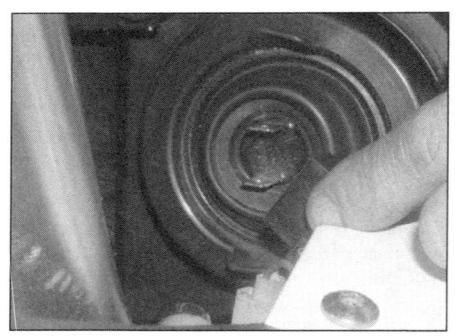

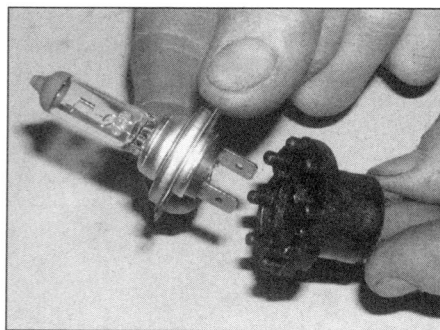

7.1a Disconnect the wiring connector...

7.1b ...and remove the dust cover

7.2a Release the clip...

7.2b ...and remove the bulb

7.3 Separate the bulb from its adapter and fit a new one

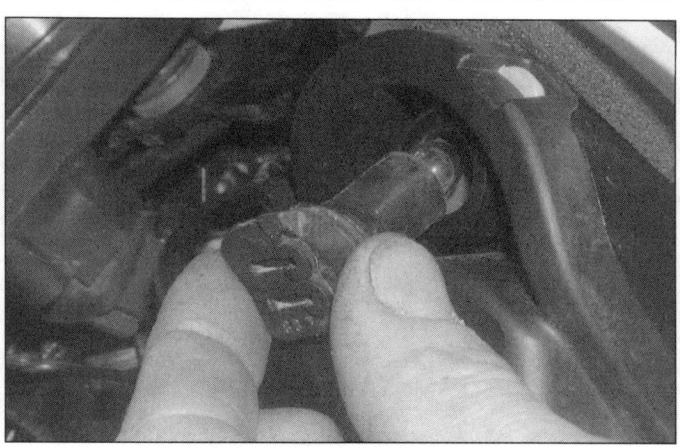

Electrical system 9•7

7.7a Draw out the bulbholder . . .

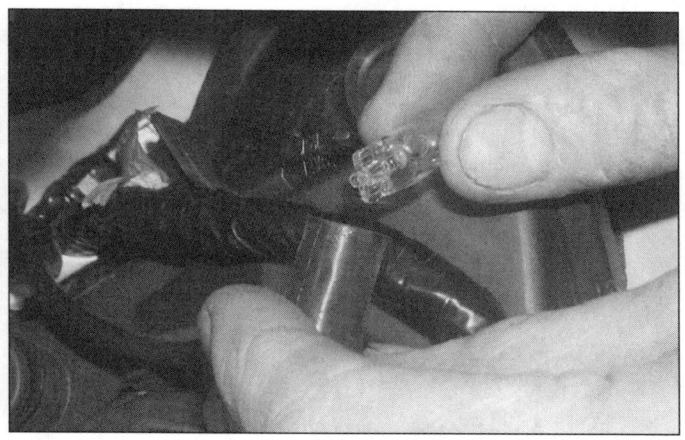

7.7b . . . and remove the bulb

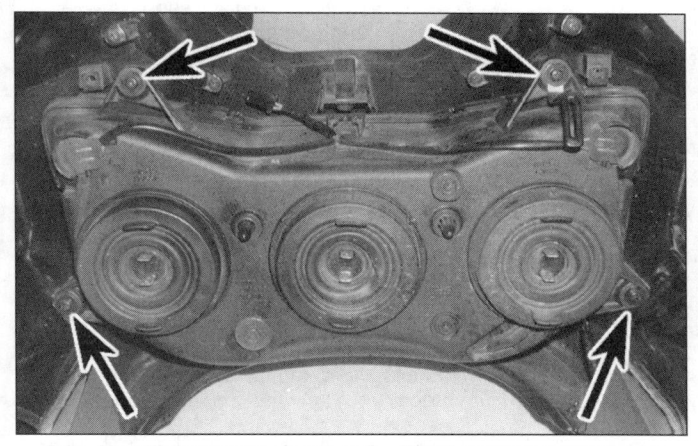

8.2a Headlight screws (arrowed) – RR-Y and RR-1 (2000 and 2001) models

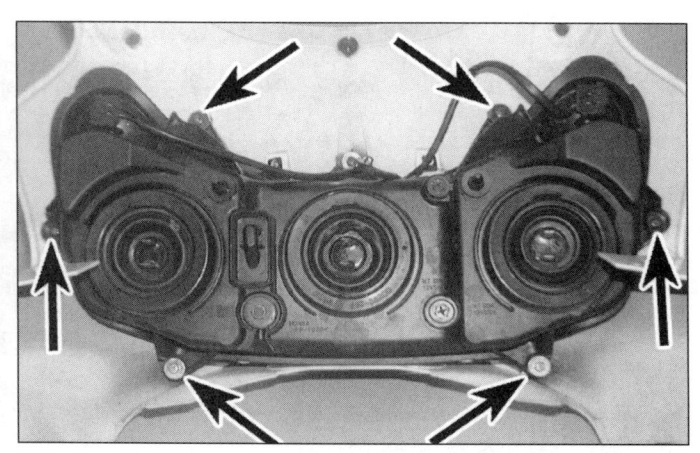

8.2b Headlight screws (arrowed) – RR-2 and RR-3 (2002 and 2003) models

Sidelight

7 Pull the bulbholder out of its socket in the side of the headlight **(see illustration)**. Carefully pull the bulb out of the holder **(see illustration)**.
8 Fit the new bulb in the bulbholder, then install the holder. Make sure it is correctly seated.
9 Check the operation of the sidelight.

8 Headlight assembly – removal and installation

Removal

1 Remove the fairing (see Chapter 8). Pull the sidelight bulbholders out of the headlight.
2 Undo the screws securing the headlight assembly to the fairing and lift it out **(see illustrations)**.

Installation

3 Installation is the reverse of removal. Make sure all the wiring is correctly connected and secured. Check the operation of the headlight and sidelight. Check the headlight aim (see Chapter 1).

9 Brake/tail light bulb and license plate bulb – replacement

Note: *It is a good idea to use a paper towel or dry cloth when handling the new bulb to prevent injury if it breaks, and to increase bulb life.*

Brake/tail light bulb – RR-Y and RR-1 models

1 Raise the passenger seat (see Chapter 8).
2 Turn the bulbholder anti-clockwise and withdraw it from the tail light **(see illustration)**.
3 Carefully pull the bulb out of the socket **(see illustration)**. Check the socket terminals for corrosion and clean them if necessary. Install the new bulb by pushing

9.2 Remove the bulbholder from the taillight . . .

9.3 . . . and the bulb from the holder

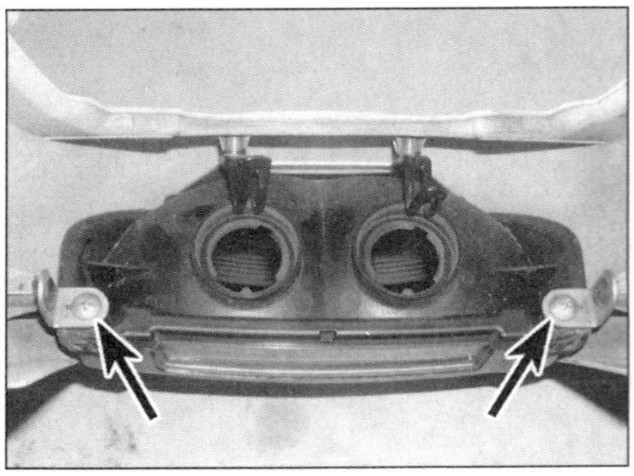

9•8 Electrical system

10.2a Undo the screws (arrowed) and remove the light . . .

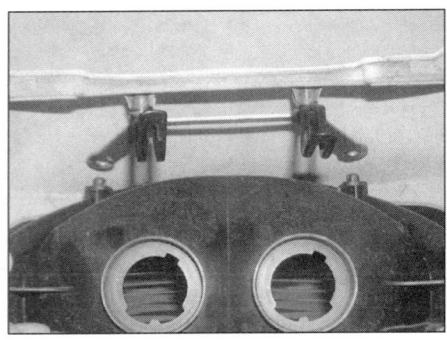

10.2b . . . noting how the pegs locate

10.3 Undo the screw (arrowed) on each side

it into the socket – it can be installed either way round.
4 Fit the bulbholder into the tail light and turn it clockwise to secure it.
5 Lower the seat.

Brake/tail light LEDs – RR-2 and RR-3 models

6 If one or more of the LEDs within the brake or tail light unit has failed, replace the entire tail light assembly with a new one – individual LEDs are not available (see Section 10).

License plate light bulb – RR-2 and RR-3 models

7 Undo the screws securing the lens and remove it.
8 Carefully pull the bulb out of its socket and replace it with a new one.
9 Check that the rubber seal is correctly seated, and is not damaged, deformed or deteriorated. Fit a new one if necessary.
10 Fit the lens back onto the light and secure it with the screws – do not overtighten them as it is easy to crack the lens or strip the threads.

10 Tail light assembly – removal and installation

Removal

1 Remove the seat cowling (see Chapter 8).
2 On RR-Y and RR-1 (2000 and 2001) models undo the two screws securing the tail light to its side brackets in the seat cowling and remove it, noting how the pegs on the top locate in the holes in the top bracket **(see illustrations)**. Remove the mounting brackets if required by undoing their screws.
3 On RR-2 and RR-3 (2002 and 2003) models undo the two screws securing the tail light side brackets in the seat cowling and draw the taillight forwards, noting how the pegs on the back locate in the grommets in the top bracket **(see illustration)**. Remove the side brackets by unscrewing the nuts. Remove the top bracket from the seat cowling if required by undoing its screws. Check the condition of

the rubber mounts and replace them with new ones if they are damaged, deformed or deteriorated

Installation

4 Installation is the reverse of removal. Check the operation of the tail and brake lights.

11 Turn signal circuit – check

Note 1: *On US and Canada models the front turn signals also function as running lights and have dual filament bulbs. When checking for faults, refer to the wiring diagram at the end of this Chapter.*
Note 2: *On RR-2 and RR-3 (2002 and 2003) models, the turn signal relay is incorporated in the instrument cluster.*

1 Most turn signal problems are the result of a burned out bulb or corroded socket. This is especially true when the turn signals function properly in one direction, but fail to flash in the other direction. If this is the case, first check the bulbs, the sockets and the wiring connectors. If all the turn signals fail to work, first check the fuse (see Section 5), and then on RR-Y and RR-1 (2000 and 2001) models the relay (see below), or on RR-2 and RR-3 (2002 and 2003) models the instrument cluster (see below). If they are good, the problem lies in the wiring or connectors, or the switch. Refer to Section 20 for the switch testing procedures, and also to the wiring diagrams at the end of this Chapter.

RR-Y and RR-1 models

2 To check the relay, remove the fairing (see Chapter 8) – the relay is mounted in front of the instrument cluster on the right-hand side **(see illustration)**. The easiest way to tell if the relay is faulty is to substitute it with another one, if available. If the turn signals then work, replace the faulty relay with a new one.
3 If a substitute is not available, or if it does not solve the problem, disconnect the relay wiring connector. Check for battery voltage at the white/green wire terminal on the loom side of the connector with the ignition ON. If no

voltage was present, check the wiring from the relay to the ignition (main) switch (via the fusebox) for continuity.
4 If voltage was present, short between the white/green and grey wire terminals on the connector using a jumper wire. Turn the ignition ON and operate the turn signal switch. If the lights come on, the relay is faulty and must be replaced with a new one.
5 If the lights do not come on, check the green wire from the connector for continuity to earth (ground), and the grey wire for continuity to the left-hand switch housing, and repair or renew the wiring or connectors as required.
6 If all is good so far, or if the lights came illuminated on one side but not the other, check the wiring between the left-hand switch housing and the turn signals themselves. Repair or renew the wiring or connectors as necessary.

RR-2 and RR-3 models

7 To check the relay, remove the fairing (see Chapter 8).
8 Disconnect the instrument cluster wiring connector **(see illustration 15.2)**. Check for battery voltage at the white/green wire terminal on the loom side of the connector with the ignition ON. If no voltage was present, check the wiring from the relay to the ignition (main) switch (via the fusebox) for continuity.
9 If voltage was present, short between the white/green and grey wire terminals on the

11.2 Turn signal relay (arrowed) – RR-Y and RR-1 (2000 and 2001) models

Electrical system 9•9

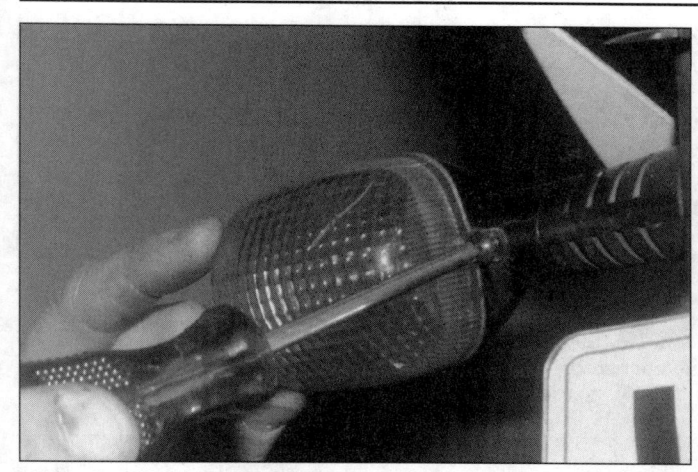

12.1a On RR-Y and RR-1 (2000 and 2001) models the screw is on the front of the housing

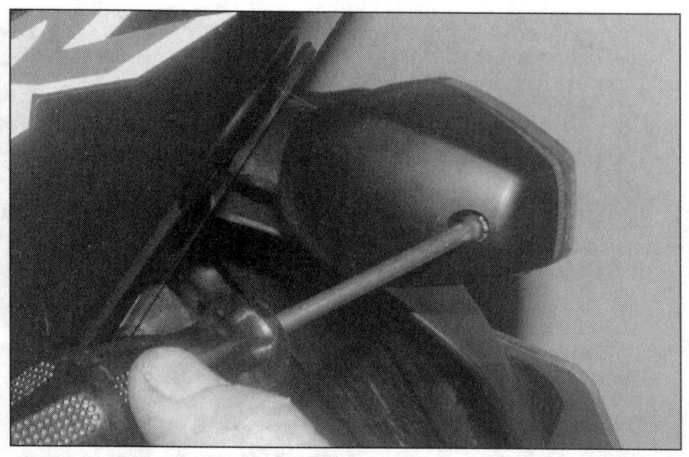

12.1b On RR-2 and RR-3 (2002 and 2003) models the screw is on the back of the housing

connector using a jumper wire. Turn the ignition ON and operate the turn signal switch. If the lights come on, the relay is faulty and must be replaced with a new one.

10 If the lights do not come on, check the grey wire for continuity to the left-hand switch housing, and repair or renew the wiring or connectors as required.

11 If all is good so far, or if the lights came on on one side but not the other, check the wiring between the left-hand switch housing and the turn signals themselves. Repair or renew the wiring or connectors as necessary.

12 Turn signal bulbs – replacement

1 Remove the screw securing the lens and detach the lens from the housing, noting how it fits **(see illustrations)**. Remove the rubber seal if it is loose, and discard it if it is damaged, deformed or deteriorated.

2 Push the bulb into the holder and twist it anti-clockwise to remove it **(see illustrations)**. Check the socket terminals for corrosion and clean them if necessary. Line up the pins of the new bulb with the slots in the socket, then push the bulb in and turn it clockwise until it locks into place. **Note:** *It is a good idea to use a paper towel or dry cloth when handling the new bulb to prevent injury if the bulb should break and to increase bulb life.*

3 Fit a new rubber seal onto the housing if required, and make sure it is properly seated and does not get pinched by the lens **(see illustration)**. On RR-Y and RR-1 (2000 and 2001) models, fit the lens onto the housing, locating the cutout on the outer end of the lens over the tab on the housing, and install the screw **(see illustration)**. On RR-2 and RR-3 (2002 and 2003) models, fit the lens onto the housing, locating the tab on the inner end of the lens into the cutout on the housing, and install the screw **(see illustration)**. Do not overtighten the screw as it is easy to strip the threads or crack the lens.

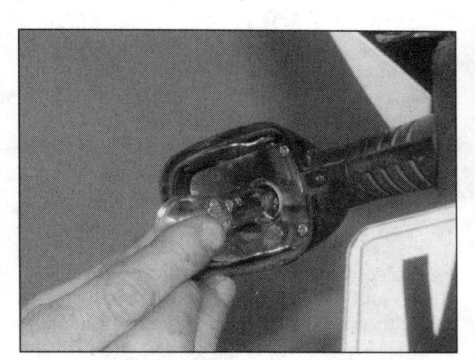

12.2a Removing the bulb – RR-Y and RR-1 (2000 and 2001) models

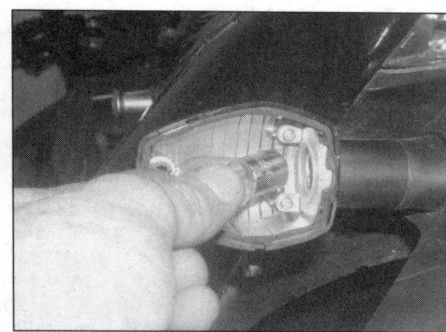

12.2b Removing the bulb – RR-2 and RR-3 (2002 and 2003) models

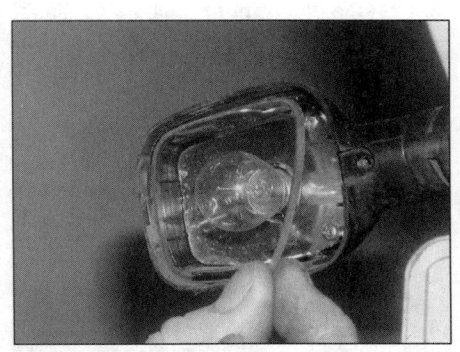

12.3a Make sure the seal seats correctly

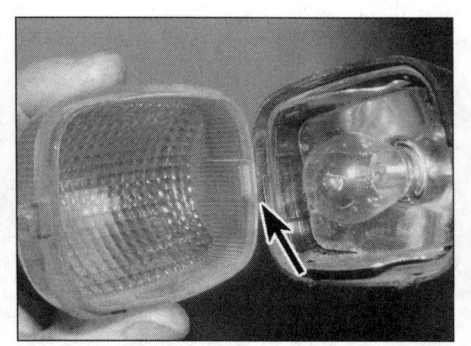

12.3b On RR-Y and RR-1 (2000 and 2001) models locate the cut-out in the lens over the tab on the housing (arrow)

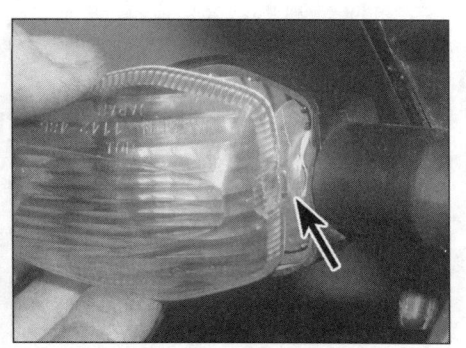

12.3c On RR-2 and RR-3 (2002 and 2003) models locate the tab on the lens in the cut-out on the housing (arrow)

9•10 Electrical system

13.2a Turn signal wiring connector – RR-Y and RR-1 (2000 and 2001) models

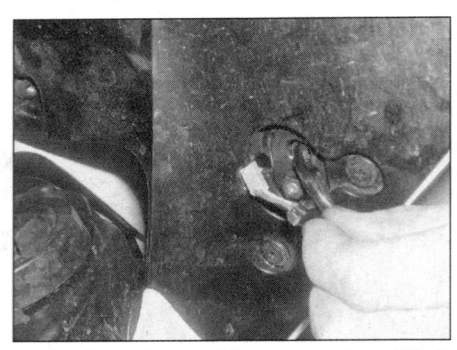

13.2b Turn signal wiring connector – RR-2 and RR-3 (2002 and 2003) models

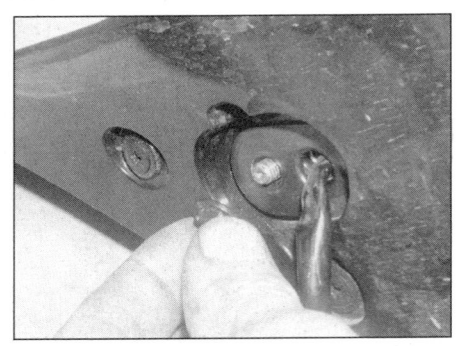

13.3a Unscrew the nut . . .

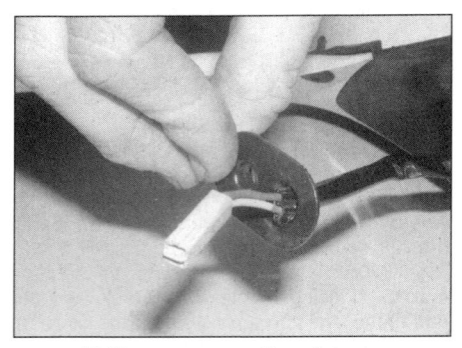

13.3b . . . remove the plate . . .

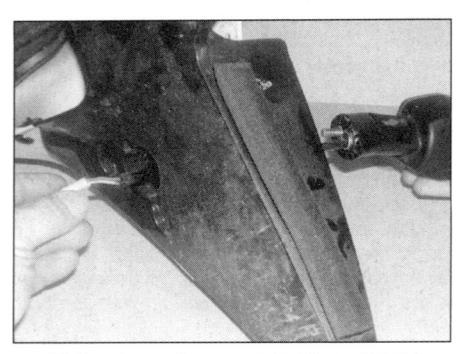

13.3c . . . and remove the turn signal

13.4a Rear turn signal wiring connectors (arrowed – RR-Y and RR-1 (2000 and 2001) models

13 Turn signal assemblies – removal and installation

Removal

Front turn signals

1 On RR-Y and RR-1 (2000 and 2001) models remove the relevant air duct cover (see Chapter 8, Section 4).
2 Disconnect the turn signal wiring connector (see illustrations).
3 Unscrew the nut securing the stem to the inside of the fairing and remove the mounting plate (see illustrations). Remove the turn signal, taking care as you draw the wiring through (see illustration).

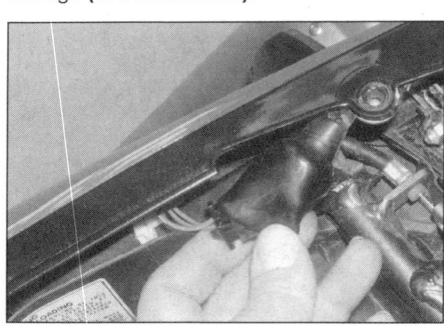

Rear turn signals

4 Raise the passenger seat (see Chapter 8). Disconnect the relevant turn signal wiring connector – on RR-Y and RR-1 (2000 and 2001) models they are easy to access (see illustration). On RR-2 and RR-3 (2002 and 2003) models they are housed inside a rubber boot tucked inside the seat cowling, so it may be necessary to remove the cowling if access is too restricted for you (see Chapter 8, Section 4) (see illustration).
5 Carefully feed the wiring through to the stem under the mudguard, taking care not to snag it.
6 Unscrew the nut securing the stem to the inside of the rear mudguard and remove the mounting plate (see illustration). Remove the turn signal, again taking care as you draw the wiring through.

Installation

7 Installation is the reverse of removal. Check the operation of the turn signals.

14 Brake light switches – check and replacement

Circuit check

1 Before checking the switches, and if not already done, check the brake light circuit (see Section 6).
2 The front brake light switch is mounted on the underside of the brake master cylinder. Disconnect the wiring connectors from the switch (see illustration). Using a continuity

13.4b On RR-2 and RR-3 (2002 and 2003) models the connectors are inside the rubber boot

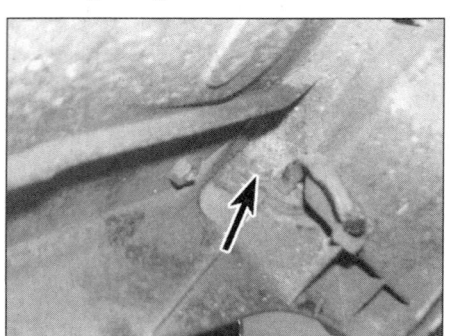

13.6 Rear turn signal nut (arrowed)

14.2 Front brake switch wiring connectors (arrowed)

Electrical system 9•11

14.3 Rear brake light switch (arrowed)

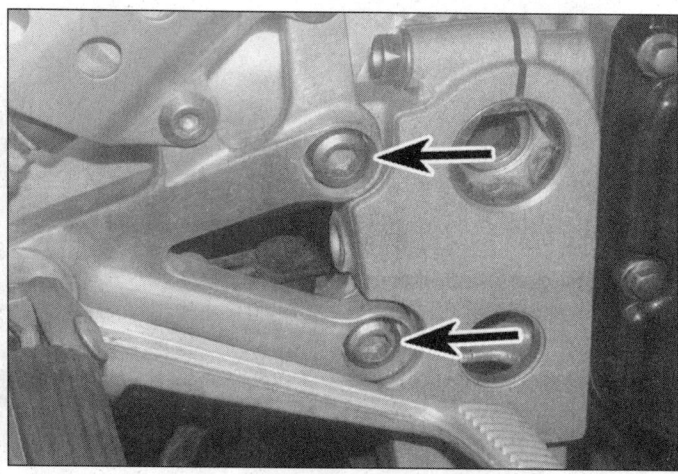

14.9 Unscrew the bolts (arrowed) and displace the bracket

tester, connect the probes to the terminals of the switch. With the brake lever at rest, there should be no continuity. With the brake lever applied, there should be continuity. If the switch does not behave as described, replace it with a new one.

3 The rear brake light switch is mounted on the inside of the right-hand rider's footrest bracket, above the brake pedal **(see illustration)**. Remove the rider's seat to access the wiring connector (see Chapter 8). Trace the wiring from the switch and disconnect it at the connector. Using a continuity tester, connect the probes to the terminals on the switch side of the wiring connector. With the brake pedal at rest, there should be no continuity. With the brake pedal applied, there should be continuity. If the switch does not behave as described, replace it with a new one, although check first that switch is adjusted correctly (see Step 11).

4 If the switches are good, check for voltage at the black/green (front) or white/green (rear) wire on the connector with the ignition switch ON – there should be battery voltage. If there's no voltage present, check the wiring between the switch and the ignition switch via the fusebox (see the wiring diagrams at the end of this Chapter). If voltage is present, check the black/yellow (front) or green/yellow (rear) wire for continuity to the brake light bulb wiring connector, referring to the relevant wiring diagram. Repair or renew the wiring as necessary.

Switch replacement

Front brake lever switch

5 The switch is mounted on the underside of the brake master cylinder. Disconnect the wiring connectors from the switch **(see illustration 14.2)**.

6 Remove the single screw securing the switch to the master cylinder and remove the switch.

7 Installation is the reverse of removal. Make sure the switch is correctly located before tightening its screw. The switch isn't adjustable.

Rear brake pedal switch

8 The rear brake light switch is mounted on the inside of the right-hand rider's footrest bracket, above the brake pedal **(see illustration 14.3)**. Remove the rider's seat to access the wiring connector (see Chapter 8). Trace the wiring from the switch and disconnect it at the connector. Feed the wiring down to the switch, noting its routing and releasing it from any ties.

9 Unscrew the footrest bracket mounting bolts and displace the bracket so that you can access the back of it **(see illustration)**.

10 Detach the rear end of the switch spring from the brake pedal **(see illustration)**. Either thread the switch out of its adjustment nut, leaving the nut in the mounting, or unscrew the bolt securing the switch bracket and remove the whole assembly, and separate them afterwards.

11 Installation is the reverse of removal. Tighten the footrest bracket bolts to the torque setting specified at the beginning of the Chapter. Make sure the brake light is activated just before the rear brake pedal takes effect. If adjustment is necessary, hold the switch body and turn the adjustment nut as required until the brake light is activated correctly – if the brake light comes on too late or not at all, turn the ring clockwise (when looked at from the front) so the switch threads out of the bracket. If the brake light comes on too soon or is permanently on, turn the ring anti-clockwise so the switch threads into the bracket. If the footrest bracket was displaced, tighten the bolts to the torque setting specified at the beginning of the Chapter.

15 Instrument cluster – removal and installation

Removal

1 Remove the fairing (see Chapter 8). On RR-2 and RR-3 (2002 and 2003) models remove the lean angle sensor (see Chapter 4, Section 6).

2 Pull back the rubber boot and disconnect the instrument cluster wiring connector **(see illustration)**.

3 Undo the bolt (RR-Y and RR-1 (2000 and

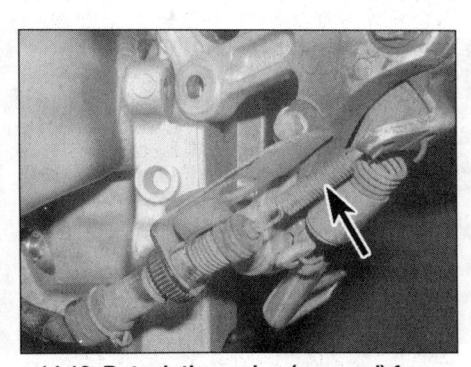

14.10 Detach the spring (arrowed) from the pedal then remove the switch as described

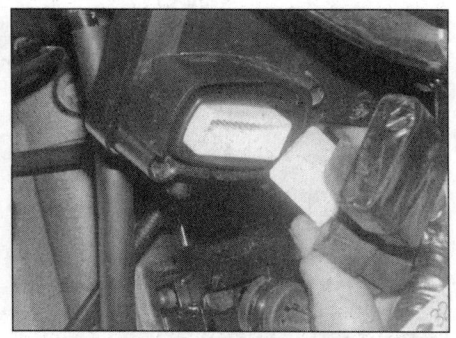

15.2 Disconnect the wiring connector

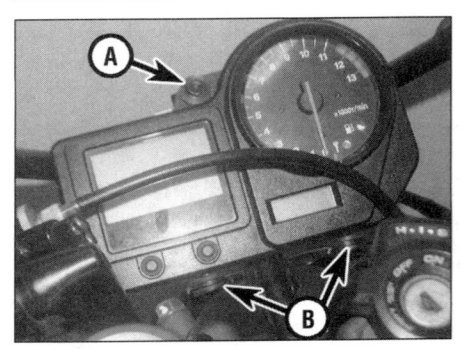

9•12 Electrical system

15.3a Instrument cluster bolt (A) – RR-Y and RR-1 (2000 and 2001) models. Note how the pegs locate in the grommets (B)

15.3b Instrument cluster screw (arrowed) – RR-2 and RR-3 (2002 and 2003) models . . .

15.3c . . . note how the pegs locate in the grommets (arrow)

2001) models) or screw (RR-2 and RR-3 (2002 and 2003) models) and lift the instrument cluster from the bracket, noting how it locates **(see illustrations)**. Note the rubber grommets fitted in the mounts.

Installation

4 Installation is the reverse of removal. Check the rubber grommets for damage, deformation and deterioration and replace them with new ones if necessary. Make sure the pegs locate correctly in the grommets. Make sure that the wiring connector is secure.

16 Instruments and speed sensor – check and replacement

Instrument cluster

Power check

1 If none of the instruments or displays are working, first check the fuse (see Section 5).
2 If the fuse is good, remove the fairing (see Chapter 8) then pull back the rubber boot and disconnect the instrument cluster wiring connector **(see illustration 15.2)**. Check the connector for loose or broken connections.
3 To check the power input wire, check for battery voltage between the black/brown wire terminal on the wiring loom side of the connector and a good earth (ground) with the ignition switch ON. There should be battery voltage. If there is no voltage, refer to the wiring diagrams and check the black/brown wire between the instrument cluster and the fusebox for loose or broken connections or a damaged wire.
4 To check the back-up power wire, check for battery voltage between the red/green wire terminal on the wiring loom side of the connector and a good earth (ground) with the ignition switch OFF. There should be battery voltage. If there is no voltage, refer to the wiring diagrams and check the red/green wire between the instrument cluster and the fusebox for loose or broken connections or a damaged wire, then check the red wire from the fusebox to the main fuse and battery.

5 If there is voltage, and to check the earth (ground) wires, check for continuity between each green/black wire terminal on the loom side of the wiring connector and earth (ground), and between the green wire terminal and earth. If there is no continuity, check the circuit for loose or broken connections or a damaged wire and repair as necessary.
6 If all power input and earth wires are good, but there is no display or instrument function, then either the PCB or the LCD unit in the instrument cluster is faulty. Individual components are not available, so replace the instrument cluster with a new one.

Speedometer and speed sensor

Check

7 First check the fuse (see Section 5).
8 Raise or remove the fuel tank (see Chapter 4). Trace the wiring from the speed sensor, which is mounted on the crankcase behind the cylinders, and disconnect it at the 3-pin connector **(see illustration)**. Check the connector for loose terminals. With the ignition switch ON, check for battery voltage between the black/brown wire terminal on the wiring loom side of the connector and a good earth (ground). If there is no voltage, refer to the wiring diagrams and check the black/brown circuit for loose or broken connections or a damaged wire.
9 If there is voltage, reconnect the wiring connector, then place the machine on an

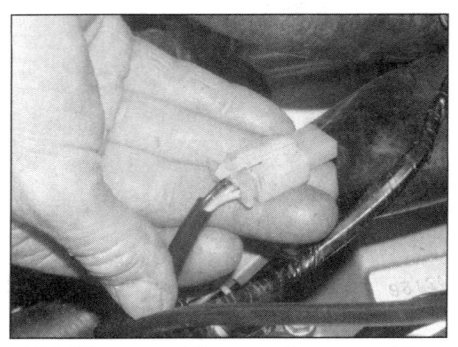

16.8 Disconnect the speed sensor wiring connector

auxiliary stand so the rear wheel is off the ground. Connect a voltmeter between the pink/green and green/black terminals on the wiring loom side of the connector. With the ignition switch ON, the transmission in gear and the clutch lever pulled in, turn the rear wheel by hand and check that a fluctuating voltage reading between 0 and 5 volts is obtained. If no reading is obtained, refer to Step 5 and check the instrument cluster earth wires. If the wiring is good, then the speed sensor is faulty.
10 If a reading is obtained, remove the fairing (see Chapter 8), then pull back the rubber boot and disconnect the instrument cluster wiring connector **(see illustration 15.2)**. Check the connector for loose or broken connections, then repeat the check in Step 8, but connect the meter between the pink/green and green/black wire terminals on the loom side of the cluster wiring connector. If no fluctuating voltage is obtained, check for continuity in the wiring between the speed sensor wiring connector and the instrument cluster wiring connector, referring to the wiring diagrams.
11 If the wiring is all good, either the PCB or the LCD unit in the instrument cluster is faulty. Individual components are not available, so replace the instrument cluster with a new one.

Speedometer replacement

12 Remove the instrument cluster (see Section 15), and replace it with a new one – individual components are not available.
13 Installation is the reverse of removal.

Speed sensor replacement

14 The speed sensor is mounted on the crankcase behind the cylinders on the left-hand side. Depending on the tools you have available, the size of your hands and your general dexterity, you might be able to access the sensor from the left-hand side of the engine after removing the fairing side panel and on RR-Y and RR-1 (2000 and 2001) models the lower fairing (see Chapter 8). Otherwise you will have to raise or remove the fuel tank (see Chapter 4) and access the sensor from the top.
15 Trace the wiring from the sensor and disconnect it at the 3-pin connector **(see illustration 16.8)**.

Electrical system 9•13

16.16a Undo the bolts (arrowed) and remove the sensor

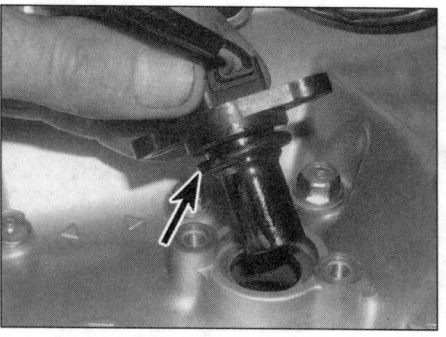

16.16b Replace the O-ring (arrowed) with a new one

18.3 Pull back the rubber then remove the terminal screw and detach the wiring

16 Unscrew the sensor mounting bolts and remove the sensor (see illustration). Remove and discard its O-ring as a new one must be used (see illustration). Plug the sensor orifice with clean rag to prevent anything falling into the engine.

17 Installation is the reverse of removal, using a new O-ring.

Tachometer

Check

18 First check the fuse (see Section 5).
19 If the fuse is good, remove the fairing (see Chapter 8), then pull back the rubber boot and disconnect the instrument cluster wiring connector (see illustration 15.2). Check the connector for loose or broken connections.
20 To check the tachometer input peak voltage Honda specify their own Imrie diagnostic tester (model 625), or the peak voltage adapter (Pt. No. 07HGJ-0020100) with an aftermarket digital multimeter having an impedance of 10 M-ohm/DCV minimum, for this test. Connect the positive (+) lead of the voltmeter and peak voltage adapter arrangement to the yellow/green wire terminal on the loom side of the connector and the negative (–) lead to a good earth (ground). Start the engine and measure the tachometer input peak voltage, which should be at least 10.5 volts. If the peak voltage is normal, check the instrument cluster earth wires (see Step 5). If they are good then the tachometer is faulty. Individual components are not available, so replace the instrument cluster with a new one.
21 If there is no reading, remove the seat (see Chapter 8), and disconnect the ECM wiring connector (see Chapter 4). Check for continuity in the yellow/green wire between the ECM wiring connector and the instrument cluster wiring connector. If there is no continuity there is a break in the wire or faulty connector. Refer to the wiring diagrams and trace and rectify the fault. If the wiring is good the ECM could be faulty (see Chapter 4, Section 6).

Replacement

22 Remove the instrument cluster (see Section 15), and replace it with a new one – individual components are not available.
23 Installation is the reverse of removal.

LCD display

Check

24 If the display is not working at all, check the instrument cluster power input (see above) and earth wires. If the wires are good, then either the printed circuit board (PCB) or the LCD display unit is faulty. Individual components are not available, so replace the instrument cluster with a new one.
25 If an individual display is not working, refer to Chapter 3 for the coolant temperature and warning display, Chapter 4 for the low fuel warning, and Section 18 for the oil pressure warning. If the clock doesn't work, first check the fuse (see Section 5). If any fault points to the LCD unit being faulty, replace the instrument cluster with a new one – individual components are not available.

Replacement

26 Remove the instrument cluster (see Section 15), and replace it with a new one – individual components are not available.
27 Installation is the reverse of removal.

17 Instrument and warning light bulbs – replacement

1 All instrument and warning lights are LEDs, which are part of the printed circuit board and are not available individually. If one of the LEDs fails the entire instrument cluster must be replaced with a new one as no individual components are available.

18 Oil pressure switch – check, removal and installation

Check

1 The oil pressure warning display should come on when the ignition (main) switch is turned ON and extinguish a few seconds after the engine is started. If the oil pressure warning light comes on whilst the engine is running, stop the engine immediately and carry out an oil level check, and if the level is correct, an oil pressure check (see Chapter 1).
2 If the oil pressure warning light does not come on when the ignition is turned ON, but the LCD display otherwise appears to be functioning, raise the fuel tank (see Chapter 4).
3 The oil pressure switch is screwed into the crankcase on the right-hand side. Pull the rubber cover off the switch and remove the screw securing the wiring connector (see illustration). With the ignition switched ON, earth (ground) the wire on the crankcase and check that the warning light comes on. If the light comes on, the switch is defective and must be replaced with a new one.
4 If the light still does not come on, check for voltage at the wire terminal. If there is no voltage present, check the wire between the switch, the instrument cluster and fusebox for continuity (see the wiring diagrams at the end of this Chapter).
5 If the warning light comes on whilst the engine is running, yet the oil pressure is satisfactory, detach the wire from the oil pressure switch (see above). With the wire detached and the ignition switched ON the light should be out. If it is illuminated, the wire between the switch and instrument cluster is earthed (grounded) at some point. If the wiring is good, the switch must be assumed faulty and replaced with a new one.

Removal

6 The oil pressure switch is screwed into the crankcase on the right-hand side. Depending on the tools you have available, the size of your hands and your general dexterity, you might be able to access the switch from the right-hand side of the engine after removing the fairing side panel and on RR-Y and RR-1 (2000 and 2001) models the lower fairing (see Chapter 8). Otherwise you will have to raise or remove the fuel tank (see Chapter 4) and access the sensor from the top.
7 Pull the rubber cover off the switch, then undo the screw securing the wiring connector (see illustration 18.3).
8 Unscrew the oil pressure switch and withdraw it from the crankcase.

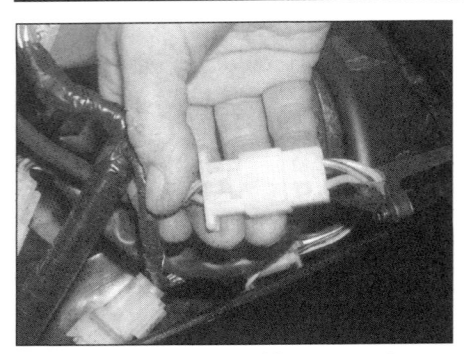

19.1 Ignition switch wiring connector – RR-Y and RR-1 (2000 and 2001) models

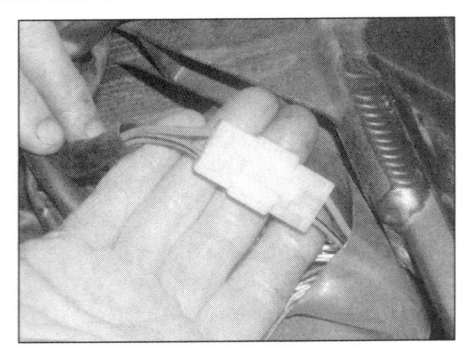

19.2 Ignition switch wiring connector – RR-2 and RR-3 (2002 and 2003) models

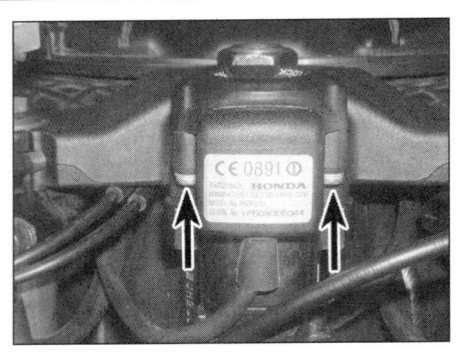

19.8 Immobiliser receiver screws (arrowed)

Installation

9 Apply a suitable sealant to the upper portion of the switch threads near the switch body, leaving the bottom 3 to 4 mm of thread clean. Install the switch in the crankcase and tighten it to the torque setting specified at the beginning of the Chapter. Attach the wiring connector and secure it with the screw, then fit the rubber cover **(see illustration 18.3)**.
10 Run the engine and check that the switch operates correctly without leakage.
11 Install the fuel tank or fairing side panel/lower fairing according to your removal method.

19 Ignition (main) switch – check, removal and installation

⚠️ **Warning: To prevent the risk of short circuits, disconnect the battery negative (–ve) lead before making any ignition (main) switch checks.**

Check

1 On RR-Y and RR-1 (2000 and 2001) models, remove the right hand air duct cover (see Chapter 8, Section 4). Trace the wiring from the ignition switch and disconnect it at the connector inside the rubber boot **(see illustration)**.
2 On RR-2 and RR-3 (2002 and 2003) models, remove the air filter housing (see Chapter 4). Trace the wiring from the ignition switch and disconnect it at the connector inside the rubber boot behind the steering stem **(see illustration)**.
3 Using an ohmmeter or a continuity tester, check the continuity of the connector terminal pairs (see the wiring diagrams at the end of this Chapter). Continuity should exist between the terminals connected by a solid line on the diagram when the switch is in the indicated position.
4 If the switch fails any of the tests, replace it with a new one.

Removal

5 Remove the instrument cluster (see Section 15).
6 On RR-Y and RR-1 (2000 and 2001) models, trace the wiring from the ignition switch and disconnect it at the connector inside the rubber boot **(see illustration 19.1)**. Feed the wiring back to the switch, freeing it from any clips and ties and noting its routing.
7 On RR-2 and RR-3 (2002 and 2003) models, remove the air filter housing (see Chapter 4). Trace the wiring from the ignition switch and disconnect it at the 4-pin connector inside the rubber boot behind the steering stem **(see illustration 19.2)**. Feed the wiring back to the switch, freeing it from any clips and ties and noting its routing.
8 On models fitted with the HISS immobiliser system, undo the screws securing the receiver around the ignition switch and displace it, noting how it fits **(see illustration)**.

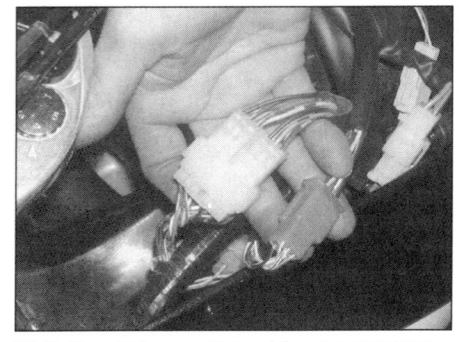

19.9 Remove the bolt (arrowed) on each side

20.3 Handlebar switch wiring connectors – RR-Y and RR-1 (2000 and 2001) models

9 Remove the switch bolts and withdraw the switch from the top yoke **(see illustration)**. Where single-use security bolts are fitted (it depends on model and country) you may have to drift the heads round using a punch – slacken the fork clamp bolts in the top yoke and unscrew the steering stem nut and remove the top yoke as a matter of precaution if required.
10 If required separate the contact plate from the bottom of the switch.

Installation

11 Installation is the reverse of removal. Tighten the ignition switch bolts (use new bolts if necessary), and where necessary the steering stem nut and fork clamp bolts (in that order) to the torque settings specified at the beginning of the Chapter. Make sure the wiring connector is correctly routed and securely connected.

20 Handlebar switches – check

1 Generally speaking, the switches are reliable and trouble-free. Most troubles, when they do occur, are caused by dirty or corroded contacts, but wear and breakage of internal parts is a possibility that should not be overlooked. If breakage does occur, the entire switch and related wiring harness will have to be replaced with a new one, as individual parts are not available.
2 The switches can be checked for continuity using an ohmmeter or a continuity test light. Always disconnect the battery negative (–ve) cable, which will prevent the possibility of a short circuit, before making the checks.
3 On RR-Y and RR-1 (2000 and 2001) models, remove the right-hand air duct cover (see Chapter 8, Section 4). Trace the wiring from the relevant switch and disconnect it at the connector inside the rubber boot **(see illustration)**.
4 On RR-2 and RR-3 (2002 and 2003) models, remove the air filter housing (see Chapter 4). Trace the wiring from the relevant switch and disconnect it at the connector(s)

Electrical system 9•15

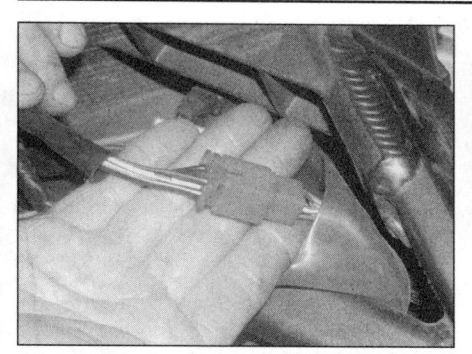

20.4a Right-hand switch wiring connector – RR-2 and RR-3 (2002 and 2003) models

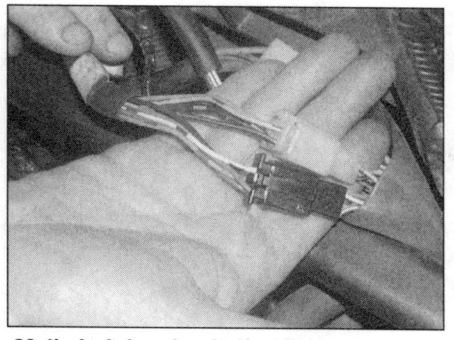

20.4b Left-hand switch wiring connectors – RR-2 and RR-3 (2002 and 2003) models

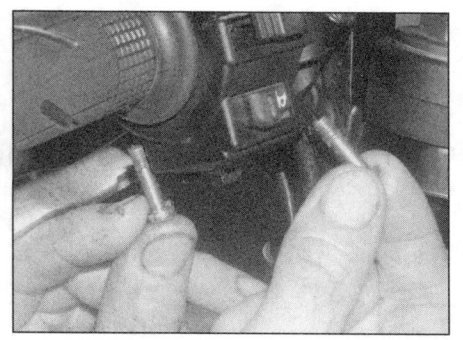

21.5a Undo the screws (arrowed) . . .

inside the rubber boot behind the steering stem **(see illustrations)**.

5 Check for continuity between the terminals of the switch connector with the switch in the various positions (i.e. switch off – no continuity, switch on – continuity) – see the wiring diagrams at the end of this Chapter. Continuity should exist between the terminals connected by a solid line on the diagram when the switch is in the indicated position.

6 If the continuity check indicates a problem exists, refer to Section 21, displace the switch housing and spray the switch contacts with electrical contact cleaner (there is no need to remove the switch completely). If they are accessible, the contacts can be scraped clean with a knife or polished with crocus cloth. If switch components are damaged or broken, it will be obvious when the switch is disassembled.

21 Handlebar switches – removal and installation

Removal

1 On RR-Y and RR-1 (2000 and 2001) models, remove the right-hand air duct cover (see Chapter 8, Section 4). Trace the wiring from the relevant switch and disconnect it at

the connector inside the rubber boot **(see illustration 20.3)**. Feed the wiring back to the switch, freeing it from any clips and ties and noting its routing.

2 On RR-2 and RR-3 (2002 and 2003) models, remove the air filter housing (see Chapter 4). Trace the wiring from the ignition switch and disconnect it at the 4-pin connector inside the rubber boot behind the steering stem **(see illustration 20.4a or 20.4b)**. Feed the wiring back to the switch, freeing it from any clips and ties and noting its routing.

3 If removing the right-hand switch disconnect the two wires from the brake light switch **(see illustration 14.2)**. If removing the left-hand switch disconnect the wires at the clutch switch and the horn **(see illustration 24.2a or 24.2b and 26.2)**.

4 When working on the right-hand switch, refer to Chapter 4 for removal of the throttle cables, which involves detaching the switch housing from the handlebars. Free the throttle cables from the twistgrip and remove them from the housing.

5 When working on the left-hand switch, unscrew the two handlebar switch screws and free the switch from the handlebar by separating the halves **(see illustrations)**.

Installation

6 Installation is the reverse of removal. Make sure the locating pin in the switch housing

locates in the hole in the handlebar **(see illustration)**. Refer to Chapter 4 for installation of the throttle cables and right-hand switch housing.

22 Neutral switch – check, removal and installation

Check

1 Before checking the electrical circuit, check the fuse (see Section 5).

2 The switch is located in the left-hand side of the transmission casing below the front sprocket cover **(see illustration)**. Detach the wiring connector from the switch. Make sure the transmission is in neutral.

3 With the connector disconnected and the ignition switch ON, the neutral light should be out. If not, the wire between the connector and instrument cluster must be earthed (grounded) at some point.

4 Check for continuity between the switch terminal and the crankcase. With the transmission in neutral, there should be continuity. With the transmission in gear, there should be no continuity. If the tests prove otherwise, then the switch is faulty.

5 If the continuity tests prove the switch is good, check for voltage at the wire terminal. If there's no voltage present, check the wire

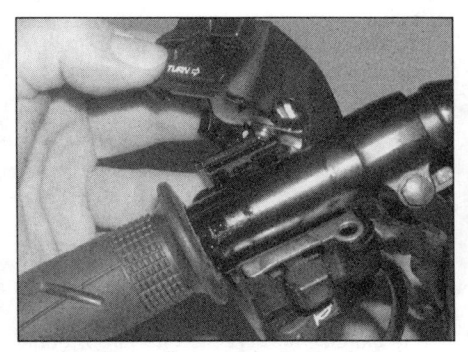

21.5b . . . then separate the halves

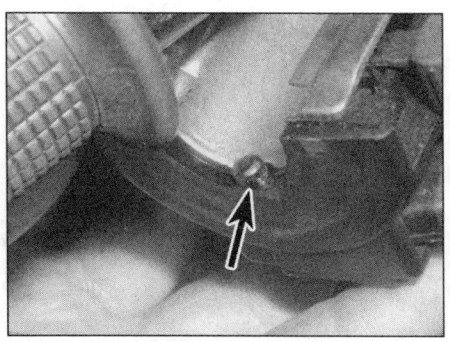

21.6 Locate the pin (arrowed) in the hole in the handlebar

22.2 Neutral switch (arrowed)

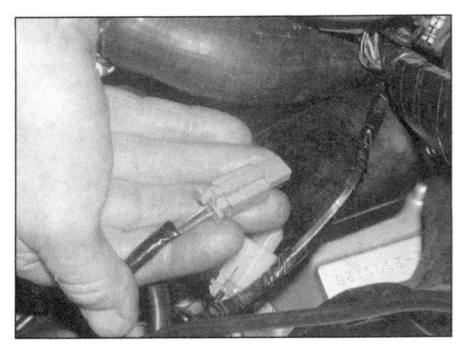

9•16 Electrical system

23.2 Sidestand switch wiring connector

23.7 Sidestand switch mounting bolt (arrowed)

between the switch, the instrument cluster and fusebox (see the wiring diagrams at the end of this Chapter). If all is good, the LCD unit in the instrument cluster could be faulty (see Section 16).

Removal

6 The switch is located in the left-hand side of the transmission casing below the front sprocket cover **(see illustration 22.2)**.
7 Detach the wiring connector from the switch.
8 Clean the area around the switch, then unscrew it from the crankcase. Discard the sealing washer as a new one should be used.

Installation

9 Install the switch using a new washer and tighten it to the torque setting specified at the beginning of the Chapter.
10 Connect the wiring connector and check the operation of the neutral light **(see illustration 22.2)**.

23 Sidestand switch – check and replacement

Check

1 The sidestand switch is mounted on the back of the sidestand. The switch is part of the starter interlock safety circuit which prevents or stops the engine running if the transmission is in gear whilst the sidestand is down, and prevents the engine from starting if the transmission is in gear unless the sidestand is up, and unless the clutch is pulled in. Before checking the electrical circuit, check the fuse (see Section 5).
2 Raise the fuel tank (see Chapter 4) to access the wiring connector. Trace the wiring back from the switch and disconnect at the green 2-pin wiring connector inside the rubber boot **(see illustration)**.
3 Check the operation of the switch using an ohmmeter or continuity test light. Connect the meter between the terminals on the switch side of the connector. With the sidestand up there should be continuity (zero resistance) between the terminals, and with the stand down there should be no continuity (infinite resistance).
4 If the switch does not perform as expected, it is faulty and must be replaced with a new one.
5 If the switch is good, check the wiring and connectors between the various components in the starter safety circuit using a continuity tester (see the wiring diagrams at the end of this Chapter). Also check for voltage at the green/white wire terminal on the loom side of the connector with the ignition ON – there should be battery voltage. Repair or renew the wiring as required.

Replacement

6 The sidestand switch is mounted on the back of the sidestand. Raise the fuel tank (see Chapter 4) to access the wiring connector. Trace the wiring back from the switch and disconnect at the green 2-pin wiring connector inside the rubber boot **(see illustration 23.2)**. Feed the wiring back to the switch, freeing it from any clips and ties and noting its routing.
7 Unscrew the switch bolt and remove the switch from the stand, noting how it fits **(see illustration)**. Note that Honda specify that the switch bolt be renewed every time it is disturbed – the new bolt has a locking compound already applied to its threads. However there is nothing to stop you cleaning up the threads on the old bolt and applying a suitable non-permanent thread locking compound of your own when you install it.
8 Fit the new switch onto the sidestand, making sure the pin locates in the hole in the sidestand, and the lug on the stand bracket locates into the cutout in the switch body. Secure the switch with a new or cleaned and threadlocked bolt and tighten it to the torque setting specified at the beginning of the Chapter **(see illustration 23.7)**.
9 Feed the wiring up to its connector, making sure it is correctly routed and secured by any clips.
10 Reconnect the wiring connector and check the operation of the sidestand switch **(see illustration 23.2)**.
11 Lower the fuel tank (see Chapter 4).

24 Clutch switch – check and replacement

Check

1 The clutch switch is mounted in or under the clutch lever bracket, depending on your model. The switch is part of the starter interlock safety circuit which prevents or stops the engine running if the transmission is in gear whilst the sidestand is down, and prevents the engine from starting if the transmission is in gear unless the sidestand is up and the clutch lever is pulled in. The switch isn't adjustable.
2 To check the switch, disconnect the wiring connectors from it **(see illustrations)**. Connect the probes of an ohmmeter or a continuity tester to the two switch terminals. With the clutch lever pulled in, continuity should be indicated. With the clutch lever out, no continuity (infinite resistance) should be indicated.
3 If the switch is good, check the other components in the starter circuit as described in the relevant sections of this Chapter. If all components are good, check the wiring between the various components (see the wiring diagrams at the end of this Chapter).

Replacement

4 The clutch switch is mounted in or under the clutch lever bracket, depending on your model.

24.2a Clutch switch wiring connectors (arrowed) – RR-Y and RR-1 (2000 and 2001) models

24.2b Clutch switch wiring connectors (arrowed) – RR-2 and RR-3 (2002 and 2003) models

Electrical system 9•17

25.2 The diode (arrowed) is in the fusebox

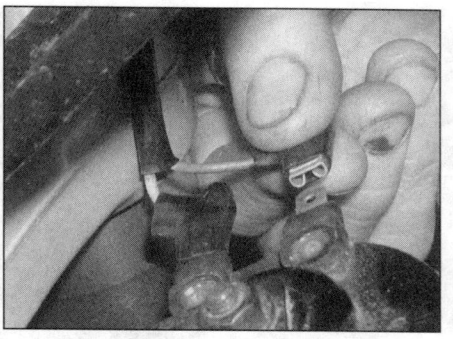

26.2 Disconnect the horn wiring connectors

27.2 Starter relay (arrowed)

5 Disconnect the wiring connectors from the switch (see illustration 24.2a or 24.2b).
6 On RR-Y and RR-1 (2000 and 2001) models release the switch from the bracket by pressing up on the small catch via the opening in the lever bracket and draw it out.
7 On RR-2 and RR-3 (2002 and 2003) models undo the single screw securing the switch and remove it, noting how it fits.
8 Installation is the reverse of removal. On RR-Y and RR-1 (2000 and 2001) models install the switch with the small catch at the bottom and press it in until the catch is felt to locate. On RR-2 and RR-3 (2002 and 2003) models make sure the switch is correctly located before tightening its screw.

25 Diode – check and replacement

Check

1 The diode is a small block that plugs into a connector in the fusebox, which is located under the rider's seat. The diode is part of the starter interlock safety circuit which prevents or stops the engine running if the transmission is in gear whilst the sidestand is down, and prevents the engine from starting if the transmission is in gear unless the sidestand is up and the clutch lever is pulled in.
2 Remove the rider's seat (see Chapter 8), open the fusebox lid and pull the diode out of its socket (see illustration).
3 Using an ohmmeter or continuity tester, connect the positive (+ve) probe to one of the outer terminals of the diode and the negative (–ve) probe to the middle terminal of the diode. The diode should show continuity. Now reverse the probes. The diode should show no continuity. Repeat the tests between the other outer terminal and the middle terminal. The same results should be achieved. If it doesn't behave as stated, replace the diode with a new one.
4 If the diode is good, check the other components in the starter circuit as described in the relevant sections of this Chapter. If all

components are good, check the wiring between the various components (see the wiring diagrams at the end of this Chapter).

Replacement

5 The diode is a small block that plugs into a connector in the fusebox, which is located under the rider's seat.
6 Remove the rider's seat (see Chapter 8), open the fusebox lid and pull the diode out of its socket (see illustration 25.2).
7 Push the diode back into position to install it.

26 Horn – check and replacement

Check

1 The horn is mounted on the bottom yoke – remove the fairing for access if required, though you should be able to get at it quite easily (see Chapter 8).
2 Disconnect the wiring connectors from the horn (see illustration). Check them for loose wires. Using two jumper wires, apply voltage from a fully-charged 12V battery directly to the terminals on the horn. If the horn doesn't sound, replace it with a new one.
3 If there is no sound check for voltage at the light green wire connector with the ignition ON and the horn button pressed. If voltage is present, check the green wire for continuity to earth.
4 If no voltage was present, check the light green wire for continuity between the horn and the switch, and the black/brown wire from the switch to the fusebox, and then to the ignition switch (see the wiring diagrams at the end of this Chapter). With the ignition switch ON, check that there is voltage at the black/brown wire to the horn button in the left-hand switch gear. If there is, the problem lies between the switch and the horn. If there isn't, the problem lies between the ignition switch and the horn switch via the fusebox.
5 If all the wiring and connectors are good, check the button contacts in the switch housing (see Section 20).

Replacement

6 The horn is mounted on the bottom yoke – remove the fairing for access if required, though you should be able to get at it quite easily (see Chapter 8).
7 Unplug the wiring connectors from the horn (see illustration 26.2). Unscrew the bolt securing the horn to the bottom yoke and remove it from the bike.
8 Install the horn and tighten the bolt. Connect the wiring to the horn. Check that it works, then install the fairing if removed (see Chapter 8).

27 Starter relay – check and replacement

Check

1 If the starter circuit is faulty, first check the fuse (see Section 5).
2 The starter relay is located under the rider's seat (see illustration) – remove the seat for access (see Chapter 8).
3 Lift the rubber terminal cover and unscrew the bolt securing the starter motor lead (see illustration 27.12b); position the lead away from the relay terminal. With the ignition switch ON, the engine kill switch in the RUN position, and the transmission in neutral, press the starter switch. The relay should be heard to click.
4 If the relay doesn't click, switch off the ignition and remove the relay as described below; test it as follows.
5 Set a multimeter to the ohms x 1 scale and connect it across the relay's starter motor and battery lead terminals. There should be no continuity. Using a fully-charged 12 volt battery and two insulated jumper wires, connect the positive (+ve) terminal of the battery to the yellow/red wire terminal of the relay, and the negative (–ve) terminal to the green/red wire terminal of the relay. At this point the relay should be heard to click and the multimeter read 0 ohms (continuity). If this is the case the relay is proved good. If the relay does not click when battery voltage is applied and indicates no

9•18 Electrical system

27.12a Disconnect the wiring connector ...

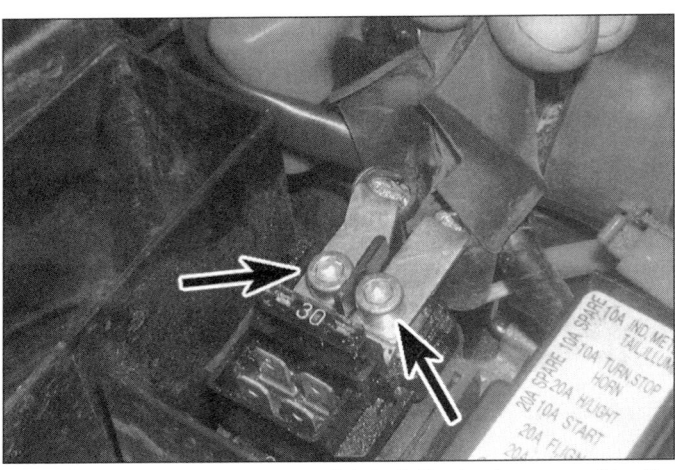

27.12b ... then lift the cover and detach the leads, secured by the bolts (arrowed)

continuity (infinite resistance) across its terminals, it is faulty and must be replaced with a new one.

6 If the relay is good, check for continuity in the main lead from the battery to the relay. Also check that the terminals and connectors at each end of the lead are tight and corrosion-free.

7 Next check for battery voltage at the yellow/red wire terminal on the relay wiring connector with the ignition ON, the kill switch in the RUN position and the starter button pressed. If there is no voltage, check the wiring between the relay wiring connector and the starter button.

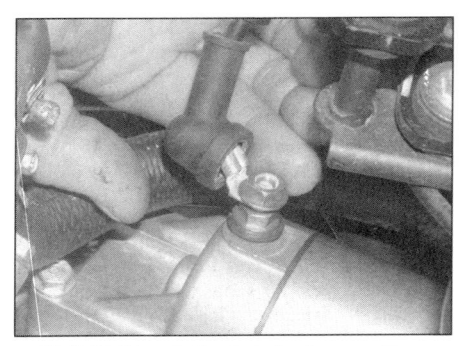

8 If voltage is present, check that there is continuity to earth in the green/red wire with the transmission in neutral (note that there will be a very slight resistance due to the diode. If not check the wiring and connectors between the relay, the fusebox and the neutral switch, then if that is good check the switch itself and the diode.

9 Now put the transmission in gear, raise the sidestand and pull the clutch lever in and check for continuity to earth again. If there is no continuity, check the clutch switch and sidestand switch as described in the relevant sections of this Chapter. If all components are good, check the wiring between the various components (see the wiring diagrams at the end of this Chapter).

Replacement

10 The starter relay is located under the rider's seat **(see illustration 27.2)** – remove the seat for access (see Chapter 8).

11 Disconnect the battery terminals, remembering to disconnect the negative (–ve) terminal first.

12 Disconnect the relay wiring connector, then lift the insulating cover and unscrew the bolts securing the starter motor and battery leads to the relay and detach the leads **(see illustrations)**. Remove the relay from its rubber sleeve. If the relay is being replaced with a new one, remove the main fuse from the relay.

13 Installation is the reverse of removal. Make sure the terminal bolts are securely tightened. Do not forget to fit the main fuse into the relay, if removed. Connect the negative (–ve) lead last when reconnecting the battery.

28 Starter motor – removal and installation

Removal

1 Remove the rider's seat (see Chapter 8). Disconnect the battery negative (–ve) lead. The starter motor is mounted on the crankcase behind the cylinders on the left-hand side.

2 For best access to the starter motor remove the thermostat housing (see Chapter 3), but note that access is possible, albeit tricky, with the fuel tank raised and the left-hand fairing side panel, and on RR-Y and RR-1 (2000 and 2001) models the lower fairing, removed (see Chapters 4 and 8).

3 Peel back the rubber terminal cover on the starter motor **(see illustration)**. Unscrew the nut securing the starter lead to the motor and detach the lead.

4 Unscrew the two bolts securing the starter motor to the crankcase, noting the earth lead secured by the rear bolts **(see illustration)**. Slide the starter motor out and remove it, from the top if the thermostat housing has been removed, otherwise from the left-hand side **(see illustration)**.

5 Remove the O-ring on the end of the starter motor and discard it as a new one must be used.

Installation

6 Fit a new O-ring onto the end of the starter motor, making sure it is seated in its groove

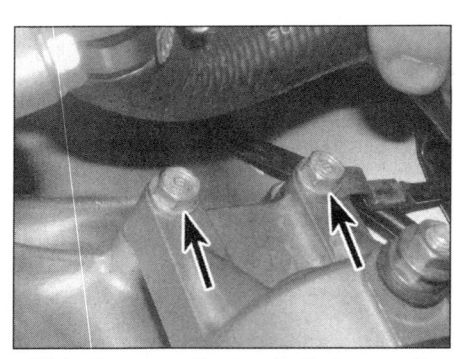

28.3 Pull back the terminal cover then unscrew the nut and detach the lead

28.4a Unscrew the two bolts (arrowed), noting the earth lead ...

28.4b ... and remove the starter motor

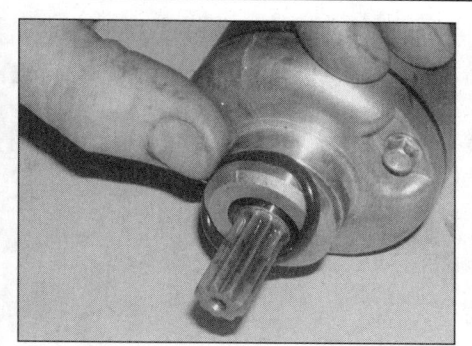

Electrical system 9•19

28.6 Fit a new O-ring and lubricate it

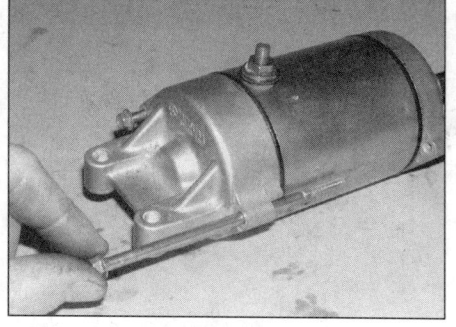

29.5 Unscrew and remove the two long bolts

29.6a Remove the front cover

(see illustration). Apply a smear of engine oil to the O-ring to aid installation.
7 Manoeuvre the motor into position and slide it into the crankcase (see illustration 28.4b). Ensure that the starter motor teeth mesh correctly with those of the starter idle/reduction gear. Install the mounting bolts, not forgetting to secure the earth lead with the rear bolt, and tighten them (see illustration 28.4a).
8 Connect the starter lead to the motor and secure it with the nut (see illustration 28.3). Fit the rubber cover over the terminal.
9 Install the thermostat housing if removed (see Chapter 3), and fill the cooling system (see Chapter 1). Otherwise install the fairing panels (see Chapter 8).
10 Connect the battery negative (–ve) lead and install the rider's seat (see Chapter 8).

29 Starter motor – check, disassembly, inspection and reassembly

Check

1 Remove the starter motor (see Section 28). Cover the body in some rag and clamp the motor in a soft-jawed vice – do not overtighten it.
2 Using a fully-charged 12 volt battery and two insulated jumper wires, connect the positive (+) terminal of the battery to the protruding terminal on the starter motor, and the negative (–) terminal to one of the motor's mounting lugs. At this point the starter motor should spin. If this is the case the motor is proved good, though it is worth disassembling it and checking it if you suspect it of not working properly under load. If the motor does not spin, disassemble it for inspection.

RR-Y and RR-1 models

Disassembly

3 Remove the starter motor (see Section 28).
4 Note any alignment marks between the main housing and the front and rear covers, or make your own if they aren't clear.

5 Unscrew the two long bolts and withdraw them from the starter motor (see illustration).
6 Wrap some insulating tape around the teeth on the end of the starter motor shaft – this will protect the oil seal from damage as the front cover is removed. Remove the front cover from the motor (see illustration). Remove the sealing ring from the main housing and discard it as a new one must be used. Remove the washer and shim(s) from the front end of the armature shaft or the inside of the front cover, noting their correct fitted locations (see illustration). Also remove the tabbed thrust washer from the front cover.
7 Remove the rear cover from the motor (see illustration). Remove the sealing ring from the main housing and discard it as a new one must be used. Remove the shims from the rear end of the armature shaft or from inside the rear cover (see illustration).
8 Withdraw the armature from the main housing, noting that you will have to pull it out against the attraction of the magnets (see illustration).
9 At this stage check for continuity between the terminal bolt and the main housing – there should be no continuity (infinite resistance). Also check for continuity between the insulated brushes and the terminal bolt – there should be continuity. Also check for continuity between the uninsulated brushes and the brushplate – there should be continuity (zero resistance). If there is no continuity when there should be or *vice versa*, identify the faulty component and replace it with a new one.
10 Slide the brushes with the yellow

29.6b Slide the insulating washer and shim(s) off the shaft

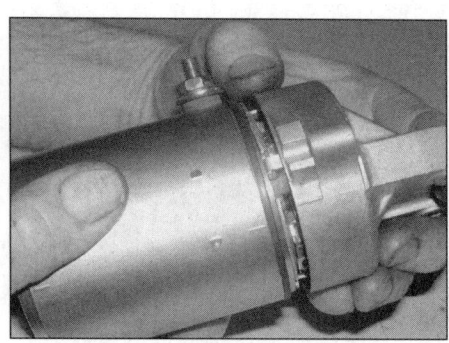

29.7a Remove the rear cover . . .

29.7b . . . and the shims (arrowed)

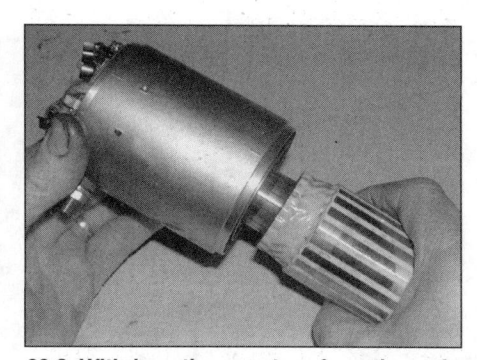

29.8 Withdraw the armature from the main housing

9•20 Electrical system

29.11a Unscrew the nut and remove the large and small insulating washers . . .

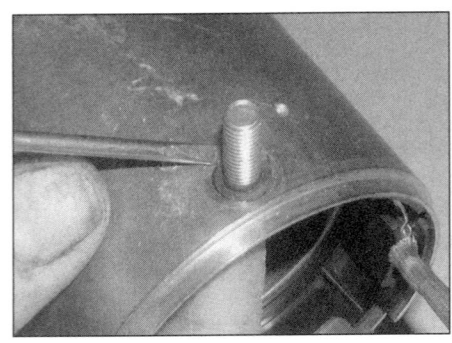

29.11b . . . and the O-ring

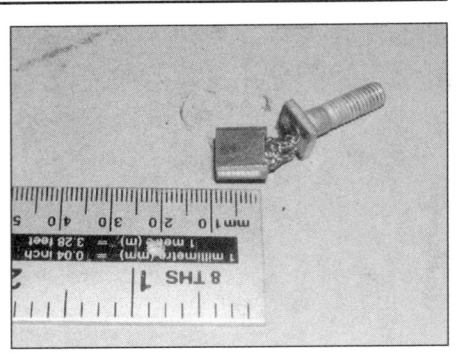

29.12 Measure the length of each brush

insulated wires out of their housings **(see illustration 29.22a)**. Remove the brushplate assembly from the main housing, noting how it locates **(see illustration 29.20)**.

11 Noting the correct fitted location of each component, unscrew the terminal nut and remove it along with its washer, the one large and two small insulating washers and the O-ring **(see illustrations)**. Withdraw the terminal bolt, then remove the insulated brush assembly from the brushplate seat, noting how it fits **(see illustrations 29.19d and 29.19c)**. Remove the insulator and the brushplate seat **(see illustrations 29.19b and 29.19a)**. Check the condition of the O-ring and renew it if it is damaged, deformed or deteriorated.

Inspection

12 The parts of the starter motor that are most likely to require attention are the brushes. Measure the length of each brush and compare the results to the length listed in this Chapter's Specifications **(see illustration)**. If any of the brushes are worn beyond the service limit, renew the brush assemblies. If the brushes are not worn excessively, nor cracked, chipped, or otherwise damaged, they may be reused.

13 Inspect the commutator bars on the armature for scoring, scratches and discoloration. The commutator can be cleaned and polished with crocus cloth, but do not use sandpaper or emery paper. After cleaning, wipe away any residue with a cloth soaked in electrical system cleaner or denatured alcohol.

14 Using an ohmmeter or a continuity test light, check for continuity between the commutator bars **(see illustration)**. Continuity should exist between each bar and all of the others. Also, check for continuity between the commutator bars and the armature shaft **(see illustration)**. There should be no continuity (infinite resistance) between the commutator and the shaft. If the checks indicate otherwise, the armature is defective.

15 Check the front end of the armature shaft for worn, cracked, chipped and broken teeth. If the shaft is damaged or worn, replace the armature.

16 Inspect the front and rear covers for signs of cracks or wear. Check the oil seal and the needle bearing in the front cover and the bush in the rear cover for wear and damage – they are not listed as being available separately so if any is found new covers must be fitted **(see illustrations)**.

17 Inspect the magnets in the main housing and the housing itself for cracks.

18 Inspect the insulating washers, O-ring, and sealing rings for signs of damage, deformation and deterioration and replace them with new ones if necessary.

Reassembly

19 Fit the brushplate seat and insulator into the main housing **(see illustrations)**. Fit the insulated brush assembly, locating the arms

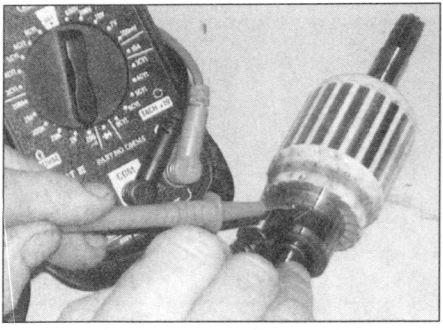

29.14a Continuity should exist between the commutator bars

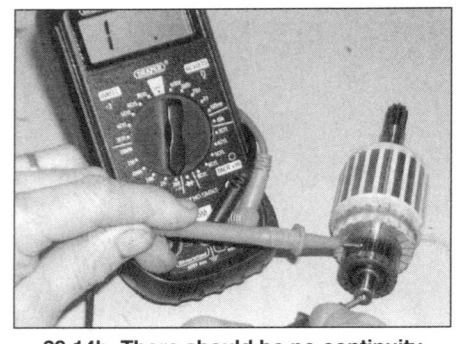

29.14b There should be no continuity between the commutator bars and the armature shaft

29.16a Check the seal and needle bearing in the front cover . . .

29.16b . . . and the bush in the rear cover

29.19a Fit the brushplate seat . . .

Electrical system 9•21

29.19b ... the insulator ...

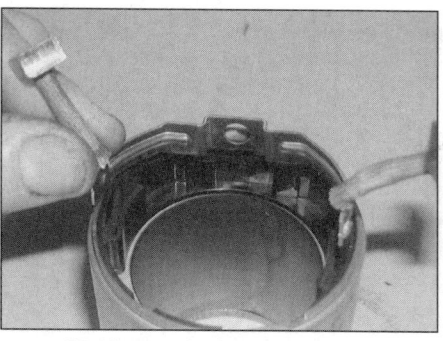

29.19c ... the insulated brush assembly ...

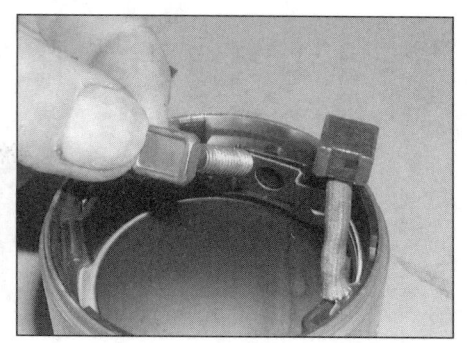

29.19d ... and the terminal bolt

into the brushplate seat **(see illustration)**. Insert the terminal bolt **(see illustration)**. Fit the O-ring, small and large insulating washers, plain washer and nut onto the bolt, and tighten the nut **(see illustrations 29.11b and 29.11a)**. Use a new O-ring if necessary.

20 Fit the brushplate assembly onto the main housing, locating the insulated brush wires in the cut-outs and making sure the tab on the plate locates in the cut-out in the housing **(see illustration)**.

21 At this stage check again for continuity between the terminal bolt and the housing – there should be no continuity (infinite resistance). Also check for continuity between the insulated brushes and the terminal bolt – there should be continuity. Also check for continuity between the uninsulated brush and the brushplate – there should be continuity (zero resistance). If there is no continuity when there should be or *vice versa*, identify the faulty component and replace it with a new one.

22 Fit each insulated brush into its housing and slide them fully in **(see illustration)**. Place each brush spring end onto the top of the outer end of its brush – this keeps the brushes held right in so they do not foul the commutator bars when the armature is installed **(see illustration)**.

23 Insert the armature into the main housing, noting that it will be forcibly drawn in by the attraction of the magnets **(see illustration 29.8)**. Locate each brush spring end onto its brush **(see illustration)**. Check that each brush is securely pressed against the commutator and is free to move in its housing.

24 Slide the shims onto the rear end of the armature **(see illustration 29.7b)**, then fit a new rear cover sealing ring **(see illustration)**. Fit the rear cover onto the housing, aligning the marks noted or made earlier, and locating the tab in the slot in the cover **(see illustration)**.

29.20 Fit the brushplate assembly, making sure the insulated brush wires and the tab locate correctly

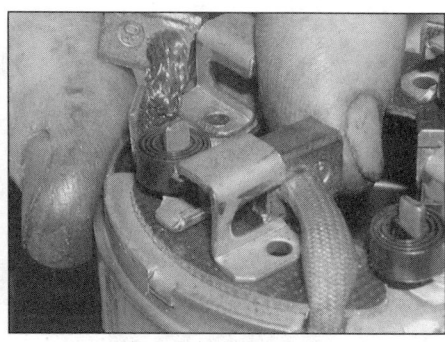

29.22a Slide the brushes into their housings ...

29.22b ... and retain them using the spring end to hold them as shown

29.23 Locate each spring onto its brush so they are pressed against the commutator

29.24a Fit a new sealing ring ...

29.24b ... then fit the rear cover

9•22 Electrical system

29.25 Fit a new sealing ring

29.29 Note the alignment marks between the housing and the covers

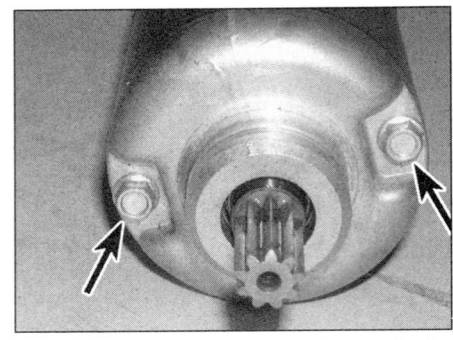

29.30a Unscrew and remove the two bolts (arrowed) . . .

25 Fit a new front cover sealing ring **(see illustration)**. Slide the shims and insulating washer onto the front of the armature **(see illustration 29.6b)**. Apply a smear of grease to the lips of the front cover oil seal. Fit the tabbed washer onto the cover, making sure the tabs locate correctly, then install the cover, aligning the marks made on removal **(see illustration 29.6a)**. Remove the protective tape from the shaft end.
26 Check the alignment marks made on removal are correctly aligned, then install the long bolts and tighten them securely **(see illustration 29.5)**.
27 Install the starter motor (see Section 28).

RR-2 and RR-3 models
Disassembly
28 Remove the starter motor (see Section 28).
29 Note any alignment marks between the main housing and the front and rear covers, or make your own if they aren't clear **(see illustration)**.
30 Unscrew the two long bolts, noting the O-rings, then remove the front cover from the motor along with its sealing ring **(see illustrations)**. Discard the sealing ring as a new one must be used. Remove the tabbed washer from the cover and slide the insulating washer and shim(s) from the front end of the armature, noting the number of shims and their correct fitted order **(see illustrations)**.

31 Remove the rear cover from the motor along with its sealing ring **(see illustration)**. Discard the sealing ring as a new one must be used. Remove the shim(s) from the rear end of the armature noting how many and their correct fitted positions **(see illustration 29.40a)**.
32 Withdraw the armature from the main housing.
33 At this stage check for continuity between the terminal bolt and the brush on the insulated base – there should be continuity (zero resistance) **(see illustration)**. Check for continuity between the terminal bolt and the cover – there should be no continuity (infinite resistance) **(see illustration)**. Also check for

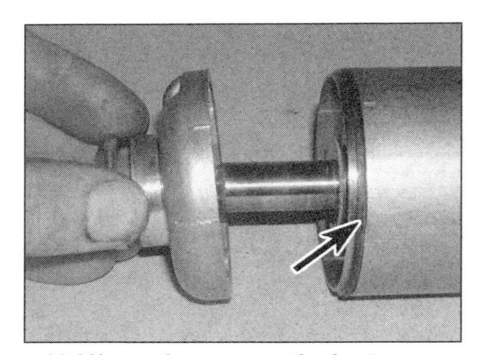

29.30b . . . then remove the front cover and sealing ring (arrowed)

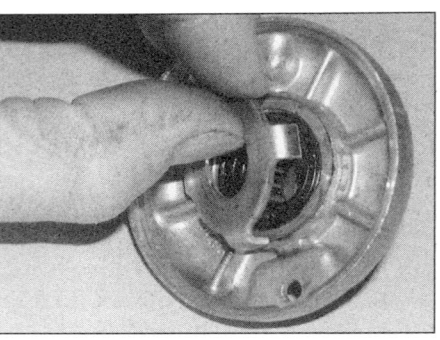

29.30c Remove the tabbed washer . . .

29.30d . . . and the insulating washer and shim(s)

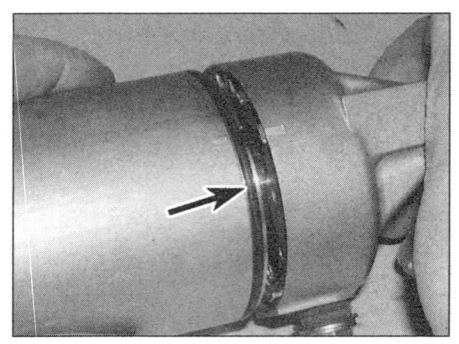

29.31 Remove the rear cover and its sealing ring (arrowed)

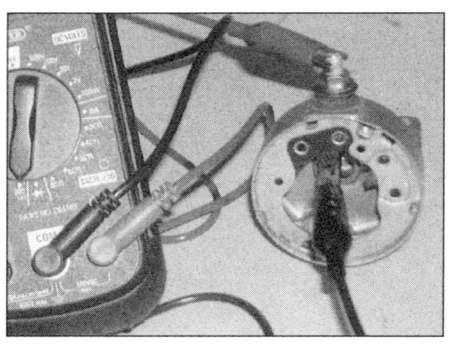

29.33a Check for continuity between the terminal and its brush . . .

29.33b . . . and between the terminal and the cover

Electrical system 9•23

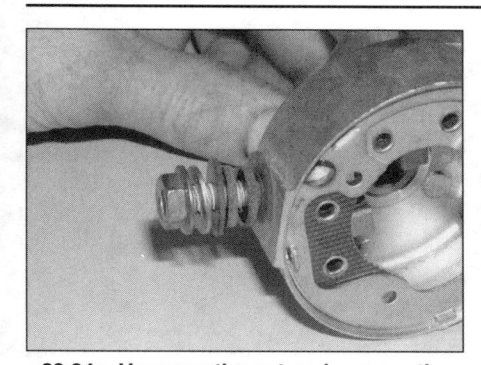

29.34a Unscrew the nut and remove the plain washer and the large and small insulating washers

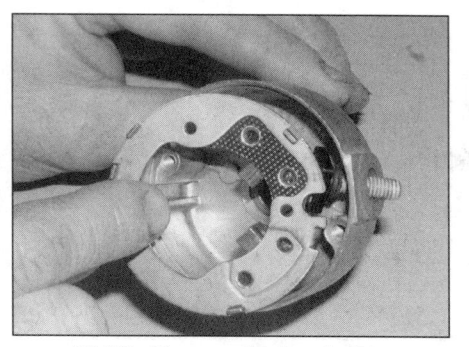

29.34b Remove the brushplate assembly . . .

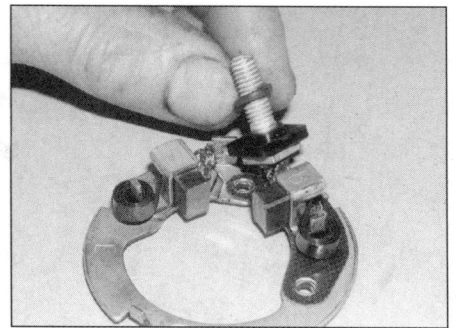

29.34c . . . then remove the O-ring . . .

continuity between the uninsulated brush and the brushplate – there should be continuity (zero resistance). If there is no continuity when there should be or *vice versa*, identify the faulty component and replace it with a new one.

34 Noting the correct fitted location of each component, unscrew the nut from the terminal bolt and remove the plain washer, the one large and two small insulating washers **(see illustration)**. Remove the brushplate assembly and terminal bolt from the main housing, noting how it locates **(see illustration)**. Remove the O-ring and the insulator from the bolt **(see illustrations)**.

35 Move each brush spring end aside and slide the brushes out **(see illustrations)**.

Inspection

36 Refer to Steps 12 to 18 above.

Reassembly

37 Slide the brushes back into position in their housings and locate the brush spring ends onto the outer ends of the brushes **(see illustration 29.35b and 29.35a)**.

38 Fit the insulator and the O-ring onto the terminal bolt **(see illustrations 29.34d and 29.34c)**. Insert the terminal bolt through its hole in the rear cover then fit the brushplate into the cover, making sure its tab is correctly

located in the groove in the cover **(see illustration 29.34b)**. Slide the small insulating washers onto the terminal bolt, followed by the large insulating washer and the plain washer **(see illustration 29.34a)**. Fit the nut onto the terminal bolt and tighten it securely.

39 At this stage check again for continuity between the terminal bolt and the brush on the insulated base – there should be continuity (zero resistance) **(see illustration 29.33a)**. Check for continuity between the terminal bolt and the cover – there should be no continuity (infinite resistance) **(see illustration 29.33b)**. Also check for continuity between the uninsulated brush and the

brushplate – there should be continuity (zero resistance). If there is no continuity when there should be or *vice versa*, identify the faulty component and replace it with a new one.

40 Fit the shims onto the rear of the armature shaft **(see illustration)**. Apply a smear of grease to the end of the shaft. Insert the armature into the brushplate at an angle so that the brushes locate against the commutator, then straighten the armature, pushing the brushes back into their housings against the springs, and slide it into the rear cover so that the shaft end locates in its bush **(see illustration)**.

41 Fit the sealing ring onto the rear of the

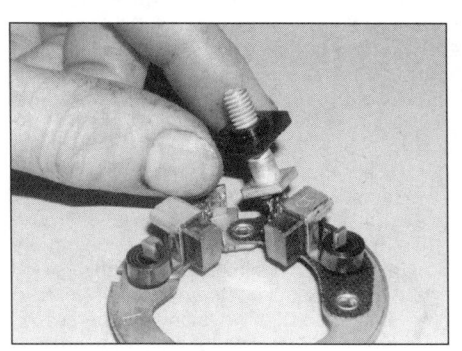

29.34d . . . and the insulator

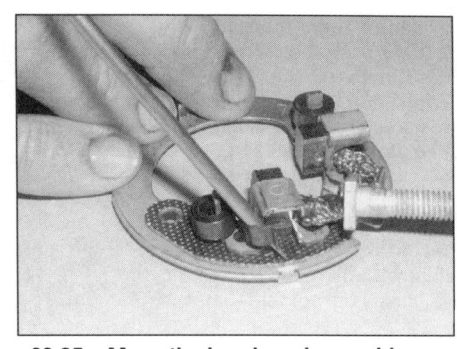

29.35a Move the brush springs aside . . .

29.35b . . . and slide the brushes out

29.40a Fit the shims onto the shaft . . .

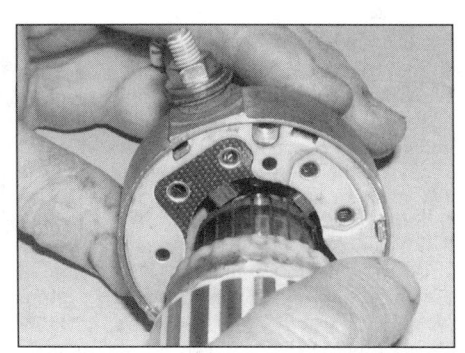

29.40b . . . then fit the armature into the rear cover making sure the brushes locate correctly onto the commutator

9•24 Electrical system

29.41a Fit a new sealing ring onto the rear of the housing...

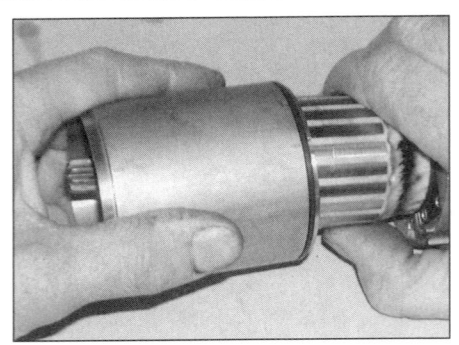

29.41b ... then fit the housing onto the armature

29.43 Fit a new sealing ring onto the front of the housing

main housing **(see illustration)**. Grasp both the armature and the rear cover in one hand and hold them together – this will prevent the armature being drawn out by the magnets in the housing. Note however that you should take care not to let the housing be drawn forcibly onto the armature by the magnets. Carefully allow the housing to be drawn onto the armature, making sure the end with the cut-out faces the rear cover and aligns with the brushplate outer tab **(see illustration)** – aligning the marks between the cover and housing (Step 29) will help.
42 Apply a smear of grease to the front cover oil seal lip. Fit the tabbed washer into the cover so that its teeth are correctly located with the cover ribs **(see illustration 29.30c)**.
43 Fit the sealing ring onto the front of the housing **(see illustration)**. Slide the shim(s) onto the front end of the armature shaft then fit the insulating washer **(see illustration 29.30d)**. Slide the front cover into position, aligning the marks made on removal **(see illustration 29.30b)**.
44 Check the marks made on removal are correctly aligned then fit the long bolts, not forgetting the O-rings (using new ones if necessary) and tighten them securely **(see illustration)**.
45 Install the starter motor (see Section 29).

30 Charging system testing – general information and precautions

1 If the performance of the charging system is suspect, the system as a whole should be checked first, followed by testing of the individual components. **Note:** *Before beginning the checks, make sure the battery is fully-charged and that all system connections are clean and tight.*
2 Checking the output of the charging system and the performance of the various components within the charging system requires the use of a multimeter (with voltage, current and resistance checking facilities).
3 When making the checks, follow the procedures carefully to prevent incorrect connections or short circuits, as irreparable damage to electrical system components may result if short circuits occur.
4 If a multimeter is not available, the job of checking the charging system should be left to a Honda dealer.

31 Charging system – leakage and output test

1 If the charging system of the machine is thought to be faulty, remove the seat (see Chapter 8) and perform the following checks.

Leakage test

Caution: Always connect an ammeter in series, never in parallel with the battery, otherwise it will be damaged. Do not turn the ignition ON or operate the starter motor when the ammeter is connected – a sudden surge in current will blow the meter's fuse.

2 Turn the ignition switch OFF and disconnect the lead from the battery negative (–ve) terminal.
3 Set the multimeter to the amps function and connect its negative (–ve) probe to the battery negative (–ve) terminal, and positive (+ve) probe to the disconnected negative (–ve) lead **(see illustration)**. Always set the meter to a high amps range initially and then bring it down to the mA (milli amps) range; if there is a high current flow in the circuit it may blow the meter's fuse.

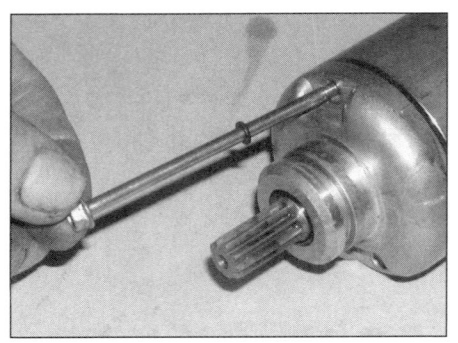

29.44 Fit the long bolts with their O-rings and tighten them securely

4 If the current leakage indicated exceeds the amount specified at the beginning of the Chapter, there is probably a short circuit in the wiring. Use the wiring diagrams at the end of this Chapter and systematically disconnect individual electrical components until the source is identified.
5 Disconnect the meter and connect the negative (–ve) lead to the battery, tightening it securely,

Output test

6 Start the engine and warm it up to normal operating temperature. Remove the rider's seat (see Chapter 8).
7 To check the voltage output, allow the engine to idle and connect a multimeter set to the 0 to 20 volts dc scale (voltmeter) across the terminals of the battery (positive (+) lead to battery positive (+) terminal, negative (–) lead to battery negative (–) terminal. Slowly increase the engine speed to 5000 rpm and note the reading obtained. The regulated voltage should be as specified at the beginning of the Chapter. If the voltage is outside these limits, check the alternator and the regulator (see Sections 32 and 33).

> **HAYNES HiNT** *Clues to a faulty regulator are constantly blowing bulbs, with brightness varying considerably with engine speed, and battery overheating.*

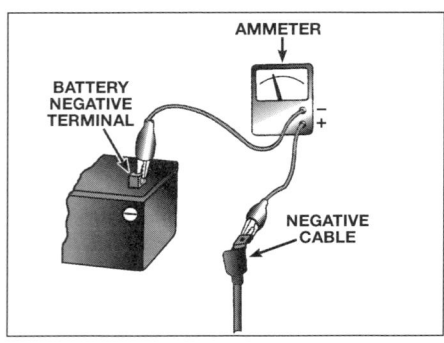

31.3 Checking the charging system leakage rate – connect the meter as shown

Electrical system 9•25

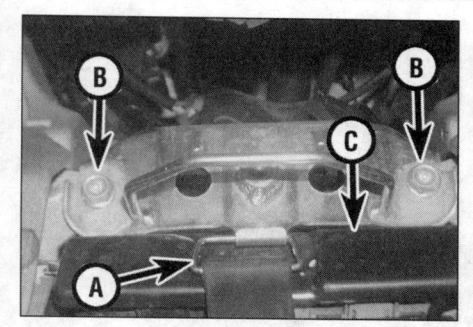

32.2 Release the strap (A), then unscrew the bolts (B) and remove the bracket, then remove the ECM cover (C)

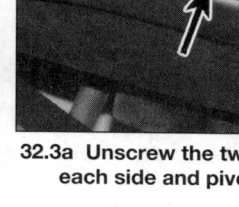

32.3a Unscrew the two bolts (arrowed) on each side and pivot the bracket up

32.3b Remove the ECM cover

32 Alternator – check, removal and installation

Check

1 Remove the rider's seat (see Chapter 8).
2 On RR-Y and RR-1 (2000 and 2001) models unscrew the fuel tank mounting bolts and raise the rear of the tank (see Chapter 4). Release the battery strap then undo the tank bracket bolts and remove the bracket **(see illustration)**. Remove the ECM cover.
3 On RR-2 and RR-3 (2002 and 2003) models unscrew the fuel tank bracket bolts and pivot the bracket up **(see illustration)**. Remove the battery (see Section 3), then remove the ECM cover **(see illustration)**.
4 Trace the wiring from the alternator and disconnect it at the white 3-pin wiring connector with the three yellow wires **(see illustration)**. Check the connector terminals for corrosion and security.
5 Using a multimeter set to the ohms x 1 (ohmmeter) scale measure the resistance between each of the yellow wires on the alternator side of the connector, taking a total of three readings, then check for continuity between each terminal and ground (earth). If the stator coil windings are in good condition the three readings should be within the range shown in the Specifications at the start of this Chapter, and there should be no continuity (infinite resistance) between any of the terminals and ground (earth). If not, the alternator stator coil assembly is at fault and should be replaced with a new one. **Note:** *Before condemning the stator coils, check the fault is not due to damaged wiring between the connector and the coils.*

Removal

6 On RR-Y and RR-1 (2000 and 2001) models remove the lower fairing and the left-hand fairing side panel (see Chapter 8). On RR-2 and RR-3 (2002 and 2003) models remove the left-hand fairing side panel (see Chapter 8). Either drain the engine oil (see Chapter 1), or place a container under the engine to catch the oil that will come out when the alternator cover is removed. If you have an auxiliary stand place the bike on it so that it is level – this minimises oil loss. If you do not have an auxiliary stand it is best to drain the oil.
7 Remove the rider's seat (see Chapter 8).
8 On RR-Y and RR-1 (2000 and 2001) models unscrew the fuel tank mounting bolts and raise the rear of the tank (see Chapter 4). Release the battery strap then undo the tank bracket bolts and remove the bracket **(see illustration 32.2)**. Remove the ECM cover.
9 On RR-2 and RR-3 (2002 and 2003) models unscrew the fuel tank bracket bolts and pivot the bracket up **(see illustration 32.3a)**. Remove the battery (see Section 3), then remove the ECM cover **(see illustration 32.3b)**.
10 Trace the wiring from the alternator and disconnect it at the white 3-pin wiring connector with the three yellow wires **(see illustration 32.4)**. Feed the wiring down to the alternator, releasing it from any ties and noting its routing.
11 Working in a criss-cross pattern, evenly slacken the alternator cover bolts, noting the cable guide secured by one of the rear bolts **(see illustration)**. Lift the cover away from the engine, noting that it will be restrained by the force of the rotor magnets, and be prepared to catch any residual oil. Remove the gasket and discard it. Remove the dowel from either the cover or the crankcase if it is loose.
12 Withdraw the idle/reduction gear shaft from the crankcase and remove the gear, noting which way round it fits **(see illustration)**.
13 To remove the rotor bolt it is necessary to stop the rotor from turning. The best way is to use a commercially available rotor strap **(see illustration)**. If one is not available, try placing

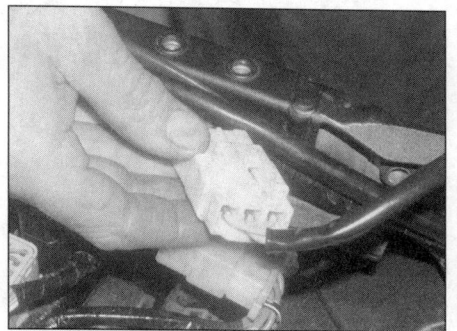

32.4 Disconnect the alternator wiring connector

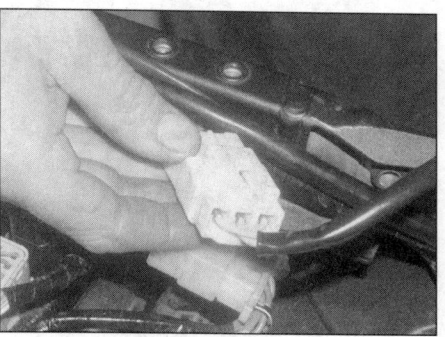

32.11 Alternator cover bolts (arrowed)

32.12 Withdraw the shaft and remove the gear

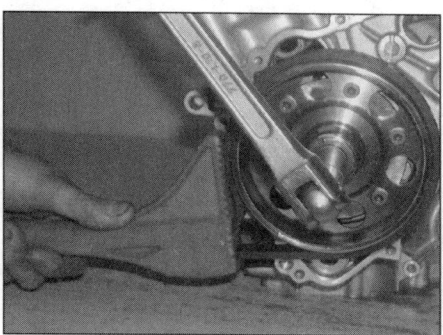

32.13 Using a rotor strap to hold the rotor while unscrewing the bolt

9•26 Electrical system

32.14a Thread the puller into the rotor . . .

32.14b . . . then hold the rotor and turn the puller, using a bar for leverage

32.14c Remove the Woodruff key if it is loose

the transmission in gear and having an assistant apply the rear brake hard. Unscrew the bolt. Note the washer fitted with the bolt.

14 To remove the rotor, complete with starter driven gear, from the shaft it is necessary to use a rotor puller. Thread the rotor puller into the centre of the rotor and turn it until the rotor is displaced from the shaft, holding the rotor to prevent the engine turning **(see illustrations)**. Remove the Woodruff key from its slot in the crankcase if it is loose **(see illustration)**. Separate the starter clutch from the rotor if required (see Chapter 2).

15 To remove the stator from the cover, unscrew its bolts, and the bolt securing the wiring clamp, then remove the assembly, noting how the rubber wiring grommet fits **(see illustration)**.

Installation

16 Fit the stator into the cover, aligning the rubber wiring grommet with the groove **(see illustration 32.15)**. Install the bolts and tighten them to the torque setting specified at the beginning of the Chapter. Apply a suitable sealant to the wiring grommet, then press it into the cut-out in the cover. Secure the wiring with its clamp and tighten the bolt.

17 If separated, fit the starter clutch onto the back of the rotor (see Chapter 2). Apply some oil to the needle bearing on the end of the crankshaft **(see illustration)**.

18 Clean the tapered end of the crankshaft and the corresponding mating surface on the inside of the rotor with a suitable solvent. Fit the Woodruff key into its slot in the crankshaft if removed **(see illustration 32.17)**. Make sure that no metal objects have attached themselves to the magnet on the inside of the rotor. Slide the rotor onto the shaft, making sure the groove on the inside of the rotor is aligned with and fits over the Woodruff key **(see illustration)**. Make sure the Woodruff key does not become dislodged when installing the rotor.

19 Apply some clean oil to the rotor bolt threads and the underside of the head **(see illustration)**. Install the rotor bolt with its washer and tighten it to the torque setting specified at the beginning of the Chapter, using the method employed on removal to prevent the rotor from turning **(see illustration)**.

20 Lubricate the idle/reduction gear shaft with clean engine oil. Position the gear in the crankcase, making sure the smaller pinion faces inwards and its teeth mesh correctly with the teeth of the starter driven gear, and the teeth of the larger pinion mesh correctly with the teeth on the starter motor, then slide the shaft into the gear **(see illustration 32.12)**.

21 Apply a smear of suitable sealant to the area around the crankcase joints. Fit the dowel into the crankcase if removed. Install the alternator cover using a new gasket,

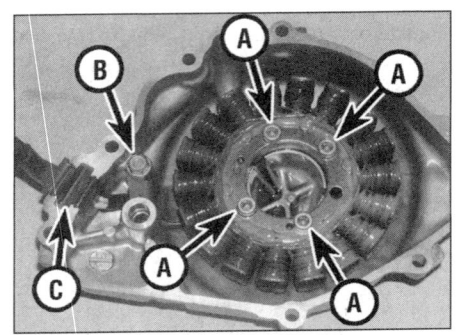

32.15 Unscrew the stator bolts (A) and the wiring clamp bolt (B) and free the grommet (C)

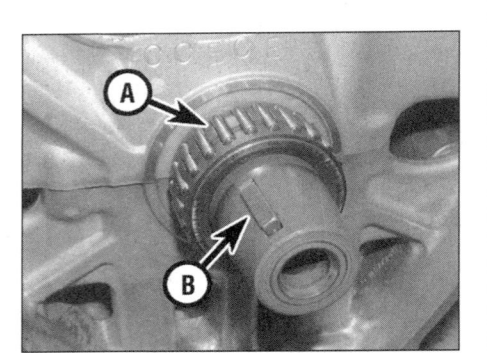

32.17 Lubricate the bearing (A) and fit the Woodruff key (B) if removed

32.18 Slide the rotor onto the shaft, aligning the cut-out with the Woodruff key

32.19a Lubricate the bolt then install it with its washer . . .

32.19b . . . and tighten it to the specified torque

Electrical system 9•27

32.21a Locate the new gasket onto the dowel (arrowed) ...

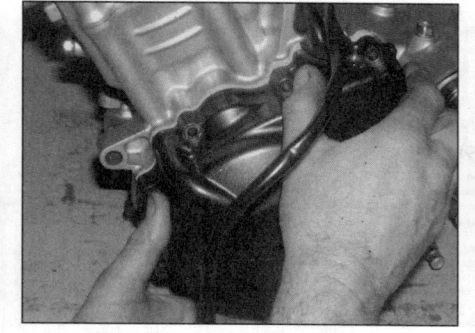

32.21b ... then install the cover

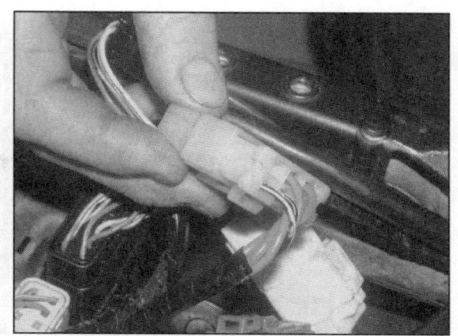

33.4 Disconnect the regulator/rectifier wiring connectors

making sure they locate onto the dowel **(see illustrations)**. Tighten the cover bolts evenly in a criss-cross sequence.

22 Reconnect the wiring at the connector and secure it with any clips or ties previously released **(see illustration 32.4)**. Fit the ECM cover, then install the battery, battery strap and tank bracket as required according to model (see Step 2 or 3).

23 Fill the engine with oil, or top it up to the correct level, according to your removal method (see Chapter 1). Install the fairing side panel and on RR-Y and RR-1 (2000 and 2001) models the lower fairing, and the seat (see Chapter 8).

33 Regulator/rectifier – check and replacement

Check

1 Remove the rider's seat (see Chapter 8).
2 On RR-Y and RR-1 (2000 and 2001) models unscrew the fuel tank mounting bolts and raise the rear of the tank (see Chapter 4). Release the battery strap then undo the tank bracket bolts and remove the bracket **(see illustration 32.2)**. Remove the ECM cover.
3 On RR-2 and RR-3 (2002 and 2003) models unscrew the fuel tank bracket bolts and pivot the bracket up **(see illustration 32.3a)**. Remove the battery (see Section 3), then remove the ECM cover **(see illustration 32.3b)**.
4 Trace the wiring from the regulator/rectifier (which is mounted on the left-hand side of the rear sub-frame behind the seat cowling) and disconnect it at the connectors **(see illustration)**. Check the connector terminals for corrosion and security.

5 Set the multimeter to the 0 to 20 dc volts setting. Connect the meter positive (+) probe to the red/white wire terminal on the loom side of the connector and the negative (–) probe to a suitable ground (earth) and check for voltage. Full battery voltage should be present at all times.
6 Switch the multimeter to the resistance (ohms) scale. Check for continuity between the green wire terminal on the loom side of the connector and ground (earth). There should be continuity.
7 Set the multimeter to the ohms x 1 (ohmmeter) scale and measure the resistance between each of the yellow wires on the alternator (loom) side of the connector, taking a total of three readings, then check for continuity between each terminal and ground (earth). The three readings should be within the range shown in the Specifications for the alternator stator coil at the start of this Chapter, and there should be no continuity (infinite resistance) between any of the terminals and ground (earth).
8 If the above checks do not provide the

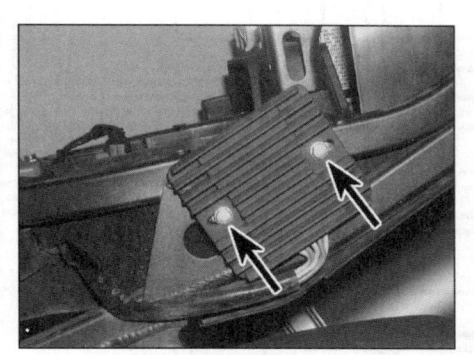

33.14 Unscrew the bolts (arrowed) and remove the regulator/rectifier

expected results check the wiring and connectors between the battery, regulator/rectifier and alternator for shorts, breaks, and loose or corroded terminals (see the wiring diagrams at the end of this chapter).
9 If the wiring checks out, the regulator/rectifier unit is probably faulty. Honda provide no test data for the unit itself. Take it to a Honda dealer for confirmation of its condition before replacing it with a new one.

Replacement

10 Remove the seat cowling (see Chapter 8).
11 On RR-Y and RR-1 (2000 and 2001) models unscrew the fuel tank mounting bolts and raise the rear of the tank (see Chapter 4). Release the battery strap then undo the tank bracket bolts and remove the bracket **(see illustration 32.2)**. Remove the ECM cover.
12 On RR-2 and RR-3 (2002 and 2003) models unscrew the fuel tank bracket bolts and pivot the bracket up **(see illustration 32.3a)**. Remove the battery (see Section 3), then remove the ECM cover **(see illustration 32.3b)**.
13 Trace the wiring from the regulator/rectifier (which is mounted on the left-hand side of the rear sub-frame) and disconnect it at the connectors **(see illustration 33.4)**. Feed the wiring back to the regulator/rectifier.
14 Unscrew the two bolts securing the regulator/rectifier and remove it **(see illustration)**.
15 Install the new unit and tighten its bolts. Connect the wiring connector.
16 Fit the ECM cover, then install the battery, battery strap and tank bracket as required according to model (see Step 2 or 3), and the seat cowling (see Chapter 8).

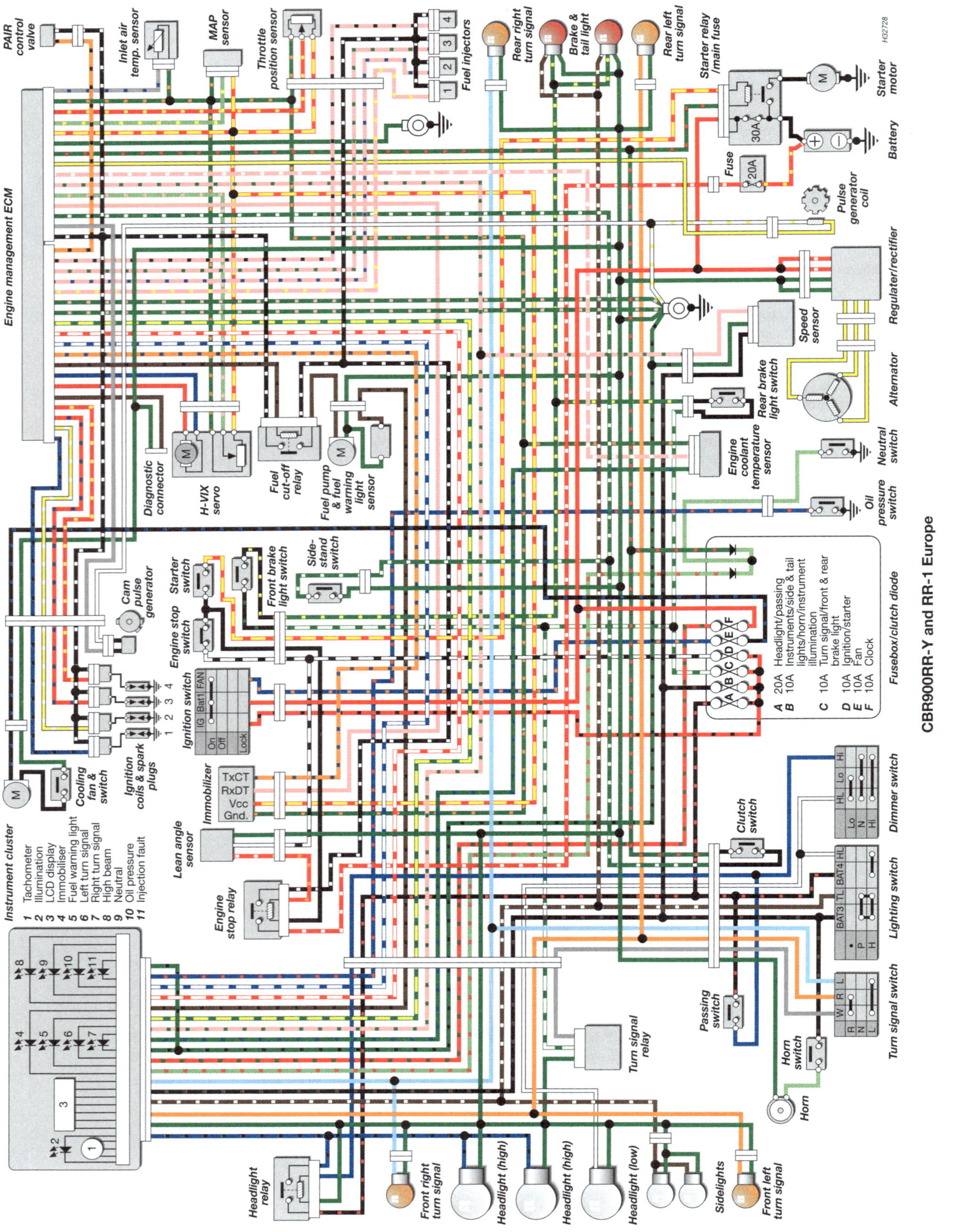
PAIR control valve
Inlet air temp. sensor
MAP sensor
Throttle position sensor
Fuel injectors
1
2
3
4
Rear right turn signal
Brake & tail light
Rear left turn signal
H32728
Starter relay /main fuse
30A
Starter motor
Battery
Fuse
20A
Pulse generator coil
Regulater/rectifier
Speed sensor
Alternator
Rear brake light switch
Neutral switch
Engine coolant temperature sensor
Oil pressure switch
Engine management ECM
Diagnostic connector
H-VIX servo
Fuel cut-off relay
Fuel pump & fuel warning light sensor
Cam pulse generator
Starter switch
Front brake light switch
Side-stand switch
Engine stop switch
Ignition switch
IG
Bat1
FAN
On
Off
Lock
Ignition coils & spark plugs
Cooling fan & switch
Immobilizer
TxCT
RxDT
Vcc
Gnd.
Lean angle sensor
Engine stop relay
Fusebox/clutch diode
A 20A Headlight/passing
B 10A Instruments/side & tail lights/horn/instrument illumination
C 10A Turn signal/front & rear brake light
D 10A Ignition/starter
E 10A Fan
F 10A Clock
CBR900RR-Y and RR-1 Europe
Instrument cluster
1 Tachometer
2 Illumination
3 LCD display
4 Immobiliser
5 Fuel warning light
6 Left turn signal
7 Right turn signal
8 High beam
9 Neutral
10 Oil pressure
11 Injection fault
Clutch switch
Dimmer switch
Lo
HL
Hi
N
Lighting switch
BAT3
TL
BAT4
P
H
Turn signal switch
W
R
L
Passing switch
Horn switch
Horn
Turn signal relay
Headlight relay
Front right turn signal
Headlight (high)
Headlight (low)
Sidelights
Front left turn signal

Wiring diagrams 9•29

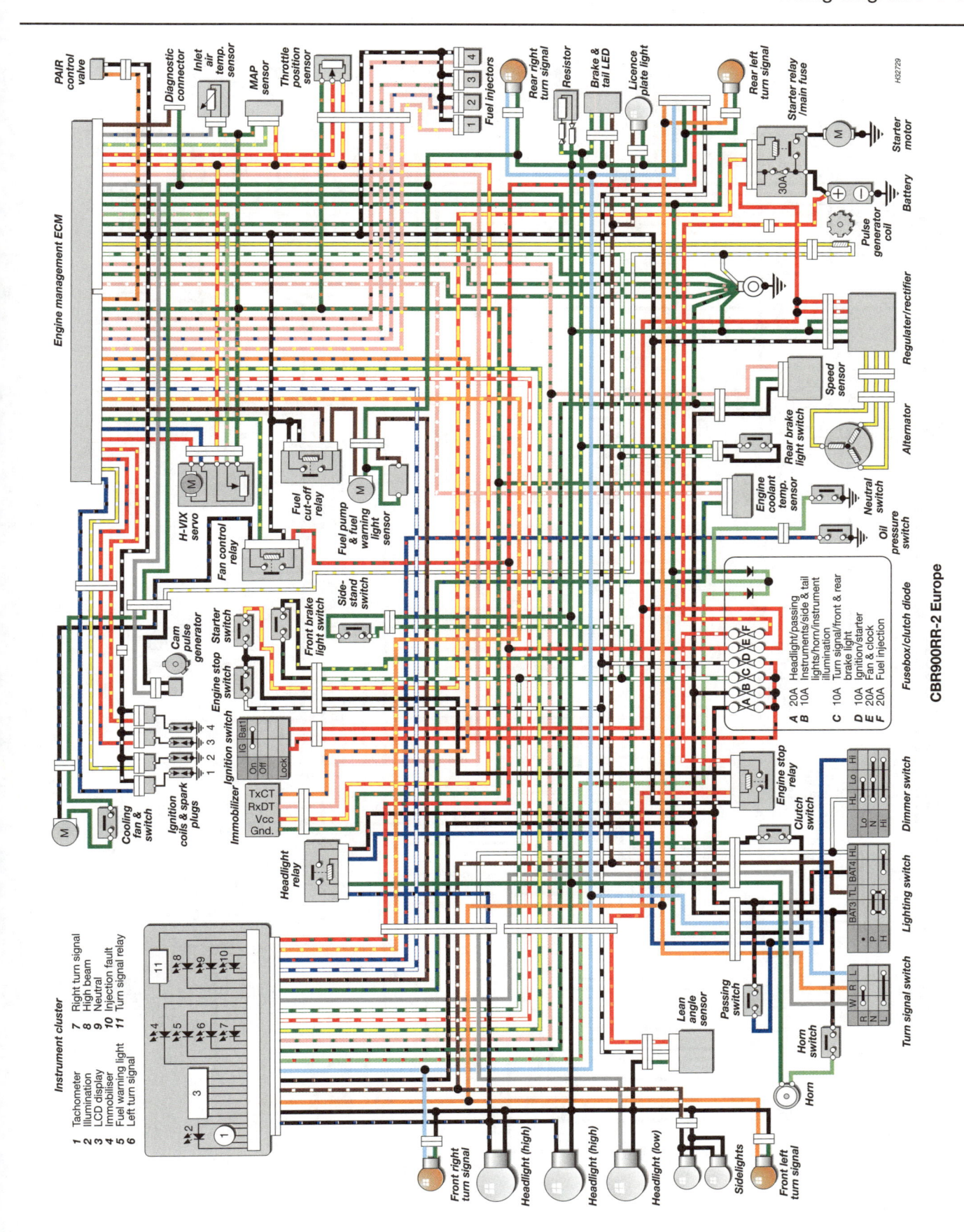

CBR900RR-2 Europe

9•30 Wiring diagrams

Wiring diagrams 9•31

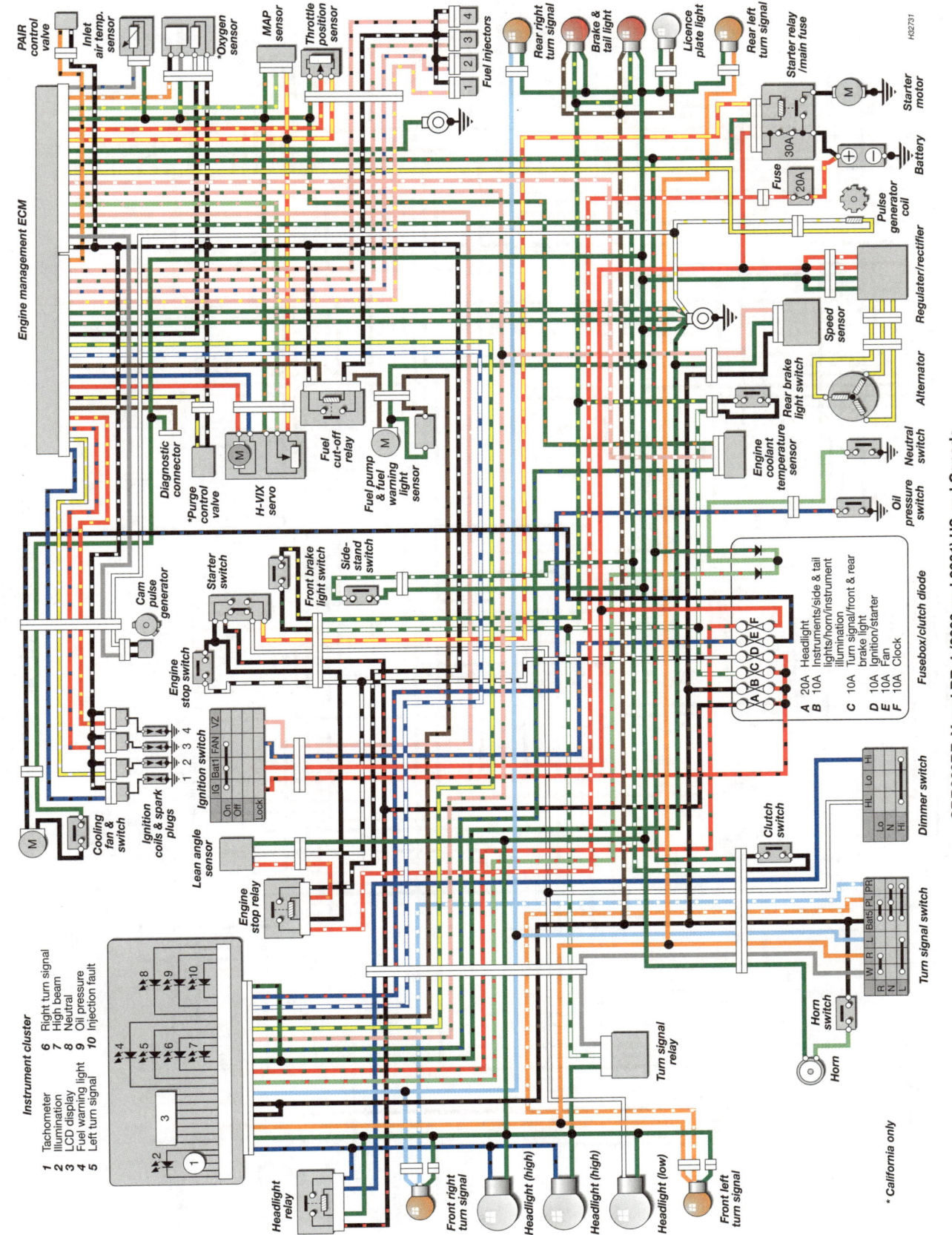

CBR929RR-Y and RR-1 (2000 and 2001) US and Canada

*California only

9•32 Wiring diagrams

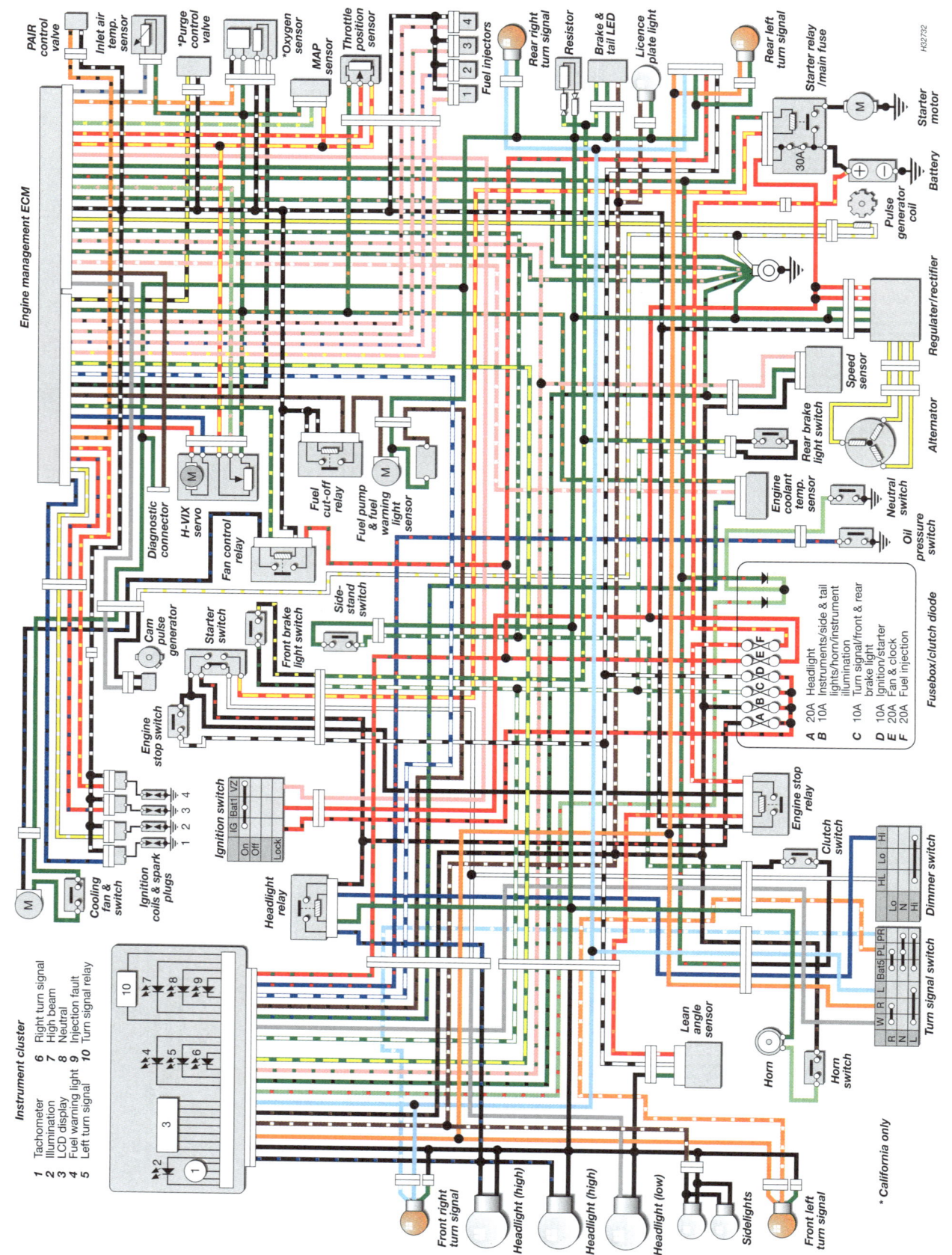

CBR954RR-2 and RR-3 (2002 and 2003) US and Canada

*California only

Reference

Tools and Workshop Tips — REF•2
- Building up a tool kit and equipping your workshop
- Using tools
- Understanding bearing, seal, fastener and chain sizes and markings
- Repair techniques

Security — REF•20
- Locks and chains
- U-locks
- Disc locks
- Alarms and immobilisers
- Security marking systems
- Tips on how to prevent bike theft

Lubricants and fluids — REF•23
- Engine oils
- Transmission (gear) oils
- Coolant/anti-freeze
- Fork oils and suspension fluids
- Brake/clutch fluids
- Spray lubes, degreasers and solvents

Conversion Factors — REF•26

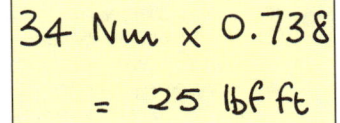

34 Nm x 0.738 = 25 lbf ft

- Formulae for conversion of the metric (SI) units used throughout the manual into Imperial measures

MOT Test Checks — REF•27
- A guide to the UK MOT test
- Which items are tested
- How to prepare your motorcycle for the test and perform a pre-test check

Storage — REF•32
- How to prepare your motorcycle for going into storage and protect essential systems
- How to get the motorcycle back on the road

Fault Finding — REF•35
- Common faults and their likely causes
- How to check engine cylinder compression
- How to make electrical tests and use test meters

Technical Terms Explained — REF•48
- Component names, technical terms and common abbreviations explained

Index — REF•52

REF•2 Tools and Workshop Tips

Buying tools

A toolkit is a fundamental requirement for servicing and repairing a motorcycle. Although there will be an initial expense in building up enough tools for servicing, this will soon be offset by the savings made by doing the job yourself. As experience and confidence grow, additional tools can be added to enable the repair and overhaul of the motorcycle. Many of the specialist tools are expensive and not often used so it may be preferable to hire them, or for a group of friends or motorcycle club to join in the purchase.

As a rule, it is better to buy more expensive, good quality tools. Cheaper tools are likely to wear out faster and need to be renewed more often, nullifying the original saving.

> ⚠ **Warning:** To avoid the risk of a poor quality tool breaking in use, causing injury or damage to the component being worked on, always aim to purchase tools which meet the relevant national safety standards.

The following lists of tools do not represent the manufacturer's service tools, but serve as a guide to help the owner decide which tools are needed for this level of work. In addition, items such as an electric drill, hacksaw, files, soldering iron and a workbench equipped with a vice, may be needed. Although not classed as tools, a selection of bolts, screws, nuts, washers and pieces of tubing always come in useful.

For more information about tools, refer to the Haynes *Motorcycle Workshop Practice TechBook* (Bk. No. 3470).

Manufacturer's service tools

Inevitably certain tasks require the use of a service tool. Where possible an alternative tool or method of approach is recommended, but sometimes there is no option if personal injury or damage to the component is to be avoided. Where required, service tools are referred to in the relevant procedure.

Service tools can usually only be purchased from a motorcycle dealer and are identified by a part number. Some of the commonly-used tools, such as rotor pullers, are available in aftermarket form from mail-order motorcycle tool and accessory suppliers.

Maintenance and minor repair tools

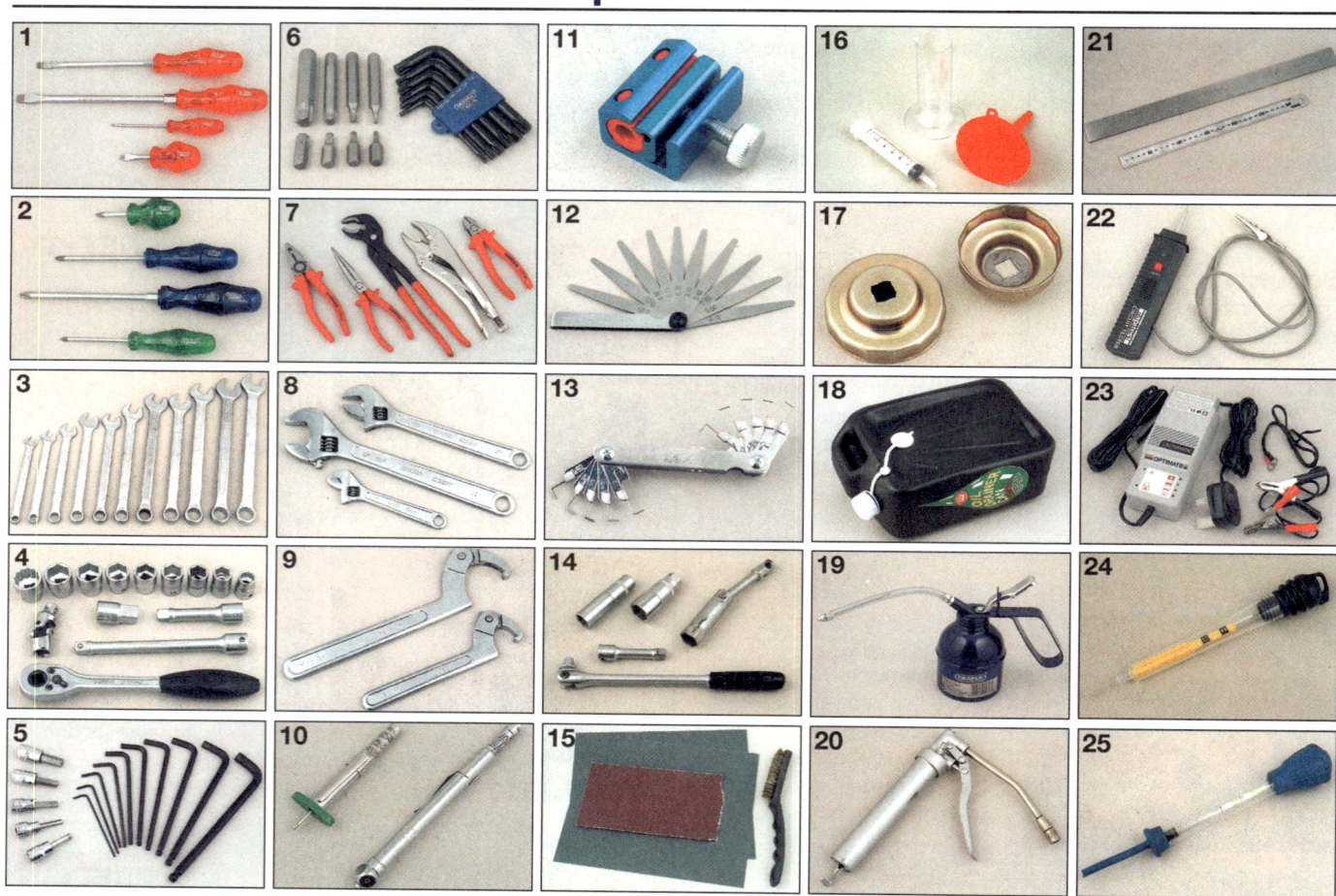

1. Set of flat-bladed screwdrivers
2. Set of Phillips head screwdrivers
3. Combination open-end and ring spanners
4. Socket set (3/8 inch or 1/2 inch drive)
5. Set of Allen keys or bits
6. Set of Torx keys or bits
7. Pliers, cutters and self-locking grips (Mole grips)
8. Adjustable spanners
9. C-spanners
10. Tread depth gauge and tyre pressure gauge
11. Cable oiler clamp
12. Feeler gauges
13. Spark plug gap measuring tool
14. Spark plug spanner or deep plug sockets
15. Wire brush and emery paper
16. Calibrated syringe, measuring vessel and funnel
17. Oil filter adapters
18. Oil drainer can or tray
19. Pump type oil can
20. Grease gun
21. Straight-edge and steel rule
22. Continuity tester
23. Battery charger
24. Hydrometer (for battery specific gravity check)
25. Anti-freeze tester (for liquid-cooled engines)

Tools and Workshop Tips REF•3

Repair and overhaul tools

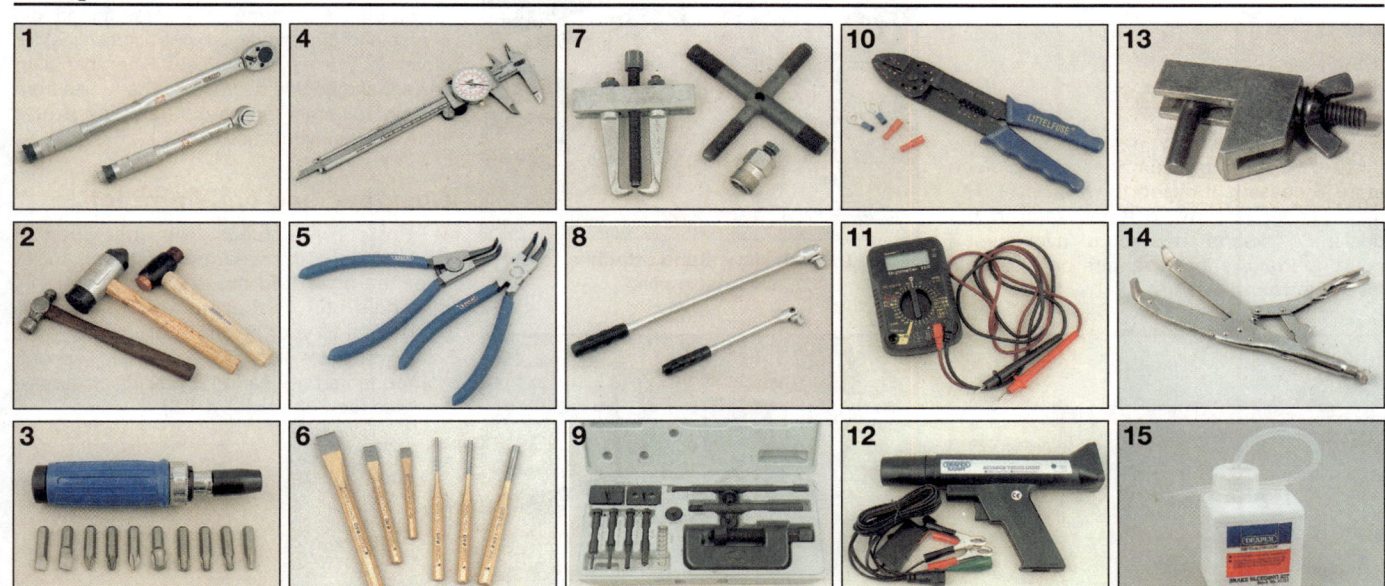

1 Torque wrench (small and mid-ranges)
2 Conventional, plastic or soft-faced hammers
3 Impact driver set
4 Vernier gauge
5 Circlip pliers (internal and external, or combination)
6 Set of cold chisels and punches
7 Selection of pullers
8 Breaker bars
9 Chain breaking/riveting tool set
10 Wire stripper and crimper tool
11 Multimeter (measures amps, volts and ohms)
12 Stroboscope (for dynamic timing checks)
13 Hose clamp (wingnut type shown)
14 Clutch holding tool
15 One-man brake/clutch bleeder kit

Specialist tools

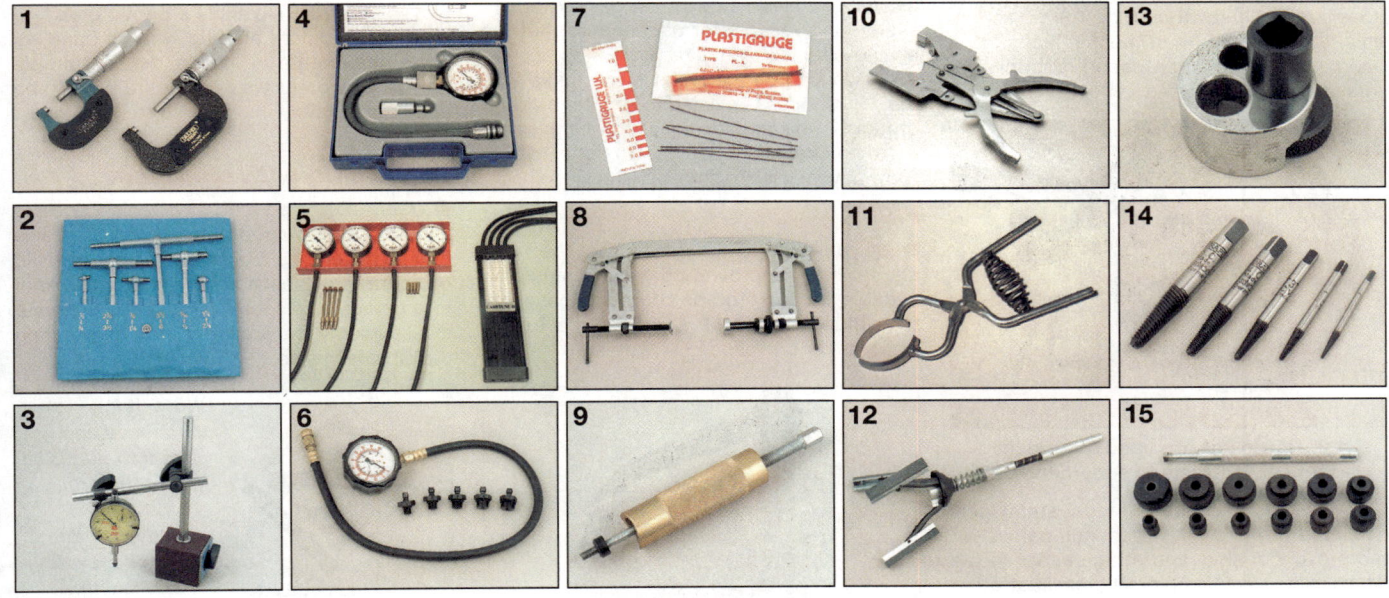

1 Micrometers (external type)
2 Telescoping gauges
3 Dial gauge
4 Cylinder compression gauge
5 Vacuum gauges (left) or manometer (right)
6 Oil pressure gauge
7 Plastigauge kit
8 Valve spring compressor (4-stroke engines)
9 Piston pin drawbolt tool
10 Piston ring removal and installation tool
11 Piston ring clamp
12 Cylinder bore hone (stone type shown)
13 Stud extractor
14 Screw extractor set
15 Bearing driver set

REF•4 Tools and Workshop Tips

1 Workshop equipment and facilities

The workbench

- Work is made much easier by raising the bike up on a ramp - components are much more accessible if raised to waist level. The hydraulic or pneumatic types seen in the dealer's workshop are a sound investment if you undertake a lot of repairs or overhauls **(see illustration 1.1)**.

1.1 Hydraulic motorcycle ramp

- If raised off ground level, the bike must be supported on the ramp to avoid it falling. Most ramps incorporate a front wheel locating clamp which can be adjusted to suit different diameter wheels. When tightening the clamp, take care not to mark the wheel rim or damage the tyre - use wood blocks on each side to prevent this.
- Secure the bike to the ramp using tie-downs **(see illustration 1.2)**. If the bike has only a sidestand, and hence leans at a dangerous angle when raised, support the bike on an auxiliary stand.

1.2 Tie-downs are used around the passenger footrests to secure the bike

- Auxiliary (paddock) stands are widely available from mail order companies or motorcycle dealers and attach either to the wheel axle or swingarm pivot **(see illustration 1.3)**. If the motorcycle has a centrestand, you can support it under the crankcase to prevent it toppling whilst either wheel is removed **(see illustration 1.4)**.

1.3 This auxiliary stand attaches to the swingarm pivot

1.4 Always use a block of wood between the engine and jack head when supporting the engine in this way

Fumes and fire

- Refer to the Safety first! page at the beginning of the manual for full details. Make sure your workshop is equipped with a fire extinguisher suitable for fuel-related fires (Class B fire - flammable liquids) - it is not sufficient to have a water-filled extinguisher.
- Always ensure adequate ventilation is available. Unless an exhaust gas extraction system is available for use, ensure that the engine is run outside of the workshop.
- If working on the fuel system, make sure the workshop is ventilated to avoid a build-up of fumes. This applies equally to fume build-up when charging a battery. Do not smoke or allow anyone else to smoke in the workshop.

Fluids

- If you need to drain fuel from the tank, store it in an approved container marked as suitable for the storage of petrol (gasoline) **(see illustration 1.5)**. Do not store fuel in glass jars or bottles.

1.5 Use an approved can only for storing petrol (gasoline)

- Use proprietary engine degreasers or solvents which have a high flash-point, such as paraffin (kerosene), for cleaning off oil, grease and dirt - never use petrol (gasoline) for cleaning. Wear rubber gloves when handling solvent and engine degreaser. The fumes from certain solvents can be dangerous - always work in a well-ventilated area.

Dust, eye and hand protection

- Protect your lungs from inhalation of dust particles by wearing a filtering mask over the nose and mouth. Many frictional materials still contain asbestos which is dangerous to your health. Protect your eyes from spouts of liquid and sprung components by wearing a pair of protective goggles **(see illustration 1.6)**.

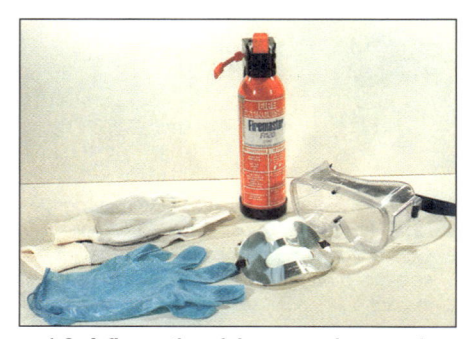

1.6 A fire extinguisher, goggles, mask and protective gloves should be at hand in the workshop

- Protect your hands from contact with solvents, fuel and oils by wearing rubber gloves. Alternatively apply a barrier cream to your hands before starting work. If handling hot components or fluids, wear suitable gloves to protect your hands from scalding and burns.

What to do with old fluids

- Old cleaning solvent, fuel, coolant and oils should not be poured down domestic drains or onto the ground. Package the fluid up in old oil containers, label it accordingly, and take it to a garage or disposal facility. Contact your local authority for location of such sites or ring the oil care hotline.

OIL CARE
OIL BANK LINE
0800 66 33 66
www.oilbankline.org.uk

Note: It is antisocial and illegal to dump oil down the drain. To find the location of your local oil recycling bank, call this number free.

In the USA, note that any oil supplier must accept used oil for recycling.

Tools and Workshop Tips REF•5

2 Fasteners - screws, bolts and nuts

Fastener types and applications

Bolts and screws

● Fastener head types are either of hexagonal, Torx or splined design, with internal and external versions of each type **(see illustrations 2.1 and 2.2)**; splined head fasteners are not in common use on motorcycles. The conventional slotted or Phillips head design is used for certain screws. Bolt or screw length is always measured from the underside of the head to the end of the item **(see illustration 2.11)**.

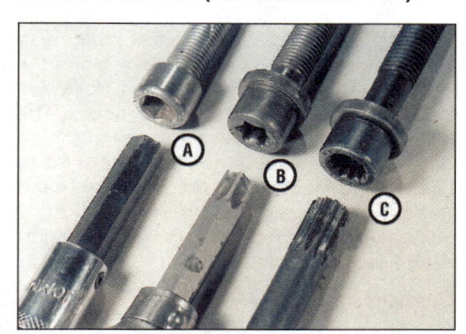

2.1 Internal hexagon/Allen (A), Torx (B) and splined (C) fasteners, with corresponding bits

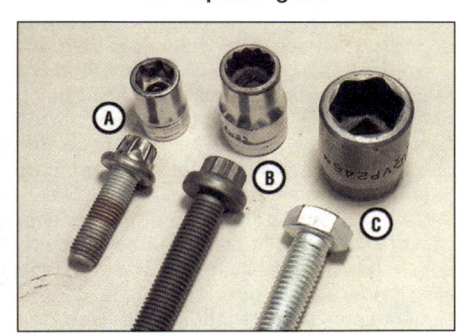

2.2 External Torx (A), splined (B) and hexagon (C) fasteners, with corresponding sockets

● Certain fasteners on the motorcycle have a tensile marking on their heads, the higher the marking the stronger the fastener. High tensile fasteners generally carry a 10 or higher marking. Never replace a high tensile fastener with one of a lower tensile strength.

Washers (see illustration 2.3)

● Plain washers are used between a fastener head and a component to prevent damage to the component or to spread the load when torque is applied. Plain washers can also be used as spacers or shims in certain assemblies. Copper or aluminium plain washers are often used as sealing washers on drain plugs.

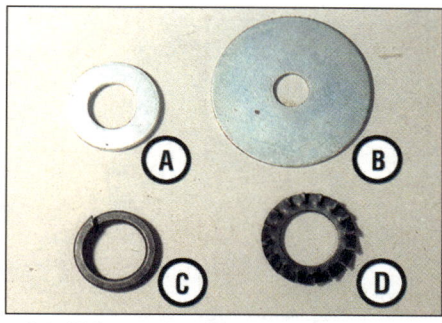

2.3 Plain washer (A), penny washer (B), spring washer (C) and serrated washer (D)

● The split-ring spring washer works by applying axial tension between the fastener head and component. If flattened, it is fatigued and must be renewed. If a plain (flat) washer is used on the fastener, position the spring washer between the fastener and the plain washer.

● Serrated star type washers dig into the fastener and component faces, preventing loosening. They are often used on electrical earth (ground) connections to the frame.

● Cone type washers (sometimes called Belleville) are conical and when tightened apply axial tension between the fastener head and component. They must be installed with the dished side against the component and often carry an OUTSIDE marking on their outer face. If flattened, they are fatigued and must be renewed.

● Tab washers are used to lock plain nuts or bolts to a shaft. A portion of the tab washer is bent up hard against one flat of the nut or bolt to prevent it loosening. Due to the tab washer being deformed in use, a new tab washer should be used every time it is disturbed.

● Wave washers are used to take up endfloat on a shaft. They provide light springing and prevent excessive side-to-side play of a component. Can be found on rocker arm shafts.

Nuts and split pins

● Conventional plain nuts are usually six-sided **(see illustration 2.4)**. They are sized by thread diameter and pitch. High tensile nuts carry a number on one end to denote their tensile strength.

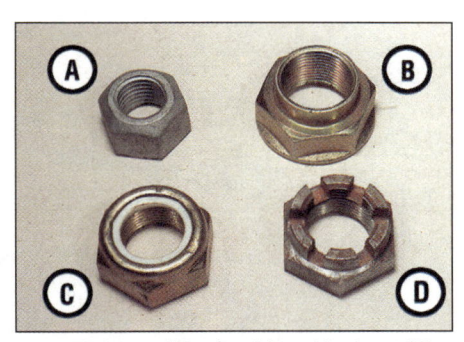

2.4 Plain nut (A), shouldered locknut (B), nylon insert nut (C) and castellated nut (D)

● Self-locking nuts either have a nylon insert, or two spring metal tabs, or a shoulder which is staked into a groove in the shaft - their advantage over conventional plain nuts is a resistance to loosening due to vibration. The nylon insert type can be used a number of times, but must be renewed when the friction of the nylon insert is reduced, ie when the nut spins freely on the shaft. The spring tab type can be reused unless the tabs are damaged. The shouldered type must be renewed every time it is disturbed.

● Split pins (cotter pins) are used to lock a castellated nut to a shaft or to prevent slackening of a plain nut. Common applications are wheel axles and brake torque arms. Because the split pin arms are deformed to lock around the nut a new split pin must always be used on installation - always fit the correct size split pin which will fit snugly in the shaft hole. Make sure the split pin arms are correctly located around the nut **(see illustrations 2.5 and 2.6)**.

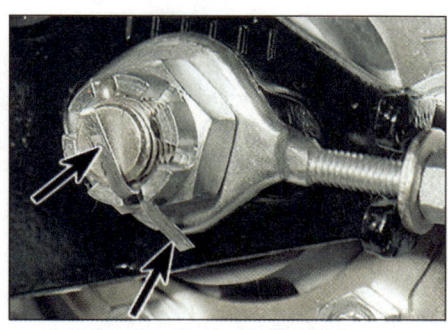

2.5 Bend split pin (cotter pin) arms as shown (arrows) to secure a castellated nut

2.6 Bend split pin (cotter pin) arms as shown to secure a plain nut

> **Caution:** If the castellated nut slots do not align with the shaft hole after tightening to the torque setting, tighten the nut until the next slot aligns with the hole - never slacken the nut to align its slot.

● R-pins (shaped like the letter R), or slip pins as they are sometimes called, are sprung and can be reused if they are otherwise in good condition. Always install R-pins with their closed end facing forwards **(see illustration 2.7)**.

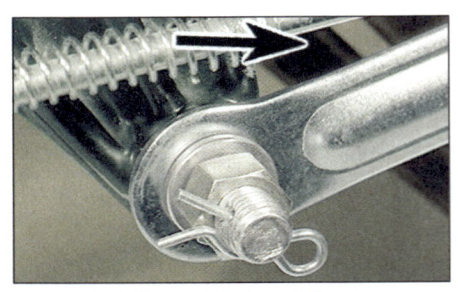

REF•6 Tools and Workshop Tips

2.7 Correct fitting of R-pin. Arrow indicates forward direction

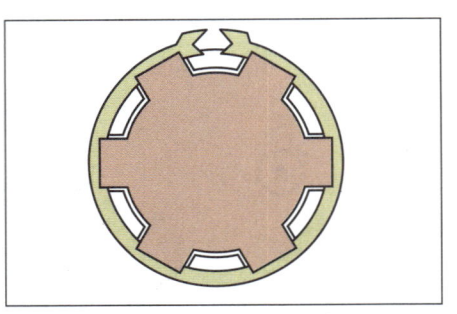

2.10 Align circlip opening with shaft channel

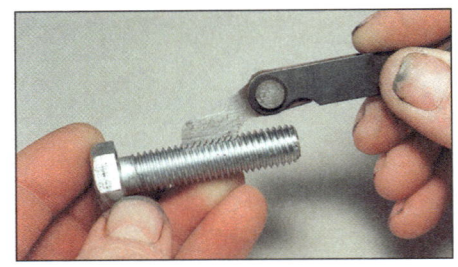

2.12 Using a thread gauge to measure pitch

Circlips (see illustration 2.8)

● Circlips (sometimes called snap-rings) are used to retain components on a shaft or in a housing and have corresponding external or internal ears to permit removal. Parallel-sided (machined) circlips can be installed either way round in their groove, whereas stamped circlips (which have a chamfered edge on one face) must be installed with the chamfer facing away from the direction of thrust load **(see illustration 2.9)**.

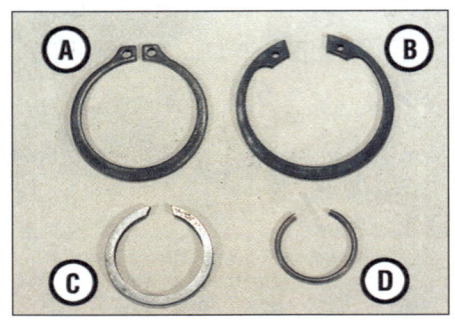

2.8 External stamped circlip (A), internal stamped circlip (B), machined circlip (C) and wire circlip (D)

● Always use circlip pliers to remove and install circlips; expand or compress them just enough to remove them. After installation, rotate the circlip in its groove to ensure it is securely seated. If installing a circlip on a splined shaft, always align its opening with a shaft channel to ensure the circlip ends are well supported and unlikely to catch **(see illustration 2.10)**.

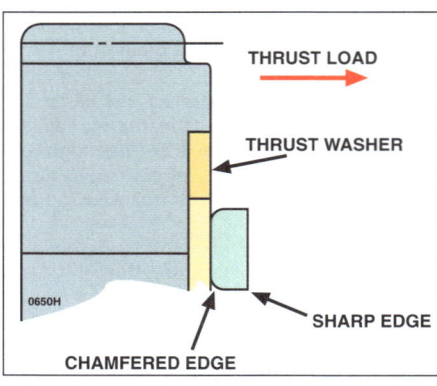

2.9 Correct fitting of a stamped circlip

● Circlips can wear due to the thrust of components and become loose in their grooves, with the subsequent danger of becoming dislodged in operation. For this reason, renewal is advised every time a circlip is disturbed.

● Wire circlips are commonly used as piston pin retaining clips. If a removal tang is provided, long-nosed pliers can be used to dislodge them, otherwise careful use of a small flat-bladed screwdriver is necessary. Wire circlips should be renewed every time they are disturbed.

Thread diameter and pitch

● Diameter of a male thread (screw, bolt or stud) is the outside diameter of the threaded portion **(see illustration 2.11)**. Most motorcycle manufacturers use the ISO (International Standards Organisation) metric system expressed in millimetres, eg M6 refers to a 6 mm diameter thread. Sizing is the same for nuts, except that the thread diameter is measured across the valleys of the nut.

● Pitch is the distance between the peaks of the thread **(see illustration 2.11)**. It is expressed in millimetres, thus a common bolt size may be expressed as 6.0 x 1.0 mm (6 mm thread diameter and 1 mm pitch). Generally pitch increases in proportion to thread diameter, although there are always exceptions.

● Thread diameter and pitch are related for conventional fastener applications and the accompanying table can be used as a guide. Additionally, the AF (Across Flats), spanner or socket size dimension of the bolt or nut **(see illustration 2.11)** is linked to thread and pitch specification. Thread pitch can be measured with a thread gauge **(see illustration 2.12)**.

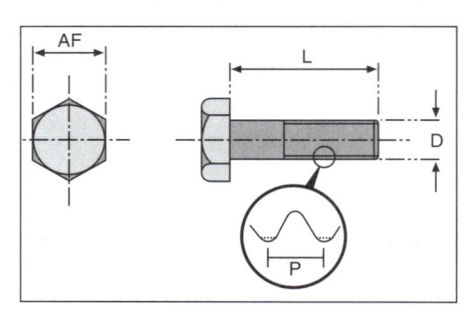

2.11 Fastener length (L), thread diameter (D), thread pitch (P) and head size (AF)

AF size	Thread diameter x pitch (mm)
8 mm	M5 x 0.8
8 mm	M6 x 1.0
10 mm	M6 x 1.0
12 mm	M8 x 1.25
14 mm	M10 x 1.25
17 mm	M12 x 1.25

● The threads of most fasteners are of the right-hand type, ie they are turned clockwise to tighten and anti-clockwise to loosen. The reverse situation applies to left-hand thread fasteners, which are turned anti-clockwise to tighten and clockwise to loosen. Left-hand threads are used where rotation of a component might loosen a conventional right-hand thread fastener.

Seized fasteners

● Corrosion of external fasteners due to water or reaction between two dissimilar metals can occur over a period of time. It will build up sooner in wet conditions or in countries where salt is used on the roads during the winter. If a fastener is severely corroded it is likely that normal methods of removal will fail and result in its head being ruined. When you attempt removal, the fastener thread should be heard to crack free and unscrew easily - if it doesn't, stop there before damaging something.

● A smart tap on the head of the fastener will often succeed in breaking free corrosion which has occurred in the threads **(see illustration 2.13)**.

● An aerosol penetrating fluid (such as WD-40) applied the night beforehand may work its way down into the thread and ease removal. Depending on the location, you may be able to make up a Plasticine well around the fastener head and fill it with penetrating fluid.

2.13 A sharp tap on the head of a fastener will often break free a corroded thread

Tools and Workshop Tips REF•7

● If you are working on an engine internal component, corrosion will most likely not be a problem due to the well lubricated environment. However, components can be very tight and an impact driver is a useful tool in freeing them **(see illustration 2.14)**.

2.14 Using an impact driver to free a fastener

● Where corrosion has occurred between dissimilar metals (eg steel and aluminium alloy), the application of heat to the fastener head will create a disproportionate expansion rate between the two metals and break the seizure caused by the corrosion. Whether heat can be applied depends on the location of the fastener - any surrounding components likely to be damaged must first be removed **(see illustration 2.15)**. Heat can be applied using a paint stripper heat gun or clothes iron, or by immersing the component in boiling water - wear protective gloves to prevent scalding or burns to the hands.

2.15 Using heat to free a seized fastener

● As a last resort, it is possible to use a hammer and cold chisel to work the fastener head unscrewed **(see illustration 2.16)**. This will damage the fastener, but more importantly extreme care must be taken not to damage the surrounding component.

> **Caution:** Remember that the component being secured is generally of more value than the bolt, nut or screw - when the fastener is freed, do not unscrew it with force, instead work the fastener back and forth when resistance is felt to prevent thread damage.

2.16 Using a hammer and chisel to free a seized fastener

Broken fasteners and damaged heads

● If the shank of a broken bolt or screw is accessible you can grip it with self-locking grips. The knurled wheel type stud extractor tool or self-gripping stud puller tool is particularly useful for removing the long studs which screw into the cylinder mouth surface of the crankcase or bolts and screws from which the head has broken off **(see illustration 2.17)**. Studs can also be removed by locking two nuts together on the threaded end of the stud and using a spanner on the lower nut **(see illustration 2.18)**.

2.17 Using a stud extractor tool to remove a broken crankcase stud

2.18 Two nuts can be locked together to unscrew a stud from a component

● A bolt or screw which has broken off below or level with the casing must be extracted using a screw extractor set. Centre punch the fastener to centralise the drill bit, then drill a hole in the fastener **(see illustration 2.19)**. Select a drill bit which is approximately half to three-quarters the diameter of the fastener and drill to a depth which will accommodate the extractor. Use the largest size extractor possible, but avoid leaving too small a wall thickness otherwise the extractor will merely force the fastener walls outwards wedging it in the casing thread.

2.19 When using a screw extractor, first drill a hole in the fastener . . .

● If a spiral type extractor is used, thread it anti-clockwise into the fastener. As it is screwed in, it will grip the fastener and unscrew it from the casing **(see illustration 2.20)**.

2.20 . . . then thread the extractor anti-clockwise into the fastener

● If a taper type extractor is used, tap it into the fastener so that it is firmly wedged in place. Unscrew the extractor (anti-clockwise) to draw the fastener out.

> **Warning:** Stud extractors are very hard and may break off in the fastener if care is not taken - ask an engineer about spark erosion if this happens.

● Alternatively, the broken bolt/screw can be drilled out and the hole retapped for an oversize bolt/screw or a diamond-section thread insert. It is essential that the drilling is carried out squarely and to the correct depth, otherwise the casing may be ruined - if in doubt, entrust the work to an engineer.

● Bolts and nuts with rounded corners cause the correct size spanner or socket to slip when force is applied. Of the types of spanner/socket available always use a six-point type rather than an eight or twelve-point type - better grip

REF•8 Tools and Workshop Tips

2.21 Comparison of surface drive ring spanner (left) with 12-point type (right)

is obtained. Surface drive spanners grip the middle of the hex flats, rather than the corners, and are thus good in cases of damaged heads (see illustration 2.21).

- Slotted-head or Phillips-head screws are often damaged by the use of the wrong size screwdriver. Allen-head and Torx-head screws are much less likely to sustain damage. If enough of the screw head is exposed you can use a hacksaw to cut a slot in its head and then use a conventional flat-bladed screwdriver to remove it. Alternatively use a hammer and cold chisel to tap the head of the fastener around to slacken it. Always replace damaged fasteners with new ones, preferably Torx or Allen-head type.

> **HAYNES HiNT**
>
> *A dab of valve grinding compound between the screw head and screwdriver tip will often give a good grip.*

Thread repair

- Threads (particularly those in aluminium alloy components) can be damaged by overtightening, being assembled with dirt in the threads, or from a component working loose and vibrating. Eventually the thread will fail completely, and it will be impossible to tighten the fastener.
- If a thread is damaged or clogged with old locking compound it can be renovated with a thread repair tool (thread chaser) **(see illustrations 2.22 and 2.23)**; special thread

2.22 A thread repair tool being used to correct an internal thread

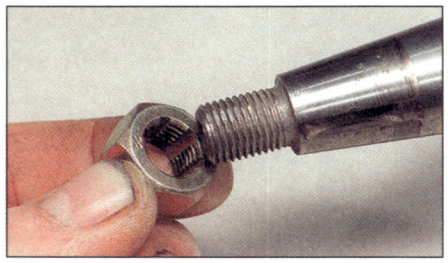

2.23 A thread repair tool being used to correct an external thread

chasers are available for spark plug hole threads. The tool will not cut a new thread, but clean and true the original thread. Make sure that you use the correct diameter and pitch tool. Similarly, external threads can be cleaned up with a die or a thread restorer file **(see illustration 2.24)**.

2.24 Using a thread restorer file

- It is possible to drill out the old thread and retap the component to the next thread size. This will work where there is enough surrounding material and a new bolt or screw can be obtained. Sometimes, however, this is not possible - such as where the bolt/screw passes through another component which must also be suitably modified, also in cases where a spark plug or oil drain plug cannot be obtained in a larger diameter thread size.
- The diamond-section thread insert (often known by its popular trade name of Heli-Coil) is a simple and effective method of renewing the thread and retaining the original size. A kit can be purchased which contains the tap, insert and installing tool **(see illustration 2.25)**. Drill out the damaged thread with the size drill specified **(see illustration 2.26)**. Carefully retap the thread **(see illustration 2.27)**. Install the

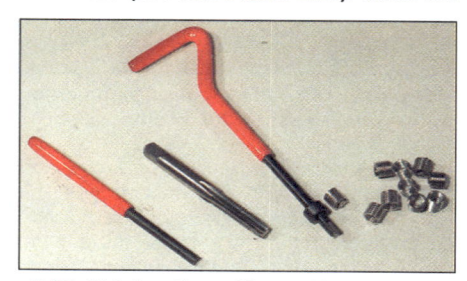

2.25 Obtain a thread insert kit to suit the thread diameter and pitch required

2.26 To install a thread insert, first drill out the original thread . . .

2.27 . . . tap a new thread . . .

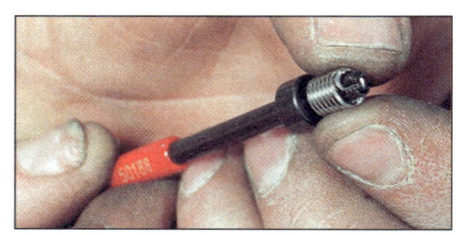

2.28 . . . fit insert on the installing tool . . .

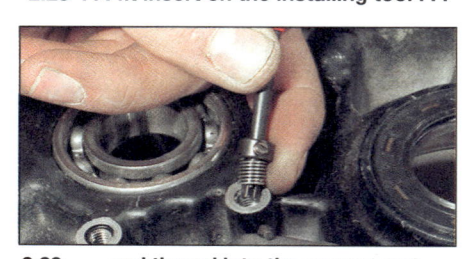

2.29 . . . and thread into the component . . .

2.30 . . . break off the tang when complete

insert on the installing tool and thread it slowly into place using a light downward pressure **(see illustrations 2.28 and 2.29)**. When positioned between a 1/4 and 1/2 turn below the surface withdraw the installing tool and use the break-off tool to press down on the tang, breaking it off **(see illustration 2.30)**.

- There are epoxy thread repair kits on the market which can rebuild stripped internal threads, although this repair should not be used on high load-bearing components.

Tools and Workshop Tips

Thread locking and sealing compounds

● Locking compounds are used in locations where the fastener is prone to loosening due to vibration or on important safety-related items which might cause loss of control of the motorcycle if they fail. It is also used where important fasteners cannot be secured by other means such as lockwashers or split pins.

● Before applying locking compound, make sure that the threads (internal and external) are clean and dry with all old compound removed. Select a compound to suit the component being secured - a non-permanent general locking and sealing type is suitable for most applications, but a high strength type is needed for permanent fixing of studs in castings. Apply a drop or two of the compound to the first few threads of the fastener, then thread it into place and tighten to the specified torque. Do not apply excessive thread locking compound otherwise the thread may be damaged on subsequent removal.

● Certain fasteners are impregnated with a dry film type coating of locking compound on their threads. Always renew this type of fastener if disturbed.

● Anti-seize compounds, such as copper-based greases, can be applied to protect threads from seizure due to extreme heat and corrosion. A common instance is spark plug threads and exhaust system fasteners.

3 Measuring tools and gauges

Feeler gauges

● Feeler gauges (or blades) are used for measuring small gaps and clearances (see illustration 3.1). They can also be used to measure endfloat (sideplay) of a component on a shaft where access is not possible with a dial gauge.

● Feeler gauge sets should be treated with care and not bent or damaged. They are etched with their size on one face. Keep them clean and very lightly oiled to prevent corrosion build-up.

3.1 Feeler gauges are used for measuring small gaps and clearances - thickness is marked on one face of gauge

● When measuring a clearance, select a gauge which is a light sliding fit between the two components. You may need to use two gauges together to measure the clearance accurately.

Micrometers

● A micrometer is a precision tool capable of measuring to 0.01 or 0.001 of a millimetre. It should always be stored in its case and not in the general toolbox. It must be kept clean and never dropped, otherwise its frame or measuring anvils could be distorted resulting in inaccurate readings.

● External micrometers are used for measuring outside diameters of components and have many more applications than internal micrometers. Micrometers are available in different size ranges, eg 0 to 25 mm, 25 to 50 mm, and upwards in 25 mm steps; some large micrometers have interchangeable anvils to allow a range of measurements to be taken. Generally the largest precision measurement you are likely to take on a motorcycle is the piston diameter.

● Internal micrometers (or bore micrometers) are used for measuring inside diameters, such as valve guides and cylinder bores. Telescoping gauges and small hole gauges are used in conjunction with an external micrometer, whereas the more expensive internal micrometers have their own measuring device.

External micrometer

Note: *The conventional analogue type instrument is described. Although much easier to read, digital micrometers are considerably more expensive.*

● Always check the calibration of the micrometer before use. With the anvils closed (0 to 25 mm type) or set over a test gauge (for

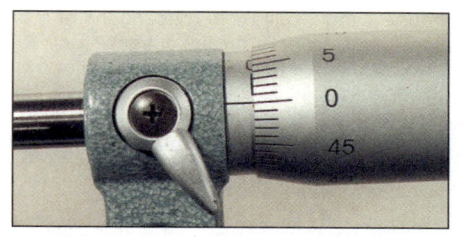

3.2 Check micrometer calibration before use

the larger types) the scale should read zero **(see illustration 3.2)**; make sure that the anvils (and test piece) are clean first. Any discrepancy can be adjusted by referring to the instructions supplied with the tool. Remember that the micrometer is a precision measuring tool - don't force the anvils closed, use the ratchet (4) on the end of the micrometer to close it. In this way, a measured force is always applied.

● To use, first make sure that the item being measured is clean. Place the anvil of the micrometer (1) against the item and use the thimble (2) to bring the spindle (3) lightly into contact with the other side of the item **(see illustration 3.3)**. Don't tighten the thimble down because this will damage the micrometer - instead use the ratchet (4) on the end of the micrometer. The ratchet mechanism applies a measured force preventing damage to the instrument.

● The micrometer is read by referring to the linear scale on the sleeve and the annular scale on the thimble. Read off the sleeve first to obtain the base measurement, then add the fine measurement from the thimble to obtain the overall reading. The linear scale on the sleeve represents the measuring range of the micrometer (eg 0 to 25 mm). The annular scale

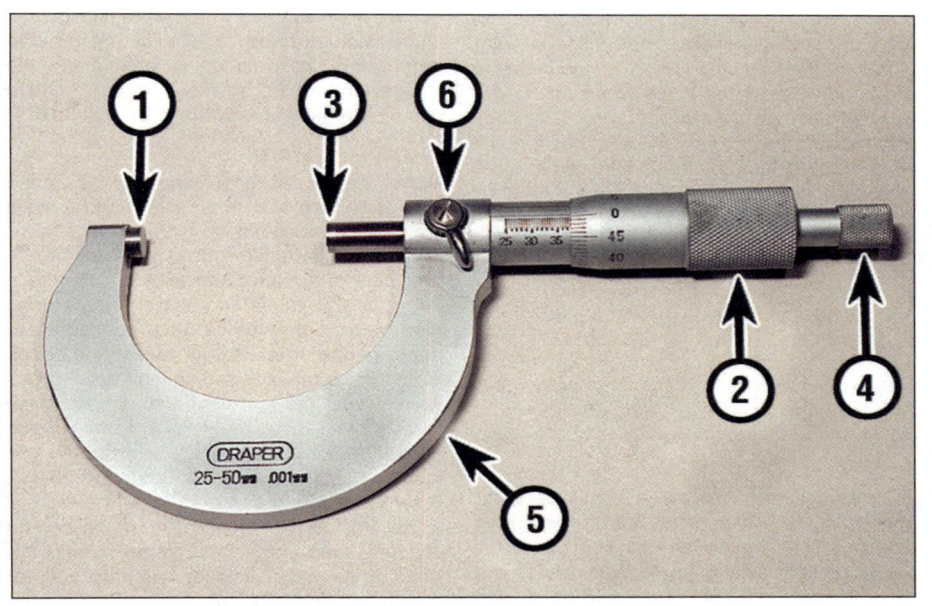

3.3 Micrometer component parts

| 1 Anvil | 3 Spindle | 5 Frame |
| 2 Thimble | 4 Ratchet | 6 Locking lever |

REF•10 Tools and Workshop Tips

on the thimble will be in graduations of 0.01 mm (or as marked on the frame) - one full revolution of the thimble will move 0.5 mm on the linear scale. Take the reading where the datum line on the sleeve intersects the thimble's scale. Always position the eye directly above the scale otherwise an inaccurate reading will result.

In the example shown the item measures 2.95 mm **(see illustration 3.4)**:

Linear scale	2.00 mm
Linear scale	0.50 mm
Annular scale	0.45 mm
Total figure	**2.95 mm**

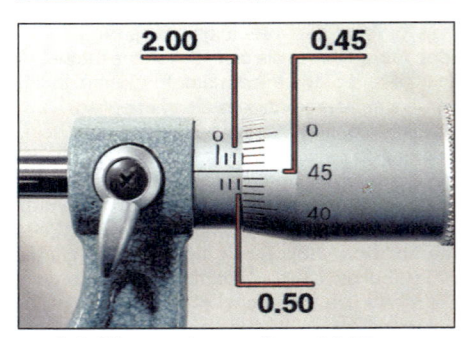

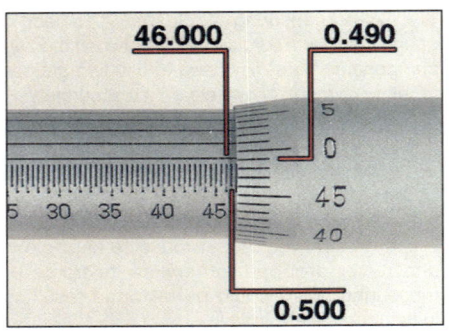

3.5 Micrometer reading of 46.99 mm on linear and annular scales . . .

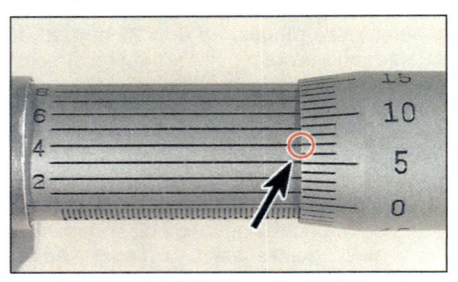

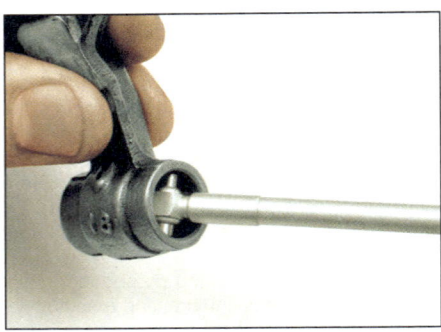

3.7 Expand the telescoping gauge in the bore, lock its position . . .

3.4 Micrometer reading of 2.95 mm

3.6 . . . and 0.004 mm on vernier scale

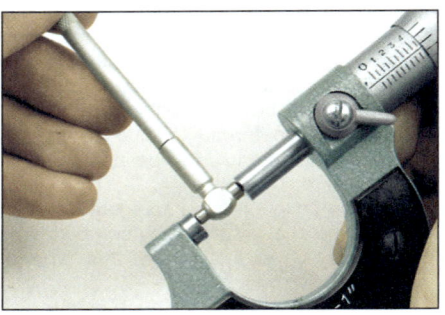

3.8 . . . then measure the gauge with a micrometer

Most micrometers have a locking lever (6) on the frame to hold the setting in place, allowing the item to be removed from the micrometer.
• Some micrometers have a vernier scale on their sleeve, providing an even finer measurement to be taken, in 0.001 increments of a millimetre. Take the sleeve and thimble measurement as described above, then check which graduation on the vernier scale aligns with that of the annular scale on the thimble **Note:** *The eye must be perpendicular to the scale when taking the vernier reading - if necessary rotate the body of the micrometer to ensure this.* Multiply the vernier scale figure by 0.001 and add it to the base and fine measurement figures.

In the example shown the item measures 46.994 mm **(see illustrations 3.5 and 3.6)**:

Linear scale (base)	46.000 mm
Linear scale (base)	00.500 mm
Annular scale (fine)	00.490 mm
Vernier scale	00.004 mm
Total figure	**46.994 mm**

Internal micrometer

• Internal micrometers are available for measuring bore diameters, but are expensive and unlikely to be available for home use. It is suggested that a set of telescoping gauges and small hole gauges, both of which must be used with an external micrometer, will suffice for taking internal measurements on a motorcycle.
• Telescoping gauges can be used to measure internal diameters of components. Select a gauge with the correct size range, make sure its ends are clean and insert it into the bore. Expand the gauge, then lock its position and withdraw it from the bore **(see illustration 3.7)**. Measure across the gauge ends with a micrometer **(see illustration 3.8)**.
• Very small diameter bores (such as valve guides) are measured with a small hole gauge. Once adjusted to a slip-fit inside the component, its position is locked and the gauge withdrawn for measurement with a micrometer **(see illustrations 3.9 and 3.10)**.

Vernier caliper

Note: *The conventional linear and dial gauge type instruments are described. Digital types are easier to read, but are far more expensive.*
• The vernier caliper does not provide the precision of a micrometer, but is versatile in being able to measure internal and external diameters. Some types also incorporate a depth gauge. It is ideal for measuring clutch plate friction material and spring free lengths.
• To use the conventional linear scale vernier, slacken off the vernier clamp screws (1) and set its jaws over (2), or inside (3), the item to be measured **(see illustration 3.11)**. Slide the jaw into contact, using the thumbwheel (4) for fine movement of the sliding scale (5) then tighten the clamp screws (1). Read off the main scale (6) where the zero on the sliding scale (5) intersects it, taking the whole number to the left of the zero; this provides the base measurement. View along the sliding scale and select the division which

3.9 Expand the small hole gauge in the bore, lock its position . . .

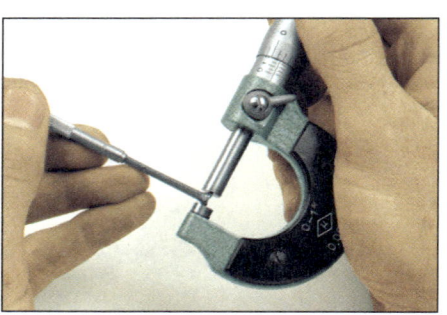

3.10 . . . then measure the gauge with a micrometer

lines up exactly with any of the divisions on the main scale, noting that the divisions usually represents 0.02 of a millimetre. Add this fine measurement to the base measurement to obtain the total reading.

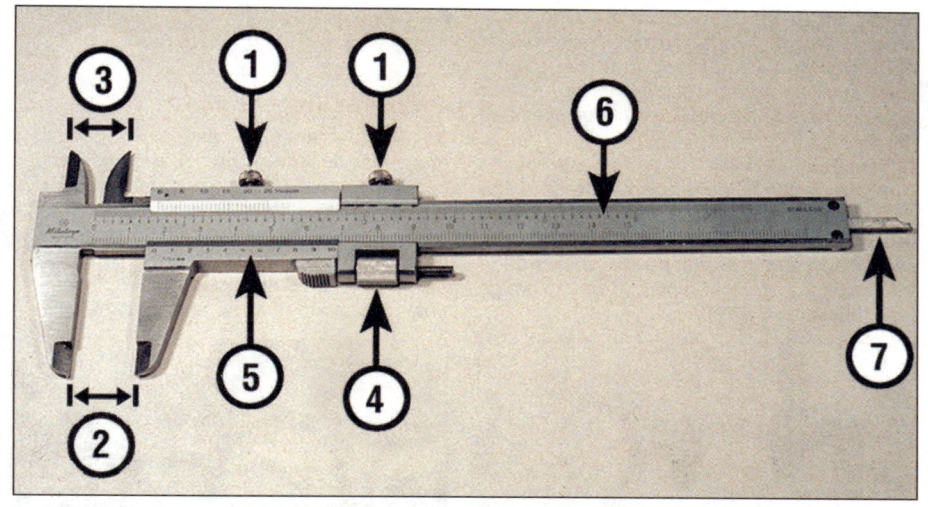

Tools and Workshop Tips REF•11

3.11 Vernier component parts (linear gauge)

1 Clamp screws
2 External jaws
3 Internal jaws
4 Thumbwheel
5 Sliding scale
6 Main scale
7 Depth gauge

In the example shown the item measures 55.92 mm **(see illustration 3.12)**:

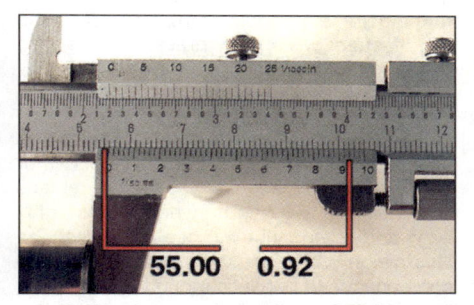

Base measurement	55.00 mm
Fine measurement	00.92 mm
Total figure	**55.92 mm**

3.12 Vernier gauge reading of 55.92 mm

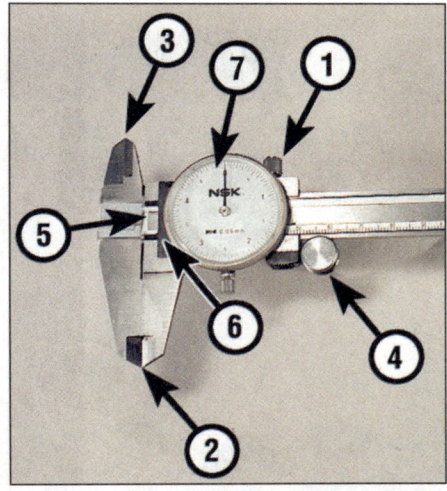

● Some vernier calipers are equipped with a dial gauge for fine measurement. Before use, check that the jaws are clean, then close them fully and check that the dial gauge reads zero. If necessary adjust the gauge ring accordingly. Slacken the vernier clamp screw (1) and set its jaws over (2), or inside (3), the item to be measured **(see illustration 3.13)**. Slide the jaws into contact, using the thumbwheel (4) for fine movement. Read off the main scale (5) where the edge of the sliding scale (6) intersects it, taking the whole number to the left of the zero; this provides the base measurement. Read off the needle position on the dial gauge (7) scale to provide the fine measurement; each division represents 0.05 of a millimetre. Add this fine measurement to the base measurement to obtain the total reading.

In the example shown the item measures 55.95 mm **(see illustration 3.14)**:

Base measurement	55.00 mm
Fine measurement	00.95 mm
Total figure	**55.95 mm**

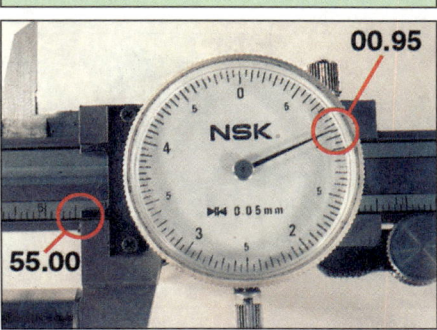

3.13 Vernier component parts (dial gauge)

1 Clamp screw
2 External jaws
3 Internal jaws
4 Thumbwheel
5 Main scale
6 Sliding scale
7 Dial gauge

3.14 Vernier gauge reading of 55.95 mm

Plastigauge

● Plastigauge is a plastic material which can be compressed between two surfaces to measure the oil clearance between them. The width of the compressed Plastigauge is measured against a calibrated scale to determine the clearance.

● Common uses of Plastigauge are for measuring the clearance between crankshaft journal and main bearing inserts, between crankshaft journal and big-end bearing inserts, and between camshaft and bearing surfaces. The following example describes big-end oil clearance measurement.

● Handle the Plastigauge material carefully to prevent distortion. Using a sharp knife, cut a length which corresponds with the width of the bearing being measured and place it carefully across the journal so that it is parallel with the shaft **(see illustration 3.15)**. Carefully install both bearing shells and the connecting rod. Without rotating the rod on the journal tighten its bolts or nuts (as applicable) to the specified torque. The connecting rod and bearings are then disassembled and the crushed Plastigauge examined.

3.15 Plastigauge placed across shaft journal

● Using the scale provided in the Plastigauge kit, measure the width of the material to determine the oil clearance **(see illustration 3.16)**. Always remove all traces of Plastigauge after use using your fingernails.

Caution: Arriving at the correct clearance demands that the assembly is torqued correctly, according to the settings and sequence (where applicable) provided by the motorcycle manufacturer.

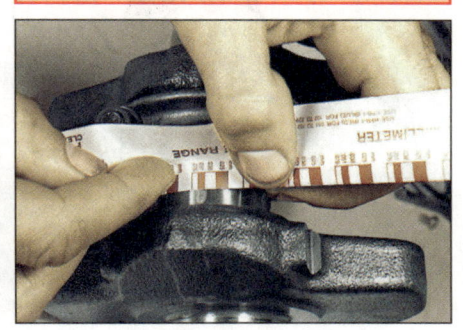

3.16 Measuring the width of the crushed Plastigauge

REF•12 Tools and Workshop Tips

Dial gauge or DTI (Dial Test Indicator)

- A dial gauge can be used to accurately measure small amounts of movement. Typical uses are measuring shaft runout or shaft endfloat (sideplay) and setting piston position for ignition timing on two-strokes. A dial gauge set usually comes with a range of different probes and adapters and mounting equipment.
- The gauge needle must point to zero when at rest. Rotate the ring around its periphery to zero the gauge.
- Check that the gauge is capable of reading the extent of movement in the work. Most gauges have a small dial set in the face which records whole millimetres of movement as well as the fine scale around the face periphery which is calibrated in 0.01 mm divisions. Read off the small dial first to obtain the base measurement, then add the measurement from the fine scale to obtain the total reading.

In the example shown the gauge reads 1.48 mm (see illustration 3.17):

Base measurement	1.00 mm
Fine measurement	0.48 mm
Total figure	**1.48 mm**

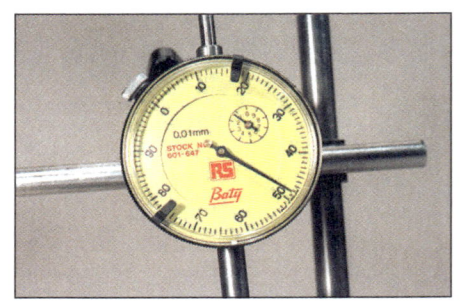

3.17 Dial gauge reading of 1.48 mm

- If measuring shaft runout, the shaft must be supported in vee-blocks and the gauge mounted on a stand perpendicular to the shaft. Rest the tip of the gauge against the centre of the shaft and rotate the shaft slowly whilst watching the gauge reading (see illustration 3.18). Take several measurements along the length of the shaft and record the maximum gauge reading as the amount of runout in the shaft. **Note:** *The reading obtained will be total runout at that point - some manufacturers specify that the runout figure is halved to compare with their specified runout limit.*
- Endfloat (sideplay) measurement requires that the gauge is mounted securely to the surrounding component with its probe touching the end of the shaft. Using hand pressure, push and pull on the shaft noting the maximum endfloat recorded on the gauge (see illustration 3.19).

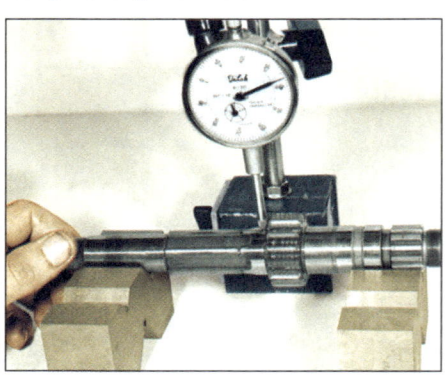

3.18 Using a dial gauge to measure shaft runout

3.19 Using a dial gauge to measure shaft endfloat

- A dial gauge with suitable adapters can be used to determine piston position BTDC on two-stroke engines for the purposes of ignition timing. The gauge, adapter and suitable length probe are installed in the place of the spark plug and the gauge zeroed at TDC. If the piston position is specified as 1.14 mm BTDC, rotate the engine back to 2.00 mm BTDC, then slowly forwards to 1.14 mm BTDC.

Cylinder compression gauges

- A compression gauge is used for measuring cylinder compression. Either the rubber-cone type or the threaded adapter type can be used. The latter is preferred to ensure a perfect seal against the cylinder head. A 0 to 300 psi (0 to 20 Bar) type gauge (for petrol/gasoline engines) will be suitable for motorcycles.
- The spark plug is removed and the gauge either held hard against the cylinder head (cone type) or the gauge adapter screwed into the cylinder head (threaded type) (see illustration 3.20). Cylinder compression is measured with the engine turning over, but not running - carry out the compression test as described in *Fault Finding Equipment*. The gauge will hold the reading until manually released.

3.20 Using a rubber-cone type cylinder compression gauge

Oil pressure gauge

- An oil pressure gauge is used for measuring engine oil pressure. Most gauges come with a set of adapters to fit the thread of the take-off point (see illustration 3.21). If the take-off point specified by the motorcycle manufacturer is an external oil pipe union, make sure that the specified replacement union is used to prevent oil starvation.

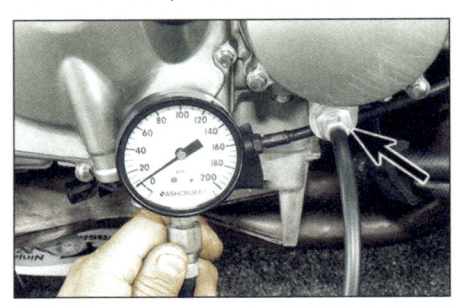

3.21 Oil pressure gauge and take-off point adapter (arrow)

- Oil pressure is measured with the engine running (at a specific rpm) and often the manufacturer will specify pressure limits for a cold and hot engine.

Straight-edge and surface plate

- If checking the gasket face of a component for warpage, place a steel rule or precision straight-edge across the gasket face and measure any gap between the straight-edge and component with feeler gauges (see illustration 3.22). Check diagonally across the component and between mounting holes (see illustration 3.23).

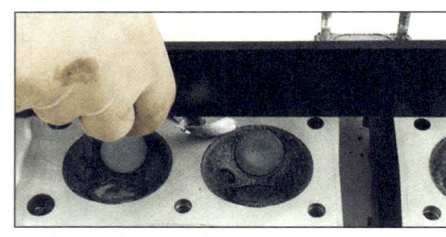

3.22 Use a straight-edge and feeler gauges to check for warpage

3.23 Check for warpage in these directions

Tools and Workshop Tips

- Checking individual components for warpage, such as clutch plain (metal) plates, requires a perfectly flat plate or piece or plate glass and feeler gauges.

4 Torque and leverage

What is torque?

- Torque describes the twisting force about a shaft. The amount of torque applied is determined by the distance from the centre of the shaft to the end of the lever and the amount of force being applied to the end of the lever; distance multiplied by force equals torque.
- The manufacturer applies a measured torque to a bolt or nut to ensure that it will not slacken in use and to hold two components securely together without movement in the joint. The actual torque setting depends on the thread size, bolt or nut material and the composition of the components being held.
- Too little torque may cause the fastener to loosen due to vibration, whereas too much torque will distort the joint faces of the component or cause the fastener to shear off. Always stick to the specified torque setting.

Using a torque wrench

- Check the calibration of the torque wrench and make sure it has a suitable range for the job. Torque wrenches are available in Nm (Newton-metres), kgf m (kilograms-force metre), lbf ft (pounds-feet), lbf in (inch-pounds). Do not confuse lbf ft with lbf in.
- Adjust the tool to the desired torque on the scale (see illustration 4.1). If your torque wrench is not calibrated in the units specified, carefully convert the figure (see *Conversion Factors*). A manufacturer sometimes gives a torque setting as a range (8 to 10 Nm) rather than a single figure - in this case set the tool midway between the two settings. The same torque may be expressed as 9 Nm ± 1 Nm. Some torque wrenches have a method of locking the setting so that it isn't inadvertently altered during use.

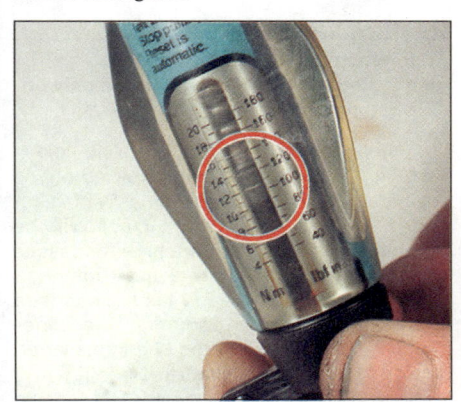

- Install the bolts/nuts in their correct location and secure them lightly. Their threads must be clean and free of any old locking compound. Unless specified the threads and flange should be dry - oiled threads are necessary in certain circumstances and the manufacturer will take this into account in the specified torque figure. Similarly, the manufacturer may also specify the application of thread-locking compound.
- Tighten the fasteners in the specified sequence until the torque wrench clicks, indicating that the torque setting has been reached. Apply the torque again to double-check the setting. Where different thread diameter fasteners secure the component, as a rule tighten the larger diameter ones first.
- When the torque wrench has been finished with, release the lock (where applicable) and fully back off its setting to zero - do not leave the torque wrench tensioned. Also, do not use a torque wrench for slackening a fastener.

Angle-tightening

- Manufacturers often specify a figure in degrees for final tightening of a fastener. This usually follows tightening to a specific torque setting.
- A degree disc can be set and attached to the socket (see illustration 4.2) or a protractor can be used to mark the angle of movement on the bolt/nut head and the surrounding casting (see illustration 4.3).

4.2 Angle tightening can be accomplished with a torque-angle gauge . . .

4.1 Set the torque wrench index mark to the setting required, in this case 12 Nm

4.3 . . . or by marking the angle on the surrounding component

Loosening sequences

- Where more than one bolt/nut secures a component, loosen each fastener evenly a little at a time. In this way, not all the stress of the joint is held by one fastener and the components are not likely to distort.
- If a tightening sequence is provided, work in the REVERSE of this, but if not, work from the outside in, in a criss-cross sequence (see illustration 4.4).

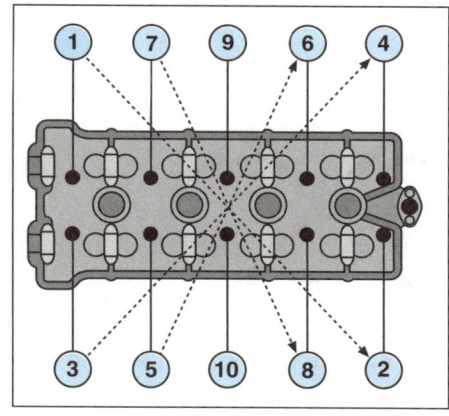

4.4 When slackening, work from the outside inwards

Tightening sequences

- If a component is held by more than one fastener it is important that the retaining bolts/nuts are tightened evenly to prevent uneven stress build-up and distortion of sealing faces. This is especially important on high-compression joints such as the cylinder head.
- A sequence is usually provided by the manufacturer, either in a diagram or actually marked in the casting. If not, always start in the centre and work outwards in a criss-cross pattern (see illustration 4.5). Start off by securing all bolts/nuts finger-tight, then set the torque wrench and tighten each fastener by a small amount in sequence until the final torque is reached. By following this practice,

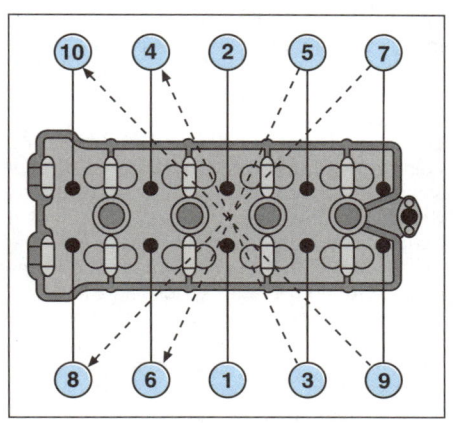

4.5 When tightening, work from the inside outwards

REF•14 Tools and Workshop Tips

the joint will be held evenly and will not be distorted. Important joints, such as the cylinder head and big-end fasteners often have two- or three-stage torque settings.

Applying leverage

● Use tools at the correct angle. Position a socket wrench or spanner on the bolt/nut so that you pull it towards you when loosening. If this can't be done, push the spanner without curling your fingers around it **(see illustration 4.6)** - the spanner may slip or the fastener loosen suddenly, resulting in your fingers being crushed against a component.

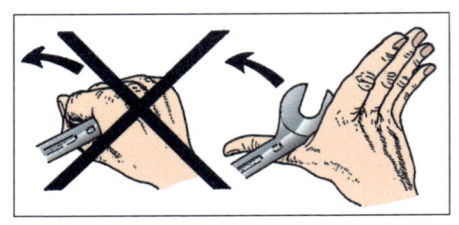

4.6 If you can't pull on the spanner to loosen a fastener, push with your hand open

● Additional leverage is gained by extending the length of the lever. The best way to do this is to use a breaker bar instead of the regular length tool, or to slip a length of tubing over the end of the spanner or socket wrench.
● If additional leverage will not work, the fastener head is either damaged or firmly corroded in place (see *Fasteners*).

5 Bearings

Bearing removal and installation
Drivers and sockets

● Before removing a bearing, always inspect the casing to see which way it must be driven out - some casings will have retaining plates or a cast step. Also check for any identifying markings on the bearing and if installed to a certain depth, measure this at this stage. Some roller bearings are sealed on one side - take note of the original fitted position.
● Bearings can be driven out of a casing using a bearing driver tool (with the correct size head) or a socket of the correct diameter. Select the driver head or socket so that it contacts the outer race of the bearing, not the balls/rollers or inner race. Always support the casing around the bearing housing with wood blocks, otherwise there is a risk of fracture. The bearing is driven out with a few blows on the driver or socket from a heavy mallet. Unless access is severely restricted (as with wheel bearings), a pin-punch is not recommended unless it is moved around the bearing to keep it square in its housing.

● The same equipment can be used to install bearings. Make sure the bearing housing is supported on wood blocks and line up the bearing in its housing. Fit the bearing as noted on removal - generally they are installed with their marked side facing outwards. Tap the bearing squarely into its housing using a driver or socket which bears only on the bearing's outer race - contact with the bearing balls/rollers or inner race will destroy it **(see illustrations 5.1 and 5.2)**.
● Check that the bearing inner race and balls/rollers rotate freely.

5.1 Using a bearing driver against the bearing's outer race

5.2 Using a large socket against the bearing's outer race

Pullers and slide-hammers

● Where a bearing is pressed on a shaft a puller will be required to extract it **(see illustration 5.3)**. Make sure that the puller clamp or legs fit securely behind the bearing and are unlikely to slip out. If pulling a bearing

5.3 This bearing puller clamps behind the bearing and pressure is applied to the shaft end to draw the bearing off

off a gear shaft for example, you may have to locate the puller behind a gear pinion if there is no access to the race and draw the gear pinion off the shaft as well **(see illustration 5.4)**.

> **Caution: Ensure that the puller's centre bolt locates securely against the end of the shaft and will not slip when pressure is applied. Also ensure that puller does not damage the shaft end.**

5.4 Where no access is available to the rear of the bearing, it is sometimes possible to draw off the adjacent component

● Operate the puller so that its centre bolt exerts pressure on the shaft end and draws the bearing off the shaft.
● When installing the bearing on the shaft, tap only on the bearing's inner race - contact with the balls/rollers or outer race with destroy the bearing. Use a socket or length of tubing as a drift which fits over the shaft end **(see illustration 5.5)**.

5.5 When installing a bearing on a shaft use a piece of tubing which bears only on the bearing's inner race

● Where a bearing locates in a blind hole in a casing, it cannot be driven or pulled out as described above. A slide-hammer with knife-edged bearing puller attachment will be required. The puller attachment passes through the bearing and when tightened expands to fit firmly behind the bearing **(see illustration 5.6)**. By operating the slide-hammer part of the tool the bearing is jarred out of its housing **(see illustration 5.7)**.
● It is possible, if the bearing is of reasonable weight, for it to drop out of its housing if the casing is heated as described opposite. If this

Tools and Workshop Tips REF•15

5.6 Expand the bearing puller so that it locks behind the bearing ...

5.7 ... attach the slide hammer to the bearing puller

method is attempted, first prepare a work surface which will enable the casing to be tapped face down to help dislodge the bearing - a wood surface is ideal since it will not damage the casing's gasket surface. Wearing protective gloves, tap the heated casing several times against the work surface to dislodge the bearing under its own weight **(see illustration 5.8)**.

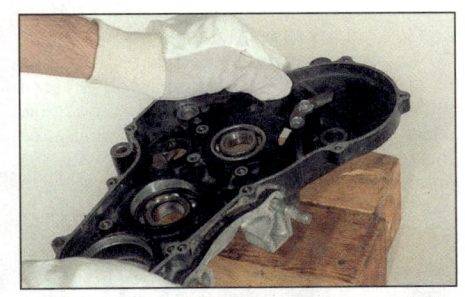

5.8 Tapping a casing face down on wood blocks can often dislodge a bearing

● Bearings can be installed in blind holes using the driver or socket method described above.

Drawbolts

● Where a bearing or bush is set in the eye of a component, such as a suspension linkage arm or connecting rod small-end, removal by drift may damage the component. Furthermore, a rubber bushing in a shock absorber eye cannot successfully be driven out of position. If access is available to a engineering press, the task is straightforward. If not, a drawbolt can be fabricated to extract the bearing or bush.

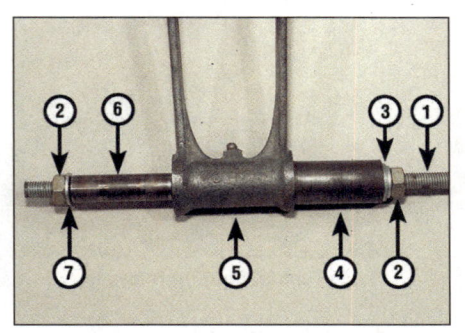

5.9 Drawbolt component parts assembled on a suspension arm

1. Bolt or length of threaded bar
2. Nuts
3. Washer (external diameter greater than tubing internal diameter)
4. Tubing (internal diameter sufficient to accommodate bearing)
5. Suspension arm with bearing
6. Tubing (external diameter slightly smaller than bearing)
7. Washer (external diameter slightly smaller than bearing)

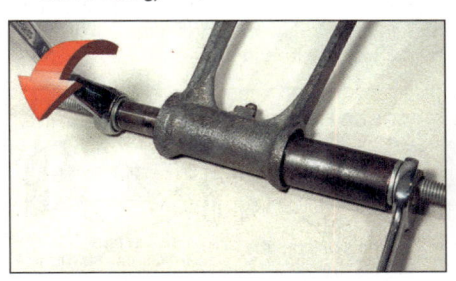

5.10 Drawing the bearing out of the suspension arm

● To extract the bearing/bush you will need a long bolt with nut (or piece of threaded bar with two nuts), a piece of tubing which has an internal diameter larger than the bearing/bush, another piece of tubing which has an external diameter slightly smaller than the bearing/bush, and a selection of washers **(see illustrations 5.9 and 5.10)**. Note that the pieces of tubing must be of the same length, or longer, than the bearing/bush.

● The same kit (without the pieces of tubing) can be used to draw the new bearing/bush back into place **(see illustration 5.11)**.

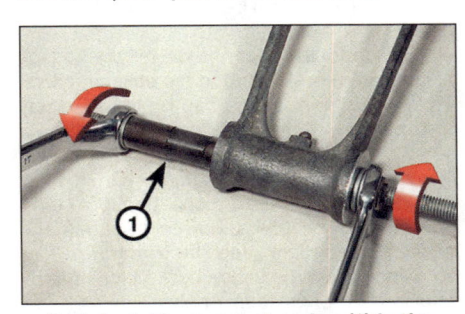

5.11 Installing a new bearing (1) in the suspension arm

Temperature change

● If the bearing's outer race is a tight fit in the casing, the aluminium casing can be heated to release its grip on the bearing. Aluminium will expand at a greater rate than the steel bearing outer race. There are several ways to do this, but avoid any localised extreme heat (such as a blow torch) - aluminium alloy has a low melting point.

● Approved methods of heating a casing are using a domestic oven (heated to 100°C) or immersing the casing in boiling water **(see illustration 5.12)**. Low temperature range localised heat sources such as a paint stripper heat gun or clothes iron can also be used **(see illustration 5.13)**. Alternatively, soak a rag in boiling water, wring it out and wrap it around the bearing housing.

> ⚠ **Warning: All of these methods require care in use to prevent scalding and burns to the hands. Wear protective gloves when handling hot components.**

5.12 A casing can be immersed in a sink of boiling water to aid bearing removal

5.13 Using a localised heat source to aid bearing removal

● If heating the whole casing note that plastic components, such as the neutral switch, may suffer - remove them beforehand.

● After heating, remove the bearing as described above. You may find that the expansion is sufficient for the bearing to fall out of the casing under its own weight or with a light tap on the driver or socket.

● If necessary, the casing can be heated to aid bearing installation, and this is sometimes the recommended procedure if the motorcycle manufacturer has designed the housing and bearing fit with this intention.

Tools and Workshop Tips

- Installation of bearings can be eased by placing them in a freezer the night before installation. The steel bearing will contract slightly, allowing easy insertion in its housing. This is often useful when installing steering head outer races in the frame.

Bearing types and markings

- Plain shell bearings, ball bearings, needle roller bearings and tapered roller bearings will all be found on motorcycles (see illustrations 5.14 and 5.15). The ball and roller types are usually caged between an inner and outer race, but uncaged variations may be found.

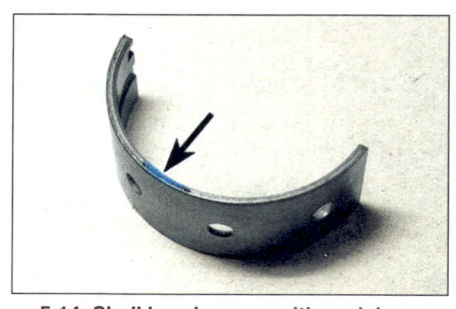

5.14 Shell bearings are either plain or grooved. They are usually identified by colour code (arrow)

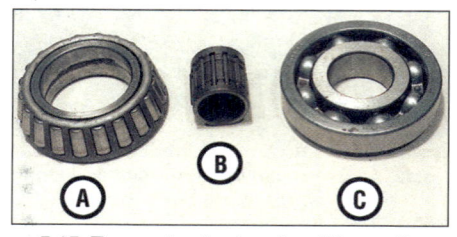

5.15 Tapered roller bearing (A), needle roller bearing (B) and ball journal bearing (C)

- Shell bearings (often called inserts) are usually found at the crankshaft main and connecting rod big-end where they are good at coping with high loads. They are made of a phosphor-bronze material and are impregnated with self-lubricating properties.
- Ball bearings and needle roller bearings consist of a steel inner and outer race with the balls or rollers between the races. They require constant lubrication by oil or grease and are good at coping with axial loads. Taper roller bearings consist of rollers set in a tapered cage set on the inner race; the outer race is separate. They are good at coping with axial loads and prevent movement along the shaft - a typical application is in the steering head.
- Bearing manufacturers produce bearings to ISO size standards and stamp one face of the bearing to indicate its internal and external diameter, load capacity and type (see illustration 5.16).
- Metal bushes are usually of phosphor-bronze material. Rubber bushes are used in suspension mounting eyes. Fibre bushes have also been used in suspension pivots.

5.16 Typical bearing marking

Bearing fault finding

- If a bearing outer race has spun in its housing, the housing material will be damaged. You can use a bearing locking compound to bond the outer race in place if damage is not too severe.
- Shell bearings will fail due to damage of their working surface, as a result of lack of lubrication, corrosion or abrasive particles in the oil (see illustration 5.17). Small particles of dirt in the oil may embed in the bearing material whereas larger particles will score the bearing and shaft journal. If a number of short journeys are made, insufficient heat will be generated to drive off condensation which has built up on the bearings.

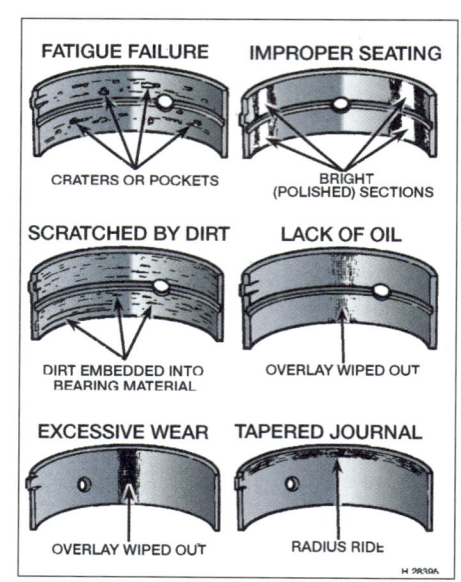

5.17 Typical bearing failures

- Ball and roller bearings will fail due to lack of lubrication or damage to the balls or rollers. Tapered-roller bearings can be damaged by overloading them. Unless the bearing is sealed on both sides, wash it in paraffin (kerosene) to remove all old grease then allow it to dry. Make a visual inspection looking to dented balls or rollers, damaged cages and worn or pitted races (see illustration 5.18).
- A ball bearing can be checked for wear by listening to it when spun. Apply a film of light oil to the bearing and hold it close to the ear - hold the outer race with one hand and spin the inner race with the other hand (see illustration 5.19). The bearing should be almost silent when spun; if it grates or rattles it is worn.

5.18 Example of ball journal bearing with damaged balls and cages

5.19 Hold outer race and listen to inner race when spun

6 Oil seals

Oil seal removal and installation

- Oil seals should be renewed every time a component is dismantled. This is because the seal lips will become set to the sealing surface and will not necessarily reseal.
- Oil seals can be prised out of position using a large flat-bladed screwdriver (see illustration 6.1). In the case of crankcase seals, check first that the seal is not lipped on the inside, preventing its removal with the crankcases joined.

6.1 Prise out oil seals with a large flat-bladed screwdriver

- New seals are usually installed with their marked face (containing the seal reference code) outwards and the spring side towards the fluid being retained. In certain cases, such as a two-stroke engine crankshaft seal, a double lipped seal may be used due to there being fluid or gas on each side of the joint.

Tools and Workshop Tips REF•17

- Use a bearing driver or socket which bears only on the outer hard edge of the seal to install it in the casing - tapping on the inner edge will damage the sealing lip.

Oil seal types and markings

- Oil seals are usually of the single-lipped type. Double-lipped seals are found where a liquid or gas is on both sides of the joint.
- Oil seals can harden and lose their sealing ability if the motorcycle has been in storage for a long period - renewal is the only solution.
- Oil seal manufacturers also conform to the ISO markings for seal size - these are moulded into the outer face of the seal **(see illustration 6.2)**.

6.2 These oil seal markings indicate inside diameter, outside diameter and seal thickness

7 Gaskets and sealants

Types of gasket and sealant

- Gaskets are used to seal the mating surfaces between components and keep lubricants, fluids, vacuum or pressure contained within the assembly. Aluminium gaskets are sometimes found at the cylinder joints, but most gaskets are paper-based. If the mating surfaces of the components being joined are undamaged the gasket can be installed dry, although a dab of sealant or grease will be useful to hold it in place during assembly.
- RTV (Room Temperature Vulcanising) silicone rubber sealants cure when exposed to moisture in the atmosphere. These sealants are good at filling pits or irregular gasket faces, but will tend to be forced out of the joint under very high torque. They can be used to replace a paper gasket, but first make sure that the width of the paper gasket is not essential to the shimming of internal components. RTV sealants should not be used on components containing petrol (gasoline).
- Non-hardening, semi-hardening and hard setting liquid gasket compounds can be used with a gasket or between a metal-to-metal joint. Select the sealant to suit the application: universal non-hardening sealant can be used on virtually all joints; semi-hardening on joint faces which are rough or damaged; hard setting sealant on joints which require a permanent bond and are subjected to high temperature and pressure. **Note:** *Check first if the paper gasket has a bead of sealant impregnated in its surface before applying additional sealant.*
- When choosing a sealant, make sure it is suitable for the application, particularly if being applied in a high-temperature area or in the vicinity of fuel. Certain manufacturers produce sealants in either clear, silver or black colours to match the finish of the engine. This has a particular application on motorcycles where much of the engine is exposed.
- Do not over-apply sealant. That which is squeezed out on the outside of the joint can be wiped off, whereas an excess of sealant on the inside can break off and clog oilways.

Breaking a sealed joint

- Age, heat, pressure and the use of hard setting sealant can cause two components to stick together so tightly that they are difficult to separate using finger pressure alone. Do not resort to using levers unless there is a pry point provided for this purpose **(see illustration 7.1)** or else the gasket surfaces will be damaged.
- Use a soft-faced hammer **(see illustration 7.2)** or a wood block and conventional hammer to strike the component near the mating surface. Avoid hammering against cast extremities since they may break off. If this method fails, try using a wood wedge between the two components.

Caution: If the joint will not separate, double-check that you have removed all the fasteners.

7.1 If a pry point is provided, apply gently pressure with a flat-bladed screwdriver

7.2 Tap around the joint with a soft-faced mallet if necessary - don't strike cooling fins

Removal of old gasket and sealant

- Paper gaskets will most likely come away complete, leaving only a few traces stuck on the sealing faces of the components. It is imperative that all traces are removed to ensure correct sealing of the new gasket.
- Very carefully scrape all traces of gasket away making sure that the sealing surfaces are not gouged or scored by the scraper **(see illustrations 7.3, 7.4 and 7.5)**. Stubborn deposits can be removed by spraying with an aerosol gasket remover. Final preparation of

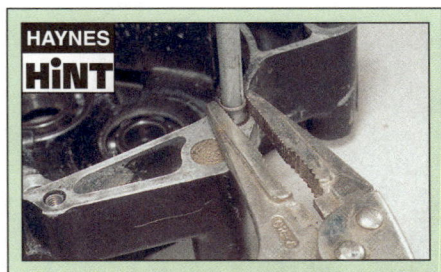

> **HAYNES HiNT**
>
> *Most components have one or two hollow locating dowels between the two gasket faces. If a dowel cannot be removed, do not resort to gripping it with pliers - it will almost certainly be distorted. Install a close-fitting socket or Phillips screwdriver into the dowel and then grip the outer edge of the dowel to free it.*

7.3 Paper gaskets can be scraped off with a gasket scraper tool . . .

7.4 . . . a knife blade . . .

7.5 . . . or a household scraper

Tools and Workshop Tips

7.6 Fine abrasive paper is wrapped around a flat file to clean up the gasket face

7.7 A kitchen scourer can be used on stubborn deposits

the gasket surface can be made with very fine abrasive paper or a plastic kitchen scourer **(see illustrations 7.6 and 7.7)**.

● Old sealant can be scraped or peeled off components, depending on the type originally used. Note that gasket removal compounds are available to avoid scraping the components clean; make sure the gasket remover suits the type of sealant used.

8 Chains

Breaking and joining final drive chains

● Drive chains for all but small bikes are continuous and do not have a clip-type connecting link. The chain must be broken using a chain breaker tool and the new chain securely riveted together using a new soft rivet-type link. Never use a clip-type connecting link instead of a rivet-type link, except in an emergency. Various chain breaking and riveting tools are available, either as separate tools or combined as illustrated in the accompanying photographs - read the instructions supplied with the tool carefully.

> **Warning: The need to rivet the new link pins correctly cannot be overstressed - loss of control of the motorcycle is very likely to result if the chain breaks in use.**

● Rotate the chain and look for the soft link. The soft link pins look like they have been deeply centre-punched instead of peened over like all the other pins **(see illustration 8.9)** and its sideplate may be a different colour. Position the soft link midway between the sprockets and assemble the chain breaker tool over one of the soft link pins **(see illustration 8.1)**. Operate the tool to push the pin out through the chain **(see illustration 8.2)**. On an O-ring chain, remove the O-rings **(see illustration 8.3)**. Carry out the same procedure on the other soft link pin.

> **Caution: Certain soft link pins (particularly on the larger chains) may require their ends to be filed or ground off before they can be pressed out using the tool.**

● Check that you have the correct size and strength (standard or heavy duty) new soft link - do not reuse the old link. Look for the size marking on the chain sideplates **(see illustration 8.10)**.

● Position the chain ends so that they are engaged over the rear sprocket. On an O-ring chain, install a new O-ring over each pin of the link and insert the link through the two chain ends **(see illustration 8.4)**. Install a new O-ring over the end of each pin, followed by the sideplate (with the chain manufacturer's marking facing outwards) **(see illustrations 8.5 and 8.6)**. On an unsealed chain, insert the link through the two chain ends, then install the sideplate with the chain manufacturer's marking facing outwards.

● Note that it may not be possible to install the sideplate using finger pressure alone. If using a joining tool, assemble it so that the plates of the tool clamp the link and press the sideplate over the pins **(see illustration 8.7)**. Otherwise, use two small sockets placed over

8.1 Tighten the chain breaker to push the pin out of the link . . .

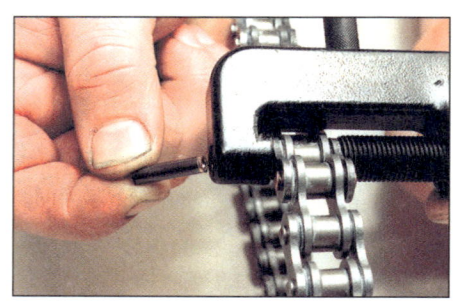

8.2 . . . withdraw the pin, remove the tool . . .

8.3 . . . and separate the chain link

8.4 Insert the new soft link, with O-rings, through the chain ends . . .

8.5 . . . install the O-rings over the pin ends . . .

8.6 . . . followed by the sideplate

8.7 Push the sideplate into position using a clamp

Tools and Workshop Tips REF•19

8.8 Assemble the chain riveting tool over one pin at a time and tighten it fully

8.9 Pin end correctly riveted (A), pin end unriveted (B)

the rivet ends and two pieces of the wood between a G-clamp. Operate the clamp to press the sideplate over the pins.
● Assemble the joining tool over one pin (following the maker's instructions) and tighten the tool down to spread the pin end securely **(see illustrations 8.8 and 8.9)**. Do the same on the other pin.

> ⚠️ **Warning: Check that the pin ends are secure and that there is no danger of the sideplate coming loose. If the pin ends are cracked the soft link must be renewed.**

Final drive chain sizing

● Chains are sized using a three digit number, followed by a suffix to denote the chain type **(see illustration 8.10)**. Chain type is either standard or heavy duty (thicker sideplates), and also unsealed or O-ring/X-ring type.
● The first digit of the number relates to the pitch of the chain, ie the distance from the centre of one pin to the centre of the next pin **(see illustration 8.11)**. Pitch is expressed in eighths of an inch, as follows:

8.10 Typical chain size and type marking

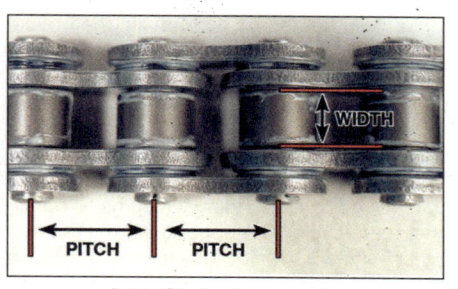

8.11 Chain dimensions

Sizes commencing with a 4 (eg 428) have a pitch of 1/2 inch (12.7 mm)

Sizes commencing with a 5 (eg 520) have a pitch of 5/8 inch (15.9 mm)

Sizes commencing with a 6 (eg 630) have a pitch of 3/4 inch (19.1 mm)

● The second and third digits of the chain size relate to the width of the rollers, again in imperial units, eg the 525 shown has 5/16 inch (7.94 mm) rollers **(see illustration 8.11)**.

9 Hoses

Clamping to prevent flow

● Small-bore flexible hoses can be clamped to prevent fluid flow whilst a component is worked on. Whichever method is used, ensure that the hose material is not permanently distorted or damaged by the clamp.
a) A brake hose clamp available from auto accessory shops **(see illustration 9.1)**.
b) A wingnut type hose clamp **(see illustration 9.2)**.

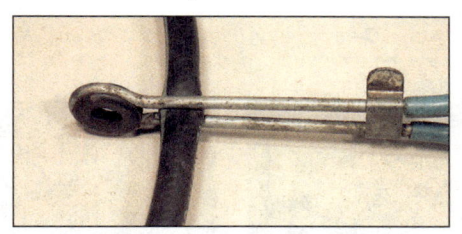

9.1 Hoses can be clamped with an automotive brake hose clamp . . .

9.2 . . . a wingnut type hose clamp . . .

c) Two sockets placed each side of the hose and held with straight-jawed self-locking grips **(see illustration 9.3)**.
d) Thick card each side of the hose held between straight-jawed self-locking grips **(see illustration 9.4)**.

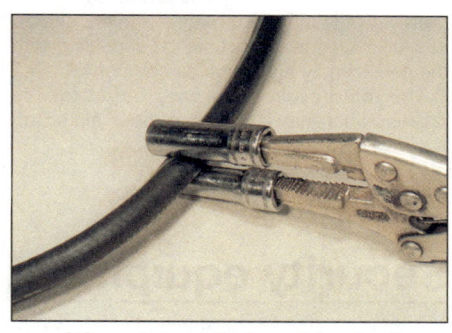

9.3 . . . two sockets and a pair of self-locking grips . . .

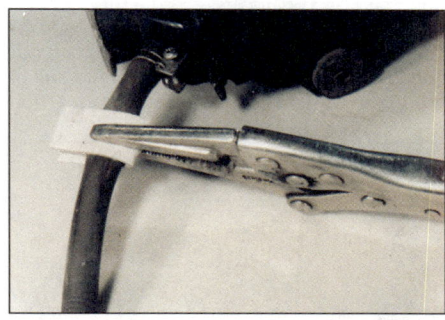

9.4 . . . or thick card and self-locking grips

Freeing and fitting hoses

● Always make sure the hose clamp is moved well clear of the hose end. Grip the hose with your hand and rotate it whilst pulling it off the union. If the hose has hardened due to age and will not move, slit it with a sharp knife and peel its ends off the union **(see illustration 9.5)**.
● Resist the temptation to use grease or soap on the unions to aid installation; although it helps the hose slip over the union it will equally aid the escape of fluid from the joint. It is preferable to soften the hose ends in hot water and wet the inside surface of the hose with water or a fluid which will evaporate.

9.5 Cutting a coolant hose free with a sharp knife

REF•20 Security

Introduction

In less time than it takes to read this introduction, a thief could steal your motorcycle. Returning only to find your bike has gone is one of the worst feelings in the world. Even if the motorcycle is insured against theft, once you've got over the initial shock, you will have the inconvenience of dealing with the police and your insurance company.

The motorcycle is an easy target for the professional thief and the joyrider alike and the official figures on motorcycle theft make for depressing reading; on average a motorcycle is stolen every 16 minutes in the UK!

Motorcycle thefts fall into two categories, those stolen 'to order' and those taken by opportunists. The thief stealing to order will be on the look out for a specific make and model and will go to extraordinary lengths to obtain that motorcycle. The opportunist thief on the other hand will look for easy targets which can be stolen with the minimum of effort and risk.

Whilst it is never going to be possible to make your machine 100% secure, it is estimated that around half of all stolen motorcycles are taken by opportunist thieves. Remember that the opportunist thief is always on the look out for the easy option: if there are two similar motorcycles parked side-by-side, they will target the one with the lowest level of security. By taking a few precautions, you can reduce the chances of your motorcycle being stolen.

Security equipment

There are many specialised motorcycle security devices available and the following text summarises their applications and their good and bad points.

Once you have decided on the type of security equipment which best suits your needs, we recommended that you read one of the many equipment tests regularly carried out by the motorcycle press. These tests compare the products from all the major manufacturers and give impartial ratings on their effectiveness, value-for-money and ease of use.

No one item of security equipment can provide complete protection. It is highly recommended that two or more of the items described below are combined to increase the security of your motorcycle (a lock and chain plus an alarm system is just about ideal). The more security measures fitted to the bike, the less likely it is to be stolen.

Lock and chain

Pros: *Very flexible to use; can be used to secure the motorcycle to almost any immovable object. On some locks and chains, the lock can be used on its own as a disc lock (see below).*

Cons: *Can be very heavy and awkward to carry on the motorcycle, although some types will be supplied with a carry bag which can be strapped to the pillion seat.*

● Heavy-duty chains and locks are an excellent security measure **(see illustration 1)**. Whenever the motorcycle is parked, use the lock and chain to secure the machine to a solid, immovable object such as a post or railings. This will prevent the machine from being ridden away or being lifted into the back of a van.

● When fitting the chain, always ensure the chain is routed around the motorcycle frame or swingarm **(see illustrations 2 and 3)**. Never merely pass the chain around one of the wheel rims; a thief may unbolt the wheel and lift the rest of the machine into a van, leaving you with just the wheel! Try to avoid having excess chain free, thus making it difficult to use cutting tools, and keep the chain and lock off the ground to prevent thieves attacking it with a cold chisel. Position the lock so that its lock barrel is facing downwards; this will make it harder for the thief to attack the lock mechanism.

1 Ensure the lock and chain you buy is of good quality and long enough to shackle your bike to a solid object

2 Pass the chain through the bike's frame, rather than just through a wheel . . .

3 . . . and loop it around a solid object

Security REF•21

U-locks

Pros: *Highly effective deterrent which can be used to secure the bike to a post or railings. Most U-locks come with a carrier which allows the lock to be easily carried on the bike.*

Cons: *Not as flexible to use as a lock and chain.*

● These are solid locks which are similar in use to a lock and chain. U-locks are lighter than a lock and chain but not so flexible to use. The length and shape of the lock shackle limit the objects to which the bike can be secured **(see illustration 4)**.

Disc locks

Pros: *Small, light and very easy to carry; most can be stored underneath the seat.*

Cons: *Does not prevent the motorcycle being lifted into a van. Can be very embarrassing if you forget to remove the lock before attempting to ride off!*

● Disc locks are designed to be attached to the front brake disc. The lock passes through one of the holes in the disc and prevents the wheel rotating by jamming against the fork/brake caliper **(see illustration 5)**. Some are equipped with an alarm siren which sounds if the disc lock is moved; this not only acts as a theft deterrent but also as a handy reminder if you try to move the bike with the lock still fitted.

● Combining the disc lock with a length of cable which can be looped around a post or railings provides an additional measure of security **(see illustration 6)**.

Alarms and immobilisers

Pros: *Once installed it is completely hassle-free to use. If the system is 'Thatcham' or 'Sold Secure-approved', insurance companies may give you a discount.*

Cons: *Can be expensive to buy and complex to install. No system will prevent the motorcycle from being lifted into a van and taken away.*

● Electronic alarms and immobilisers are available to suit a variety of budgets. There are three different types of system available: pure alarms, pure immobilisers, and the more expensive systems which are combined alarm/immobilisers **(see illustration 7)**.

● An alarm system is designed to emit an audible warning if the motorcycle is being tampered with.

● An immobiliser prevents the motorcycle being started and ridden away by disabling its electrical systems.

● When purchasing an alarm/immobiliser system, check the cost of installing the system unless you are able to do it yourself. If the motorcycle is not used regularly, another consideration is the current drain of the system. All alarm/immobiliser systems are powered by the motorcycle's battery; purchasing a system with a very low current drain could prevent the battery losing its charge whilst the motorcycle is not being used.

U-locks can be used to secure the bike to a solid object – ensure you purchase one which is long enough

A typical disc lock attached through one of the holes in the disc

A disc lock combined with a security cable provides additional protection

A typical alarm/immobiliser system

REF•22 Security

Indelible markings can be applied to most areas of the bike – always apply the manufacturer's sticker to warn off thieves

Chemically-etched code numbers can be applied to main body panels . . .

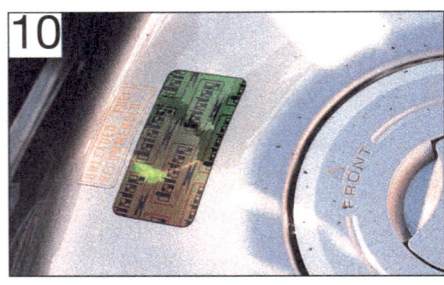

. . . again, always ensure that the kit manufacturer's sticker is applied in a prominent position

Security marking kits

Pros: *Very cheap and effective deterrent. Many insurance companies will give you a discount on your insurance premium if a recognised security marking kit is used on your motorcycle.*

Cons: *Does not prevent the motorcycle being stolen by joyriders.*

● There are many different types of security marking kits available. The idea is to mark as many parts of the motorcycle as possible with a unique security number **(see illustrations 8, 9 and 10)**. A form will be included with the kit to register your personal details and those of the motorcycle with the kit manufacturer. This register is made available to the police to help them trace the rightful owner of any motorcycle or components which they recover should all other forms of identification have been removed. Always apply the warning stickers provided with the kit to deter thieves.

Ground anchors, wheel clamps and security posts

Pros: *An excellent form of security which will deter all but the most determined of thieves.*

Cons: *Awkward to install and can be expensive.*

● Whilst the motorcycle is at home, it is a good idea to attach it securely to the floor or a solid wall, even if it is kept in a securely locked garage. Various types of ground anchors, security posts and wheel clamps are available for this purpose **(see illustration 11)**. These security devices are either bolted to a solid concrete or brick structure or can be cemented into the ground.

Permanent ground anchors provide an excellent level of security when the bike is at home

Security at home

A high percentage of motorcycle thefts are from the owner's home. Here are some things to consider whenever your motorcycle is at home:

✔ Where possible, always keep the motorcycle in a securely locked garage. Never rely solely on the standard lock on the garage door, these are usual hopelessly inadequate. Fit an additional locking mechanism to the door and consider having the garage alarmed. A security light, activated by a movement sensor, is also a good investment.

✔ Always secure the motorcycle to the ground or a wall, even if it is inside a securely locked garage.
✔ Do not regularly leave the motorcycle outside your home, try to keep it out of sight wherever possible. If a garage is not available, fit a motorcycle cover over the bike to disguise its true identity.
✔ It is not uncommon for thieves to follow a motorcyclist home to find out where the bike is kept. They will then return at a later date. Be aware of this whenever you are returning home on your motorcycle. If you suspect you are being followed, do not return home, instead ride to a garage or shop and stop as a precaution.
✔ When selling a motorcycle, do not provide your home address or the location where the bike is normally kept. Arrange to meet the buyer at a location away from your home. Thieves have been known to pose as potential buyers to find out where motorcycles are kept and then return later to steal them.

Security away from the home

As well as fitting security equipment to your motorcycle here are a few general rules to follow whenever you park your motorcycle.
✔ Park in a busy, public place.
✔ Use car parks which incorporate security features, such as CCTV.
✔ At night, park in a well-lit area, preferably directly underneath a street light.
✔ Engage the steering lock.
✔ Secure the motorcycle to a solid, immovable object such as a post or railings with an additional lock. If this is not possible, secure the bike to a friend's motorcycle. Some public parking places provide security loops for motorcycles.
✔ Never leave your helmet or luggage attached to the motorcycle. Take them with you at all times.

Lubricants and fluids

A wide range of lubricants, fluids and cleaning agents is available for motor-cycles. This is a guide as to what is available, its applications and properties.

Four-stroke engine oil

● Engine oil is without doubt the most important component of any four-stroke engine. Modern motorcycle engines place a lot of demands on their oil and choosing the right type is essential. Using an unsuitable oil will lead to an increased rate of engine wear and could result in serious engine damage. Before purchasing oil, always check the recommended oil specification given by the manufacturer. The manufacturer will state a recommended 'type or classification' and also a specific 'viscosity' range for engine oil.

● The oil 'type or classification' is identified by its API (American Petroleum Institute) rating. The API rating will be in the form of two letters, e.g. SG. The S identifies the oil as being suitable for use in a petrol (gasoline) engine (S stands for spark ignition) and the second letter, ranging from A to J, identifies the oil's performance rating. The later this letter, the higher the specification of the oil; for example API SG oil exceeds the requirements of API SF oil. **Note:** *On some oils there may also be a second rating consisting of another two letters, the first letter being C, e.g. API SF/CD. This rating indicates the oil is also suitable for use in a diesel engines (the C stands for compression ignition) and is thus of no relevance for motorcycle use.*

● The 'viscosity' of the oil is identified by its SAE (Society of Automotive Engineers) rating. All modern engines require multigrade oils and the SAE rating will consist of two numbers, the first followed by a W, e.g. 10W/40. The first number indicates the viscosity rating of the oil at low temperatures (W stands for winter – tested at –20°C) and the second number represents the viscosity of the oil at high temperatures (tested at 100°C). The lower the number, the thinner the oil. For example an oil with an SAE 10W/40 rating will give better cold starting and running than an SAE 15W/40 oil.

● As well as ensuring the 'type' and 'viscosity' of the oil match the recommendations, another consideration to make when buying engine oil is whether to purchase a standard mineral-based oil, a semi-synthetic oil (also known as a synthetic blend or synthetic-based oil) or a fully-synthetic oil. Although all oils will have a similar rating and viscosity, their cost will vary considerably; mineral-based oils are the cheapest, the fully-synthetic oils the most expensive with the semi-synthetic oils falling somewhere in-between. This decision is very much up to the owner, but it should be noted that modern synthetic oils have far better lubricating and cleaning qualities than traditional mineral-based oils and tend to retain these properties for far longer. Bearing in mind the operating conditions inside a modern, high-revving motorcycle engine it is highly recommended that a fully synthetic oil is used. The extra expense at each service could save you money in the long term by preventing premature engine wear.

● As a final note always ensure that the oil is specifically designed for use in motorcycle engines. Engine oils designed primarily for use in car engines sometimes contain additives or friction modifiers which could cause clutch slip on a motorcycle fitted with a wet-clutch.

Two-stroke engine oil

● Modern two-stroke engines, with their high power outputs, place high demands on their oil. If engine seizure is to be avoided it is essential that a high-quality oil is used. Two-stroke oils differ hugely from four-stroke oils. The oil lubricates only the crankshaft and piston(s) (the transmission has its own lubricating oil) and is used on a total-loss basis where it is burnt completely during the combustion process.

● The Japanese have recently introduced a classification system for two-stroke oils, the JASO rating. This rating is in the form of two letters, either FA, FB or FC – FA is the lowest classification and FC the highest. Ensure the oil being used meets or exceeds the recommended rating specified by the manufacturer.

● As well as ensuring the oil rating matches the recommendation, another consideration to make when buying engine oil is whether to purchase a standard mineral-based oil, a semi-synthetic oil (also known as a synthetic blend or synthetic-based oil) or a fully-synthetic oil. The cost of each type of oil varies considerably; mineral-based oils are the cheapest, the fully-synthetic oils the most expensive with the semi-synthetic oils falling somewhere in-between. This decision is very much up to the owner, but it should be noted that modern synthetic oils have far better lubricating properties and burn cleaner than traditional mineral-based oils. It is therefore recommended that a fully synthetic oil is used. The extra expense could save you money in the long term by preventing premature engine wear, engine performance will be improved, carbon deposits and exhaust smoke will be reduced.

Lubricants and fluids

- Always ensure that the oil is specifically designed for use in an injector system. Many high quality two-stroke oils are designed for competition use and need to be pre-mixed with fuel. These oils are of a much higher viscosity and are not designed to flow through the injector pumps used on road-going two-stroke motorcycles.

Transmission (gear) oil

- On a two-stroke engine, the transmission and clutch are lubricated by their own separate oil bath which must be changed in accordance with the Maintenance Schedule.
- Although the engine and transmission units of most four-strokes use a common lubrication supply, there are some exceptions where the engine and gearbox have separate oil reservoirs and a dry clutch is used.
- Motorcycle manufacturers will either recommend a monograde transmission oil or a four-stroke multigrade engine oil to lubricate the transmission.
- Transmission oils, or gear oils as they are often called, are designed specifically for use in transmission systems. The viscosity of these oils is represented by an SAE number, but the scale of measurement applied is different to that used to grade engine oils. As a rough guide a SAE90 gear oil will be of the same viscosity as an SAE50 engine oil.

Shaft drive oil

- On models equipped with shaft final drive, the shaft drive gears are will have their own oil supply. The manufacturer will state a recommended 'type or classification' and also a specific 'viscosity' range in the same manner as for four-stroke engine oil.
- Gear oil classification is given by the number which follows the API GL (GL standing for gear lubricant) rating, the higher the number, the higher the specification of the oil, e.g. API GL5 oil is a higher specification than API GL4 oil. Ensure the oil meets or exceeds the classification specified and is of the correct viscosity. The viscosity of gear oils is also represented by an SAE number but the scale of measurement used is different to that used to grade engine oils. As a rough guide an SAE90 gear oil will be of the same viscosity as an SAE50 engine oil.
- If the use of an EP (Extreme Pressure) gear oil is specified, ensure the oil purchased is suitable.

Fork oil and suspension fluid

- Conventional telescopic front forks are hydraulic and require fork oil to work. To ensure the forks function correctly, the fork oil must be changed in accordance with the Maintenance Schedule.
- Fork oil is available in a variety of viscosities, identified by their SAE rating; fork oil ratings vary from light (SAE 5) to heavy (SAE 30). When purchasing fork oil, ensure the viscosity rating matches that specified by the manufacturer.
- Some lubricant manufacturers also produce a range of high-quality suspension fluids which are very similar to fork oil but are designed mainly for competition use. These fluids may have a different viscosity rating system which is not to be confused with the SAE rating of normal fork oil. Refer to the manufacturer's instructions if in any doubt.

Brake and clutch fluid

- All disc brake systems and some clutch systems are hydraulically operated. To ensure correct operation, the hydraulic fluid must be changed in accordance with the Maintenance Schedule.
- Brake and clutch fluid is classified by its DOT rating with most motorcycle manufacturers specifying DOT 3 or 4 fluid. Both fluid types are glycol-based and can be mixed together without adverse effect; DOT 4 fluid exceeds the requirements of DOT 3 fluid. Although it is safe to use DOT 4 fluid in a system designed for use with DOT 3 fluid, never use DOT 3 fluid in a system which specifies the use of DOT 4 as this will adversely affect the system's performance. The type required for the system will be marked on the fluid reservoir cap.
- Some manufacturers also produce a DOT 5 hydraulic fluid. DOT 5 hydraulic fluid is silicone-based and is not compatible with the glycol-based DOT 3 and 4 fluids. Never mix DOT 5 fluid with DOT 3 or 4 fluid as this will seriously affect the performance of the hydraulic system.

Coolant/antifreeze

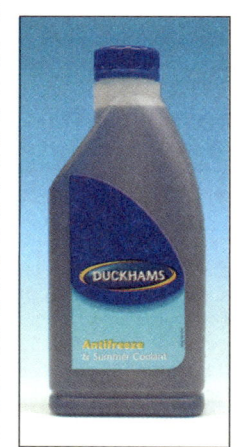

- When purchasing coolant/antifreeze, always ensure it is suitable for use in an aluminium engine and contains corrosion inhibitors to prevent possible blockages of the internal coolant passages of the system. As a general rule, most coolants are designed to be used neat and should not be diluted whereas antifreeze can be mixed with distilled water to provide a coolant solution of the required strength. Refer to the manufacturer's instructions on the bottle.
- Ensure the coolant is changed in accordance with the Maintenance Schedule.

Chain lube

- Chain lube is an aerosol-type spray lubricant specifically designed for use on motorcycle final drive chains. Chain lube has two functions, to minimise friction between the final drive chain and sprockets and to prevent corrosion of the chain. Regular use of a good-quality chain lube will extend the life of the drive chain and sprockets and thus maximise the power being transmitted from the transmission to the rear wheel.
- When using chain lube, always allow some time for the solvents in the lube to evaporate before riding the motorcycle. This will minimise the amount of lube which will

Lubricants and fluids

'fling' off from the chain when the motorcycle is used. If the motorcycle is equipped with an 'O-ring' chain, ensure the chain lube is labelled as being suitable for use on 'O-ring' chains.

Degreasers and solvents

● There are many different types of solvents and degreasers available to remove the grime and grease which accumulate around the motorcycle during normal use. Degreasers and solvents are usually available as an aerosol-type spray or as a liquid which you apply with a brush. Always closely follow the manufacturer's instructions and wear eye protection during use. Be aware that many solvents are flammable and may give off noxious fumes; take adequate precautions when using them (see Safety First!).

● For general cleaning, use one of the many solvents or degreasers available from most motorcycle accessory shops. These solvents are usually applied then left for a certain time before being washed off with water.

Brake cleaner is a solvent specifically designed to remove all traces of oil, grease and dust from braking system components. Brake cleaner is designed to evaporate quickly and leaves behind no residue.

Carburettor cleaner is an aerosol-type solvent specifically designed to clear carburettor blockages and break down the hard deposits and gum often found inside carburettors during overhaul.

Contact cleaner is an aerosol-type solvent designed for cleaning electrical components. The cleaner will remove all traces of oil and dirt from components such as switch contacts or fouled spark plugs and then dry, leaving behind no residue.

Gasket remover is an aerosol-type solvent designed for removing stubborn gaskets from engine components during overhaul. Gasket remover will minimise the amount of scraping required to remove the gasket and therefore reduce the risk of damage to the mating surface.

Spray lubricants

● Aerosol-based spray lubricants are widely available and are excellent for lubricating lever pivots and exposed cables and switches. Try to use a lubricant which is of the dry-film type as the fluid evaporates, leaving behind a dry-film of lubricant. Lubricants which leave behind an oily residue will attract dust and dirt which will increase the rate of wear of the cable/lever.

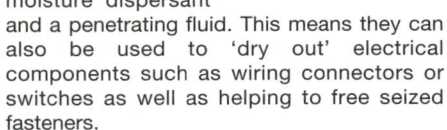

● Most lubricants also act as a moisture dispersant and a penetrating fluid. This means they can also be used to 'dry out' electrical components such as wiring connectors or switches as well as helping to free seized fasteners.

Greases

● Grease is used to lubricate many of the pivot-points. A good-quality multi-purpose grease is suitable for most applications but some manufacturers will specify the use of specialist greases for use on components such as swingarm and suspension linkage bushes. These specialist greases can be purchased from most motorcycle (or car) accessory shops; commonly specified types include molybdenum disulphide grease, lithium-based grease, graphite-based grease, silicone-based grease and high-temperature copper-based grease.

Gasket sealing compounds

● Gasket sealing compounds can be used in conjunction with gaskets, to improve their sealing capabilities, or on their own to seal metal-to-metal joints. Depending on their type, sealing compounds either set hard or stay relatively soft and pliable.

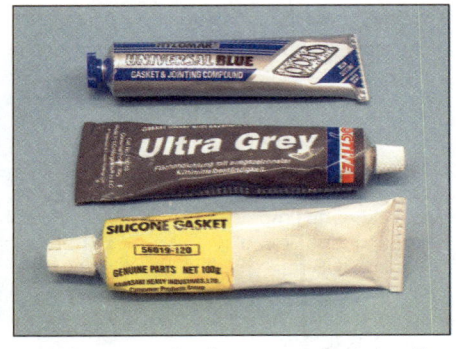

● When purchasing a gasket sealing compound, ensure that it is designed specifically for use on an internal combustion engine. General multi-purpose sealants available from DIY stores may appear visibly similar but they are not designed to withstand the extreme heat or contact with fuel and oil encountered when used on an engine (see 'Tools and Workshop Tips' for further information).

Thread locking compound

● Thread locking compounds are used to secure certain threaded fasteners in position to prevent them from loosening due to vibration. Thread locking compounds can be purchased from most motorcycle (and car) accessory shops. Ensure the threads of the both components are completely clean and dry before sparingly applying the locking compound (see 'Tools and Workshop Tips' for further information).

Fuel additives

● Fuel additives which protect and clean the fuel system components are widely available. These additives are designed to remove all traces of deposits that build up on the carburettors/injectors and prevent wear, helping the fuel system to operate more efficiently. If a fuel additive is being used, check that it is suitable for use with your motorcycle, especially if your motorcycle is equipped with a catalytic converter.

● Octane boosters are also available. These additives are designed to improve the performance of highly-tuned engines being run on normal pump-fuel and are of no real use on standard motorcycles.

Conversion Factors

Length (distance)
Inches (in)	x 25.4	= Millimetres (mm)	x 0.0394	= Inches (in)
Feet (ft)	x 0.305	= Metres (m)	x 3.281	= Feet (ft)
Miles	x 1.609	= Kilometres (km)	x 0.621	= Miles

Volume (capacity)
Cubic inches (cu in; in³)	x 16.387	= Cubic centimetres (cc; cm³)	x 0.061	= Cubic inches (cu in; in³)
Imperial pints (Imp pt)	x 0.568	= Litres (l)	x 1.76	= Imperial pints (Imp pt)
Imperial quarts (Imp qt)	x 1.137	= Litres (l)	x 0.88	= Imperial quarts (Imp qt)
Imperial quarts (Imp qt)	x 1.201	= US quarts (US qt)	x 0.833	= Imperial quarts (Imp qt)
US quarts (US qt)	x 0.946	= Litres (l)	x 1.057	= US quarts (US qt)
Imperial gallons (Imp gal)	x 4.546	= Litres (l)	x 0.22	= Imperial gallons (Imp gal)
Imperial gallons (Imp gal)	x 1.201	= US gallons (US gal)	x 0.833	= Imperial gallons (Imp gal)
US gallons (US gal)	x 3.785	= Litres (l)	x 0.264	= US gallons (US gal)

Mass (weight)
Ounces (oz)	x 28.35	= Grams (g)	x 0.035	= Ounces (oz)
Pounds (lb)	x 0.454	= Kilograms (kg)	x 2.205	= Pounds (lb)

Force
Ounces-force (ozf; oz)	x 0.278	= Newtons (N)	x 3.6	= Ounces-force (ozf; oz)
Pounds-force (lbf; lb)	x 4.448	= Newtons (N)	x 0.225	= Pounds-force (lbf; lb)
Newtons (N)	x 0.1	= Kilograms-force (kgf; kg)	x 9.81	= Newtons (N)

Pressure
Pounds-force per square inch (psi; lbf/in²; lb/in²)	x 0.070	= Kilograms-force per square centimetre (kgf/cm²; kg/cm²)	x 14.223	= Pounds-force per square inch (psi; lbf/in²; lb/in²)
Pounds-force per square inch (psi; lbf/in²; lb/in²)	x 0.068	= Atmospheres (atm)	x 14.696	= Pounds-force per square inch (psi; lbf/in²; lb/in²)
Pounds-force per square inch (psi; lbf/in²; lb/in²)	x 0.069	= Bars	x 14.5	= Pounds-force per square inch (psi; lbf/in²; lb/in²)
Pounds-force per square inch (psi; lbf/in²; lb/in²)	x 6.895	= Kilopascals (kPa)	x 0.145	= Pounds-force per square inch (psi; lbf/in²; lb/in²)
Kilopascals (kPa)	x 0.01	= Kilograms-force per square centimetre (kgf/cm²; kg/cm²)	x 98.1	= Kilopascals (kPa)
Millibar (mbar)	x 100	= Pascals (Pa)	x 0.01	= Millibar (mbar)
Millibar (mbar)	x 0.0145	= Pounds-force per square inch (psi; lbf/in²; lb/in²)	x 68.947	= Millibar (mbar)
Millibar (mbar)	x 0.75	= Millimetres of mercury (mmHg)	x 1.333	= Millibar (mbar)
Millibar (mbar)	x 0.401	= Inches of water (inH$_2$O)	x 2.491	= Millibar (mbar)
Millimetres of mercury (mmHg)	x 0.535	= Inches of water (inH$_2$O)	x 1.868	= Millimetres of mercury (mmHg)
Inches of water (inH$_2$O)	x 0.036	= Pounds-force per square inch (psi; lbf/in²; lb/in²)	x 27.68	= Inches of water (inH$_2$O)

Torque (moment of force)
Pounds-force inches (lbf in; lb in)	x 1.152	= Kilograms-force centimetre (kgf cm; kg cm)	x 0.868	= Pounds-force inches (lbf in; lb in)
Pounds-force inches (lbf in; lb in)	x 0.113	= Newton metres (Nm)	x 8.85	= Pounds-force inches (lbf in; lb in)
Pounds-force inches (lbf in; lb in)	x 0.083	= Pounds-force feet (lbf ft; lb ft)	x 12	= Pounds-force inches (lbf in; lb in)
Pounds-force feet (lbf ft; lb ft)	x 0.138	= Kilograms-force metres (kgf m; kg m)	x 7.233	= Pounds-force feet (lbf ft; lb ft)
Pounds-force feet (lbf ft; lb ft)	x 1.356	= Newton metres (Nm)	x 0.738	= Pounds-force feet (lbf ft; lb ft)
Newton metres (Nm)	x 0.102	= Kilograms-force metres (kgf m; kg m)	x 9.804	= Newton metres (Nm)

Power
Horsepower (hp)	x 745.7	= Watts (W)	x 0.0013	= Horsepower (hp)

Velocity (speed)
Miles per hour (miles/hr; mph)	x 1.609	= Kilometres per hour (km/hr; kph)	x 0.621	= Miles per hour (miles/hr; mph)

Fuel consumption*
Miles per gallon (mpg)	x 0.354	= Kilometres per litre (km/l)	x 2.825	= Miles per gallon (mpg)

Temperature

Degrees Fahrenheit = (°C x 1.8) + 32   Degrees Celsius (Degrees Centigrade; °C) = (°F - 32) x 0.56

It is common practice to convert from miles per gallon (mpg) to litres/100 kilometres (l/100km), where mpg x l/100 km = 282

MOT Test Checks REF•27

About the MOT Test

In the UK, all vehicles more than three years old are subject to an annual test to ensure that they meet minimum safety requirements. A current test certificate must be issued before a machine can be used on public roads, and is required before a road fund licence can be issued. Riding without a current test certificate will also invalidate your insurance.

For most owners, the MOT test is an annual cause for anxiety, and this is largely due to owners not being sure what needs to be checked prior to submitting the motorcycle for testing. The simple answer is that a fully roadworthy motorcycle will have no difficulty in passing the test.

This is a guide to getting your motorcycle through the MOT test. Obviously it will not be possible to examine the motorcycle to the same standard as the professional MOT tester, particularly in view of the equipment required for some of the checks. However, working through the following procedures will enable you to identify any problem areas before submitting the motorcycle for the test.

It has only been possible to summarise the test requirements here, based on the regulations in force at the time of printing. Test standards are becoming increasingly stringent, although there are some exemptions for older vehicles. More information about the MOT test can be obtained from the TSO publications, *How Safe is your Motorcycle* and *The MOT Inspection Manual for Motorcycle Testing*.

Many of the checks require that one of the wheels is raised off the ground. If the motorcycle doesn't have a centre stand, note that an auxiliary stand will be required. Additionally, the help of an assistant may prove useful.

Certain exceptions apply to machines under 50 cc, machines without a lighting system, and Classic bikes - if in doubt about any of the requirements listed below seek confirmation from an MOT tester prior to submitting the motorcycle for the test.

Check that the frame number is clearly visible.

> **HAYNES HiNT** *If a component is in borderline condition, the tester has discretion in deciding whether to pass or fail it. If the motorcycle presented is clean and evidently well cared for, the tester may be more inclined to pass a borderline component than if the motorcycle is scruffy and apparently neglected.*

Electrical System

Lights, turn signals, horn and reflector

✔ With the ignition on, check the operation of the following electrical components. **Note:** *The electrical components on certain small-capacity machines are powered by the generator, requiring that the engine is run for this check.*

a) Headlight and tail light. Check that both illuminate in the low and high beam switch positions.
b) Position lights. Check that the front position (or sidelight) and tail light illuminate in this switch position.
c) Turn signals. Check that all flash at the correct rate, and that the warning light(s) function correctly. Check that the turn signal switch works correctly.
d) Hazard warning system (where fitted). Check that all four turn signals flash in this switch position.
e) Brake stop light. Check that the light comes on when the front and rear brakes are independently applied. Models first used on or after 1st April 1986 must have a brake light switch on each brake.
f) Horn. Check that the sound is continuous and of reasonable volume.

✔ Check that there is a red reflector on the rear of the machine, either mounted separately or as part of the tail light lens.
✔ Check the condition of the headlight, tail light and turn signal lenses.

Headlight beam height

✔ The MOT tester will perform a headlight beam height check using specialised beam setting equipment **(see illustration 1)**. This equipment will not be available to the home mechanic, but if you suspect that the headlight is incorrectly set or may have been maladjusted in the past, you can perform a rough test as follows.
✔ Position the bike in a straight line facing a brick wall. The bike must be off its stand, upright and with a rider seated. Measure the height from the ground to the centre of the headlight and mark a horizontal line on the wall at this height. Position the motorcycle 3.8 metres from the wall and draw a vertical line up the wall central to the centreline of the motorcycle. Switch to dipped beam and check that the beam pattern falls slightly lower than the horizontal line and to the left of the vertical line **(see illustration 2)**.

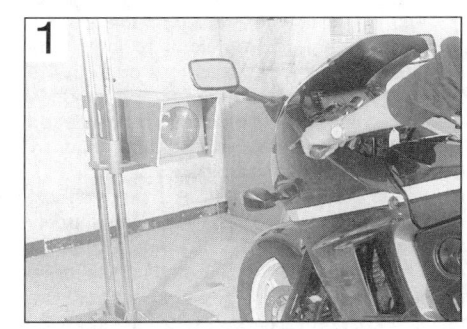

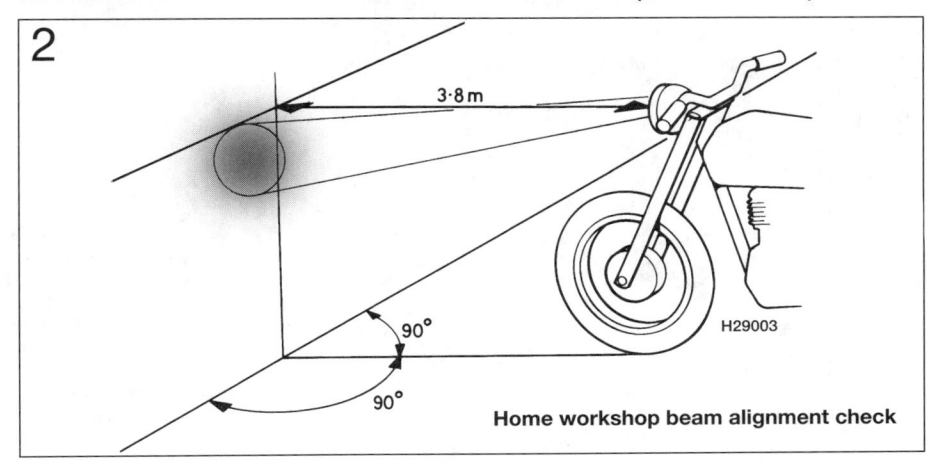

1 Headlight beam height checking equipment

2 Home workshop beam alignment check

REF•28 MOT Test Checks

Exhaust System and Final Drive

Exhaust

✔ Check that the exhaust mountings are secure and that the system does not foul any of the rear suspension components.
✔ Start the motorcycle. When the revs are increased, check that the exhaust is neither holed nor leaking from any of its joints. On a linked system, check that the collector box is not leaking due to corrosion.
✔ Note that the exhaust decibel level ("loudness" of the exhaust) is assessed at the discretion of the tester. If the motorcycle was first used on or after 1st January 1985 the silencer must carry the BSAU 193 stamp, or a marking relating to its make and model, or be of OE (original equipment) manufacture. If the silencer is marked NOT FOR ROAD USE, RACING USE ONLY or similar, it will fail the MOT.

Final drive

✔ On chain or belt drive machines, check that the chain/belt is in good condition and does not have excessive slack. Also check that the sprocket is securely mounted on the rear wheel hub. Check that the chain/belt guard is in place.
✔ On shaft drive bikes, check for oil leaking from the drive unit and fouling the rear tyre.

Steering and Suspension

Steering

✔ With the front wheel raised off the ground, rotate the steering from lock to lock. The handlebar or switches must not contact the fuel tank or be close enough to trap the rider's hand. Problems can be caused by damaged lock stops on the lower yoke and frame, or by the fitting of non-standard handlebars.
✔ When performing the lock to lock check, also ensure that the steering moves freely without drag or notchiness. Steering movement can be impaired by poorly routed cables, or by overtight head bearings or worn bearings. The tester will perform a check of the steering head bearing lower race by mounting the front wheel on a surface plate, then performing a lock to lock check with the weight of the machine on the lower bearing (see illustration 3).
✔ Grasp the fork sliders (lower legs) and attempt to push and pull on the forks (see illustration 4). Any play in the steering head bearings will be felt. Note that in extreme cases, wear of the front fork bushes can be misinterpreted for head bearing play.
✔ Check that the handlebars are securely mounted.
✔ Check that the handlebar grip rubbers are secure. They should by bonded to the bar left end and to the throttle cable pulley on the right end.

Front wheel mounted on a surface plate for steering head bearing lower race check

Front suspension

✔ With the motorcycle off the stand, hold the front brake on and pump the front forks up and down (see illustration 5). Check that they are adequately damped.

Checking the steering head bearings for freeplay

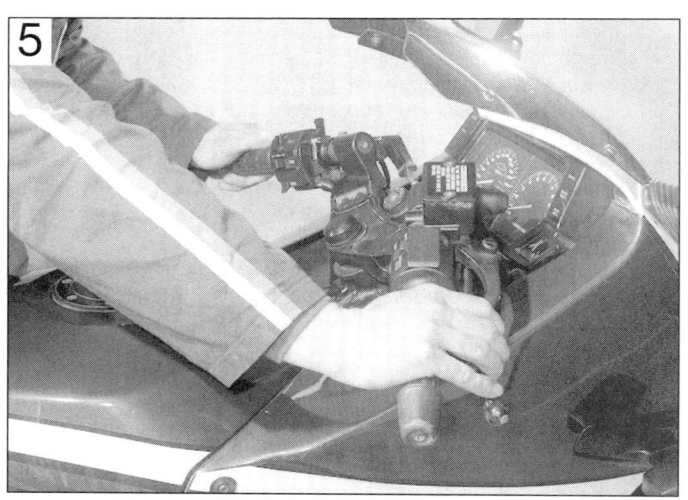

Hold the front brake on and pump the front forks up and down to check operation

MOT Test Checks REF•29

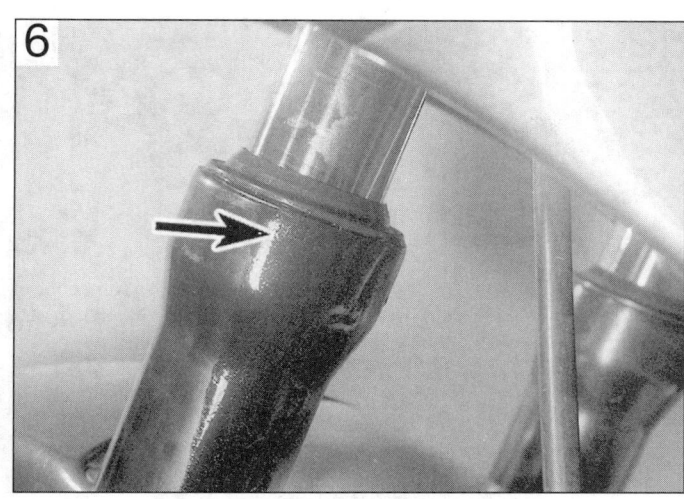

Inspect the area around the fork dust seal for oil leakage (arrow)

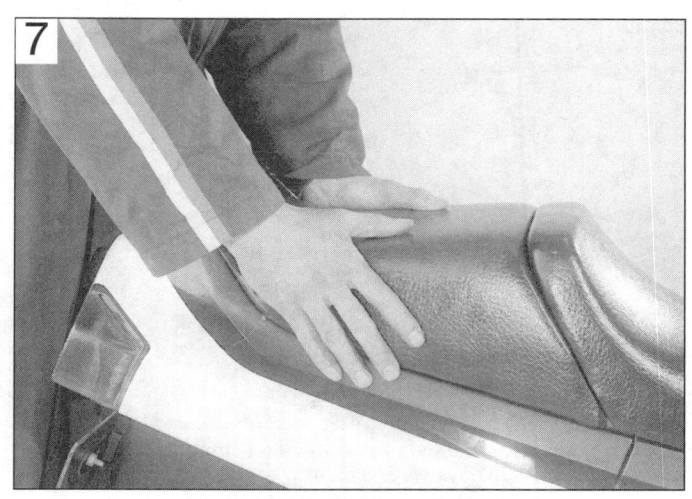

Bounce the rear of the motorcycle to check rear suspension operation

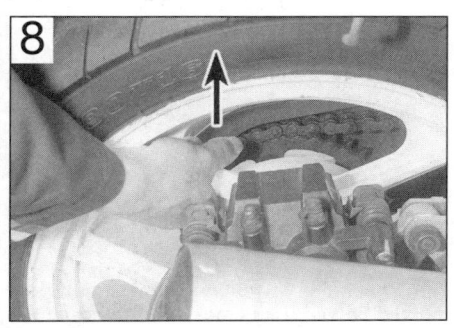

Checking for rear suspension linkage play

✔ Inspect the area above and around the front fork oil seals **(see illustration 6)**. There should be no sign of oil on the fork tube (stanchion) nor leaking down the slider (lower leg). On models so equipped, check that there is no oil leaking from the anti-dive units.
✔ On models with swingarm front suspension, check that there is no freeplay in the linkage when moved from side to side.

Rear suspension

✔ With the motorcycle off the stand and an assistant supporting the motorcycle by its handlebars, bounce the rear suspension **(see illustration 7)**. Check that the suspension components do not foul on any of the cycle parts and check that the shock absorber(s) provide adequate damping.
✔ Visually inspect the shock absorber(s) and check that there is no sign of oil leakage from its damper. This is somewhat restricted on certain single shock models due to the location of the shock absorber.
✔ With the rear wheel raised off the ground, grasp the wheel at the highest point and attempt to pull it up **(see illustration 8)**. Any play in the swingarm pivot or suspension linkage bearings will be felt as movement.
Note: *Do not confuse play with actual suspension movement.* Failure to lubricate suspension linkage bearings can lead to bearing failure **(see illustration 9)**.
✔ With the rear wheel raised off the ground, grasp the swingarm ends and attempt to move the swingarm from side to side and forwards and backwards - any play indicates wear of the swingarm pivot bearings **(see illustration 10)**.

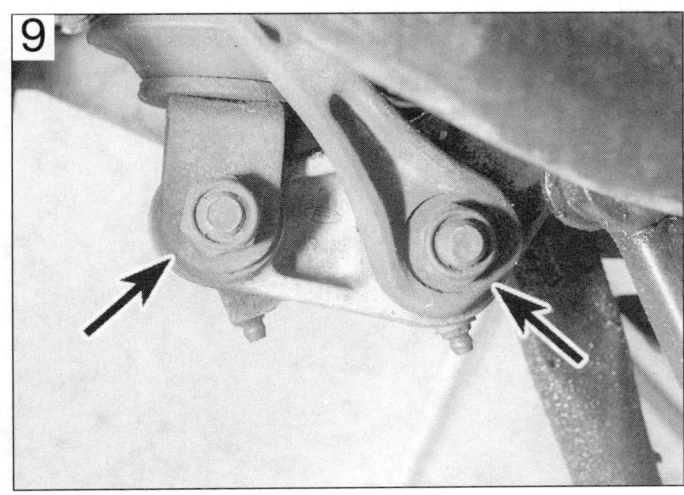

Worn suspension linkage pivots (arrows) are usually the cause of play in the rear suspension

Grasp the swingarm at the ends to check for play in its pivot bearings

REF•30 MOT Test Checks

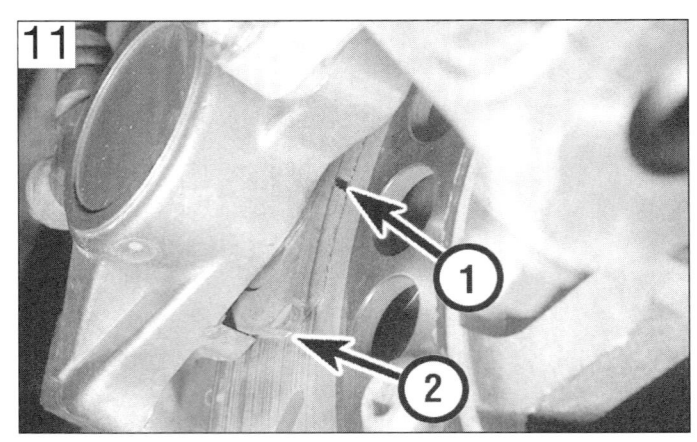

Brake pad wear can usually be viewed without removing the caliper. Most pads have wear indicator grooves (1) and some also have indicator tangs (2)

On drum brakes, check the angle of the operating lever with the brake fully applied. Most drum brakes have a wear indicator pointer and scale.

Brakes, Wheels and Tyres

Brakes

✔ With the wheel raised off the ground, apply the brake then free it off, and check that the wheel is about to revolve freely without brake drag.

✔ On disc brakes, examine the disc itself. Check that it is securely mounted and not cracked.

✔ On disc brakes, view the pad material through the caliper mouth and check that the pads are not worn down beyond the limit **(see illustration 11)**.

✔ On drum brakes, check that when the brake is applied the angle between the operating lever and cable or rod is not too great **(see illustration 12)**. Check also that the operating lever doesn't foul any other components.

✔ On disc brakes, examine the flexible hoses from top to bottom. Have an assistant hold the brake on so that the fluid in the hose is under pressure, and check that there is no sign of fluid leakage, bulges or cracking. If there are any metal brake pipes or unions, check that these are free from corrosion and damage. Where a brake-linked anti-dive system is fitted, check the hoses to the anti-dive in a similar manner.

✔ Check that the rear brake torque arm is secure and that its fasteners are secured by self-locking nuts or castellated nuts with split-pins or R-pins **(see illustration 13)**.

✔ On models with ABS, check that the self-check warning light in the instrument panel works.

✔ The MOT tester will perform a test of the motorcycle's braking efficiency based on a calculation of rider and motorcycle weight. Although this cannot be carried out at home, you can at least ensure that the braking systems are properly maintained. For hydraulic disc brakes, check the fluid level, lever/pedal feel (bleed of air if its spongy) and pad material. For drum brakes, check adjustment, cable or rod operation and shoe lining thickness.

Wheels and tyres

✔ Check the wheel condition. Cast wheels should be free from cracks and if of the built-up design, all fasteners should be secure. Spoked wheels should be checked for broken, corroded, loose or bent spokes.

✔ With the wheel raised off the ground, spin the wheel and visually check that the tyre and wheel run true. Check that the tyre does not foul the suspension or mudguards.

✔ With the wheel raised off the ground, grasp the wheel and attempt to move it about the axle (spindle) **(see illustration 14)**. Any play felt here indicates wheel bearing failure.

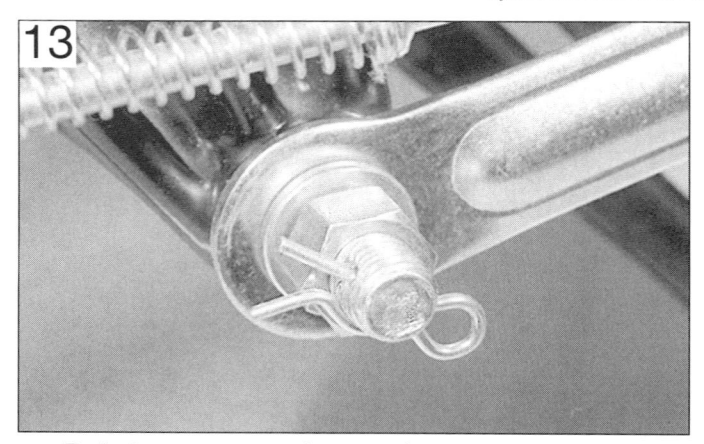

Brake torque arm must be properly secured at both ends

Check for wheel bearing play by trying to move the wheel about the axle (spindle)

MOT Test Checks REF•31

Checking the tyre tread depth

Tyre direction of rotation arrow can be found on tyre sidewall

Castellated type wheel axle (spindle) nut must be secured by a split pin or R-pin

Two straightedges are used to check wheel alignment

✔ Check the tyre tread depth, tread condition and sidewall condition **(see illustration 15)**.
✔ Check the tyre type. Front and rear tyre types must be compatible and be suitable for road use. Tyres marked NOT FOR ROAD USE, COMPETITION USE ONLY or similar, will fail the MOT.

✔ If the tyre sidewall carries a direction of rotation arrow, this must be pointing in the direction of normal wheel rotation **(see illustration 16)**.
✔ Check that the wheel axle (spindle) nuts (where applicable) are properly secured. A self-locking nut or castellated nut with a split-pin or R-pin can be used **(see illustration 17)**.
✔ Wheel alignment is checked with the motorcycle off the stand and a rider seated. With the front wheel pointing straight ahead, two perfectly straight lengths of metal or wood and placed against the sidewalls of both tyres **(see illustration 18)**. The gap each side of the front tyre must be equidistant on both sides. Incorrect wheel alignment may be due to a cocked rear wheel (often as the result of poor chain adjustment) or in extreme cases, a bent frame.

General checks and condition

✔ Check the security of all major fasteners, bodypanels, seat, fairings (where fitted) and mudguards.

✔ Check that the rider and pillion footrests, handlebar levers and brake pedal are securely mounted.

✔ Check for corrosion on the frame or any load-bearing components. If severe, this may affect the structure, particularly under stress.

Sidecars

A motorcycle fitted with a sidecar requires additional checks relating to the stability of the machine and security of attachment and swivel joints, plus specific wheel alignment (toe-in) requirements. Additionally, tyre and lighting requirements differ from conventional motorcycle use. Owners are advised to check MOT test requirements with an official test centre.

Preparing for storage

Before you start

If repairs or an overhaul is needed, see that this is carried out now rather than left until you want to ride the bike again.

Give the bike a good wash and scrub all dirt from its underside. Make sure the bike dries completely before preparing for storage.

Engine

● Remove the spark plug(s) and lubricate the cylinder bores with approximately a teaspoon of motor oil using a spout-type oil can (see illustration 1). Reinstall the spark plug(s). Crank the engine over a couple of times to coat the piston rings and bores with oil. If the bike has a kickstart, use this to turn the engine over. If not, flick the kill switch to the OFF position and crank the engine over on the starter (see illustration 2). If the nature on the ignition system prevents the starter operating with the kill switch in the OFF position, remove the spark plugs and fit them back in their caps; ensure that the plugs are earthed (grounded) against the cylinder head when the starter is operated (see illustration 3).

⚠️ **Warning:** *It is important that the plugs are earthed (grounded) away from the spark plug holes otherwise there is a risk of atomised fuel from the cylinders igniting.*

HAYNES HiNT: *On a single cylinder four-stroke engine, you can seal the combustion chamber completely by positioning the piston at TDC on the compression stroke.*

● Drain the carburettor(s) otherwise there is a risk of jets becoming blocked by gum deposits from the fuel (see illustration 4).

● If the bike is going into long-term storage, consider adding a fuel stabiliser to the fuel in the tank. If the tank is drained completely, corrosion of its internal surfaces may occur if left unprotected for a long period. The tank can be treated with a rust preventative especially for this purpose. Alternatively, remove the tank and pour half a litre of motor oil into it, install the filler cap and shake the tank to coat its internals with oil before draining off the excess. The same effect can also be achieved by spraying WD40 or a similar water-dispersant around the inside of the tank via its flexible nozzle.

● Make sure the cooling system contains the correct mix of antifreeze. Antifreeze also contains important corrosion inhibitors.

● The air intakes and exhaust can be sealed off by covering or plugging the openings. Ensure that you do not seal in any condensation; run the engine until it is hot,

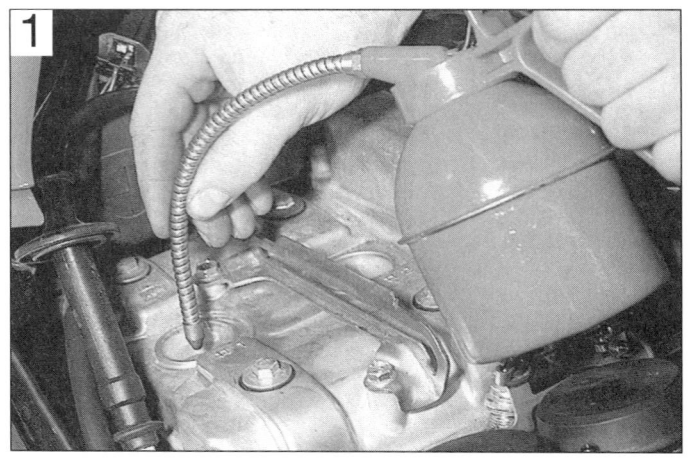

1 Squirt a drop of motor oil into each cylinder

2 Flick the kill switch to OFF . . .

3 . . . and ensure that the metal bodies of the plugs (arrows) are earthed against the cylinder head

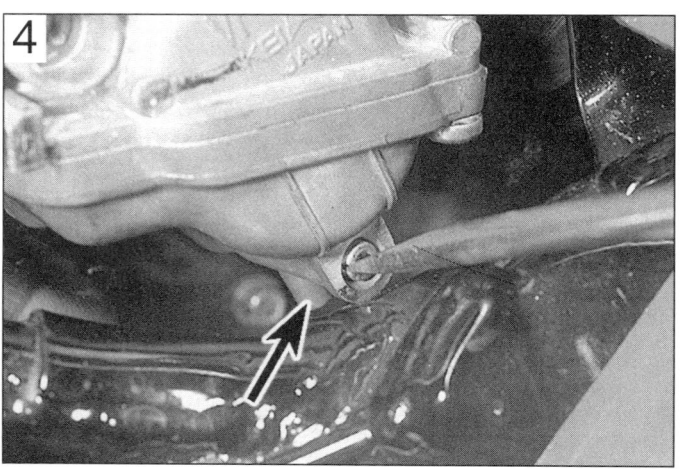

4 Connect a hose to the carburettor float chamber drain stub (arrow) and unscrew the drain screw

Storage   REF•33

Exhausts can be sealed off with a plastic bag

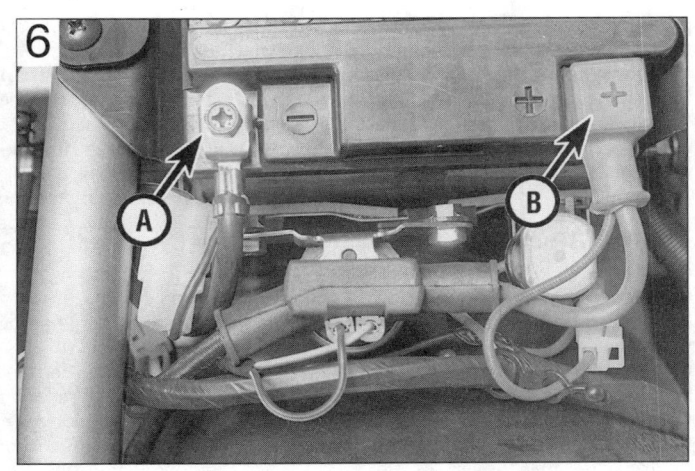

Disconnect the negative lead (A) first, followed by the positive lead (B)

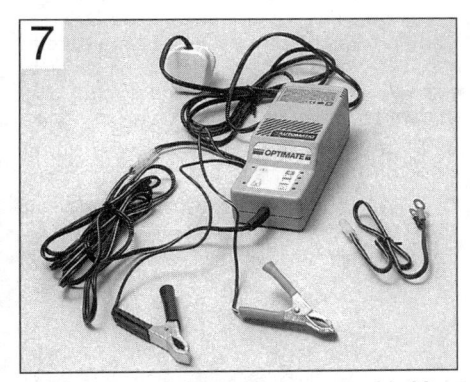

Use a suitable battery charger - this kit also assess battery condition

then switch off and allow to cool. Tape a piece of thick plastic over the silencer end(s) **(see illustration 5)**. Note that some advocate pouring a tablespoon of motor oil into the silencer(s) before sealing them off.

Battery

● Remove it from the bike - in extreme cases of cold the battery may freeze and crack its case **(see illustration 6)**.

● Check the electrolyte level and top up if necessary (conventional refillable batteries). Clean the terminals.
● Store the battery off the motorcycle and away from any sources of fire. Position a wooden block under the battery if it is to sit on the ground.
● Give the battery a trickle charge for a few hours every month **(see illustration 7)**.

Tyres

● Place the bike on its centrestand or an auxiliary stand which will support the motorcycle in an upright position. Position wood blocks under the tyres to keep them off the ground and to provide insulation from damp. If the bike is being put into long-term storage, ideally both tyres should be off the ground; not only will this protect the tyres, but will also ensure that no load is placed on the steering head or wheel bearings.
● Deflate each tyre by 5 to 10 psi, no more or the beads may unseat from the rim, making subsequent inflation difficult on tubeless tyres.

Pivots and controls

● Lubricate all lever, pedal, stand and footrest pivot points. If grease nipples are fitted to the rear suspension components, apply lubricant to the pivots.
● Lubricate all control cables.

Cycle components

● Apply a wax protectant to all painted and plastic components. Wipe off any excess, but don't polish to a shine. Where fitted, clean the screen with soap and water.
● Coat metal parts with Vaseline (petroleum jelly). When applying this to the fork tubes, do not compress the forks otherwise the seals will rot from contact with the Vaseline.
● Apply a vinyl cleaner to the seat.

Storage conditions

● Aim to store the bike in a shed or garage which does not leak and is free from damp.
● Drape an old blanket or bedspread over the bike to protect it from dust and direct contact with sunlight (which will fade paint). This also hides the bike from prying eyes. Beware of tight-fitting plastic covers which may allow condensation to form and settle on the bike.

Getting back on the road

Engine and transmission

● Change the oil and replace the oil filter. If this was done prior to storage, check that the oil hasn't emulsified - a thick whitish substance which occurs through condensation.
● Remove the spark plugs. Using a spout-type oil can, squirt a few drops of oil into the cylinder(s). This will provide initial lubrication as the piston rings and bores comes back into contact. Service the spark plugs, or fit new ones, and install them in the engine.

● Check that the clutch isn't stuck on. The plates can stick together if left standing for some time, preventing clutch operation. Engage a gear and try rocking the bike back and forth with the clutch lever held against the handlebar. If this doesn't work on cable-operated clutches, hold the clutch lever back against the handlebar with a strong elastic band or cable tie for a couple of hours **(see illustration 8)**.
● If the air intakes or silencer end(s) were blocked off, remove the bung or cover used.
● If the fuel tank was coated with a rust

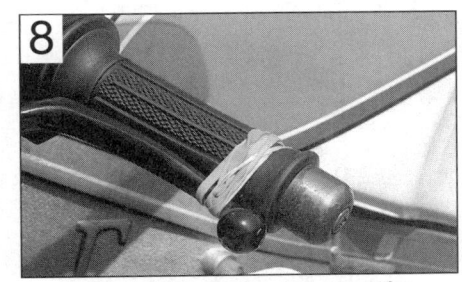

Hold clutch lever back against the handlebar with elastic bands or a cable tie

Storage

preventative, oil or a stabiliser added to the fuel, drain and flush the tank and dispose of the fuel sensibly. If no action was taken with the fuel tank prior to storage, it is advised that the old fuel is disposed of since it will go off over a period of time. Refill the fuel tank with fresh fuel.

Frame and running gear

- Oil all pivot points and cables.
- Check the tyre pressures. They will definitely need inflating if pressures were reduced for storage.
- Lubricate the final drive chain (where applicable).
- Remove any protective coating applied to the fork tubes (stanchions) since this may well destroy the fork seals. If the fork tubes weren't protected and have picked up rust spots, remove them with very fine abrasive paper and refinish with metal polish.
- Check that both brakes operate correctly. Apply each brake hard and check that it's not possible to move the motorcycle forwards, then check that the brake frees off again once released. Brake caliper pistons can stick due to corrosion around the piston head, or on the sliding caliper types, due to corrosion of the slider pins. If the brake doesn't free after repeated operation, take the caliper off for examination. Similarly drum brakes can stick due to a seized operating cam, cable or rod linkage.
- If the motorcycle has been in long-term storage, renew the brake fluid and clutch fluid (where applicable).
- Depending on where the bike has been stored, the wiring, cables and hoses may have been nibbled by rodents. Make a visual check and investigate disturbed wiring loom tape.

Battery

- If the battery has been previously removal and given top up charges it can simply be reconnected. Remember to connect the positive cable first and the negative cable last.
- On conventional refillable batteries, if the battery has not received any attention, remove it from the motorcycle and check its electrolyte level. Top up if necessary then charge the battery. If the battery fails to hold a charge and a visual checks show heavy white sulphation of the plates, the battery is probably defective and must be renewed. This is particularly likely if the battery is old. Confirm battery condition with a specific gravity check.
- On sealed (MF) batteries, if the battery has not received any attention, remove it from the motorcycle and charge it according to the information on the battery case - if the battery fails to hold a charge it must be renewed.

Starting procedure

- If a kickstart is fitted, turn the engine over a couple of times with the ignition OFF to distribute oil around the engine. If no kickstart is fitted, flick the engine kill switch OFF and the ignition ON and crank the engine over a couple of times to work oil around the upper cylinder components. If the nature of the ignition system is such that the starter won't work with the kill switch OFF, remove the spark plugs, fit them back into their caps and earth (ground) their bodies on the cylinder head. Reinstall the spark plugs afterwards.
- Switch the kill switch to RUN, operate the choke and start the engine. If the engine won't start don't continue cranking the engine - not only will this flatten the battery, but the starter motor will overheat. Switch the ignition off and try again later. If the engine refuses to start, go through the fault finding procedures in this manual. **Note:** *If the bike has been in storage for a long time, old fuel or a carburettor blockage may be the problem. Gum deposits in carburettors can block jets - if a carburettor cleaner doesn't prove successful the carburettors must be dismantled for cleaning.*
- Once the engine has started, check that the lights, turn signals and horn work properly.
- Treat the bike gently for the first ride and check all fluid levels on completion. Settle the bike back into the maintenance schedule.

Fault Finding   REF•35

This Section provides an easy reference-guide to the more common faults that are likely to afflict your machine. Obviously, the opportunities are almost limitless for faults to occur as a result of obscure failures, and to try and cover all eventualities would require a book. Indeed, a number have been written on the subject.

Successful troubleshooting is not a mysterious 'black art' but the application of a bit of knowledge combined with a systematic and logical approach to the problem. Approach any troubleshooting by first accurately identifying the symptom and then checking through the list of possible causes, starting with the simplest or most obvious and progressing in stages to the most complex.

Take nothing for granted, but above all apply liberal quantities of common sense.

The main symptom of a fault is given in the text as a major heading below which are listed the various systems or areas which may contain the fault. Details of each possible cause for a fault and the remedial action to be taken are given, in brief, in the paragraphs below each heading. Further information should be sought in the relevant Chapter.

1 Engine doesn't start or is difficult to start
- [ ] Starter motor doesn't rotate
- [ ] Starter motor rotates but engine does not turn over
- [ ] Starter works but engine won't turn over (seized)
- [ ] No fuel flow
- [ ] Engine flooded
- [ ] No spark or weak spark
- [ ] Compression low
- [ ] Stalls after starting
- [ ] Rough idle

2 Poor running at low speeds
- [ ] Spark weak
- [ ] Fuel/air mixture incorrect
- [ ] Compression low
- [ ] Poor acceleration

3 Poor running or no power at high speed
- [ ] Firing incorrect
- [ ] Fuel/air mixture incorrect
- [ ] Compression low
- [ ] Knocking or pinking
- [ ] Miscellaneous causes

4 Overheating
- [ ] Engine overheats
- [ ] Firing incorrect
- [ ] Fuel/air mixture incorrect
- [ ] Compression too high
- [ ] Engine load excessive
- [ ] Lubrication inadequate
- [ ] Miscellaneous causes

5 Clutch problems
- [ ] Clutch slipping
- [ ] Clutch not disengaging completely

6 Gearchange problems
- [ ] Doesn't go into gear, or lever doesn't return
- [ ] Jumps out of gear
- [ ] Overshifts

7 Abnormal engine noise
- [ ] Knocking or pinking
- [ ] Piston slap or rattling
- [ ] Valve noise
- [ ] Other noise

8 Abnormal driveline noise
- [ ] Clutch noise
- [ ] Transmission noise
- [ ] Final drive noise

9 Abnormal frame and suspension noise
- [ ] Front end noise
- [ ] Rear suspension noise
- [ ] Brake noise

10 Oil pressure warning light comes on
- [ ] Engine lubrication system
- [ ] Electrical system

11 Excessive exhaust smoke
- [ ] White smoke
- [ ] Black smoke
- [ ] Brown smoke

12 Poor handling or stability
- [ ] Handlebar hard to turn
- [ ] Handlebar shakes or vibrates excessively
- [ ] Handlebar pulls to one side
- [ ] Poor shock absorbing qualities

13 Braking problems
- [ ] Brakes are spongy, or lack power
- [ ] Brake lever or pedal pulsates
- [ ] Brakes drag

14 Electrical problems
- [ ] Battery dead or weak
- [ ] Battery overcharged

1 Engine doesn't start or is difficult to start

Starter motor doesn't rotate
- [ ] Engine kill switch OFF.
- [ ] Fuse blown. Check main fuse and ignition/starter circuit fuse (Chapter 9).
- [ ] Battery voltage low. Check and recharge battery (Chapter 9).
- [ ] Starter motor defective. Make sure the wiring to the starter is secure. Make sure the starter relay clicks when the start button is pushed. If the relay clicks, then the fault is in the wiring or motor.
- [ ] Starter relay faulty. Check it according to the procedure in Chapter 9.
- [ ] Starter switch not contacting. The contacts could be wet, corroded or dirty. Disassemble and clean the switch (Chapter 9).
- [ ] Wiring open or shorted. Check all wiring connections and harnesses to make sure that they are dry, tight and not corroded. Also check for broken or frayed wires that can cause a short to ground (earth) (see wiring diagrams, Chapter 9).
- [ ] Ignition (main) switch defective. Check the switch according to the procedure in Chapter 9. Replace the switch with a new one if it is defective.
- [ ] Engine kill switch defective. Check for wet, dirty or corroded contacts. Clean or replace the switch as necessary (Chapter 9).
- [ ] Faulty neutral, side stand or clutch switch, or diode. Check the wiring to each switch and the switch itself according to the procedures in Chapter 9. Also check the diode.

Starter motor rotates but engine does not turn over
- [ ] Starter clutch defective. Inspect and repair or replace (Chapter 2).
- [ ] Damaged idle/reduction or starter gears. Inspect and renew the damaged parts (Chapter 2).

Starter works but engine won't turn over (seized)
- [ ] Seized engine caused by one or more internally damaged components. Failure due to wear, abuse or lack of lubrication. Damage can include seized valves, followers, camshafts, pistons, crankshaft, connecting rod bearings, or transmission gears or bearings. Refer to Chapter 2 for engine disassembly.

No fuel flow
- [ ] No fuel in tank.
- [ ] Fuel tank breather hose obstructed.
- [ ] Fuel pump assembly filter clogged. Remove the tap or pump and clean or renew the filter (Chapter 1 and 4).
- [ ] Fuel hose clogged. Remove the fuel tank and check the hoses.
- [ ] Faulty fuel cut-off relay. Check the relay (Chapter 4).
- [ ] Fuel pump faulty. Check the fuel pump flow rate and fuel pressure (Chapter 4).

Engine flooded
- [ ] Faulty pressure regulator – if it is stuck closed there could be excessive pressure in the fuel rail. Check as described in Chapter 4.
- [ ] Injector(s) stuck open, allowing a constant flow of fuel into the engine. Check as described in Chapter 4.
- [ ] Starting technique incorrect. The throttle should be fully closed when starting the engine – a fully open throttle will cause the fuel supply to be shut off by the ECM. If the engine is flooded, set the engine kill switch to RUN and turn the ignition ON. Open the throttle fully and operate the starter motor for 5 seconds. The engine can then be started normally.

No spark or weak spark
- [ ] Engine kill switch turned to the OFF position.
- [ ] Battery voltage low. Check and recharge the battery as necessary (Chapter 9).
- [ ] Spark plugs dirty, defective or worn out. Locate reason for fouled plugs using spark plug condition chart and follow the plug maintenance procedures (Chapter 1).
- [ ] Spark plug caps/ignition coils faulty or not making good contact over the spark plugs. Check condition. Replace if cracks or deterioration are evident (Chapter 5).
- [ ] ECM defective. Check the unit, referring to Chapter 4 for details.
- [ ] Pulse generator defective. Check the unit, referring to Chapter 4 for details.
- [ ] Ignition or kill switch shorted. This is usually caused by water, corrosion, damage or excessive wear. The switches can be disassembled and cleaned with electrical contact cleaner. If cleaning does not help, renew the switches (Chapter 9).
- [ ] Wiring shorted or broken between:
    a) Ignition (main) switch and engine kill switch (or blown fuse)
    b) ECM and engine kill switch
    c) ECM and ignition HT coils
    d) ECM and pulse generator
- [ ] Make sure that all wiring connections are clean, dry and tight. Look for chafed and broken wires (Chapters 4, 5 and 9).

Compression low
- [ ] Spark plugs loose. Remove the plugs and inspect their threads. Reinstall and tighten to the specified torque (Chapter 1).
- [ ] Cylinder head not sufficiently tightened down. If a cylinder head is suspected of being loose, then there's a chance that the gasket or head is damaged if the problem has persisted for any length of time. The head bolts should be tightened to the proper torque in the correct sequence (Chapter 2).
- [ ] Incorrect valve clearance. This means that the valve is not closing completely and compression pressure is leaking past the valve. Check and adjust the valve clearances (Chapter 1).
- [ ] Cylinder and/or piston worn. Excessive wear will cause compression pressure to leak past the rings. This is usually accompanied by worn rings as well. A top-end overhaul is necessary (Chapter 2).
- [ ] Piston rings worn, weak, broken, or sticking. Broken or sticking piston rings usually indicate a lubrication or fuelling problem that causes excess carbon deposits or seizures to form on the pistons and rings. Top-end overhaul is necessary (Chapter 2).
- [ ] Piston ring-to-groove clearance excessive. This is caused by excessive wear of the piston ring lands. Piston replacement is necessary (Chapter 2).
- [ ] Cylinder head gasket damaged. If a head is allowed to become loose, or if excessive carbon build-up on the piston crown and combustion chamber causes extremely high compression, the head gasket may leak. Retorquing the head is not always sufficient to restore the seal, so gasket renewal is necessary (Chapter 2).
- [ ] Cylinder head warped. This is caused by overheating or improperly tightened head bolts. Machine shop resurfacing or head replacement is necessary (Chapter 2).
- [ ] Valve spring broken or weak. Caused by component failure or wear; the springs must be renewed (Chapter 2).
- [ ] Valve not seating properly. This is caused by a bent valve (from over-revving or improper valve adjustment), burned valve or seat (improper fuelling) or an accumulation of carbon deposits on the seat (from fuelling or lubrication problems). The valves must be cleaned and/or renewed and the seats serviced if possible (Chapter 2).

Stalls after starting
- [ ] Ignition malfunction (Chapter 5).
- [ ] Fuel injection system malfunction (Chapter 4).
- [ ] Fuel contaminated. The fuel can be contaminated with either dirt or water, or can change chemically if the machine is allowed to sit for several months or more. Drain the tank and fuel rail (Chapter 4). Also check that the fuel flows freely and is not being restricted.

Fault Finding   REF•37

1 Engine doesn't start or is difficult to start (continued)

- ☐ Intake air leak. Check for loose throttle body-to-intake manifold connections, loose or missing vacuum gauge adapter screws or hoses (Chapter 4).
- ☐ Engine idle speed incorrect. Turn idle adjusting screw until the engine idles at the specified rpm (Chapter 1). Also check the fast idle system wax unit and starter valves (Chapter 4).

Rough idle
- ☐ Ignition malfunction (Chapter 5).
- ☐ Idle speed incorrect (Chapter 1).

- ☐ Throttle body starter valves not synchronised. Adjust with vacuum gauge or manometer set as described in Chapter 4.
- ☐ Throttle body starter valve or fuel injection system malfunction (Chapter 4).
- ☐ Fuel contaminated. The fuel can be contaminated with either dirt or water, or can change chemically if the machine is allowed to sit for several months or more. Drain the tank and fuel rail (Chapter 4).
- ☐ Intake air leak. Check for loose throttle body-to-intake manifold connections, loose or missing vacuum gauge adapter screws or hoses (Chapter 4).
- ☐ Air filter clogged. Clean or renew the air filter element (Chapter 1).

2 Poor running at low speeds

Spark weak
- ☐ Battery voltage low. Check and recharge battery (Chapter 9).
- ☐ Spark plugs fouled, defective or worn out. Refer to Chapter 1 for spark plug maintenance.
- ☐ Incorrect spark plugs. Wrong type, heat range or cap configuration. Check and install correct plugs listed in Chapter 1.
- ☐ ECM defective. Check the unit, referring to Chapter 4 for details.
- ☐ Pulse generator defective (Chapter 4).
- ☐ Ignition HT coils/spark plug caps defective or not making good contact with spark plugs (Chapter 5).

Fuel/air mixture incorrect
- ☐ Fuel injection system malfunction (Chapter 4).
- ☐ Fuel injector clogged (Chapter 4).
- ☐ Fuel pump or pressure regulator faulty (Chapter 4).
- ☐ Throttle body intake manifolds loose. Check for cracks, breaks, tears or loose clamps. Renew the rubber intake manifold joints if split or perished.
- ☐ Air filter clogged, poorly sealed or missing (Chapter 1).
- ☐ Air filter housing poorly sealed. Look for cracks, holes or loose clamps and renew or repair defective parts.
- ☐ Fuel tank breather hose obstructed.

Compression low
- ☐ Spark plugs loose. Remove the plugs and inspect their threads. Reinstall and tighten to the specified torque (Chapter 1).
- ☐ Cylinder head not sufficiently tightened down. If a cylinder head is suspected of being loose, then there's a chance that the gasket and head are damaged if the problem has persisted for any length of time. The head bolts should be tightened to the proper torque in the correct sequence (Chapter 2).
- ☐ Incorrect valve clearance. This means that the valve is not closing completely and compression pressure is leaking past the valve. Check and adjust the valve clearances (Chapter 1).
- ☐ Cylinder and/or piston worn. Excessive wear will cause compression pressure to leak past the rings. This is usually accompanied by worn rings as well. A top-end overhaul is necessary (Chapter 2).
- ☐ Piston rings worn, weak, broken, or sticking. Broken or sticking piston rings usually indicate a lubrication or fuelling problem that causes excess carbon deposits or seizures to form on the pistons and rings. Top-end overhaul is necessary (Chapter 2).
- ☐ Piston ring-to-groove clearance excessive. This is caused by excessive wear of the piston ring lands. Piston renewal is necessary (Chapter 2).
- ☐ Cylinder head gasket damaged. If a head is allowed to become loose, or if excessive carbon build-up on the piston crown and combustion chamber causes extremely high compression, the head gasket may leak. Retorquing the head is not always sufficient to restore the seal, so gasket renewal is necessary (Chapter 2).
- ☐ Cylinder head warped. This is caused by overheating or improperly tightened head bolts. Machine shop resurfacing or head renewal is necessary (Chapter 2).
- ☐ Valve spring broken or weak. Caused by component failure or wear; the springs must be renewed (Chapter 2).
- ☐ Valve not seating properly. This is caused by a bent valve (from over-revving or improper valve adjustment), burned valve or seat (improper fuelling) or an accumulation of carbon deposits on the seat (from fuelling, lubrication problems). The valves must be cleaned and/or renewed and the seats serviced if possible (Chapter 2).

Poor acceleration
- ☐ Throttle bodies leaking or dirty. Overhaul them (Chapter 4).
- ☐ Fuel injection system malfunction, faulty fuel pump, or pressure regulator (Chapter 4).
- ☐ Timing not advancing. The pulse generator or ECM may be defective. If so, they must be renewed, as they can't be repaired. Check them (Chapter 4).
- ☐ Throttle bodies (starter valves) not synchronised. Adjust them with a vacuum gauge set or manometer as described in Chapter 4.
- ☐ Engine oil viscosity too high. Using a heavier oil than that recommended in Chapter 1 can damage the oil pump or lubrication system and cause drag on the engine.
- ☐ Brakes dragging. Usually caused by debris which has entered the brake piston seals, or from a warped disc or bent axle. Repair as necessary (Chapter 7).

3 Poor running or no power at high speed

Firing incorrect
- ☐ Air filter restricted. Clean or renew filter (Chapter 1).
- ☐ Spark plugs fouled, defective or worn out. See Chapter 1 for spark plug maintenance.
- ☐ Incorrect spark plugs. Wrong type, heat range or cap configuration. Check and install correct plugs listed in Chapter 1.
- ☐ ECM defective (Chapter 4).
- ☐ Ignition HT coils/spark plug caps defective or not making good contact with spark plugs (Chapter 5).

Fuel/air mixture incorrect
- ☐ Fuel injection system malfunction (Chapter 4).
- ☐ Fuel injector clogged (Chapter 4).
- ☐ Fuel pump or pressure regulator faulty (Chapter 4).
- ☐ Throttle body intake manifolds loose. Check for cracks, breaks, tears or loose clamps. Renew the rubber intake manifold joints if split or perished.
- ☐ Air filter clogged, poorly sealed or missing (Chapter 1).
- ☐ Air filter housing poorly sealed. Look for cracks, holes or loose clamps and renew or repair defective parts.
- ☐ Fuel tank breather hose obstructed.

Compression low
- ☐ Spark plugs loose. Remove the plugs and inspect their threads. Reinstall and tighten to the specified torque (Chapter 1).
- ☐ Cylinder head not sufficiently tightened down. If a cylinder head is suspected of being loose, then there's a chance that the gasket and head are damaged if the problem has persisted for any length of time. The head bolts should be tightened to the proper torque in the correct sequence (Chapter 2).
- ☐ Incorrect valve clearance. This means that the valve is not closing completely and compression pressure is leaking past the valve. Check and adjust the valve clearances (Chapter 1).
- ☐ Cylinder and/or piston worn. Excessive wear will cause compression pressure to leak past the rings. This is usually accompanied by worn rings as well. A top-end overhaul is necessary (Chapter 2).
- ☐ Piston rings worn, weak, broken, or sticking. Broken or sticking piston rings usually indicate a lubrication problem that causes excess carbon deposits or seizures to form on the pistons and rings. A top-end overhaul is necessary (Chapter 2).
- ☐ Piston ring-to-groove clearance excessive. This is caused by excessive wear of the piston ring lands. Piston renewal is necessary (Chapter 2).
- ☐ Cylinder head gasket damaged. If a head is allowed to become loose, or if excessive carbon build-up on the piston crown and combustion chamber causes extremely high compression, the head gasket may leak. Retorquing the head is not always sufficient to restore the seal, so gasket renewal is necessary (Chapter 2).
- ☐ Cylinder head warped. This is caused by overheating or improperly tightened head bolts. Machine shop resurfacing or head renewal is necessary (Chapter 2).
- ☐ Valve spring broken or weak. Caused by component failure or wear; the springs must be renewed (Chapter 2).
- ☐ Valve not seating properly. This is caused by a bent valve (from over-revving or improper valve adjustment), burned valve or seat (improper fuelling) or an accumulation of carbon deposits on the seat (from fuelling or lubrication problems). The valves must be cleaned and/or renewed and the seats serviced if possible (Chapter 2).

Knocking or pinking
- ☐ Carbon build-up in combustion chamber. Use of a fuel additive that will dissolve the adhesive bonding the carbon particles to the crown and chamber is the easiest way to remove the build-up. Otherwise, the cylinder head will have to be removed and decarbonised (Chapter 2).
- ☐ Incorrect or poor quality fuel. Old or improper grades of fuel can cause detonation. This causes the piston to rattle, thus the knocking or pinking sound. Drain old fuel and always use the recommended fuel grade.
- ☐ Spark plug heat range incorrect. Uncontrolled detonation indicates the plug heat range is too hot. The plug in effect becomes a glow plug, raising cylinder temperatures. Install the proper heat range plug (Chapter 1).
- ☐ Improper air/fuel mixture. This will cause the cylinders to run hot, which leads to detonation. Clogged injectors or an air leak can cause this imbalance (Chapter 4).

Miscellaneous causes
- ☐ Throttle valve doesn't open fully. Adjust the throttle grip freeplay (Chapter 1).
- ☐ Clutch slipping. May be caused by loose or worn clutch components. Refer to Chapter 2 for clutch overhaul procedures.
- ☐ Timing not advancing – faulty ECM.
- ☐ Engine oil viscosity too high. Using a heavier oil than the one recommended in Chapter 1 can damage the oil pump or lubrication system and cause drag on the engine.
- ☐ Brakes dragging. Usually caused by debris which has entered the brake piston seals, or from a warped disc or bent axle. Repair as necessary (Chapter 7).

4 Overheating

Engine overheats
- [ ] Coolant level low. Check and add coolant (Chapter 1).
- [ ] Leak in cooling system. Check cooling system hoses and radiator for leaks and other damage. Repair or renew parts as necessary (Chapter 3).
- [ ] Thermostat sticking open or closed. Test as described in Chapter 3.
- [ ] Faulty radiator cap. Remove the cap and have it pressure tested by a dealer.
- [ ] Coolant passages clogged. Have the entire system drained and flushed, then refill with fresh coolant.
- [ ] Water pump defective. Remove the pump and check the components (Chapter 3).
- [ ] Clogged radiator fins. Clean them by blowing compressed air through the fins from the rear side of the radiator.
- [ ] Cooling fan or fan switch fault (Chapter 3).

Firing incorrect
- [ ] Spark plugs fouled, defective or worn out. See Chapter 1 for spark plug maintenance.
- [ ] Incorrect spark plugs.
- [ ] ECM defective (Chapter 4).
- [ ] Pulse generator faulty (Chapter 4).
- [ ] Faulty HT ignition coils/spark plug caps (Chapter 5).

Fuel/air mixture incorrect
- [ ] Fuel injection system malfunction (Chapter 4).
- [ ] Fuel injector clogged (Chapter 4).
- [ ] Fuel pump or pressure regulator faulty (Chapter 4).
- [ ] Throttle body intake manifolds loose. Check for cracks, breaks, tears or loose clamps. Renew the rubber intake manifold joints if split or perished.
- [ ] Air filter clogged, poorly sealed or missing (Chapter 1).
- [ ] Air filter housing poorly sealed. Look for cracks, holes or loose clamps and renew or repair defective parts.
- [ ] Fuel tank breather hose obstructed.

Compression too high
- [ ] Carbon build-up in combustion chamber. Use of a fuel additive that will dissolve the adhesive bonding the carbon particles to the piston crown and chamber is the easiest way to remove the build-up. Otherwise, the cylinder head will have to be removed and decarbonised (Chapter 2).

Engine load excessive
- [ ] Clutch slipping. Can be caused by damaged, loose or worn clutch components. Refer to Chapter 2 for overhaul procedures.
- [ ] Engine oil level too high. The addition of too much oil will cause pressurisation of the crankcase and inefficient engine operation. Check Specifications and drain to proper level (Chapter 1 and *Daily (pre-ride) checks*).
- [ ] Engine oil viscosity too high. Using a heavier oil than the one recommended in Chapter 1 can damage the oil pump or lubrication system as well as cause drag on the engine.
- [ ] Brakes dragging. Usually caused by debris which has entered the brake piston seals, or from a warped disc or bent axle. Repair as necessary (Chapter 7).

Lubrication inadequate
- [ ] Engine oil level too low. Friction caused by intermittent lack of lubrication or from oil that is overworked can cause overheating. The oil provides a definite cooling function in the engine. Check the oil level (*Daily (pre-ride) checks*).
- [ ] Poor quality engine oil or incorrect viscosity or type. Oil is rated not only according to viscosity but also according to type. Some oils are not rated high enough for use in this engine. Check the Specifications section and change to the correct oil (Chapter 1).

Miscellaneous causes
- [ ] Modification to exhaust system. Most aftermarket exhaust systems cause the engine to run leaner, which make them run hotter.

5 Clutch problems

Clutch slipping
- [ ] Clutch cable incorrectly adjusted. Check and adjust (Chapter 1).
- [ ] Friction plates worn or warped. Overhaul the clutch assembly (Chapter 2).
- [ ] Plain plates warped (Chapter 2).
- [ ] Clutch springs broken or weak. Old or heat-damaged (from slipping clutch) springs should be replaced with new ones (Chapter 2).
- [ ] Clutch release mechanism defective. Replace any defective parts (Chapter 2).
- [ ] Clutch centre or housing unevenly worn. This causes improper engagement of the plates. Renew the damaged or worn parts (Chapter 2).

Clutch not disengaging completely
- [ ] Clutch cable incorrectly adjusted. Check and adjust (Chapter 1).
- [ ] Clutch plates warped or damaged. This will cause clutch drag, which in turn will cause the machine to creep. Overhaul the clutch assembly (Chapter 2).
- [ ] Clutch spring tension uneven. Usually caused by a sagged or broken spring. Check and renew the springs as a set (Chapter 2).
- [ ] Engine oil deteriorated. Old, thin, worn out oil will not provide proper lubrication for the plates, causing the clutch to drag. Renew the oil and filter (Chapter 1).
- [ ] Engine oil viscosity too high. Using a heavier oil than recommended in Chapter 1 can cause the plates to stick together, putting a drag on the engine. Change to the correct weight oil (Chapter 1).
- [ ] Clutch housing guide seized on gearbox input shaft. Lack of lubrication, severe wear or damage can cause the guide to seize on the shaft. Overhaul of the clutch, and perhaps transmission, may be necessary to repair the damage (Chapter 2).
- [ ] Clutch release mechanism defective.
- [ ] Loose clutch centre nut. Causes housing and centre misalignment putting a drag on the engine. Engagement adjustment continually varies. Overhaul the clutch assembly (Chapter 2).

REF•40 Fault Finding

6 Gearchange problems

Doesn't go into gear or lever doesn't return
- [ ] Clutch not disengaging. See above.
- [ ] Selector fork(s) bent or seized. Often caused by dropping the machine or from lack of lubrication. Overhaul the transmission (Chapter 2).
- [ ] Gear(s) stuck on shaft. Most often caused by a lack of lubrication or excessive wear in transmission bearings and bushings. Overhaul the transmission (Chapter 2).
- [ ] Selector drum binding. Caused by lubrication failure or excessive wear. Renew the drum and bearing (Chapter 2).
- [ ] Gearchange lever return spring weak or broken (Chapter 2).
- [ ] Gearchange lever broken. Splines stripped out of lever or shaft, caused by allowing the lever to get loose or from accident damage. Renew necessary parts (Chapter 2).
- [ ] Gearchange mechanism stopper arm broken or worn. Full engagement and rotary movement of selector drum results. Renew the arm (Chapter 2).
- [ ] Stopper arm spring broken. Allows arm to float, causing sporadic shift operation. Renew spring (Chapter 2).

Jumps out of gear
- [ ] Selector fork(s) worn. Overhaul the transmission (Chapter 2).
- [ ] Gear groove(s) worn. Overhaul the transmission (Chapter 2).
- [ ] Gear dogs or dog slots worn or damaged. The gears should be inspected and renewed if necessary.

Overshifts
- [ ] Stopper arm spring weak or broken (Chapter 2).
- [ ] Gearchange shaft return spring post broken or distorted (Chapter 2).

7 Abnormal engine noise

Knocking or pinking
- [ ] Carbon build-up in combustion chamber. Use of a fuel additive that will dissolve the adhesive bonding the carbon particles to the piston crown and chamber is the easiest way to remove the build-up. Otherwise, the cylinder head will have to be removed and decarbonised (Chapter 2).
- [ ] Incorrect or poor quality fuel. Old or improper fuel can cause detonation. This causes the pistons to rattle, thus the knocking or pinking sound. Drain the old fuel and always use the recommended grade fuel (Chapter 4).
- [ ] Spark plug heat range incorrect. Uncontrolled detonation indicates that the plug heat range is too hot. The plug in effect becomes a glow plug, raising cylinder temperatures. Install the proper heat range plug (Chapter 1).
- [ ] Improper fuel/air mixture. This will cause the cylinders to run hot and lead to detonation. Clogged injectors or an air leak can cause this imbalance (Chapter 4).

Piston slap or rattling
- [ ] Cylinder-to-piston clearance excessive. Caused by improper assembly. Inspect and overhaul top-end parts (Chapter 2).
- [ ] Connecting rod bent. Caused by over-revving, trying to start a badly flooded engine or from ingesting a foreign object into the combustion chamber. Renew the damaged parts (Chapter 2).
- [ ] Piston pin or piston pin bore worn or seized from wear or lack of lubrication. Renew damaged parts (Chapter 2).
- [ ] Piston ring(s) worn, broken or sticking. Overhaul the top-end (Chapter 2).
- [ ] Piston seizure damage. Usually from lack of lubrication or overheating. Renew the pistons and bore the cylinders, as necessary (Chapter 2).
- [ ] Connecting rod upper or lower end clearance excessive. Caused by excessive wear or lack of lubrication. Renew worn parts.

Valve noise
- [ ] Incorrect valve clearances. Adjust the clearances by referring to Chapter 1.
- [ ] Valve spring broken or weak. Check and renew weak valve springs (Chapter 2).
- [ ] Camshaft or cylinder head worn or damaged. Lack of lubrication at high rpm is usually the cause of damage. Insufficient oil or failure to change the oil at the recommended intervals are the chief causes. Since there are no replaceable bearings in the head, the head and camshaft holder will have to be renewed if there is excessive wear or damage (Chapter 2).

Other noise
- [ ] Cylinder head gasket leaking.
- [ ] Exhaust pipe leaking at cylinder head connection. Caused by improper fit of pipe(s) or loose exhaust flange. All exhaust fasteners should be tightened evenly and carefully. Failure to do this will lead to a leak.
- [ ] Crankshaft runout excessive. Caused by a bent crankshaft (from over-revving) or damage from an upper cylinder component failure. Can also be attributed to dropping the machine on either of the crankshaft ends.
- [ ] Engine mounting bolts loose. Tighten all engine mount bolts (Chapter 2).
- [ ] Crankshaft bearings worn (Chapter 2).
- [ ] Cam chain tensioner defective, cam chain or guide blades worn. Renew according to the procedure in Chapter 2.

Fault Finding   REF•41

8 Abnormal driveline noise

Clutch noise
☐ Clutch housing/friction plate clearance excessive (Chapter 2).
☐ Loose or damaged clutch pressure plate and/or bolts (Chapter 2).

Transmission noise
☐ Bearings worn. Also includes the possibility that the shafts are worn. Overhaul the transmission (Chapter 2).
☐ Gears worn or chipped (Chapter 2).
☐ Metal chips jammed in gear teeth. Probably pieces from a broken clutch, gear or selector mechanism that were picked up by the gears. This will cause early bearing failure (Chapter 2).
☐ Engine oil level too low. Causes a howl from transmission. Also affects engine power and clutch operation (*Daily (pre-ride) checks*).

Final drive noise
☐ Chain not adjusted properly (Chapter 1).
☐ Front or rear sprocket loose. Tighten fasteners (Chapter 6).
☐ Sprockets worn. Renew sprockets (Chapter 6).
☐ Rear sprocket warped. Renew sprockets (Chapter 6).
☐ Rubber dampers in rear wheel hub worn. Check and renew (Chapter 7).

9 Abnormal frame and suspension noise

Front end noise
☐ Low fluid level or improper viscosity oil in forks. This can sound like spurting and is usually accompanied by irregular fork action (Chapter 6).
☐ Spring weak or broken. Makes a clicking or scraping sound. Fork oil, when drained, will have a lot of metal particles in it (Chapter 6).
☐ Steering head bearings loose or damaged. Clicks when braking. Check and adjust or replace as necessary (Chapters 1 and 6).
☐ Fork yokes loose. Make sure all clamp pinch bolts are tightened to the specified torque (Chapter 6).
☐ Fork tube bent. Good possibility if machine has been dropped. Replace tube with a new one (Chapter 6).
☐ Front axle bolt or axle clamp bolts loose. Tighten them to the specified torque (Chapter 7).
☐ Loose or worn wheel bearings. Check and renew as necessary (Chapter 7).

Rear suspension noise
☐ Fluid level incorrect. Indicates a leak caused by defective seal. Shock will be covered with oil. Renew shock or seek advice on repair from a Honda dealer or suspension specialist (Chapter 6).
☐ Defective shock absorber with internal damage. This is in the body of the shock and can't be remedied. The shock must be replaced with a new one (Chapter 6).
☐ Bent or damaged shock body. Replace the shock with a new one (Chapter 6).
☐ Loose or worn suspension linkage or swingarm components. Check and renew as necessary (Chapter 6).

Brake noise
☐ Squeal caused by pad shim not installed or positioned correctly (where fitted) (Chapter 7).
☐ Squeal caused by dust on brake pads. Usually found in combination with glazed pads. Clean using brake cleaning solvent (Chapter 7).
☐ Contamination of brake pads. Oil, brake fluid or dirt causing brake to chatter or squeal. Replace pads (Chapter 7).
☐ Pads glazed. Renew the pads (Chapter 7).
☐ Disc warped. Can cause a chattering, clicking or intermittent squeal. Usually accompanied by a pulsating lever and uneven braking. Renew the disc (Chapter 7).
☐ Loose or worn wheel bearings. Check (Chapter 1) and renew (Chapter 7) as necessary.

10 Oil pressure warning light comes on

Engine lubrication system
☐ Engine oil pump defective, blocked oil strainer gauze or failed relief valve. Carry out an oil pressure check (Chapter 1).
☐ Engine oil level low. Inspect for leak or other problem causing low oil level and add recommended oil (*Daily (pre-ride) checks*).
☐ Engine oil viscosity too low. Very old, thin oil or an improper weight of oil used in the engine. Change to correct oil (Chapter 1).
☐ Camshaft or journals worn. Excessive wear causing drop in oil pressure. Measure oil clearance (Chapter 2). Abnormal wear could be caused by oil starvation at high rpm from low oil level or improper weight or type of oil (Chapter 1).
☐ Crankshaft and/or bearings worn. Same problems as above. Measure oil connecting rod and main bearing oil clearance (Chapter 2).

Electrical system
☐ Oil pressure switch defective. Check the switch according to the procedure in Chapter 9. Replace it if it is defective.
☐ Oil pressure warning light circuit defective. Check for pinched, shorted, disconnected or damaged wiring (Chapter 9).

REF•42 Fault Finding

11 Excessive exhaust smoke

White smoke
- [ ] Piston oil ring worn. The ring may be broken or damaged, causing oil from the crankcase to be pulled past the piston into the combustion chamber. Replace the rings with new ones (Chapter 2).
- [ ] Cylinders worn, cracked, or scored. Caused by overheating or oil starvation. The cylinders will have to be rebored and new pistons installed.
- [ ] Valve oil seal damaged or worn. Replace oil seals with new ones (Chapter 2).
- [ ] Valve guide worn. Perform a complete valve job (Chapter 2).
- [ ] Engine oil level too high, which causes the oil to be forced past the rings. Drain oil to the proper level (Chapter 1).
- [ ] Head gasket broken between oil return and cylinder. Causes oil to be pulled into the combustion chamber. Renew the head gasket and check the head for warpage (Chapter 2).
- [ ] Abnormal crankcase pressurisation, which forces oil past the rings. Clogged breather is usually the cause.

Black smoke
- [ ] Fuel injection system malfunction (Chapter 4).
- [ ] Air filter clogged (Chapter 1).

Brown smoke
- [ ] Fuel injection system malfunction (Chapter 4).
- [ ] Faulty fuel pump or pressure regulator (Chapter 4).
- [ ] Air filter poorly sealed or not installed (Chapter 1).

12 Poor handling or stability

Handlebars hard to turn
- [ ] Steering head bearing adjuster nut too tight. Check adjustment as described in Chapter 1.
- [ ] Bearings damaged. Roughness can be felt as the bars are turned from side-to-side. Renew bearings and races (Chapter 6).
- [ ] Races dented or worn. Denting results from wear in only one position (e.g. straight ahead), from a collision or hitting a pothole or from dropping the machine. Renew bearings (Chapter 6).
- [ ] Steering stem lubrication inadequate. Causes are grease getting hard from age or being washed out by high pressure car washes. Disassemble steering head and repack bearings with fresh grease (Chapter 6).
- [ ] Steering stem bent. Caused by a collision, hitting a pothole or by dropping the machine. Renew damaged part. Don't try to straighten the steering stem (Chapter 6).
- [ ] Front tyre air pressure too low (Chapter 1).

Handlebar shakes or vibrates excessively
- [ ] Tyres worn or out of balance (Chapter 7).
- [ ] Swingarm bearings worn. Renew worn bearings (Chapter 6).
- [ ] Wheel rim(s) warped or damaged. Inspect wheels for runout (Chapter 7).
- [ ] Wheel bearings worn. Worn front or rear wheel bearings can cause poor tracking. Worn front bearings will cause wobble (Chapter 7).
- [ ] Handlebar clamp bolts loose (Chapter 6).
- [ ] Fork yoke bolts loose. Tighten them to the specified torque (Chapter 6).
- [ ] Engine mounting bolts loose. Will cause excessive vibration with increased engine rpm (Chapter 2).

Handlebar pulls to one side
- [ ] Frame bent. Definitely suspect this if the machine has been dropped. May or may not be accompanied by cracking near the bend. Renew the frame (Chapter 6).
- [ ] Wheels out of alignment. Caused by improper location of axle spacers or from bent steering stem or frame (Chapter 6).
- [ ] Swingarm bent or twisted from accident damage. Renew the swingarm (Chapter 6).
- [ ] Steering stem bent. Caused by impact damage or by dropping the motorcycle. Renew the steering stem (Chapter 6).
- [ ] Fork tube bent. Disassemble the forks and renew the damaged parts (Chapter 6).
- [ ] Fork oil level uneven. Check and add or drain as necessary (Chapter 6).

Poor shock absorbing qualities
- [ ] Too hard:
    a) *Fork oil level excessive (Chapter 6).*
    b) *Fork oil viscosity too high. Use a lighter oil (see the Specifications in Chapter 6).*
    c) *Fork tube bent. Causes a harsh, sticking feeling (Chapter 6).*
    d) *Rear shock shaft or body bent or damaged (Chapter 6).*
    e) *Fork internal damage (Chapter 6).*
    f) *Shock internal damage.*
    g) *Tyre pressure too high (Chapter 1).*
    h) *Incorrect adjustment settings (Chapter 6).*
- [ ] Too soft:
    a) *Fork or shock oil insufficient and/or leaking (Chapter 6).*
    b) *Fork oil level too low (Chapter 6).*
    c) *Fork oil viscosity too light (Chapter 6).*
    d) *Fork springs weak or broken (Chapter 6).*
    e) *Shock internal damage or leakage (Chapter 6).*
    f) *Incorrect adjustment settings (Chapter 6).*

Fault Finding REF•43

13 Braking problems

Brakes are spongy, or lack power
- [ ] Air in brake line. Caused by inattention to master cylinder fluid level or by leakage. Locate problem and bleed brakes (Chapter 7).
- [ ] Pad or disc worn (Chapters 1 and 7).
- [ ] Brake fluid leak. Locate problem and bleed brakes (Chapter 7).
- [ ] Contaminated pads. Caused by contamination with oil, grease, brake fluid, etc. Renew pads (Chapter 7). Clean disc thoroughly with brake cleaner (Chapter 7).
- [ ] Brake fluid deteriorated. Fluid is old or contaminated. Drain system, replenish with new fluid and bleed the system (Chapter 7).
- [ ] Master cylinder internal parts worn or damaged causing fluid to bypass (Chapter 7).
- [ ] Master cylinder bore scratched by foreign material or broken spring. Repair or renew master cylinder (Chapter 7).

Brake lever or pedal pulsates
- [ ] Disc warped. Renew disc (Chapter 7).
- [ ] Wheel axle bent. Renew axle (Chapter 7).
- [ ] Brake caliper bolts loose (Chapter 7).
- [ ] Brake caliper sliders damaged or sticking (rear caliper), causing caliper to bind. Lubricate the sliders or renew them if they are corroded or bent (Chapter 7).
- [ ] Wheel warped or otherwise damaged (Chapter 7).
- [ ] Wheel bearings damaged or worn (Chapters 1 and 7).

Brakes drag
- [ ] Master cylinder piston seized. Caused by wear or damage to piston or cylinder bore (Chapter 7).
- [ ] Lever balky or stuck. Check pivot and lubricate (Chapter 7).
- [ ] Brake caliper binds on bracket (rear caliper). Caused by inadequate lubrication or damage to caliper slider pins (Chapter 7).
- [ ] Brake caliper piston seized in bore. Caused by wear or ingestion of dirt past deteriorated seal (Chapter 7).
- [ ] Brake pad damaged. Renew pads (Chapter 7).
- [ ] Pads improperly installed (Chapter 7).

14 Electrical problems

Battery dead or weak
- [ ] Battery faulty. Caused by sulphated plates which are shorted through sedimentation. Also, broken battery terminal making only occasional contact (Chapter 9).
- [ ] Battery cables making poor contact (Chapter 9).
- [ ] Load excessive. Caused by addition of high wattage lights or other electrical accessories.
- [ ] Ignition (main) switch defective. Switch either earths (grounds) internally or fails to shut off system. Renew the switch (Chapter 9).
- [ ] Regulator/rectifier defective (Chapter 9).
- [ ] Alternator stator coil open or shorted (Chapter 9).
- [ ] Wiring faulty. Wiring earthed (grounded) or connections loose in ignition, charging or lighting circuits (Chapter 9).

Battery overcharged
- [ ] Regulator/rectifier defective. Overcharging is noticed when battery gets excessively warm (Chapter 9).
- [ ] Battery defective. Replace battery with a new one (Chapter 9).
- [ ] Battery amperage too low, wrong type or size. Install manufacturer's specified amp-hour battery to handle charging load (Chapter 9).

REF•44 Fault Finding Equipment

Checking engine compression

- Low compression will result in exhaust smoke, heavy oil consumption, poor starting and poor performance. A compression test will provide useful information about an engine's condition and if performed regularly, can give warning of trouble before any other symptoms become apparent.
- A compression gauge will be required, along with an adapter to suit the spark plug hole thread size. Note that the screw-in type gauge/adapter set up is preferable to the rubber cone type.
- Before carrying out the test, first check the valve clearances as described in Chapter 1.

1 Run the engine until it reaches normal operating temperature, then stop it and remove the spark plug(s), taking care not to scald your hands on the hot components.
2 Install the gauge adapter and compression gauge in No. 1 cylinder spark plug hole (see illustration 1).

Screw the compression gauge adapter into the spark plug hole, then screw the gauge into the adapter

3 On kickstart-equipped motorcycles, make sure the ignition switch is OFF, then open the throttle fully and kick the engine over a couple of times until the gauge reading stabilises.
4 On motorcycles with electric start only, the procedure will differ depending on the nature of the ignition system. Flick the engine kill switch (engine stop switch) to OFF and turn the ignition switch ON; open the throttle fully and crank the engine over on the starter motor for a couple of revolutions until the gauge reading stabilises. If the starter will not operate with the kill switch OFF, turn the ignition switch OFF and refer to the next paragraph.
5 Install the plugs back in their caps and arrange the plug electrodes so that their metal bodies are earthed (grounded) against the cylinder head; this is essential to prevent damage to the ignition system (see illustration 2). Position the plugs well away from the plug holes otherwise there is a risk of

All spark plugs must be earthed (grounded) against the cylinder head

atomised fuel escaping from the plug holes and igniting. As a safety precaution, cover the cylinder head cover with rag and disconnect the fuel pump wiring connector (see Chapter 4). Turn the ignition switch and kill switch ON, open the throttle fully and crank the engine over on the starter motor for a couple of revolutions until the gauge reading stabilises.
6 After one or two revolutions the pressure should build up to a maximum figure and then stabilise. Take a note of this reading and on multi-cylinder engines repeat the test on the remaining cylinders.
7 The correct pressures are given in Chapter 1 Specifications. If the results fall within the specified range and on multi-cylinder engines all are relatively equal, the engine is in good condition. If there is a marked difference between the readings, or if the readings are lower than specified, inspection of the top-end components will be required.
8 Low compression pressure may be due to worn cylinder bores, pistons or rings, failure of the cylinder head gasket, worn valve seals, or poor valve seating.
9 To distinguish between cylinder/piston wear and valve leakage, pour a small quantity of oil into the bore to temporarily seal the piston rings, then repeat the compression tests (see illustration 3). If the readings show

Bores can be temporarily sealed with a squirt of motor oil

a noticeable increase in pressure this confirms that the cylinder bore, piston, or rings are worn. If, however, no change is indicated, the cylinder head gasket or valves should be examined.
10 High compression pressure indicates excessive carbon build-up in the combustion chamber and on the piston crown. If this is the case the cylinder head should be removed and the deposits removed. Note that excessive carbon build-up is less likely with the used on modern fuels.

Checking battery open-circuit voltage

Warning: The gases produced by the battery are explosive - never smoke or create any sparks in the vicinity of the battery. Never allow the electrolyte to contact your skin or clothing - if it does, wash it off and seek immediate medical attention.

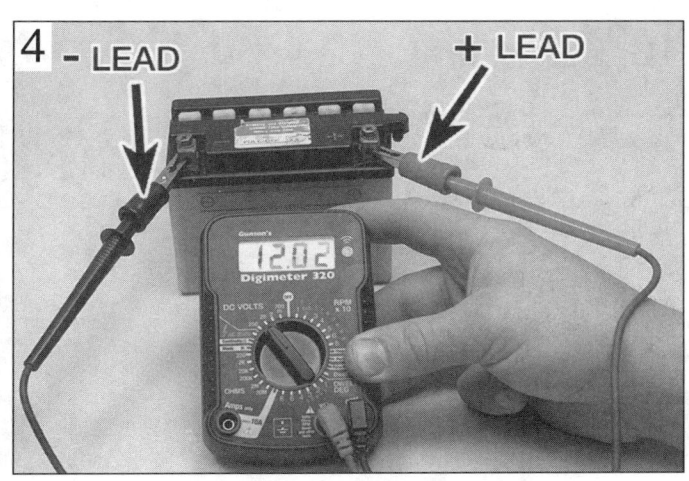

Fault Finding Equipment   REF•45

Measuring open-circuit battery voltage

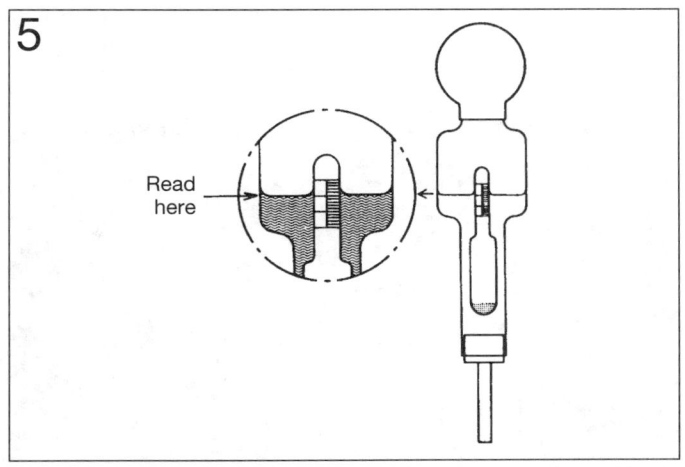

Float-type hydrometer for measuring battery specific gravity

- Before any electrical fault is investigated the battery should be checked.
- You'll need a dc voltmeter or multimeter to check battery voltage. Check that the leads are inserted in the correct terminals on the meter, red lead to positive (+ve), black lead to negative (-ve). Incorrect connections can damage the meter.
- A sound fully-charged 12 volt battery should produce between 12.3 and 12.6 volts across its terminals (12.8 volts for a maintenance-free battery). On machines with a 6 volt battery, voltage should be between 6.1 and 6.3 volts.

1 Set a multimeter to the 0 to 20 volts dc range and connect its probes across the battery terminals. Connect the meter's positive (+ve) probe, usually red, to the battery positive (+ve) terminal, followed by the meter's negative (-ve) probe, usually black, to the battery negative terminal (-ve) **(see illustration 4)**.

2 If battery voltage is low (below 10 volts on a 12 volt battery or below 4 volts on a six volt battery), charge the battery and test the voltage again. If the battery repeatedly goes flat, investigate the motorcycle's charging system.

Checking battery specific gravity (SG)

> ⚠️ **Warning:** *The gases produced by the battery are explosive - never smoke or create any sparks in the vicinity of the battery. Never allow the electrolyte to contact your skin or clothing - if it does, wash it off and seek immediate medical attention.*

- The specific gravity check gives an indication of a battery's state of charge.
- A hydrometer is used for measuring specific gravity. Make sure you purchase one which has a small enough hose to insert in the aperture of a motorcycle battery.
- Specific gravity is simply a measure of the electrolyte's density compared with that of water. Water has an SG of 1.000 and fully-charged battery electrolyte is about 26% heavier, at 1.260.
- Specific gravity checks are not possible on maintenance-free batteries. Testing the open-circuit voltage is the only means of determining their state of charge.

1 To measure SG, remove the battery from the motorcycle and remove the first cell cap. Draw some electrolyte into the hydrometer and note the reading **(see illustration 5)**. Return the electrolyte to the cell and install the cap.

2 The reading should be in the region of 1.260 to 1.280. If SG is below 1.200 the battery needs charging. Note that SG will vary with temperature; it should be measured at 20°C (68°F). Add 0.007 to the reading for every 10°C above 20°C, and subtract 0.007 from the reading for every 10°C below 20°C. Add 0.004 to the reading for every 10°F above 68°F, and subtract 0.004 from the reading for every 10°F below 68°F.

3 When the check is complete, rinse the hydrometer thoroughly with clean water.

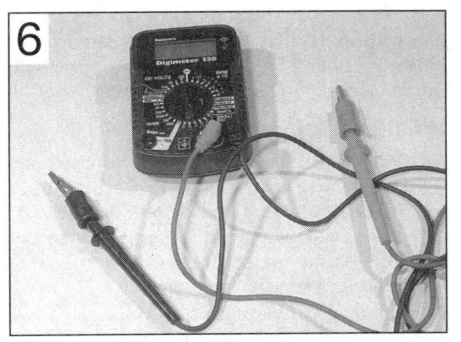

Checking for continuity

- The term continuity describes the uninterrupted flow of electricity through an electrical circuit. A continuity check will determine whether an **open-circuit** situation exists.
- Continuity can be checked with an ohmmeter, multimeter, continuity tester or battery and bulb test circuit **(see illustrations 6, 7 and 8)**.

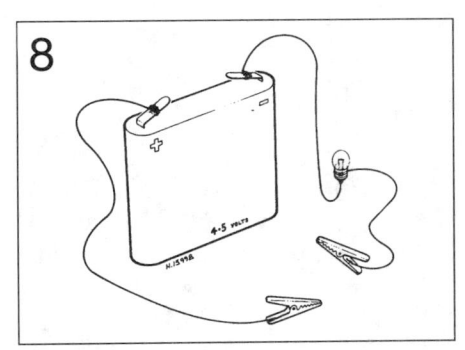

Digital multimeter can be used for all electrical tests

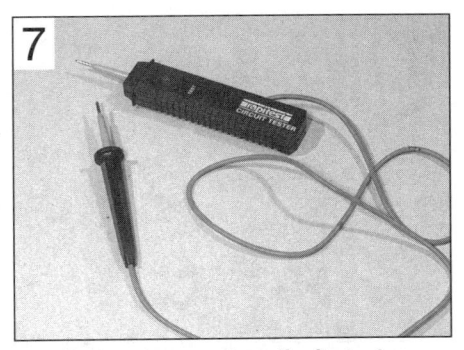

Battery-powered continuity tester

Battery and bulb test circuit

REF•46 Fault Finding Equipment

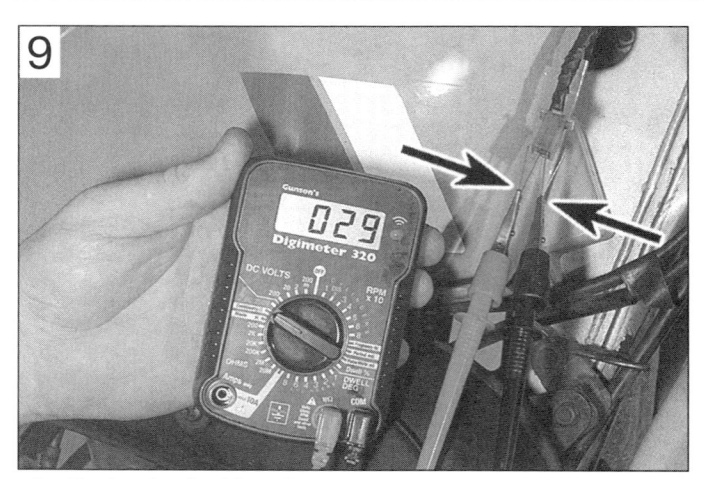

Continuity check of front brake light switch using a meter - note split pins used to access connector terminals

Continuity check of rear brake light switch using a continuity tester

- All of these instruments are self-powered by a battery, therefore the checks are made with the ignition OFF.
- As a safety precaution, always disconnect the battery negative (-ve) lead before making checks, particularly if ignition switch checks are being made.
- If using a meter, select the appropriate ohms scale and check that the meter reads infinity (∞). Touch the meter probes together and check that meter reads zero; where necessary adjust the meter so that it reads zero.
- After using a meter, always switch it OFF to conserve its battery.

Switch checks

1 If a switch is at fault, trace its wiring up to the wiring connectors. Separate the wire connectors and inspect them for security and condition. A build-up of dirt or corrosion here will most likely be the cause of the problem - clean up and apply a water dispersant such as WD40.
2 If using a test meter, set the meter to the ohms x 10 scale and connect its probes across the wires from the switch (see illustration 9). Simple ON/OFF type switches, such as brake light switches, only have two wires whereas combination switches, like the ignition switch, have many internal links. Study the wiring diagram to ensure that you are connecting across the correct pair of wires. Continuity (low or no measurable resistance - 0 ohms) should be indicated with the switch ON and no continuity (high resistance) with it OFF.
3 Note that the polarity of the test probes doesn't matter for continuity checks, although care should be taken to follow specific test procedures if a diode or solid-state component is being checked.
4 A continuity tester or battery and bulb circuit can be used in the same way. Connect its probes as described above (see illustration 10). The light should come on to indicate continuity in the ON switch position, but should extinguish in the OFF position.

Wiring checks

- Many electrical faults are caused by damaged wiring, often due to incorrect routing or chaffing on frame components.
- Loose, wet or corroded wire connectors can also be the cause of electrical problems, especially in exposed locations.

1 A continuity check can be made on a single length of wire by disconnecting it at each end and connecting a meter or continuity tester across both ends of the wire (see illustration 11).
2 Continuity (low or no resistance - 0 ohms) should be indicated if the wire is good. If no continuity (high resistance) is shown, suspect a broken wire.

Checking for voltage

- A voltage check can determine whether current is reaching a component.
- Voltage can be checked with a dc voltmeter, multimeter set on the dc volts scale, test light or buzzer (see illustrations 12 and 13). A meter has the advantage of being able to measure actual voltage.
- When using a meter, check that its leads are inserted in the correct terminals on the meter, red to positive (+ve), black to negative (-ve). Incorrect connections can damage the meter.
- A voltmeter (or multimeter set to the dc volts scale) should always be connected in parallel (across the load). Connecting it in series will not harm the meter, but the reading will not be meaningful.
- Voltage checks are made with the ignition ON.

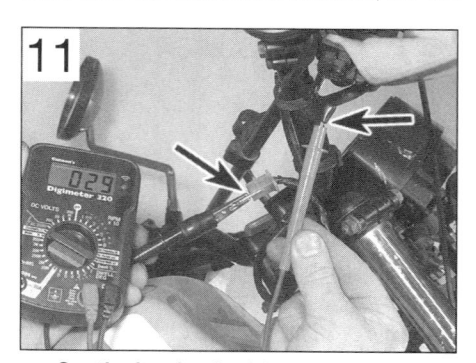

Continuity check of front brake light switch sub-harness

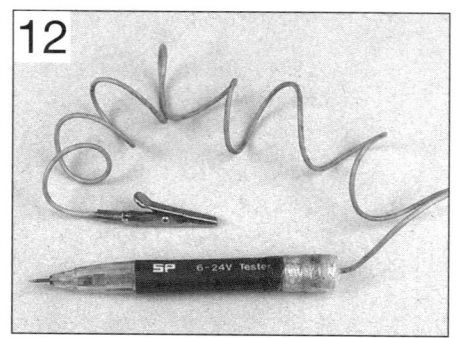

A simple test light can be used for voltage checks

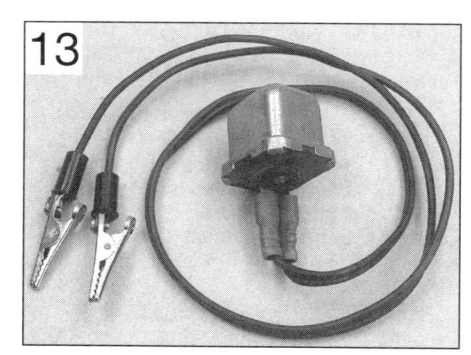

A buzzer is useful for voltage checks

Fault Finding Equipment   REF•47

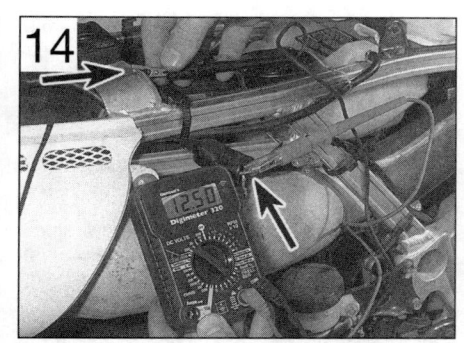

Checking for voltage at the rear brake light power supply wire using a meter . . .

1 First identify the relevant wiring circuit by referring to the wiring diagram at the end of this manual. If other electrical components share the same power supply (ie are fed from the same fuse), take note whether they are working correctly - this is useful information in deciding where to start checking the circuit.
2 If using a meter, check first that the meter leads are plugged into the correct terminals on the meter (see above). Set the meter to the dc volts function, at a range suitable for the battery voltage. Connect the meter red probe (+ve) to the power supply wire and the black probe to a good metal earth (ground) on the motorcycle's frame or directly to the battery negative (-ve) terminal **(see illustration 14)**. Battery voltage should be shown on the meter

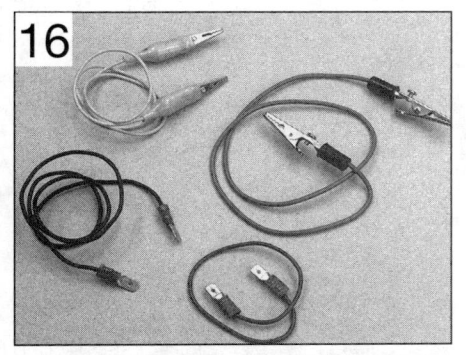

A selection of jumper wires for making earth (ground) checks

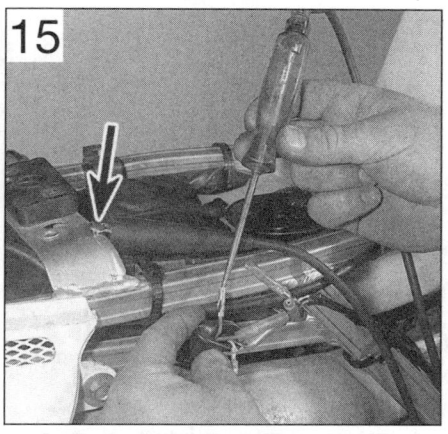

. . . or a test light - note the earth connection to the frame (arrow)

with the ignition switched ON.
3 If using a test light or buzzer, connect its positive (+ve) probe to the power supply terminal and its negative (-ve) probe to a good earth (ground) on the motorcycle's frame or directly to the battery negative (-ve) terminal **(see illustration 15)**. With the ignition ON, the test light should illuminate or the buzzer sound.
4 If no voltage is indicated, work back towards the fuse continuing to check for voltage. When you reach a point where there is voltage, you know the problem lies between that point and your last check point.

Checking the earth (ground)

● Earth connections are made either directly to the engine or frame (such as sensors, neutral switch etc. which only have a positive feed) or by a separate wire into the earth circuit of the wiring harness. Alternatively a short earth wire is sometimes run directly from the component to the motorcycle's frame.
● Corrosion is often the cause of a poor earth connection.
● If total failure is experienced, check the security of the main earth lead from the negative (-ve) terminal of the battery and also the main earth (ground) point on the wiring harness. If corroded, dismantle the connection and clean all surfaces back to bare metal.
1 To check the earth on a component, use an insulated jumper wire to temporarily bypass its earth connection **(see illustration 16)**. Connect one end of the jumper wire between the earth terminal or metal body of the component and the other end to the motorcycle's frame.
2 If the circuit works with the jumper wire installed, the original earth circuit is faulty. Check the wiring for open-circuits or poor connections. Clean up direct earth connections, removing all traces of corrosion and remake the joint. Apply petroleum jelly to the joint to prevent future corrosion.

Tracing a short-circuit

● A short-circuit occurs where current shorts to earth (ground) bypassing the circuit components. This usually results in a blown fuse.

● A short-circuit is most likely to occur where the insulation has worn through due to wiring chafing on a component, allowing a direct path to earth (ground) on the frame.

1 Remove any bodypanels necessary to access the circuit wiring.
2 Check that all electrical switches in the circuit are OFF, then remove the circuit fuse and connect a test light, buzzer or voltmeter (set to the dc scale) across the fuse terminals. No voltage should be shown.
3 Move the wiring from side to side whilst observing the test light or meter. When the test light comes on, buzzer sounds or meter shows voltage, you have found the cause of the short. It will usually shown up as damaged or burned insulation.
4 Note that the same test can be performed on each component in the circuit, even the switch.

Technical Terms Explained

A

ABS (Anti-lock braking system) A system, usually electronically controlled, that senses incipient wheel lockup during braking and relieves hydraulic pressure at wheel which is about to skid.

Aftermarket Components suitable for the motorcycle, but not produced by the motorcycle manufacturer.

Allen key A hexagonal wrench which fits into a recessed hexagonal hole.

Alternating current (ac) Current produced by an alternator. Requires converting to direct current by a rectifier for charging purposes.

Alternator Converts mechanical energy from the engine into electrical energy to charge the battery and power the electrical system.

Ampere (amp) A unit of measurement for the flow of electrical current. Current = Volts ÷ Ohms.

Ampere-hour (Ah) Measure of battery capacity.

Angle-tightening A torque expressed in degrees. Often follows a conventional tightening torque for cylinder head or main bearing fasteners **(see illustration)**.

Angle-tightening cylinder head bolts

Antifreeze A substance (usually ethylene glycol) mixed with water, and added to the cooling system, to prevent freezing of the coolant in winter. Antifreeze also contains chemicals to inhibit corrosion and the formation of rust and other deposits that would tend to clog the radiator and coolant passages and reduce cooling efficiency.

Anti-dive System attached to the fork lower leg (slider) to prevent fork dive when braking hard.

Anti-seize compound A coating that reduces the risk of seizing on fasteners that are subjected to high temperatures, such as exhaust clamp bolts and nuts.

API American Petroleum Institute. A quality standard for 4-stroke motor oils.

Asbestos A natural fibrous mineral with great heat resistance, commonly used in the composition of brake friction materials. Asbestos is a health hazard and the dust created by brake systems should never be inhaled or ingested.

ATF Automatic Transmission Fluid. Often used in front forks.

ATU Automatic Timing Unit. Mechanical device for advancing the ignition timing on early engines.

ATV All Terrain Vehicle. Often called a Quad.

Axial play Side-to-side movement.

Axle A shaft on which a wheel revolves. Also known as a spindle.

B

Backlash The amount of movement between meshed components when one component is held still. Usually applies to gear teeth.

Ball bearing A bearing consisting of a hardened inner and outer race with hardened steel balls between the two races.

Bearings Used between two working surfaces to prevent wear of the components and a build-up of heat. Four types of bearing are commonly used on motorcycles: plain shell bearings, ball bearings, tapered roller bearings and needle roller bearings.

Bevel gears Used to turn the drive through 90°. Typical applications are shaft final drive and camshaft drive **(see illustration)**.

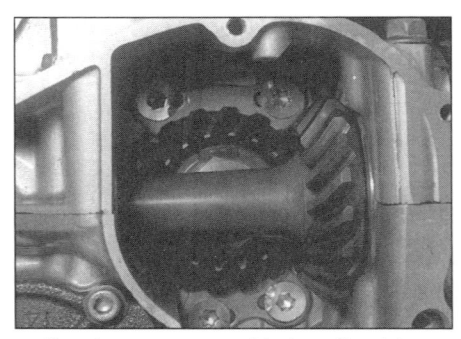

Bevel gears are used to turn the drive through 90°

BHP Brake Horsepower. The British measurement for engine power output. Power output is now usually expressed in kilowatts (kW).

Bias-belted tyre Similar construction to radial tyre, but with outer belt running at an angle to the wheel rim.

Big-end bearing The bearing in the end of the connecting rod that's attached to the crankshaft.

Bleeding The process of removing air from an hydraulic system via a bleed nipple or bleed screw.

Bottom-end A description of an engine's crankcase components and all components contained there-in.

BTDC Before Top Dead Centre in terms of piston position. Ignition timing is often expressed in terms of degrees or millimetres BTDC.

Bush A cylindrical metal or rubber component used between two moving parts.

Burr Rough edge left on a component after machining or as a result of excessive wear.

C

Cam chain The chain which takes drive from the crankshaft to the camshaft(s).

Canister The main component in an evaporative emission control system (California market only); contains activated charcoal granules to trap vapours from the fuel system rather than allowing them to vent to the atmosphere.

Castellated Resembling the parapets along the top of a castle wall. For example, a castellated wheel axle or spindle nut.

Catalytic converter A device in the exhaust system of some machines which converts certain pollutants in the exhaust gases into less harmful substances.

Charging system Description of the components which charge the battery, ie the alternator, rectifer and regulator.

Circlip A ring-shaped clip used to prevent endwise movement of cylindrical parts and shafts. An internal circlip is installed in a groove in a housing; an external circlip fits into a groove on the outside of a cylindrical piece such as a shaft. Also known as a snap-ring.

Clearance The amount of space between two parts. For example, between a piston and a cylinder, between a bearing and a journal, etc.

Coil spring A spiral of elastic steel found in various sizes throughout a vehicle, for example as a springing medium in the suspension and in the valve train.

Compression Reduction in volume, and increase in pressure and temperature, of a gas, caused by squeezing it into a smaller space.

Compression damping Controls the speed the suspension compresses when hitting a bump.

Compression ratio The relationship between cylinder volume when the piston is at top dead centre and cylinder volume when the piston is at bottom dead centre.

Continuity The uninterrupted path in the flow of electricity. Little or no measurable resistance.

Continuity tester Self-powered bleeper or test light which indicates continuity.

Cp Candlepower. Bulb rating commonly found on US motorcycles.

Crossply tyre Tyre plies arranged in a criss-cross pattern. Usually four or six plies used, hence 4PR or 6PR in tyre size codes.

Cush drive Rubber damper segments fitted between the rear wheel and final drive sprocket to absorb transmission shocks **(see illustration)**.

Cush drive rubbers dampen out transmission shocks

D

Degree disc Calibrated disc for measuring piston position. Expressed in degrees.

Dial gauge Clock-type gauge with adapters for measuring runout and piston position. Expressed in mm or inches.

Diaphragm The rubber membrane in a master cylinder or carburettor which seals the upper chamber.

Diaphragm spring A single sprung plate often used in clutches.

Direct current (dc) Current produced by a dc generator.

Technical Terms Explained

Decarbonisation The process of removing carbon deposits - typically from the combustion chamber, valves and exhaust port/system.
Detonation Destructive and damaging explosion of fuel/air mixture in combustion chamber instead of controlled burning.
Diode An electrical valve which only allows current to flow in one direction. Commonly used in rectifiers and starter interlock systems.
Disc valve (or rotary valve) A induction system used on some two-stroke engines.
Double-overhead camshaft (DOHC) An engine that uses two overhead camshafts, one for the intake valves and one for the exhaust valves.
Drivebelt A toothed belt used to transmit drive to the rear wheel on some motorcycles. A drivebelt has also been used to drive the camshafts. Drivebelts are usually made of Kevlar.
Driveshaft Any shaft used to transmit motion. Commonly used when referring to the final driveshaft on shaft drive motorcycles.

E

Earth return The return path of an electrical circuit, utilising the motorcycle's frame.
ECU (Electronic Control Unit) A computer which controls (for instance) an ignition system, or an anti-lock braking system.
EGO Exhaust Gas Oxygen sensor. Sometimes called a Lambda sensor.
Electrolyte The fluid in a lead-acid battery.
EMS (Engine Management System) A computer controlled system which manages the fuel injection and the ignition systems in an integrated fashion.
Endfloat The amount of lengthways movement between two parts. As applied to a crankshaft, the distance that the crankshaft can move side-to-side in the crankcase.
Endless chain A chain having no joining link. Common use for cam chains and final drive chains.
EP (Extreme Pressure) Oil type used in locations where high loads are applied, such as between gear teeth.
Evaporative emission control system Describes a charcoal filled canister which stores fuel vapours from the tank rather than allowing them to vent to the atmosphere. Usually only fitted to California models and referred to as an EVAP system.
Expansion chamber Section of two-stroke engine exhaust system so designed to improve engine efficiency and boost power.

F

Feeler blade or gauge A thin strip or blade of hardened steel, ground to an exact thickness, used to check or measure clearances between parts.
Final drive Description of the drive from the transmission to the rear wheel. Usually by chain or shaft, but sometimes by belt.
Firing order The order in which the engine cylinders fire, or deliver their power strokes, beginning with the number one cylinder.
Flooding Term used to describe a high fuel level in the carburettor float chambers, leading to fuel overflow. Also refers to excess fuel in the combustion chamber due to incorrect starting technique.

Free length The no-load state of a component when measured. Clutch, valve and fork spring lengths are measured at rest, without any preload.
Freeplay The amount of travel before any action takes place. The looseness in a linkage, or an assembly of parts, between the initial application of force and actual movement. For example, the distance the rear brake pedal moves before the rear brake is actuated.
Fuel injection The fuel/air mixture is metered electronically and directed into the engine intake ports (indirect injection) or into the cylinders (direct injection). Sensors supply information on engine speed and conditions.
Fuel/air mixture The charge of fuel and air going into the engine. See **Stoichiometric ratio**.
Fuse An electrical device which protects a circuit against accidental overload. The typical fuse contains a soft piece of metal which is calibrated to melt at a predetermined current flow (expressed as amps) and break the circuit.

G

Gap The distance the spark must travel in jumping from the centre electrode to the side electrode in a spark plug. Also refers to the distance between the ignition rotor and the pickup coil in an electronic ignition system.
Gasket Any thin, soft material - usually cork, cardboard, asbestos or soft metal - installed between two metal surfaces to ensure a good seal. For instance, the cylinder head gasket seals the joint between the block and the cylinder head.
Gauge An instrument panel display used to monitor engine conditions. A gauge with a movable pointer on a dial or a fixed scale is an analogue gauge. A gauge with a numerical readout is called a digital gauge.
Gear ratios The drive ratio of a pair of gears in a gearbox, calculated on their number of teeth.
Glaze-busting see **Honing**
Grinding Process for renovating the valve face and valve seat contact area in the cylinder head.
Gudgeon pin The shaft which connects the connecting rod small-end with the piston. Often called a piston pin or wrist pin.

H

Helical gears Gear teeth are slightly curved and produce less gear noise that straight-cut gears. Often used for primary drives.

Installing a Helicoil thread insert in a cylinder head

Helicoil A thread insert repair system. Commonly used as a repair for stripped spark plug threads **(see illustration)**.
Honing A process used to break down the glaze on a cylinder bore (also called glaze-busting). Can also be carried out to roughen a rebored cylinder to aid ring bedding-in.
HT (High Tension) Description of the electrical circuit from the secondary winding of the ignition coil to the spark plug.
Hydraulic A liquid filled system used to transmit pressure from one component to another. Common uses on motorcycles are brakes and clutches.
Hydrometer An instrument for measuring the specific gravity of a lead-acid battery.
Hygroscopic Water absorbing. In motorcycle applications, braking efficiency will be reduced if DOT 3 or 4 hydraulic fluid absorbs water from the air - care must be taken to keep new brake fluid in tightly sealed containers.

I

lbf ft Pounds-force feet. An imperial unit of torque. Sometimes written as ft-lbs.
lbf in Pound-force inch. An imperial unit of torque, applied to components where a very low torque is required. Sometimes written as in-lbs.
IC Abbreviation for Integrated Circuit.
Ignition advance Means of increasing the timing of the spark at higher engine speeds. Done by mechanical means (ATU) on early engines or electronically by the ignition control unit on later engines.
Ignition timing The moment at which the spark plug fires, expressed in the number of crankshaft degrees before the piston reaches the top of its stroke, or in the number of millimetres before the piston reaches the top of its stroke.
Infinity (∞) Description of an open-circuit electrical state, where no continuity exists.
Inverted forks (upside down forks) The sliders or lower legs are held in the yokes and the fork tubes or stanchions are connected to the wheel axle (spindle). Less unsprung weight and stiffer construction than conventional forks.

J

JASO Quality standard for 2-stroke oils.
Joule The unit of electrical energy.
Journal The bearing surface of a shaft.

K

Kickstart Mechanical means of turning the engine over for starting purposes. Only usually fitted to mopeds, small capacity motorcycles and off-road motorcycles.
Kill switch Handebar-mounted switch for emergency ignition cut-out. Cuts the ignition circuit on all models, and additionally prevent starter motor operation on others.
km Symbol for kilometre.
kmh Abbreviation for kilometres per hour.

L

Lambda (λ) sensor A sensor fitted in the exhaust system to measure the exhaust gas oxygen content (excess air factor).

Technical Terms Explained

Lapping see **Grinding**.
LCD Abbreviation for Liquid Crystal Display.
LED Abbreviation for Light Emitting Diode.
Liner A steel cylinder liner inserted in a aluminium alloy cylinder block.
Locknut A nut used to lock an adjustment nut, or other threaded component, in place.
Lockstops The lugs on the lower triple clamp (yoke) which abut those on the frame, preventing handlebar-to-fuel tank contact.
Lockwasher A form of washer designed to prevent an attaching nut from working loose.
LT Low Tension Description of the electrical circuit from the power supply to the primary winding of the ignition coil.

M

Main bearings The bearings between the crankshaft and crankcase.
Maintenance-free (MF) battery A sealed battery which cannot be topped up.
Manometer Mercury-filled calibrated tubes used to measure intake tract vacuum. Used to synchronise carburettors on multi-cylinder engines.
Micrometer A precision measuring instrument that measures component outside diameters **(see illustration)**.

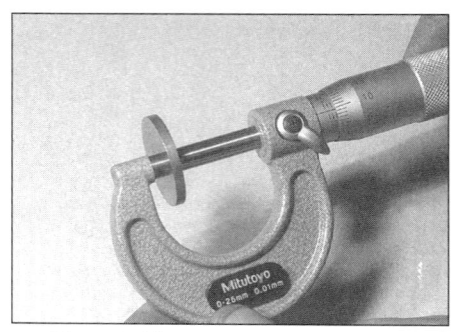

Tappet shims are measured with a micrometer

MON (Motor Octane Number) A measure of a fuel's resistance to knock.
Monograde oil An oil with a single viscosity, eg SAE80W.
Monoshock A single suspension unit linking the swingarm or suspension linkage to the frame.
mph Abbreviation for miles per hour.
Multigrade oil Having a wide viscosity range (eg 10W40). The W stands for Winter, thus the viscosity ranges from SAE10 when cold to SAE40 when hot.
Multimeter An electrical test instrument with the capability to measure voltage, current and resistance. Some meters also incorporate a continuity tester and buzzer.

N

Needle roller bearing Inner race of caged needle rollers and hardened outer race. Examples of uncaged needle rollers can be found on some engines. Commonly used in rear suspension applications and in two-stroke engines.
Nm Newton metres.
NOx Oxides of Nitrogen. A common toxic pollutant emitted by petrol engines at higher temperatures.

O

Octane The measure of a fuel's resistance to knock.
OE (Original Equipment) Relates to components fitted to a motorcycle as standard or replacement parts supplied by the motorcycle manufacturer.
Ohm The unit of electrical resistance. Ohms = Volts ÷ Current.
Ohmmeter An instrument for measuring electrical resistance.
Oil cooler System for diverting engine oil outside of the engine to a radiator for cooling purposes.
Oil injection A system of two-stroke engine lubrication where oil is pump-fed to the engine in accordance with throttle position.
Open-circuit An electrical condition where there is a break in the flow of electricity - no continuity (high resistance).
O-ring A type of sealing ring made of a special rubber-like material; in use, the O-ring is compressed into a groove to provide the sealing action.
Oversize (OS) Term used for piston and ring size options fitted to a rebored cylinder.
Overhead cam (sohc) engine An engine with single camshaft located on top of the cylinder head.
Overhead valve (ohv) engine An engine with the valves located in the cylinder head, but with the camshaft located in the engine block or crankcase.
Oxygen sensor A device installed in the exhaust system which senses the oxygen content in the exhaust and converts this information into an electric current. Also called a Lambda sensor.

P

Plastigauge A thin strip of plastic thread, available in different sizes, used for measuring clearances. For example, a strip of Plastigauge is laid across a bearing journal. The parts are assembled and dismantled; the width of the crushed strip indicates the clearance between journal and bearing.
Polarity Either negative or positive earth (ground), determined by which battery lead is connected to the frame (earth return). Modern motorcycles are usually negative earth.
Pre-ignition A situation where the fuel/air mixture ignites before the spark plug fires. Often due to a hot spot in the combustion chamber caused by carbon build-up. Engine has a tendency to 'run-on'.
Pre-load (suspension) The amount a spring is compressed when in the unloaded state. Preload can be applied by gas, spacer or mechanical adjuster.
Premix The method of engine lubrication on older two-stroke engines. Engine oil is mixed with the petrol in the fuel tank in a specific ratio. The fuel/oil mix is sometimes referred to as "petroil".
Primary drive Description of the drive from the crankshaft to the clutch. Usually by gear or chain.
PS Pfedestärke - a German interpretation of BHP.
PSI Pounds-force per square inch. Imperial measurement of tyre pressure and cylinder pressure measurement.
PTFE Polytetrafluoroethylene. A low friction substance.
Pulse secondary air injection system A process of promoting the burning of excess fuel present in the exhaust gases by routing fresh air into the exhaust ports.

Q

Quartz halogen bulb Tungsten filament surrounded by a halogen gas. Typically used for the headlight **(see illustration)**.

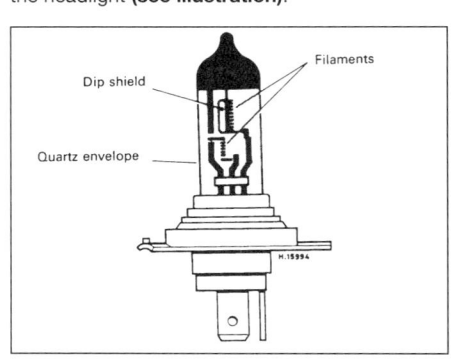

Quartz halogen headlight bulb construction

R

Rack-and-pinion A pinion gear on the end of a shaft that mates with a rack (think of a geared wheel opened up and laid flat). Sometimes used in clutch operating systems.
Radial play Up and down movement about a shaft.
Radial ply tyres Tyre plies run across the tyre (from bead to bead) and around the circumference of the tyre. Less resistant to tread distortion than other tyre types.
Radiator A liquid-to-air heat transfer device designed to reduce the temperature of the coolant in a liquid cooled engine.
Rake A feature of steering geometry - the angle of the steering head in relation to the vertical **(see illustration)**.

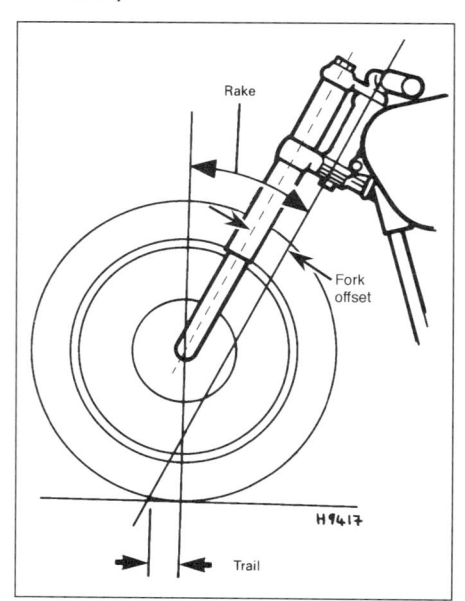

Steering geometry

Technical Terms Explained REF•51

Rebore Providing a new working surface to the cylinder bore by boring out the old surface. Necessitates the use of oversize piston and rings.
Rebound damping A means of controlling the oscillation of a suspension unit spring after it has been compressed. Resists the spring's natural tendency to bounce back after being compressed.
Rectifier Device for converting the ac output of an alternator into dc for battery charging.
Reed valve An induction system commonly used on two-stroke engines.
Regulator Device for maintaining the charging voltage from the generator or alternator within a specified range.
Relay A electrical device used to switch heavy current on and off by using a low current auxiliary circuit.
Resistance Measured in ohms. An electrical component's ability to pass electrical current.
RON (Research Octane Number) A measure of a fuel's resistance to knock.
rpm revolutions per minute.
Runout The amount of wobble (in-and-out movement) of a wheel or shaft as it's rotated. The amount a shaft rotates 'out-of-true'. The out-of-round condition of a rotating part.

S

SAE (Society of Automotive Engineers) A standard for the viscosity of a fluid.
Sealant A liquid or paste used to prevent leakage at a joint. Sometimes used in conjunction with a gasket.
Service limit Term for the point where a component is no longer useable and must be renewed.
Shaft drive A method of transmitting drive from the transmission to the rear wheel.
Shell bearings Plain bearings consisting of two shell halves. Most often used as big-end and main bearings in a four-stroke engine. Often called bearing inserts.
Shim Thin spacer, commonly used to adjust the clearance or relative positions between two parts. For example, shims inserted into or under tappets or followers to control valve clearances. Clearance is adjusted by changing the thickness of the shim.
Short-circuit An electrical condition where current shorts to earth (ground) bypassing the circuit components.
Skimming Process to correct warpage or repair a damaged surface, eg on brake discs or drums.
Slide-hammer A special puller that screws into or hooks onto a component such as a shaft or bearing; a heavy sliding handle on the shaft bottoms against the end of the shaft to knock the component free.
Small-end bearing The bearing in the upper end of the connecting rod at its joint with the gudgeon pin.
Spalling Damage to camshaft lobes or bearing journals shown as pitting of the working surface.
Specific gravity (SG) The state of charge of the electrolyte in a lead-acid battery. A measure of the electrolyte's density compared with water.
Straight-cut gears Common type gear used on gearbox shafts and for oil pump and water pump drives.
Stanchion The inner sliding part of the front forks, held by the yokes. Often called a fork tube.

Stoichiometric ratio The optimum chemical air/fuel ratio for a petrol engine, said to be 14.7 parts of air to 1 part of fuel.
Sulphuric acid The liquid (electrolyte) used in a lead-acid battery. Poisonous and extremely corrosive.
Surface grinding (lapping) Process to correct a warped gasket face, commonly used on cylinder heads.

T

Tapered-roller bearing Tapered inner race of caged needle rollers and separate tapered outer race. Examples of taper roller bearings can be found on steering heads.
Tappet A cylindrical component which transmits motion from the cam to the valve stem, either directly or via a pushrod and rocker arm. Also called a cam follower.
TCS Traction Control System. An electronically-controlled system which senses wheel spin and reduces engine speed accordingly.
TDC Top Dead Centre denotes that the piston is at its highest point in the cylinder.
Thread-locking compound Solution applied to fastener threads to prevent slackening. Select type to suit application.
Thrust washer A washer positioned between two moving components on a shaft. For example, between gear pinions on gearshaft.
Timing chain See **Cam Chain**.
Timing light Stroboscopic lamp for carrying out ignition timing checks with the engine running.
Top-end A description of an engine's cylinder block, head and valve gear components.
Torque Turning or twisting force about a shaft.
Torque setting A prescribed tightness specified by the motorcycle manufacturer to ensure that the bolt or nut is secured correctly. Undertightening can result in the bolt or nut coming loose or a surface not being sealed. Overtightening can result in stripped threads, distortion or damage to the component being retained.
Torx key A six-point wrench.
Tracer A stripe of a second colour applied to a wire insulator to distinguish that wire from another one with the same colour insulator. For example, Br/W is often used to denote a brown insulator with a white tracer.
Trail A feature of steering geometry. Distance from the steering head axis to the tyre's central contact point.
Triple clamps The cast components which extend from the steering head and support the fork stanchions or tubes. Often called fork yokes.
Turbocharger A centrifugal device, driven by exhaust gases, that pressurises the intake air. Normally used to increase the power output from a given engine displacement.
TWI Abbreviation for Tyre Wear Indicator. Indicates the location of the tread depth indicator bars on tyres.

U

Universal joint or U-joint (UJ) A double-pivoted connection for transmitting power from a driving to a driven shaft through an angle. Typically found in shaft drive assemblies.
Unsprung weight Anything not supported by the bike's suspension (ie the wheel, tyres, brakes, final drive and bottom (moving) part of the suspension).

V

Vacuum gauges Clock-type gauges for measuring intake tract vacuum. Used for carburettor synchronisation on multi-cylinder engines.
Valve A device through which the flow of liquid, gas or vacuum may be stopped, started or regulated by a moveable part that opens, shuts or partially obstructs one or more ports or passageways. The intake and exhaust valves in the cylinder head are of the poppet type.
Valve clearance The clearance between the valve tip (the end of the valve stem) and the rocker arm or tappet/follower. The valve clearance is measured when the valve is closed. The correct clearance is important - if too small the valve won't close fully and will burn out, whereas if too large noisy operation will result.
Valve lift The amount a valve is lifted off its seat by the camshaft lobe.
Valve timing The exact setting for the opening and closing of the valves in relation to piston position.
Vernier caliper A precision measuring instrument that measures inside and outside dimensions. Not quite as accurate as a micrometer, but more convenient.
VIN Vehicle Identification Number. Term for the bike's engine and frame numbers.
Viscosity The thickness of a liquid or its resistance to flow.
Volt A unit for expressing electrical "pressure" in a circuit. Volts = current x ohms.

W

Water pump A mechanically-driven device for moving coolant around the engine.
Watt A unit for expressing electrical power. Watts = volts x current.
Wear limit see **Service limit**
Wet liner A liquid-cooled engine design where the pistons run in liners which are directly surrounded by coolant **(see illustration)**.

Wet liner arrangement

Wheelbase Distance from the centre of the front wheel to the centre of the rear wheel.
Wiring harness or loom Describes the electrical wires running the length of the motorcycle and enclosed in tape or plastic sheathing. Wiring coming off the main harness is usually referred to as a sub harness.
Woodruff key A key of semi-circular or square section used to locate a gear to a shaft. Often used to locate the alternator rotor on the crankshaft.
Wrist pin Another name for gudgeon or piston pin.

Index

Note: *References throughout this index are in the form - "Chapter number" • "Page number"*

A

Air duct covers – 8•3, 8•6
Air filter – 1•19
Air filter housing – 4•5
Alignment (wheel) – 7•15
Alternator – 9•25

B

Battery – 1•12, 9•3, REF•44, REF•45
Body panels – 8•3
Brake light
   bulb – 9•7
   check – 9•5
   switches – 1•14, 9•10
Brakes
   bleeding – 7•14
   calipers – 7•6, 7•7
   check – 1•14
   discs – 7•5
   fault finding – REF•43
   fluid change – 1•19, 7•15
   fluid level check – 0•16
   hoses – 1•25, 7•13
   lever (front) – 6•8
   master cylinders – 7•9, 7•11
   pads – 1•8, 7•3
   pedal (rear) – 6•3
   seals – 1•25
   specifications – 7•1
Bulbs
   brake/tail light – 9•7
   headlight – 9•6
   license plate light – 9•8
   sidelight – 9•7
   turn signal – 9•9
   wattages – 9•2

C

Cables
   clutch – 1•8, 2•34
   H-VIX – 1•21
   lubrication – 1•10
   throttle – 1•12, 4•22
Caliper (brake) – 1•25, 7•6, 7•7
Cam chain tensioner – 2•15
Cam chain, tensioner blade and guide blades – 2•20
Cam pulse generator – 4•13

Camshafts and followers – 2•16
Catalytic converter – 4•32
Chain
   cam – 2•20
   final drive – 1•6, 6•27
Charging system – 9•24
Clutch
   fault finding – REF•39
   removal, inspection and installation – 2•27
   specifications – 2•3
Clutch cable
   check and adjustment – 1•8
   renewal – 2•34
Clutch lever – 6•8
Clutch switch – 9•16
Coolant
   change – 1•22
   level check – 0•15
   reservoir – 3•8
Cooling system
   check – 1•13
   draining, flushing and refilling – 1•22
   fan and fan switch/relay – 3•2
   hoses and unions – 3•8
   pressure cap – 1•14, 3•2
   radiator – 3•6
   specifications – 3•1
   temperature display and sensor – 3•4, 4•12
   thermostat – 3•4
   water pump – 3•6
Connecting rods – 2•47
Conversion factors – REF•26
Crankcase halves – 2•43, 2•46
Crankshaft – 2•53
Cylinder compression check – 1•24, REF•44
Cylinder bores – 2•46
Cylinder head – 2•21, 2•22

D

Dimensions – 0•10
Diode – 9•17
Disc (brake) – 7•5
Drive chain
   adjustment – 1•6
   checks – 0•14, 1•6
   cleaning and lubrication – 1•7
   removal, cleaning and installation – 6•27
   specifications – 1•2, 6•2
   sprockets – 6•28

Index REF•53

E

Electrical system
  alternator – 9•25
  battery – 1•12, 9•3,
  brake light switches – 1•14, 9•10
  brake/tail light – 9•7, 9•8
  charging system – 9•24
  clutch switch – 9•16
  diode – 9•17
  fault finding – 9•2, REF•41, REF•43, REF•44
  fuses – 9•4
  handlebar switches – 9•14, 9•15
  headlight – 1•15, 9•6, 9•7
  horn – 9•17
  ignition (main) switch – 9•14
  instrument cluster – 9•11, 9•12, 9•13
  license plate light – 9•7
  lighting system – 9•5
  neutral switch – 9•15
  oil pressure switch – 9•13
  regulator/rectifier – 9•27
  sidelight – 9•7
  sidestand switch – 1•15, 9•16
  specifications – 9•1
  starter motor – 9•18, 9•19
  starter relay – 9•17
  turn signals – 9•8, 9•9, 9•10
  wiring diagrams – 9•28

Engine
  cam chain tensioner – 2•15
  cam chain tensioner blade and guide blades – 2•20
  camshafts and followers – 2•16
  connecting rods – 2•47
  crankcase halves – 2•43, 2•46
  crankshaft – 2•53
  cylinder compression check – 1•24, REF•44
  cylinder bores – 2•46
  cylinder head – 2•21, 2•22
  fault finding – REF•36
  idle speed – 1•8
  main and connecting rod bearings – 2•47
  oil and filter change – 1•11
  oil cooler – 2•12
  oil level check – 0•14
  oil pressure check – 1•24
  oil pump – 2•37
  oil sump, strainer and pressure relief valve – 2•35
  piston rings – 2•51
  pistons – 2•49
  removal and installation – 2•7
  running-in – 2•64
  specifications – 1•1, 2•1
  starter clutch – 2•26
  timing rotor – 2•42
  valve clearances – 1•19
  valve cover – 2•13
  valves – 2•22

Engine control module (ECM) – 4•15
Engine coolant temperature sensor (ECT) – 4•12
Engine number – 0•12
Engine stop relay – 4•14
EVAP (Evaporative emission control system) – 1•19, 4•31
Exhaust system
  catalytic converter – 4•32
  H-VIX system – 1•21, 4•25
  oxygen sensor – 4•16
  removal and installation – 4•24

F

Fairing – 8•3
Fan and switch (cooling) – 3•2
Fast idle wax unit – 4•20
Fault finding – 4•9, 9•2, REF•35, REF• 44
Filter
  air – 1•19, 4•5
  fuel – 1•25
  oil – 1•11
Final drive
  chain – 1•6, 6•27
  specifications – 1•2, 6•2
  sprockets – 6•28
Footrests – 6•3
Frame – 6•3
Frame number – 0•12
Front brake
  calipers – 7•6
  discs – 7•5
  lever – 6•8
  master cylinder – 7•9
  pads – 1•8, 7•3
Front mudguard – 8•7
Front suspension (forks)
  adjustment – 6•23
  check – 1•15
  disassembly, inspection and reassembly – 6•12
  oil change – 1•25, 6•10
  removal and installation – 6•9
  specifications – 6•1
Front wheel
  bearings – 1•24, 7•18
  removal and installation – 7•16
Fuel injection system
  cam pulse generator – 4•13
  component checks – 4•10
  engine control module (ECM) – 4•15
  engine coolant temperature sensor (ECT) – 4•12
  engine stop relay – 4•14
  fast idle wax unit – 4•20
  fault diagnosis and codes – 4•9
  fuel cut-off relay – 4•15
  fuel pressure regulator – 4•11
  fuel rail and injectors – 4•10
  general information – 4•7
  ignition pulse generator – 4•13
  intake air temperature (IAT) sensor – 4•12
  lean angle sensor – 4•14
  manifold absolute pressure (MAP) sensor – 4•12
  oxygen sensor – 4•16
  throttle body – 4•16
  throttle body starter valves – 1•25, 4•18
  throttle position sensor (TPS) – 4•11
Fuel supply system
  check – 1•12
  filter – 1•25
  fuel flow rate check – 4•20
  fuel pressure check – 4•21
  hoses – 1•25
  injection system components – 4•7, 4•9, 4•10
  level warning light and sensor – 4•22
  pump – 4•21
  tank and tap – 4•2
Fuses – 9•2, 9•4

Index

G

Gearchange
 fault finding – REF•40
 lever and linkage – 6•4
 mechanism – 2•38
 selector drum and forks – 2•40
Gearshafts and bearings – 2•56, 2•57

H

Handlebar switches – 9•14, 9•15
Handlebars and levers – 6•5
Handling and stability fault finding – REF•42
Headlight
 aim – 1•15
 assembly – 9•7
 bulb – 9•6
 check and relay – 9•5
HISS (Honda Ignition Security System) – 5•4
Horn – 9•17
HT coils – 5•2
H-VIX (Honda variable intake and exhaust) system
 check and cable adjustment – 1•21
 components – 4•25

I

Idle speed – 1•8
Ignition (main) switch – 9•14
Ignition system
 check – 5•2
 engine control module (ECM) – 4•15
 HT coils – 5•2
 ignition pulse generator – 4•13
 spark plugs – 1•9, 1•10
 specifications – 5•1
 timing – 5•3
 timing rotor – 2•42
Immobiliser system – 5•4
Injection system (fuel) – 4•7, 4•9, 4•10
Injectors (fuel) – 4•10
Instrument cluster – 9•11, 9•12, 9•13
Intake air temperature (IAT) sensor – 4•12

L

LCD display – 9•13
Lean angle sensor – 4•14
Legal checks – 0•14
Lighting system – 9•5
License plate light
 bulb – 9•8
 check – 9•6
Lubricants and fluids
 general information – REF•23
 recommended – 1•2
Lubrication (maintenance) – 1•10

M

Main bearings – 2•47, 2•53
Maintenance schedule – 1•4, 1•5
Manifold absolute pressure (MAP) sensor – 4•12
Master cylinder (brake) – 1•25, 7•9, 7•11
Mirrors – 8•2
MOT test checks – REF•27
Mudguard (front) – 8•7

N

Neutral switch – 9•15

O

Oil (engine/transmission)
 level check – 0•14
 oil and filter change – 1•11
Oil (front forks) – 1•25, 6•10
Oil cooler – 2•12
Oil pressure check – 1•24
Oil pressure switch – 9•13
Oil pump – 2•37
Oil sump, strainer and pressure relief valve – 2•35
Oxygen sensor – 4•16

P

Pads (brake) – 1•8, 7•3
PAIR (Pulse secondary air injection system) – 1•18, 4•29
Parts buying – 0•12
Performance data – 0•11
PGM-FI system – 4•7, 4•9, 4•10
Piston rings – 2•51
Pistons – 2•49
Pressure check (fuel) – 4•21
Pressure check (oil) – 1•24
Pressure relief valve (oil) – 2•35
Pressure regulator (fuel) – 4•11
Pressure switch (oil) – 9•13
Pump
 fuel – 4•21
 water – 3•6

R

Radiator – 3•6
Radiator pressure cap – 1•14, 3•2
Rear brake
 caliper – 7•7
 disc – 7•5
 master cylinder – 7•11
 pads – 1•8, 7•4
 pedal – 6•3
Rear suspension
 adjustment – 6•24
 bearings – 1•25
 check – 1•15
 linkage – 6•22
 shock absorber – 6•20
 swingarm – 6•25, 6•27
Rear view mirrors – 8•2
Rear wheel
 bearings – 1•24, 7•19
 removal and installation – 7•17
Regulator/rectifier – 9•27

Index

Relay
  cooling fan – 3•3
  engine stop – 4•14
  fuel cut-off – 4•15
  headlight – 9•5
  starter – 9•17
  turn signal – 9•8
Rubber dampers (sprocket coupling) – 6•29
Running-in – 2•64

S

Safety – 0•13, 0•14, 4•2
Seat – 8•2
Seat cowling – 8•5, 8•7
Security advice – REF•20
Selector drum and forks – 2•40
Shock absorber (rear) – 6•20, 6•24
Sidelight – 9•7
Sidestand – 6•4
Sidestand switch – 1•15, 9•16
Spare parts buying – 0•12
Spark plugs – 1•9, 1•10
Specifications
  brake – 7•1
  clutch – 2•3
  cooling system – 3•1
  electrical system – 9•1
  engine – 1•1, 2•1
  final drive – 6•2
  front forks – 6•1
  fuel system – 4•1
  ignition system – 5•1
  transmission – 2•5
  wheels and tyres – 7•2
Speedometer and speed sensor – 9•12
Sprocket coupling (rear) – 6•29, 7•20
Sprockets – 1•6, 6•28
Starter clutch – 2•26
Starter interlock circuit check – 1•15
Starter motor – 9•18, 9•19
Starter relay – 9•17
Steering
  checks – 0•14
  head bearings – 1•16, 1•25, 6•19
  stem – 6•17
Storage advice – REF•32
Suspension
  adjustments – 6•23
  checks – 0•14, 1•15
  fault finding – REF•41, REF•42
  front forks – 1•25, 6•9, 6•10, 6•12
  rear linkage – 6•22
  rear shock absorber – 6•20
  swingarm – 6•25, 6•27
Swingarm – 6•25, 6•27

T

Tachometer – 9•13
Tail light
  assembly – 9•8
  bulb – 9•7
  check – 9•5
Tank and tap (fuel) – 4•2
Temperature display and sensor – 3•4
Tensioner (cam chain) – 2•15
Thermostat – 3•4
Throttle body – 4•16
Throttle body starter valves – 1•25, 4•18
Throttle cables – 1•12, 4•22
Throttle position sensor (TPS) – 4•11
Timing (ignition) – 5•3
Timing rotor – 2•42
Tools – REF•2
Torque settings – 1•2, 2•6, 3•1, 4•2, 5•1, 6•2, 7•2, 9•2
Transmission
  fault finding – REF•40
  gearchange lever and linkage – 6•4
  gearchange mechanism – 2•38
  gearshafts and bearings – 2•56, 2•57
  selector drum and forks – 2•40
  specifications – 2•5
Turn signals – 9•8, 9•9, 9•10
Tyre
  checks and pressure – 0•15
  general information and fitting – 7•21
  sizes – 7•2

V

Valve clearances – 1•19
Valve cover – 2•13
VIN (Vehicle Identification Number) – 0•12

W

Water pump – 3•6
Weights – 0•10
Wheel
  alignment – 7•15
  bearings – 1•24, 7•18
  check – 1•18
  inspection and repair – 7•15
  removal and installation – 7•16, 7•17
  specifications – 7•2
Windshield – 8•7
Wiring diagrams – 9•28
Workshop tips – REF•2

Notes

Haynes Motorcycle Manuals – The Complete List

Title	Book No
BMW	
BMW 2-valve Twins (70 - 96)	0249
BMW K100 & 75 2-valve Models (83 - 96)	1373
BMW R850 & R1100 4-valve Twins (93 - 97)	3466
BSA	
BSA Bantam (48 - 71)	0117
BSA Unit Singles (58 - 72)	0127
BSA Pre-unit Singles (54 - 61)	0326
BSA A7 & A10 Twins (47 - 62)	0121
BSA A50 & A65 Twins (62 - 73)	0155
DUCATI	
Ducati MK III & Desmo Singles (69 - 76)	0445
Ducati 600, 750 & 900 2-valve V-Twins (91 - 96)	3290
Ducati 748, 916 & 996 4-valve V-Twins (94 - 01)	3756
HARLEY-DAVIDSON	
Harley-Davidson Sportsters (70 - 01)	0702
Harley-Davidson Big Twins (70 - 99)	0703
HONDA	
Honda NB, ND, NP & NS50 Melody (81 - 85)	◇ 0622
Honda NE/NB50 Vision & SA50 Vision Met-in (85 - 95)	◇ 1278
Honda MB, MBX, MT & MTX50 (80 - 93)	0731
Honda C50, C70 & C90 (67 - 99)	0324
Honda XR80R & XR100R (85 - 96)	2218
Honda XL/XR 80, 100, 125, 185 & 200 2-valve Models (78 - 87)	0566
Honda H100 & H100S Singles (80 - 92)	◇ 0734
Honda CB/CD125T & CM125C Twins (77 - 88)	◇ 0571
Honda CG125 (76 - 00)	◇ 0433
Honda NS125 (86 - 93)	◇ 3056
Honda MBX/MTX125 & MTX200 (83 - 93)	◇ 1132
Honda CD/CM185 200T & CM250C 2-valve Twins (77 - 85)	0572
Honda XL/XR 250 & 500 (78 - 84)	0567
Honda XR250L, XR250R & XR400R (86 - 01)	2219
Honda CB250 & CB400N Super Dreams (78 - 84)	◇ 0540
Honda CR Motocross Bikes (86 - 01)	2222
Honda Elsinore 250 (73 - 75)	0217
Honda CBR400RR Fours (88 - 99)	3552
Honda VFR400 (NC30) & RVF400 (NC35) V-Fours (89 - 98)	3496
Honda CB500 (93 - 01)	3753
Honda CB400 & CB550 Fours (73 - 77)	0262
Honda CX/GL500 & 650 V-Twins (78 - 86)	0442
Honda CBX550 Four (82 - 86)	◇ 0940
Honda XL600R & XR600R (83 - 00)	2183
Honda XL600/650V Transalp & XRV750 Africa Twin (87 - 02)	3919
Honda CBR600F1 & 1000F Fours (87 - 96)	1730
Honda CBR600F2 & F3 Fours (91 - 98)	2070
Honda CBR600F4 (99 - 02)	3911
Honda CB600F Hornet (98 - 02)	3915
Honda CB650 sohc Fours (78 - 84)	0665
Honda NTV600/650/Deauville V-Twins (88 - 01)	3243
Honda Shadow VT600 & 750 (USA) (88 - 00)	2312
Honda CB750 sohc Four (69 - 79)	0131
Honda V45/65 Sabre & Magna (82 - 88)	0820
Honda VFR750 & 700 V-Fours (86 - 97)	2101
Honda VFR800 V-Fours (97 - 01)	3703
Honda VTR1000 (FireStorm, Super Hawk) & XL1000V (Varadero) (97 - 00)	3744
Honda CB750 & CB900 dohc Fours (78 - 84)	0535
Honda CBR900RR FireBlade (92 - 99)	2161
Honda CBR900RR FireBlade (00 - 03)	4060
Honda CBR1100XX Super Blackbird (97 - 02)	3901
Honda ST1100 Pan European V-Fours (90 - 01)	3384
Honda Shadow VT1100 (USA) (85 - 98)	2313
Honda GL1000 Gold Wing (75 - 79)	0309
Honda GL1100 Gold Wing (79 - 81)	0669
Honda Gold Wing 1200 (USA) (84 - 87)	2199
Honda Gold Wing 1500 (USA) (88 - 00)	2225
KAWASAKI	
Kawasaki AE/AR 50 & 80 (81 - 95)	1007
Kawasaki KC, KE & KH100 (75 - 99)	1371
Kawasaki KMX125 & 200 (86 - 02)	◇ 3046
Kawasaki 250, 350 & 400 Triples (72 - 79)	0134
Kawasaki 400 & 440 Twins (74 - 81)	0281
Kawasaki 400, 500 & 550 Fours (79 - 91)	0910
Kawasaki EN450 & 500 Twins (Ltd/Vulcan) (85 - 93)	2053
Kawasaki EX & ER500 (GPZ500S & ER-5) Twins (87 - 99)	2052
Kawasaki ZX600 (Ninja ZX-6, ZZ-R600) Fours (90 - 00)	2146
Kawasaki ZX-6R Ninja Fours (95 - 98)	3541
Kawasaki ZX600 (GPZ600R, GPX600R, Ninja 600R & RX) & ZX750 (GPX750R, Ninja 750R) Fours (85 - 97)	1780
Kawasaki 650 Four (76 - 78)	0373
Kawasaki Vulcan 700/750 & 800 (85 - 01)	2457
Kawasaki 750 Air-cooled Fours (80 - 91)	0574
Kawasaki ZR550 & 750 Zephyr Fours (90 - 97)	3382
Kawasaki ZX750 (Ninja ZX-7 & ZXR750) Fours (89 - 96)	2054
Kawasaki Ninja ZX-7R & ZX-9R (ZX750P, ZX900B/C/D/E) (94 - 00)	3721
Kawasaki 900 & 1000 Fours (73 - 77)	0222
Kawasaki ZX900, 1000 & 1100 Liquid-cooled Fours (83 - 97)	1681
MOTO GUZZI	
Moto Guzzi 750, 850 & 1000 V-Twins (74 - 78)	0339
MZ	
MZ ETZ Models (81 - 95)	◇ 1680
NORTON	
Norton 500, 600, 650 & 750 Twins (57 - 70)	0187
Norton Commando (68 - 77)	0125
PEUGEOT	
Peugeot Speedfight, Trekker & Vivacity Scooters (96 - 02)	3920
PIAGGIO	
Piaggio (Vespa) Scooters (91 - 98)	3492
SUZUKI	
Suzuki GT, ZR & TS50 (77 - 90)	◇ 0799
Suzuki TS50X (84 - 00)	◇ 1599
Suzuki 100, 125, 185 & 250 Air-cooled Trail bikes (79 - 89)	0797
Suzuki GP100 & 125 Singles (78 - 93)	◇ 0576
Suzuki GS, GN, GZ & DR125 Singles (82 - 99)	◇ 0888
Suzuki 250 & 350 Twins (68 - 78)	0120
Suzuki GT250X7, GT200X5 & SB200 Twins (78 - 83)	◇ 0469
Suzuki GS/GSX250, 400 & 450 Twins (79 - 85)	0736
Suzuki GS500 Twin (89 - 02)	3238
Suzuki GS550 (77 - 82) & GS750 Fours (76 - 79)	0363
Suzuki GS/GSX550 4-valve Fours (83 - 88)	1133
Suzuki SV650 (99 - 02)	3912
Suzuki GSX-R600 & 750 (96 - 99)	3553
Suzuki GSX-R600 (01 - 02), GSX-R750 (00 - 02) & GSX-R1000 (01 - 02)	3986
Suzuki GSF600 & 1200 Bandit Fours (95 - 01)	3367
Suzuki GS850 Fours (78 - 88)	0536
Suzuki GS1000 Four (77 - 79)	0484
Suzuki GSX-R750, GSX-R1100 (85 - 92), GSX600F, GSX750F, GSX1100F (Katana) Fours (88 - 96)	2055
Suzuki GSX600/750F & GSX750 (98 - 02)	3987
Suzuki GS/GSX1000, 1100 & 1150 4-valve Fours (79 - 88)	0737
TRIUMPH	
Triumph Tiger Cub & Terrier (52 - 68)	0414
Triumph 350 & 500 Unit Twins (58 - 73)	0137
Triumph Pre-Unit Twins (47 - 62)	0251
Triumph 650 & 750 2-valve Unit Twins (63 - 83)	0122
Triumph Trident & BSA Rocket 3 (69 - 75)	0136
Triumph Fuel Injected Triples (97 - 00)	3755
Triumph Triples & Fours (carburettor engines) (91 - 99)	2162
VESPA	
Vespa P/PX125, 150 & 200 Scooters (78 - 95)	0707
Vespa Scooters (59 - 78)	0126
YAMAHA	
Yamaha DT50 & 80 Trail Bikes (78 - 95)	◇ 0800
Yamaha T50 & 80 Townmate (83 - 95)	◇ 1247
Yamaha YB100 Singles (73 - 91)	◇ 0474
Yamaha RS/RXS100 & 125 Singles (74 - 95)	0331
Yamaha RD & DT125LC (82 - 87)	◇ 0887
Yamaha TZR125 (87 - 93) & DT125R (88 - 02)	◇ 1655
Yamaha TY50, 80, 125 & 175 (74 - 84)	◇ 0464
Yamaha XT & SR125 (82 - 02)	1021
Yamaha Trail Bikes (81 - 00)	2350
Yamaha 250 & 350 Twins (70 - 79)	0040
Yamaha XS250, 360 & 400 sohc Twins (75 - 84)	0378
Yamaha RD250 & 350LC Twins (80 - 82)	0803
Yamaha RD350 YPVS Twins (83 - 95)	1158
Yamaha RD400 Twin (75 - 79)	0333
Yamaha XT, TT & SR500 Singles (75 - 83)	0342
Yamaha XZ550 Vision V-Twins (82 - 85)	0821
Yamaha FJ, FZ, XJ & YX600 Radian (84 - 92)	2100
Yamaha XJ600S (Diversion, Seca II) & XJ600N Fours (92 - 99)	2145
Yamaha YZF600R Thundercat & FZS600 Fazer (96 - 01)	3702
Yamaha YZF-R6 (98 - 02)	3900
Yamaha 650 Twins (70 - 83)	0341
Yamaha XJ650 & 750 Fours (80 - 84)	0738
Yamaha XS750 & 850 Triples (76 - 85)	0340
Yamaha TDM850, TRX850 & XTZ750 (89 - 99)	3540
Yamaha YZF750R & YZF1000R Thunderace (93 - 00)	3720
Yamaha FZR600, 750 & 1000 Fours (87 - 96)	2056
Yamaha XV V-Twins (81 - 96)	0802
Yamaha XJ900F Fours (83 - 94)	3239
Yamaha XJ900S Diversion (94 - 01)	3739
Yamaha YZF-R1 (98 - 01)	3754
Yamaha FJ1100 & 1200 Fours (84 - 96)	2057
Yamaha XJR1200 & 1300 (95 - 03)	3981
ATVs	
Honda ATC70, 90, 110, 185 & 200 (71 - 85)	0565
Honda TRX300 Shaft Drive ATVs (88 - 00)	2125
Honda TRX300EX & TRX400EX ATVs (93 - 99)	2318
Honda Foreman 400 and 450 ATVs (95 - 02)	2465
Kawasaki Bayou 220/300 & Prairie 300 ATVs (86 - 01)	2351
Polaris ATVs (85 to 97)	2302
Yamaha YT, YFM, YTM & YTZ ATVs (80 - 85)	1154
Yamaha YFS200 Blaster ATV (88 - 98)	2317
Yamaha YFB250 Timberwolf ATV (92 - 96)	2217
Yamaha YFM350 (ER and Big Bear) ATVs (87 - 99)	2126
Yamaha Warrior and Banshee ATVs (87 - 99)	2314
ATV Basics	10450
MOTORCYCLE TECHBOOKS	
Motorcycle Basics TechBook (2nd Edition)	3515
Motorcycle Electrical TechBook (3rd Edition)	3471
Motorcycle Fuel Systems TechBook	3514
Motorcycle Workshop Practice TechBook (2nd Edition)	3470

◇ = not available in the USA   **Bold type** = Superbike

The manuals on this page are available through good motorcyle dealers and accessory stores.
In case of difficulty, contact: **Haynes Publishing**
(UK) +44 1963 442030   (USA) +1 805 498 6703
(FR) +33 1 47 17 66 29   (SV) +46 18 124016
(Australia/New Zealand) +61 3 9763 8100

MCL14.4/03

Preserving Our Motoring Heritage

The Model J Duesenberg Derham Tourster. Only eight of these magnificent cars were ever built – this is the only example to be found outside the United States of America

Almost every car you've ever loved, loathed or desired is gathered under one roof at the Haynes Motor Museum. Over 300 immaculately presented cars and motorbikes represent every aspect of our motoring heritage, from elegant reminders of bygone days, such as the superb Model J Duesenberg to curiosities like the bug-eyed BMW Isetta. There are also many old friends and flames. Perhaps you remember the 1959 Ford Popular that you did your courting in? The magnificent 'Red Collection' is a spectacle of classic sports cars including AC, Alfa Romeo, Austin Healey, Ferrari, Lamborghini, Maserati, MG, Riley, Porsche and Triumph.

A Perfect Day Out

Each and every vehicle at the Haynes Motor Museum has played its part in the history and culture of Motoring. Today, they make a wonderful spectacle and a great day out for all the family. Bring the kids, bring Mum and Dad, but above all bring your camera to capture those golden memories for ever. You will also find an impressive array of motoring memorabilia, a comfortable 70 seat video cinema and one of the most extensive transport book shops in Britain. The Pit Stop Cafe serves everything from a cup of tea to wholesome, home-made meals or, if you prefer, you can enjoy the large picnic area nestled in the beautiful rural surroundings of Somerset.

John Haynes O.B.E., Founder and Chairman of the museum at the wheel of a Haynes Light 12.

The 1936 490cc sohc-engined International Norton – well known for its racing success

The Museum is situated on the A359 Yeovil to Frome road at Sparkford, just off the A303 in Somerset. It is about 40 miles south of Bristol, and 25 minutes drive from the M5 intersection at Taunton.
Open 9.30am - 5.30pm (10.00am - 4.00pm Winter) 7 days a week, *except Christmas Day, Boxing Day and New Years Day*
Special rates available for schools, coach parties and outings Charitable Trust No. 292048